Low-luminosity Stars

A

Low-luminosity Stars

Proceedings of the Symposium on
Low-luminosity Stars
University of Virginia, 1968

Edited by

SHIV S. KUMAR

Leander McCormick Observatory
University of Virginia

GORDON AND BREACH SCIENCE PUBLISHERS
New York   London   Paris

QB
843
.D9S9
1968

Editorial office for the United Kingdom:

Gordon and Breach Science Publishers Ltd.
12 Bloomsbury Way
London W.C.1

Editorial office for France:

Gordon & Breach
7–9 rue Emile Dubois
Paris 14e

Distributed in Canada by:

The Ryerson Press
299 Queen Street West
Toronto 2B, Ontario

PRINTED IN GREAT BRITAIN BY PAGE BROS (NORWICH) LTD.

Dedicated to
Alexander T. Vyssotsky
and Emma Williams-Vyssotsky

Preface

AN INTERNATIONAL SYMPOSIUM on low-luminosity stars was held at the University of Virginia on March 28–30, 1968. The proceedings of the symposium are being published in this volume. The symposium was sponsored by the National Science Foundation, the American Astronomical Society, and the University of Virginia. I am grateful to the members of the Organizing Committee and to Miss M. King for their help in the organization of the symposium. The publication of this book has been made possible by a grant from the National Science Foundation.

I would also like to take this opportunity to thank my colleagues at the University of Virginia who helped me in organizing the symposium and in putting together the proceedings. I am particularly grateful to Mrs. K. Orfanedes and Miss M. Kerr for their help in typing the manuscript. Finally, I want to thank all the contributors for their cooperation in getting the proceedings published in such a short time.

The discussions reported in this book were tape recorded during the sessions, and I have tried to extract as much information out of the tapes as possible. Since all the discussions were *not* checked for accuracy by individual authors, the editor should be held responsible for any mistakes inadvertently reported in these discussions!

SHIV S. KUMAR

Charlottesville, Virginia
February 1st, 1969

Contents

Contents

Address of welcome

FRANK L. HEREFORD, Jr.

Provost, University of Virginia

IT IS A great pleasure to welcome all of you to the University of Virginia for the Symposium on Low-luminosity Stars, an area of astronomy in which this institution has been active for many years.

I cannot resist making a few comments about astronomy at the University of Virginia. It had its beginning during the 1820's, when records indicate that the first observations were made. During the 1830's Professor William Barton Rogers gave a series of lectures on astronomy which are famous for the numbers of students who were attracted and for the great enthusiasm with which the lectures were greeted. Professor Rogers later left the University of Virginia, moved to Massachusetts, and founded the Massachusetts Institute of Technology.

Following the establishment of several small and temporary observatories in the Grounds, the Leander McCormick Observatory was opened in 1883 on Mt. Jefferson, a site which had been recommended earlier by Thomas Jefferson, the founder of the University. Under Professors Ormond Stone and Samuel A. Mitchell the McCormick Observatory became well known throughout the world.

With this early beginning, our Department of Astronomy is today an active and thriving one. In recent years it has gained added strength through the establishment of the scientific headquarters of the National Radio Astronomy Observatory here at the University which has brought about a close interaction between the University and NRAO astronomers.

Although I am only an interested physicist spectator, I cannot help but share in the excitement of astronomy and cosmology today. The rapidity with which the discovery of the quasars was followed by that of the three degree black body radiation and now the pulsars would excite anyone.

Late last night, while deciding what I should say to you this morning, I believe that I achieved a profound insight into one aspect of the pulsars' behavior. I believe that I should share it with you even before publication!

As nearly as I could remember at 11:00 p.m., the period of the observed pulses is 1.37 sec. I began my deliberations with Dr. Hewish (h) and then recalled that he and his group are at Cambridge (*c*). Then I recalled that my friends in Electrical Engineering have told me that there would be no radio astronomy today but for the distinguished and outstanding work which they have done through the years. Being a physicist, I was unwilling to give them top drawer recognition, and I placed them beneath *h* and *c* with the notation, e^2.

Finally, I considered the physics of the problem—always the most important ingredient. Every first-year physics graduate student knows that there is always a factor of $1/2\pi$. Thus, I arrived at

$$\frac{1}{2\pi} \quad \frac{hc}{e^2}$$

I am sure that all of you recognize the above expression as 137. The fact that it differs by two orders of magnitude from the period of the pulsars I can only attribute to the lateness of the hour when this work was performed.

To be serious once again, we are delighted that you are here with us in Charlottesville and confident that your discussions and deliberations during the next few days will be productive. We at the University of Virginia hope that we can make them enjoyable.

Introductory address

JESSE L. GREENSTEIN

Mount Wilson and Palomar Observatories

ONE OF THE most difficult problems in science is to learn when you and your favorite subject of study have become outmoded. I have had it happen to me, now, so often that I have become optimistic: all I have to do is wait. In a more serious vein, the evolution of topics of critical importance for astronomy is at least as complex as the evolution of a star. The individual scientist, who must take an instantaneous slice of science, from his own point of view in space and time, leads a life in which he is comforted or discouraged by the degree of fashionableness of his own specialty. For a recent example, the introduction of the spectrograph, and with it the application of atomic physics to the atmospheres of the stars, had a shattering effect on the popularity and apparent significance of classical meridian-circle work, double star observations, parallax and proper-motion measurements. Similarly, photography had killed off the need for observers with good eyes and long patience. But now the spectrograph itself has become outmoded. Facts may even become unnecessary, since the computer permits the theoretical astrophysicist to specify what the universe *must* be like, at every instant, from the first big bang till the last star will fade away. The dinosaur, the dodo, and the large telescope should and may already be extinct. If so, even fossilized remains of disciplines like astrometry, stellar parallax and motion, determinations of mass, temperature, and composition should be crumbling into dry dust.

But, fortunately, the perverse originality of nature has kept the observer and the data collector alive and happy in old and new fields. Radio sources, quasars, the cosmic background form one end of the spectrum of surprises with X-ray flare stars and pulsars at the other. These new objects born from observation, not theory, as their names show, remind us of the absolute need to be modest.

The symposium we are starting today is about stars of small mass, undramatic, petit-bourgeois members of the lower middle class of inhabitants of our Galaxy. But supergiants and giants will come and go; the quasars will run out of energy; the pulsars will stop ticking, but the dwarf M stars will go on forever (or if not forever, for a few trillion years). In a hundred billion years, the sun will have faded into a dead, small, infrared degenerate star, probably invisible from the earth. All presently existing stars except late *K* and *M* dwarfs will have burnt out, and our only youthful neighbor will be a randomly passing *M* dwarf, probably too faint to be seen without a telescope. But that *M* dwarf will still have its full youthful vigor, with 95 per cent of its long life ahead! We had better start now, by laying a foundation of knowledge about them, to guarantee employment for the astronomer of those future days. By default, at least, those who study poor, small stars shall inherit the skies.

It is of course no coincidence that the International Symposium on Low-luminosity Stars meets at the University of Virginia. The Observatory here has long shown outstanding devotion to gathering fundamental data about small stars, *M* dwarfs, invisible companions, double star orbits, mass ratios, stellar motions and distances. The titles of the many invited and contributed papers to be presented show that related data are also being obtained in many newer institutions. And they also suggest that the theoretical astrophysicist can find as fascinating and difficult problems in little stars as anywhere in the bigger universe. Why are late-type dwarfs unstable and active? How do young stars reach the main sequence and how do old stars die? How many faint red stars are there on the main sequence, how many white dwarfs are there, and do stars of intermediate luminosity exist? Do neutron stars exist? Where is the border-line between stars and planets, and how many stars have faint, dark, quasi-planetary companions occupied by our own boring colleagues? Is the relative rate of formation of stars of different masses the same in old and new clusters, and in the galactic field? How much invisible mass is there in the Galaxy? What is the interior structure of a cool, low-mass star and what happened to the initial rotational angular momentum and magnetic fields it might have had? Can theoretical physicists deduce the equation of state for cool, partially degenerate, partially solidified matter? Is the relative frequency of wide and close binaries the same for the low and high mass stars, and for young population I and old halo population II? Do degenerate stars of very low mass and high hydrogen content exist and how can they be recognized?

To answer some of these questions, many of the apparently oldest-fashioned techniques of astronomy are required and have again been reborn and become active. We can't all work on galaxies, and we shouldn't. Faint companions are being found by Van de Kamp, who worked for many years in Charlottesville. New proper-motion surveys by Giclas and collaborators (at Lowell) and by Luyten (now repeating second-epoch Palomar schmidt plates) have been extremely productive. Mechanization of the search process by electronic techniques is underway at Minnesota and elsewhere, and of course the computer has enormously simplified reductions. The fundamental coordinate system is being investigated once more, but now with reference to galaxies at Lick, where the astrographic plate-measurement process has been successfully automated. The first new, large parallax telescope in generations is working actively at the U.S. Naval Observatory at Flagstaff. Similar developments are under way elsewhere, and especially in the U.S.S.R.; we may hope that southern-hemisphere programs will soon be started. Photoelectric photometry has been very useful for faint stars; the UBV system needs to be supplemented with infrared data for late-type dwarfs; the apparent, as well as intrinsic faintness of these objects makes this a difficult problem even for moderately large telescopes. As for the spectroscopy of faint stars it has been exclusively a large telescope problem, and our ignorance, therefore, remains enormous. After Kuiper and Vyssotsky's early classification work, the *M* dwarfs have been magnificently neglected. Their atmospheres present fascinating problems, because of molecular opacity, but so far not even one has had an accurate abundance analysis. Important problems are unexplored like the metal to hydrogen, and helium to hydrogen ratios. The complex effect of changes in oxygen to carbon ratio are unstudied in dwarfs, and we have recently found carbon-rich dwarfs. The late *K* and *M* dwarfs, essentially unevolved, because of their more completely convective structure, may be more homogeneous in composition than are solar type stars. One very important detail for the Hayashi contraction theory is the lithium content, never as yet studied in the completely convective *M* dwarfs. There are also new types of problems. How can some *dKe* and *dMe* stars show different periods of light and velocity variation? Are these star spots, as in Castor C? What happens in an *M* dwarf super-flare? To give a personal touch, I observed one of our nearest neighbors in space, Wolf 359 (= Giclas 45–20) a *dM*6*e* star of moderate velocity, just one week before this conference, using the image-tube spectrograph at Palomar. The first plate of a stellar spectrum taken with this incredibly efficient device showed a violent spectroscopic flare, a blue continuum and H and He I emission lines greatly enhanced and widened! We have reason to hope that this, and other modern technological advances, will permit spectroscopic and spectrophotometric observations of faint stars of small mass at many observatories. We can be sure that every type of freak star imagined by our theoretical colleagues will be found. But we also know, from past experience with more luminous objects, that more surprising and wonderful objects have been invented by nature than by the computer.

Let me make one speculative remark about non-thermal energy release. In faint stars the nuclear energy supply is ample, but collision rates are low and insensitive to temperature. But the potential energy supply in a small star is not much less than in the sun. With a mass–luminosity law $L \propto M^{7/2}$, we can derive that the ratio of potential energies, for homologous structures of star and sun is:

$$(\Omega/\Omega_\odot) = (M/M_\odot)^{1/4}(T_e/T_{e\odot})^2.$$

Thus, even M dwarfs with $T_e = 1/2\,T_{e\odot}$, and $M/M_\odot = 0.1$, have potential energy about 0.15 of that of the sun; this is enormous when compared to the ratio of luminosity, which is less than 0.001. Let us assume that some type of equipartition may exist between magnetic and gravitational energies. If potential energy changes occur in a magnetized plasma because of internal structural readjustments, as in the solar cycle, large magnetic energy releases might occur. Therefore, the stabilization of an internally convecting structure, with release of rotational and magnetic energy, will for a long time produce variability, flares and intense stellar winds, on a relatively large scale compared to the nuclear energy release.

Naturally, I must close with my own favorite topic, the white dwarfs. Here in the celestial mortuary are found, as in Forest Lawn cemetery in Los Angeles, the record of the peculiarity of the star's youth. Gravitational separation of the elements can produce the hydrogen-rich, metal-poor, surface composition of most of the white dwarfs. But extremes of nuclear evolution are demonstrated to the eye by the white dwarfs with helium and carbon-rich surfaces. We can discover any desired number of blue white dwarfs from the lists by Giclas *et al.*, and of Luyten, and these can easily be confirmed by photoelectric and spectroscopic observations down to the 17th magnitude. But what we really need, instead, are more examples of the pathological types of white dwarfs. An important fact is the rarity of objects of blue color but low luminosity. Is there an effective upper limit to the central density? The extraordinarily interesting possibility of vibration leading to gravitational collapse in high-density white dwarfs has suggested that a new type of energetic event may be found. We need, even more badly, as you will see from work by Eggen and by myself, discovery methods for fainter, red degenerate stars. And most of all, we need parallax measurements, and more parallax measurements.

I know we will have a good meeting. I hope the observers from Leander McCormick, Sproul, Allegheny, Lowell, and the Naval Observatory who do the dirty work of collecting astrometric data will listen to and enjoy contact with the other speculative world of theoretical astrophysics. And with astrometric patience, I hope theoreticians can wait the decades needed to get data and appreciate the art and skill required. It is only from a synthesis of fact and fancy by both groups that we can understand our strange world.

Section I

Subluminous stars

O. J. EGGEN

Mount Stromlo and Siding Spring Observatories
Australian National University

Abstract

Photometric observations of 150 stars with large proper motion indicate that either (a) the space density of late-type degenerate stars is at least an order of magnitude greater than the early-type degenerates or (b) between the degenerate stars and the main sequence there is a population of stars that show an ultraviolet excess larger than expected from abundance effects alone. The available luminosity estimates indicate that the alternate (b) may be correct.

IN THE present context the term subluminous will be used for stars more than two or three magnitudes below the main sequence. These stars might be called white dwarfs except that, strictly, the latter designation is reserved for blue degenerate objects and some of the stars discussed here may not be wholly degenerate and certainly are not blue.

Photometry of 347, or 75 per cent, of the 461 stars with color classes 0 or -1 found in the first 180 fields of the Lowell proper motion survey has been discussed elsewhere (Eggen 1968a; Paper I). It was found that 75 per cent of the stars of these color classes are probably white dwarfs of the DA type ($B-V$ between about 0 and $+0.41$): this becomes 90% if we consider only stars with proper motions exceeding $0''.2$/year. A minimum space density of about 1×10^{-3} ps^{-3} was derived for these objects. An additional, smaller sample of 75 southern stars with proper motion exceeding $0''.2$/year and given a color class of a or b by Luyten, confirmed this result.

A third sample of suspected white dwarfs is given in Table I. These 59 stars represent 50 per cent of the 119 objects brighter photovisual magnitude than $Pv = 15^{m}$ in Luyten's Bruce Proper Motion Catalogue, south of declination $-45°$ and, from photographic colors, considered by Luyten (1958) as probable white dwarfs. The position of these stars in the $(U-B, B-V)$ plane

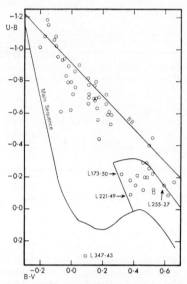

FIGURE 1   Position of the stars in Table I in the $(U - B, B - V)$ plane.

is shown in Figure 1 together with the normal relations for Hyades main-sequence stars and for blackbodies (**BB**). Also, the domain of the subdwarfs is outlined. Seventy per cent of these stars are almost certainly white dwarfs and at least three of the objects lying in the subdwarf region are, in fact, also white dwarfs if we accept the evidence that the space velocity of stars near the sun does not exceed 600 km/sec: L173–50, L221–49 and L255–27. The differences $Pv - V_E$ show a dispersion of nearly one magnitude but the correlation between the photographic colors, C, and $U - V$ is, as seen in Figure 2, considerably better. As suspected by Luyten (1958) the values of C correlate better with $U - V$ than with $B - V$ because of the ultraviolet included in the photographic magnitudes. As a result there is a very natural division at $U - V = C = 0.0$ and the bluer stars are all white dwarfs. The minimum number of white dwarfs of the DA type south of declination $-45°$ and brighter than $Pv = 15^{m}$ is then about 50. If, as before (Eggen 1968a), we assume a value of $M_V = +12.5$ for those predominately DA type stars, the space density of about 2×10^{-3} pc^{-3} agrees reasonably well with that derived from the deeper (limiting magnitude 17) survey of the Lowell proper motion stars which covered twice the area. The minimum proper motions in both surveys is near 0″.05/year. The distribution of the tangential velocities of the white dwarfs in Table I is shown in Figure 3 (hatched) superimposed on that for the white dwarfs in the Lowell survey (Eggen 1968a).

TABLE I Photometry of white dwarf suspects south of −45°.

Name	α (1950)		δ		P_v	C	μ	V_E	$B - V$	$U - B$	n
L170-27	0	24.3	−55	42	15ᵐ.06	−0ᵐ.16	0ᵛ.59	15ᵐ.21	+0ᵐ.15	−0ᵐ.68	1
L291-66*	0	39.4	−47	45	14.28	+0.09	0.09	14.47	+0.50	−0.255	4
L219-48	0	47.8	−52	25	14.43	−0.18	0.11	14.20	−0.02	−0.79	2
L221-49	1	07.2	−53	14	16.17	−0.20	0.23	15.04	+0.38	−0.09	3
L171-50	1	08.3	−58	08	14.54	+0.06	0.12	14.20	+0.48	−0.29	2
L222-53	1	26.1	−53	17	14.86	−0.30	0.20	14.48	+0.01	−0.72	2
L173-50	2	00.0	−58	12	14.48	−0.02	0.10	14.25	+0.32	−0.22	3
L54-5	2	55.8	−70	34	14.30	−0.06	0.67	14.08	+0.23	−0.59	3
L175-34	3	08.5	−56	34	13.71	−0.50	0.11	14.07	−0.11	−1.06	4
L227-140	3	11.3	−54	18	14.43	−0.05	0.08	14.71	+0.16	−0.55	3
L230-204	4	18.1	−53	58	14.58	−0.12	0.11	15.32	−0.10	−1.08	2
L302-89	4	19.6	−48	46	14.87	−0.19	0.56	14.36	+0.52	−0.45	2
L31-99	4	46.5	−78	57	13.50	−0.40	0.06	13.47	−0.10	−1.02	2
L182-61*	6	15.5	−59	11	13.80	−0.30	0.30	14.09	−0.09	−0.95	6
L184-75	7	01.5	−58	46	14.81	−0.41	0.19	14.46	+0.22	−0.72	2
L384-25	7	32.1	−42	47	14.37	−0.45	0.66	14.16	+0.11	−0.66	1
L97-12	7	52.8	−67	38	14.34	+0.29	2.05	14.09	+0.66	−0.17	5
L97-3	8	06.4	−66	09	13.82	−0.02	0.47	13.92	+0.05	−0.90	2
L243-50	8	30.4	−53	30	14.39	−0.44	0.17	14.47	−0.15	−1.15	2
L139-26	8	50.0	−61	43	14.67	−0.59	0.09	14.73	−0.02	−0.89	1
L64-40	9	28.6	−71	20	15.83	−0.23	0.43	15.44	+0.175	−0.435	2
L189-39	9	50.0	−57	12	14.87	−0.22	0.19	15.29	−0.07	−0.61	1
L64-27	9	54.6	−71	02	13.83	−0.19	0.17	13.48	+0.16	−0.69	1
L247-17	10	04.4	−50	55	11.95	+0.01	0.14	11.90	+0.49	−0.20	2
L101-80	10	42.6	−69	02	12.45	−0.30	0.28	13.09	−0.04	−0.84	2
L250-52	10	53.3	−55	04	13.81	+0.02	0.40	14.32	+0.095	−0.62	4
L251-24	11	21.5	−50	44	14.73	−0.35	0.12	15.00	+0.02	−0.86	2

continued

TABLE I—*(cont.)*

Name	α (1950)		δ		Pv	C	μ	V_E	B−V	U−B	n
L145–141	11	42.9	−64	34	11.42	−0.24	2.68	11.63	+0.22	−0.66	2
L325–214	11	53.7	−48	24	13.02	−0.86	0.05	12.85	−0.20	−1.01	2
L255–27	12	23.8	−53	16	13.05	+0.05	0.30	12.36	+0.595	−0.15	3
L327–186	12	36.1	−49	33	13.53	−0.14	0.56	13.96	+0.18	−0.70	2
L257–47	13	23.0	−51	26	14.65	−0.44	0.50	14.60	0.00	−0.79	2
L19–2	14	25.4	−81	07	13.21	−0.08	0.45	13.75	+0.25	−0.53	1
L72–91	15	24.3	−74	55	15.93	−0.31	0.44	15.93	−0.06	−0.93	1
L299–41	16	57.8	−52	43	15.42	−0.35	0.34	15.85	−0.155	−1.185	2
L203–131	17	09.8	−57	34	14.75	−0.16	0.21	15.10	+0.03	−0.76	3
L270–137	17	33.2	−54	24	15.25	+0.35	0.45	15.90	+0.50	−0.40	2
L342–53	17	41.3	−46	41	13.51	−0.09	0.12	13.36	+0.62	−0.08	1
L342–40	17	44.5	−46	03	14.00	+0.01	0.20	13.90	+0.53	−0.11	3
L158–53	18	37.6	−61	56	14.67	−0.13	0.39	14.94	+0.11	−0.70	6
L347–43	19	19.8	−46	52	13.42	−0.36	0.04	13.69	+0.09	+0.29	3
L160–108	19	44.8	−63	07	10.27	+0.19	0.44	10.89	+0.42	−0.21	3
L277–125	20	08.0	−52	21	11.94	+0.16	0.21	10.90	+0.52	−0.22	2
L210–114	20	14.9	−57	31	13.42	−0.67	0.08	13.73	−0.17	−1.08	4
L279–25	20	34.5	−53	16	14.35	−0.51	0.20	14.46	−0.05	−0.94	3
L116–79	20	39.6	−68	16	13.02	−0.07	0.25	13.53	+0.11	−0.78	2
L24–52	21	05.2	−82	01	13.61	−0.02	0.37	13.62	+0.24	−0.61	2
L212–19	21	15.8	−56	03	14.24	−0.12	0.45	14.28	+0.26	−0.59	2
L213–30	21	46.2	−56	18	13.17	+0.12	0.10	13.30	+0.46	−0.12	1
L283–7*	21	54.4	−51	14	14.89	−0.08	0.41	14.68	+0.16	−0.80	4
L48–15	21	59.8	−75	28	14.83	−0.13	0.52	15.06	+0.16	−0.63	4
L356–71	22	10.6	−47	03	14.13	+0.22	0.10	14.02	+0.41	−0.15	1
L119–34	22	16.2	−65	44	14.66	−0.18	0.65	14.43	+0.135	−0.77	2
L214–57	22	32.0	−57	31	14.53	−0.19	0.20	14.96	+0.105	−0.815	2

L285–14	22	48.0	−50	26	14.62	−0.22	0.21	14.92	+0.11	−0.72	3
L385–17	22	55.0	−46	33	14.50	−0.11	0.18	13.97	+0.465	−0.29	3
L390–96	23	33.4	−46	13	14.35	+0.09	0.09	14.06	+0.38	−0.18	4
L26–23	23	37.7	−76	03	14.69	−0.12	0.26	13.53	+0.53	−0.11	1
L289–12	23	46.3	−50	43	14.52	+0.15	0.06	14.49	+0.47	−0.20	3

NOTES TO TABLE

*L291–66 Cpm V_E = 14^m.80, B − V = +1.32, U − B = +0.80
L182–61 Cpm V_E = 6.44, B − V = +0.58, U − B = +0.13
L283–7 Cpm V_E = 10.36, B − V = +1.51, U − B = +0.83

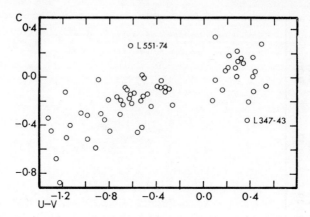

FIGURE 2 Comparison between the photoelectric and photographic colors.

In an attempt to estimate the number of later type subluminous stars, observations of all stars of color class 1 in four Lowell proper motion fields (45 stars) and of 104 southern objects with proper motion greater than $0''.2$/year and color classes f, g and k were also discussed in Paper I (Eggen 1958a). If we place an upper limit of 600 km/sec on the space velocity, as indicated by the distribution of space velocities of some 150 known subdwarfs, and assume that ultraviolet excesses that place stars outside the domain of subdwarfs in Figure 1 are indicative of subluminous stars, it was found that these late-type objects are an order of magnitude more numerous than the white dwarfs. This result was tested in Paper II (Eggen 1968b) with photometric observations of 145 randomly selected proper motion stars including those of color classes 1 and 2 in the Lowell proper motion fields (G99 and 160) and stars of color classes g and k in the LTT catalogues (Luyten 1957, 1961). Applying only the condition that the space motions shall not exceed 600 km/sec, 19 or 13 per cent of these stars are probably subluminous.

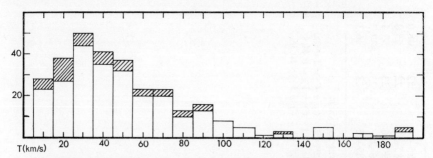

FIGURE 3 Distribution of the tangential velocities of the white dwarfs.

An unexpected result for these late-type stars was the predominance of large ultraviolet excesses. The position of the stars in Lowell fields G99 and G160, from Paper II, are shown as crosses in the $(U - B, B - V)$ plane of Figure 4 where the normal relations for Hyades main-sequence stars and for

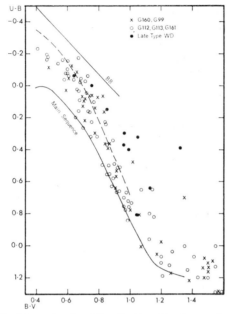

FIGURE 4 Positions in the $(U - B, B - V)$ plane of some proper motion stars and known, late type subluminous objects.

blackbodies (BB) are also indicated. When corrections for the 'abundance effect' on the (UBV) colors were first computed (cf. Sandage and Eggen 1959) it was recognized that the slope of the blanketing vectors in the $(U - B, B - V)$ plane increased with increasing color and near $B - V = +0.8$ they became nearly parallel with the intrinsic $(U - B, B - V)$ relation. Because of the extreme difficulty in measuring the effect of line blocking in stars this cool (cf. Wildey *et al.* 1962) there is no precise estimate of the maximum ultraviolet excess expected from the abundance effect. An estimate of this maximum can be obtained from the members of the σ Pup group (Eggen 1964) which have only slightly greater metal abundance than most of the globular cluster stars and has members bright enough for accurate photometry. The relation between the observed colors $(B - V)_0$, and the ultraviolet excess, δ, is shown in Figure 5. Many observations of field sub-dwarfs of types F and G as well as some observations in globular clusters, indicate that $\delta = +0.30$ is a conservative estimate of the maximum value at $(B - V)_0$ near $+0.5$. If the attenuation with increasing color follows that of the members of the σ Pup group, the broken line in Figure 5 should represent the maximum values of the excess that are expected from the abundance effect. The filled circles in Figure 5 represent the few, known late-type subdwarfs, including Groombridge 1830, ADS 2757A and B and the pair HD 134439/40. Assuming that the broken curve in Figure 5 represents the maximum values of δ from the abundance effect, the broken curve in Figure 4 then constitutes the upper envelope for subdwarfs in the $(U - B, B - V)$ plane.

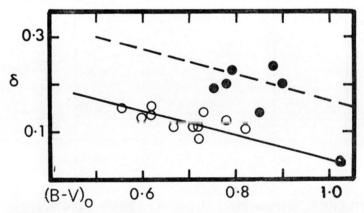

FIGURE 5.   The ultraviolet excesses for members of the Sigma Puppis group (open circles) and some known, late type subdwarfs (dots).

New observations of all the stars of color classes 1 and 2 in the three Lowell proper motions fields G112, G113 and G161 (88 stars) are given in Table II; other sources of proper motion and overlapping Lowell fields are also noted. These stars are shown in the $(U - B, B - V)$ plane of Figure 4 by open circles. For comparison, the known late-type subluminous stars (Eggen and Greenstein 1965, 1968; Eggen 1968a) are shown in the figure as filled circles. At least a half dozen of the stars in Table II probably have tangential velocities exceeding 600 km/sec, as subdwarfs, regardless of the interpretation of the large ultraviolet excess (G112–1, 112–28, 113–28, 161–56, 161–79). Also, because of the similarity of their colors with those of known subluminous stars (filled circles), G112–26, 113–6, 113–36, and 113–48

TABLE II Stars of color +1 or +2 in Lowell proper motion fields G112, 113 and 161.

Name	α/δ (1950)	Pg	C	μ	θ	V_E	B−V	U−B	n	δ $(0^m.001)$
G112–1	7 09.5 +6 12	$15^m.1$	+1	$0''.27$	235°	$14^m.30$	$+0^m.67$	$-0^m.16$	3	+37
G112–3	7 13.9 −2 00	12.3	+2	0.30	159	10.71	+0.99	+0.76	1	+11
G112–5	7 14.8 −3 06	15.0	+2	0.27	170	12.58	+1.41	+1.07	1	—
G112–6	7 16.1 +3 47	12.5	+1	0.39	167	11.11	+0.685	−0.045	2	+24
G112–7 −4° 1914	7 18.2 −4 45	10.5	+2	0.29	340	9.91	+0.87	+0.55	1	+3
G112–12	7 19.6 −1 46	10.2	+2	0.56	255	8.75	+0.79	+0.32	1	+9
LTT2813		10.0	KO	0.42	305					
G112–13	7 19.7 +9 00	11.0	—	0.31	155	9.35	+0.93	+0.66	2	+4
G89–13		10.8	+2	0.32	152					
LTT12008		8.9	F2	0.38	150					
G112–14	7 19.8 +8 55	11.0	+1	0.31	155	10.24	+0.47	−0.16	2	+17
G89–14		11.3	+1	0.32	152					
G112–19	7 25.8 +8 15	14.4	+1	0.28	170	13.26	+0.66	+0.02	1	+19
G89–21		14.4	+1	0.27	163					
G112–21	7 26.2 −3 11	13.2	+2	0.95	151	11.49	+1.55	+1.13	1	—
G112–23	7 27.0 +8 33	13.5	+1	0.38	185	12.58	+0.99	+0.72	2	+10
G89–24		14.3	+2	0.34	176					
G112–26	7 29.2 −0 56	15.0	+1	0.31	153	13.29	+0.71	−0.025	2	+30
G112–28	7 33.2 +3 37	15.1	+2	1.01	169	13.72	+1.15	+0.82	2	+20

continued

TABLE II—(*cont.*)

Name	α/δ (1950)	Pg	C	μ	θ	V_E	B−V	U−B	n	δ $(0^{\rm m}.001)$
G112–32	7 35.2 / +5 50	11.3	+1	0.33	123	10.28	+0.65	+0.30	1	0
G89–33		10.8	+1	0.32	124					
G112–33	7 36.0 / +6 06	11.2	+2	0.28	258	9.68	+1.01	+0.73	2	+13
LTT12052		9.8	K2	0.21	256					
G89–35		11.3	+2	0.28	259					
G112–36	7 37.3 / −1 24	10.1	+2	0.33	152	9.22	+0.84	+0.245	2	+28
LTT2909		10.2	G0	0.27	150					
G112–43	7 41.2 / +0 04	11.5	+1	0.33	202	10.20	+0.48	−0.18	1	+20
G112–44		12.5	+1	0.33	202	11.27	+0.57	−0.16	1	+26
G112–45	7 41.6 / +0 55	13.3	+2	0.30	95	11.81	+1.20	+1.11	1	0
G112–47	7 43.3 / +0 59	10.7	+2	0.28	194	9.06	+0.68	+0.09	1	+14
G113–3	7 51.0 / −6 36	13.4	+2	0.49	163	12.07	+1.26	+1.12	1	0
LTT2970		13.5	k−m	0.43	162					
G112–54	7 52.0 / −1 17	8.9	+2	0.31	275	7.43	+0.73	+0.16	3	+16
G113–4		8.9	+1	0.36	264					
LTT2974		8.3	G5	0.27	258					
G113–6	7 54.1 / −7 36	14.7	+2	0.27	130	13.54	+0.73	−0.06	2	+36
LTT2993		14.6	k	0.24	139					
G113–9	7 58.4 / −3 57	12.3	+2	0.31	164	10.98	+0.47	−0.19	2	+20
G113–11	7 59.8 / −7 26	15.1	+1	0.32	56	13.98	+0.84	+0.23	2	+29
G113–14	8 00.9 / −1 01	9.3	+2	0.27	198	8.26	+0.75	+0.32	1	+2
LTT3038		9.1	G5	0.23	198					

Name	RA	Dec									
G113-15	8 03.8	+0 02	11.5	+2	0.29	154	9.92	+0.91	+0.66	2	0
+0°2186	8 10.5	−7 25	15.8	+2	0.49	146	13.75	+1.44	—	1	—
G113-17											
G113-18	8 10.6	−4 22	12.4	+2	0.29	288	10.58	+0.96	+0.70	1	+6
LTT3081			12.8	m	0.32	292					
G113-21			10.6	+2	0.27	302					
G113-22	8 13.6	−5 05	10.2	+1	0.33	123	9.36	+0.59	−0.08	1	+20
G113-24	8 14.4	+0 10	8.7	+1	0.54	197	9.69	+0.595	−0.08	2	+20
LTT3098			8.2	F8	0.48	200					
G113-25			15.1	+2	0.27	147					
G113-28	8 15.0	−3 50	15.5	+1	0.27	163	7.74	+0.58	−0.04	1	+15
G113-33	8 16.0	−6 06	10.3	+1	0.27	173	13.68	+1.40	—	1	—
G113-34	8 16.2	−2 06	12.7	+2	0.27	202	14.78	+0.77	+0.10	2	+28
G113-35	8 21.0	+2 21	12.9	+2	0.34	136	9.31	+0.70	+0.26	1	−1
LTT3125			13.3	k	0.28	136					
G113-36			16.5	+2	0.30	141					
G113-38	8 23.1	−2 15	16.1	+2	0.48	221	10.83	+0.85	+0.56	1	−2
G52-13	8 23.2	−6 44	16.0	+3	0.55	219	12.18	+0.97	+0.78	1	0
G113-40	8 23.8	−5 48	13.4	+2	1.00	157	15.45	+1.14	+0.64:	1	(+30:)
LTT3144	8 25.1	+4 28	13.5	k − m	0.96	156	15.3:	+1.35	+1.0:	2	—
G113-41	8 27.1	−1 34	13.1	+2	0.28	222	11.97	+0.93	+0.51	3	+19
LTT3153	8 28.5	−5 00	13.0	m	0.25	214	11.44	+1.34	—	1	—

continued

TABLE II—*(cont.)*

Name	α/δ (1950)	Pg	C	μ	θ	V_E	B − V	U − B	n	δ (0^m.001)
G113-42	8 2.89	12.9	+2	0.44	256	11.21	+1.41	—	1	—
LTT3154	−5 52	12.8	m	0.44	259					
G113-43	8 28.9	14.6	+2	0.44	256	12.37	+1.47	—	1	—
LTT3155	−5 51	13.9	m	0.44	259					
G113-44	8 29.4	14.1	+2	0.39	192	12.87	+1.34	—	1	—
LTT3158	−2 22	14.6	m	0.33	196					
G113-46	8 31.9	14.3	+2	0.56	151	12.73	+1.62	—	1	—
G114-2	−0 58	14.6	+2	0.49	152					
LTT3171		14.8	m	0.47	149					
G113-47	8 31.9	15.0	+2	0.32	124	13.51	+1.20	+1.1:	1	0
G114-3	−5 32	14.9	+1	0.27	132					
LTT3172		14.8	m	0.31	147					
G113-48	8 32.5	15.6	+2	0.31	180	14.78	+1.07	+0.65	3	+32
G114-5	−4 04	15.6	+1	0.27	186					
G113-49	8 32.6	14.1	+1	0.28	136	12.56	+0.75	+0.23	1	+11
	+3 38									
G113-50	8 32.9	12.8	+2	0.27	178	12.25	+0.64	−0.04	1	+22
	+4 37									
G161-1	9 10.0	13.6	+2	0.30	285	12.37	+1.34	—	1	—
G114-46	−6 58	14.0	+2	0.28	279					
LTT3391		14.4	m	0.30	290					
G161-3	9 11.2	12.4	+1	0.30	159	10.63	+0.67	+0.16	1	+5
G114-48	−3 41	12.1	+1	0.27	165					
LTT3400		13.2	k	0.20	169					
G161-4	9 11.3	13.7	+2	0.31	137	12.03	+1.27	+1.05	1	—
LTT3401	−10 20	13.2	m	0.30	135					

Name	α (9ʰ)	δ	m	Sp	μ	θ	V	B−V	U−B	n	(Δ)
G161-6	9 11.6	+0 28	11.9	+2	0.27	259	9.70	+0.88	+0.63	1	−3
G114-50			11.4	+2	0.27	257					
G161-8	9 13.5	−6 44	15.4	+1	0.27	163	14.35	+0.975	+0.66	2	+12
G114-51			15.9	+2	0.30	180					
G161-9	9 14.4	−5 11	14.3	k	0.28	166	13.70	+0.68	−0.02	2	+25
LTT3415			14.8	+1	0.22	171					
G161-14	9 17.7	−5 09	13.5	k	0.27	159	12.07	+0.60	−0.10	1	+23
			13.5		0.32	127					
G161-16	9 20.3	−9 52	16.3	+2	0.30	130	15.06	+1.58	—	1	—
G161-17	9 20.6	−12 25	14.1	+1	0.28	128	12.88	+0.74	+0.14	1	+18
G161-18	9 20.8	−4 56	11.7	+2	0.28	272	9.76	+0.86	+0.50	1	+6
LTT3455			12.6	—	0.25	276					
G161-21	9 21.3	+0 21	14.0	+2	0.30	228	—	+1.50	+1.20	1	—
G48-10			12.5	+2	0.32	225					
G161-24	9 22.7	−12 45	10.5	—	0.78	130	10.10	+0.96	+0.76	1	0
LTT3468			10.7	K2	0.86	133					
G161-25	9 22.9	−7 07	14.6	+2	0.85	130	12.66	+1.20	+1.07	2	5
			13.8		0.79	132					
G161-26	9 23.0	+0 32	16.1	K5	0.63	179	15.12	+1.50	—	1	—
G46-40			16.3	+2	0.61	186					
G48-14			16.3	+4	0.62	192					
G161-29	9 23.2	−6 33	12.7	+2	0.35	132	11.36	+0.98	+0.845	2	−4
LTT3472			12.2	k	0.35	135					
G161-32	9 26.3	−11 57	14.1	+2	0.37	175	12.27	+1.47	—	1	—
G161-35	9 26.8	−7 32	16.6	+2	0.30	198	16.00	+1.54	+1.0:	1	—

B

Low-luminosity stars

TABLE II—(*cont.*)

Name	α/δ (1950)	Pg	C	μ	θ	V_E	B − V	U − B	n	δ (0^m.001)
G161-41 / G48-20 / LTT	9 28.3 / +0 33	14.0 / 12.7	+2 / +2	0.79	226	11.70	+1.60	+1.35	1	—
G161-43	9 28.9 / −2 28	16.0	+2	0.35	197	14.74	+1.44	+1.20	1	—
G161-45 / LTT3510	9 30.0 / −10 58	10.2 / 8.6	+2 / K0	0.30 / 0.27	273 / 276	7.88	+0.925	+0.51	2	+17
G161-46 / LTT3514	9 31.0 / −10 58	12.3 / 12.7	+2 / K2	0.28 / 0.20	238 / 234	10.30	+0.89	+0.58	1	+4
G161-53 / BPM73515	9 35.0 / −5 26	13.4	+2	0.35	280	11.96	+1.13	+0.96	2	—
G161-56	9 37.4 / −9 32	16.1	+1	0.37	152	15.90	+1.12	+0.8:	2	+20:
G161-58 / G48-29 / LTT	9 38.1 / +1 14	11.6 / 10.8	+1 / +1	0.56 / 0.48	165 / 163	10.48	+0.41	−0.23	1	+24
G161-59 / BPM73550	9 38.2 / −4 03	13.0	+1	0.27	147	11.55	+0.72	+0.05	1	+24
G161-62 / LTT3550	9 38.6 / −8 14	12.7 / 11.3	+2 / g	0.30 / 0.29	161 / 162	10.96	+0.90	+0.49	1	+15
G161-65 / LTT3557	9 39.7 / −7 32	9.8 / 9.7	+2 / G0	0.32 / 0.26	215 / 215	8.58	+0.67	+0.03	1	+18
G161-69 / LTT3566	9 41.4 / −9 46	11.0 / 9.5	+2 / G5	0.29 / 0.22	261 / 268	9.34	+0.61	−0.07	1	+21
G161-72 / LTT3577	9 42.5 / −3 30	15.3 / 14.3	+2 / m	0.30 / 0.27	117 / 112	13.40	+1.30	—	1	—

continued

Name											
G161–73	9	43.1	12.2	+1	0.31	152	10.86	+0.54	−0.14	1	+21
LTT3581	−4	26	12.2	g	0.27	150	11.40	+1.40	—	1	—
G161–77	9	46.2	14.4	+2	0.30	175	14.01	+0.85	+0.37	2	+17
LTT3600	−10	21	13.6	m	0.25	172					
G161–79	9	48.4	15.0	+1	0.35	138					
G162–2	−6	31	14.5	+1	0.32	133					
LTT3611			14.7	k	0.29	138					
G161–80	9	48.7	12.5	+2	1.90	143	10.10	+1.48	—	1	—
LTT3614	−12	05	11.4	M2	1.79	143	12.77	+1.41	—	1	—
G161–81	9	48.7	14.8	+2	0.27	317					
G162–3	−10	14	14.9	+2	0.27	317					
G161–82	9	49.0	13.1	+1	0.27	127	12.02	+0.60	−0.17	1	+30
G162–4	−4	24	13.0	+1	0.27	132					
LTT3615			12.4	g	0.34	118					
G161–83	9	49.1	11.5	+2	0.28	235	9.46	+0.82	+0.50	1	−2
G162–5	−2	39	11.7	+2	0.28	232					
LTT3618			10.4	K0	0.27	230	12.25	+0.89	+0.34	1	+28
G161–84	9	49.2	14.5	+1	0.28	278					
G162–6	−3	36	13.3	+1	0.30	281					
LTT3620			13.3	g	0.30	272					
G161–87	9	51.6	14.8	+2	0.37	155					
LTT3633	−12	01	13.7	m	0.32	159	12.50	+1.37	+1.15	1	—
G161–90	9	55.2	15.7	—	0.35	121					
G162–10	−2	35	15.7	+2	0.34	126	14.50	+1.55	+1.30	1	—

are likely prospects. A similar number of objects in the previously published
Lowell fields (crosses in Figure 4) are also probable subluminous stars by the
same arguments.

Most of the samples discussed thus far represent random selections of
proper motion stars. A different approach was made to the selection of the
60 stars in Table III. Values of $B - V = +0.6$ and $+1.0$ and $M_V = +5.5$
and $+6.5$ were assumed to apply to color classes g and k of the LTT catalogues (Luyten 1957, 1961) and the stars in Table III were selected on the
basis of a tangential velocity larger than 500 km/sec if the above calibration
of the photographic colors applies. The distribution of these stars in the
$(U - B, B - V)$ plane is shown in Figure 6 by open circles; the known
subluminous stars of late type are shown as filled circles. A dozen of these
stars are considerably redder than indicated by the photographic color class:
these are mainly M type dwarfs. Fifteen stars lie between the main-sequence
relation and the broken curve representing the maximum ultraviolet excess
expected from the abundance effect but at least nine of these (LTT 1116,
1424, 1679, 2480, 3746, 4124, 6885 and 9765) are probably subluminous

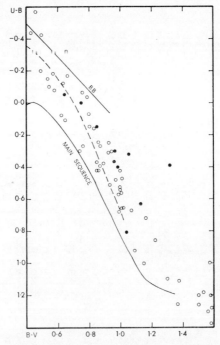

FIGURE 6 Positions of the stars in Table 3 (open circles) and some
known, subluminous late type stars (dots) in the $(U - B, B - V)$ plane.

TABLE III Proper motion stars which, on the basis of their photographic magnitudes and colors, may have tangential velocities greater than 500 km/sec.

LTT	α (1950)	δ	Pg	CC	μ	V_E	$B-V$	$U-B$	n	δ (0^m.01)	π pt (0″.001)
237	00 24.5	−32 37	13^m.3	k	0″.47	12^m.11	+0^m.99	+0^m.57	4	+25	—
329	00 34.9	−21 09	14.2	g	0.32	14.53	+0.45	−0.57	3	+57	40
375	00 39.0	−22 38	15.1	g:	0.60	14.60	+0.65	−0.32	4	+51	87, 30
378	00 39.3	−35 38	14.2	k	0.77	13.38	+1.30	+1.10	4	—	—
458	00 46.1	−17 10	14.1	g	0.26	14.04	+0.54	−0.15	2	+22	110, 40
560	00 57.1	−26 48	14.9	g	0.27	15.95	+0.495	−0.425	2	+45	42, 14
598	01 01.5	−46 03	11.8	k	1.71	11.63	+1.09	+0.92	2	+9	Note
934	01 41.6	−67 32	14.7	k	1.05	13.90	+0.43	−0.44	3	+44	100, 35
1116	02 06.5	−49 28	14.4	g	0.22	14.21	+0.545	−0.10	2	+18	100, 35
1418	02 55.8	−36 50	13.4	k	0.56	13.00	+1.01	+0.66	2	+20	—
1424	02 56.3	−29 15	14.4	k	0.50	13.56	+0.49	−0.20	3	+22	140, 47
1534	03 12.0	−56 18	14.1	g	0.30	14.71	+0.16	−0.55	3	+67	36
1561	03 15.7	−2 37	14.5	k	0.31	14.99	+0.92	+0.25	4	+43	125, 43
1611	03 21.5	−73 49	13.5	g	0.39	13.13	+0.89	+0.375	3	+24	(180, 83)
1679	03 31.3	−52 50	16.2	k	0.45	14.92	+0.74	+0.30	2	+2	100, 35
1815	03 50.2	−46 05	15.2	k	0.51	14.36	+1.22	+0.86	2	+27	sd?
1821	03 51.5	−37 11	12.8	k	1.14	12.22	+0.85	+0.42	3	+12	Note
1925	04 15.1	−26 10	13.4	g	0.62	11.77	+1.37	+1.11	2	—	Note
2008	04 28.8	−11 37	15.2	k	0.47	15.06	+0.83	+0.16	2	+34	Note
2030	04 31.9	−14 03	15.0	g	0.44	14.57	+0.80	+0.15	3	+28	110, 38
2103	04 48.1	−39 59	14.1	k	0.51	11.88	+0.86	+0.42	2	+14	—
2175	05 05.3	−23 15	14.1	k	0.40	13.71	+0.63	−0.02	3	+19	—
2236	05 16.9	−53 43	13.1	k	0.52	13.19	+1.16	+0.72	2	+38	Note
2480	06 09.4	−43 25	13.4	k	0.73	12.91	+0.125	+0.075	2	+8	220, 70
2535	06 20.3	−12 50	13.3	g	0.88	13.18	+1.00	+0.53	3	+31	265, 100
2746	07 06.9	−22 44	13.3	g	0.47	11.57	+1.50	+1.26	2	—	—

continued

TABLE III—*(cont.)*

LTT	α (1950) δ				Pg	CC	μ	V_E	B − V	U − B	n	δ (0^m.01)	π pt (0″.001)
2981	07	52.8	−67	38	15.0	g	2.05	14.09	+0.66	−0.17	5	+37	115, 38
3144	08	27.1	−01	34	13.5	k	0.96	11.97	+0.93	+0.51	3	+19	Note
3578	09	42.6	−18	00	14.3	k	1.58	12.70	+1.58	+1.38	2	−7	—
3611	09	48.4	−06	31	14.7	k	0.29	14.01	+0.85	+0.37	2	+17	—
3746	10	11.0	−57	35	13.9	g	0.37	13.12	+0.645	+0.11	2	+	190, 70
4124	11	09.1	−40	48	13.8	k	1.26	14.32	+0.99	+0.68	2	+14	165, 55
4210	11	20.1	−26	56	12.2	g:	0.31	12.78	+0.56	−0.18	3	+27	Note
4211	11	20.1	−26	56	14.8	g:	0.31	14.94	+1.00	+0.54	4	+30	
4697	12	23.4	−24	19	13.7	k	0.98	12.80	+0.97	+0.48	4	+30	300, 100
4702	12	23.4	−48	35	12.0	g	0.60	10.72	+1.15	+1.00	2	+10	Note
4806	12	35.4	−51	44	$11^m.4$	g	$1″.02$	$10^m.82$	$+1^m.53$	$+1^m.18$	3	—	Note
4984	12	59.5	−1	48	14.0	g	0.47	12.84	+0.76	−0.07	3	+43	Note
5028	13	04.8	−33	50	14.4	g	0.3	13.53	+0.58	−0.08	3	+19	210, 70
5074	13	11.9	−3	40	13.7	g	0.55	12.94	+0.95	+0.30	3	+44	185, 63
5299	13	38.1	−24	21	14.1	g	0.7	13.26	+0.78	−0.04	2	+43	190, 66
14272	14	27.6	+39	45	15.4	k	0.27	13.51	+0.86	+0.27	2	+39	
6767	16	54.4	−4	17	14.1	g − k	0.75	12.30	+1.61	+1.20	2	—	Note
6854	17	09.3	−1	48	13.1	g	0.53	11.45	+1.60	+1.28	3	+9	150, 50
6885	17	15.1	−43	23	14.6	k	1.05	13.98	+0.76	+0.27	3	—	
7048	17	39.5	−8	48	15.7	g	0.95	13.60	+1.58	+1.30	2	—	Note
7050	17	39.9	−16	27	13.8	g	0.70	13.15	+1.70	—	2	+38	sd ?
7151	17	56.1	−9	24	14.5	g	0.27	13.78	+0.925	+0.31	2	—	
7164	17	58.5	−20	35	15.5	k	0.35	14.55	+1.57	+1.00:	2	—	
7297	18	18.4	−1	04	14.7	k	1.03	12.71	+1.66	+1.12	2	—	sd ?
7381	18	33.0	−8	18	15.3	k	1.26	13.70	+1.36	+1.25	2	—	
7424	18	41.1	−20	42	14.8	g	0.31	12.86	+1.00	+0.56	2	+28	(320, 100)
9267	22	52.3	−75	42	12.6	k	1.44	10.22	+1.50	+1.20	2	—	Note

													Note
9324	22	58.5	−54	46	13.2	g	0.54	12.33	+0.86	+0.25	4	+31	
9353	23	03.2	−2	26	14.0	k	0.67	12.95	+1.01	+0.65	2	+21	sd?
9372	23	06.1	−8	02	14.2	k	0.40	13.57	+1.00	+0.42	3	+42	(210, 72)
9765	23	51.4	−19	15	14.0	g	0.75	13.55	+0.87	+0.395	4	+18	200, 70

LTT598 The star is almost certainly an extreme subdwarf with a parallax near 0″.02; the trigonometric parallax determinations are 0″.021 (wt. 8) Cape and 0″.008 (wt. 10) Yale.

LTT1821 The magnitudes and colors given in Paper I (Eggen 1968a) are incorrect. If the star is a subdwarf the tangential motion is about 550 km/sec and the parallax near 0″.01. As a white dwarf, on the upper sequence, the parallax is 0″.09. The two trigonometric determinations are 0″.052 (wt. 8) Yale and 0″.043 (wt. 7) Cape for a mean of 0″.049 and $M_v = +10^m.7$.

LTT1925 The single trigonometric parallax determination is 0″.045 (wt. 12) Yale which gives $M_v = +10^m$ and a tangential velocity of 65 km/sec.

LTT2008 An almost identical star $(V_E, B - V, U - B) = (15.08, +0.84, +0.32)$ lies 1′ North Proceeding the object. There is no certainty as to which is the proper motion star, but in either case it is probably a white dwarf with a parallax near 0″.08.

LTT2236 If the star is a subdwarf the parallax is near 0″.01. If it is a white dwarf on the upper sequence the parallax is 0″.1. The single trigonometric determination is 0″.040 (wt. 8) Cape.

LTT3144 If the star is a subdwarf the parallax is near 0″.015. If it is a white dwarf on the upper sequence the parallax is near 0″.1. The trigonometric determinations are 0″.042 (wt. 8) Van Vleck and 0″.049 (wt. 7) Cape for a mean of 0″.045 and $M_v = +10.2$.

LTT4210/11 A parallax of 0″.115 would place the bright star on the upper sequence and the fainter on the lower sequence of white dwarfs. If we ignore the extreme excess of the fainter component and consider both stars as subdwarfs the parallax is 0″.004 and the transverse velocity is near 400 km/sec.

LTT4702 Common proper motion with HR 4734; $(V_E, B - V, U - B) = (6.25, +0.67, +0.21)$. If the faint star is an extreme subdwarf, the parallax is near 0″.035 and the bright star, which shows no ultraviolet excess, is about 1^m above the main sequence. The trigonometric parallax determinations are 0″.019 (wt. 8) Yale, 0″.050 (wt. 16) Lembang and 0″.022 (wt. 7) Cape for the bright star and 0″.044 (wt. 10) Yale, 0″.023 (wt. 16) Lembang and 0″.025 (wt. 6) Cape for the fainter. The mean trigonometric parallax is 0″.034. The radial velocity of the bright star is 30 km/sec.

LTT4806 The trigonometric parallax determinations are 0″.117 (wt. 12) Yale and 0″.115 (wt. 6) Cape. The resulting value of the luminosity is $+11^m.1$ and the tangential velocity is 30 km/sec.

LTT4985 A parallax of 0″.075 would place this star on the upper sequence of white dwarfs. The only trigonometric determination is 0″.074 (wt. 8) Cape. (cf. Eggen and Greenstein 1965).

LTT6854 The trigonometric parallax is 0″.098 (wt. 12) Yale, giving $M_v = +11.4$.

LTT7050 The trigonometric parallax is 0″.050 (wt. 6) Yale, giving $M_v = +11.6$.

LTT9267 The trigonometric parallax determinations are 0″.122 (wt. 10) Yale and 0″.109 (wt. 7) Cape, giving $M_v = +10.6$.

LTT9324 The single trigonometric parallax determination is 0″.020 (wt. 7) Cape, placing the star $2^m.5$ below the main sequence. A parallax of 0″.120 would place it on the upper sequence of white dwarfs.

because the tangential velocities, as subdwarfs, exceed 600 km/sec. The remaining objects, or 55 per cent of the total in Table III, show ultraviolet excesses greater than expected from the abundance effect alone.

Photometric parallaxes have been derived for the 27 stars in Table III which are very probably subluminous and are entered in the last column of the table. Two values of the parallax derived from the assumptions that the stars lie on the upper ($M_V = 11.65 + 0.85\,(U - V)$) or lower ($M_V = 14.0 + 0.85$ $(U - V)$) sequences of white dwarfs (Eggen and Greenstein 1965) respectively are listed; single values are given for the two bluest objects based on an assumed luminosity of $M_V = +12.5$.

As detailed in the notes to Table III, 11 of the stars have trigonometric parallaxes greater than 0″.04. Six of these, unfortunately, refer to the M dwarfs, misclassified as color class g or k, but the remaining five are of earlier type. The positions of the 11 stars in the (M_V, $U - V$) plane are shown in Figure 7 by open triangles labelled with the LTT No. The main sequence

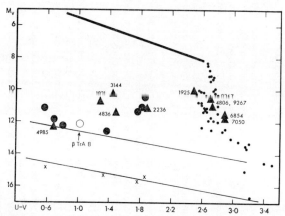

FIGURE 7   Positions in the (M_V, $U - V$) plane of some red degenerate stars (crosses), late type subluminous members of the Hyades cluster (large dots) and stars in Table 3 with trigonometric parallaxes greater than 0″.04 (triangles). The small dots indicate individual M type dwarfs with trignometric parallax greater than 0″.125.

(thick line) and the extension of the two white dwarf sequences (thin lines) are also shown together with the individual M type dwarfs (Eggen 1968c) with trigonometric parallax greater than 0″.125 (small dots) and the known late-type subluminous stars (crosses). Also, six very probable members of the Hyades cluster, discussed in Paper II (Eggen 1968b) are identified by large dots. An additional star, the common proper motion companion (LTT 6333) to $\beta\,TrA$ is shown in Figure 7 as an open circle. The observed magnitude and colors are $(V_E, \ B - V, \ U - B) = (13.50, \ +0.86, \ +0.15)$ and the single determination of the parallax of the bright star is 0″.078 (wt. 7) Cape, giving $M_V = +12.0$ for the companion.

The crosses in Figure 7 may all represent objects belonging to the lower sequence of white dwarfs; the spectrum of these stars are decidedly abnormal. The only two objects that originally defined the upper sequence in this color region were GH7–138, a member of the Hyades cluster, here shown as a filled circle, and LTT 4985 (G14–24), which was not recognized, spectroscopically, as a white dwarf (Eggen and Greenstein 1965) and shown in Figure 7 as an open triangle (Table III). 'If this is not a white dwarf it could easily be an extremely weak lined *sdK*,' and for G14–24, 'Spectrum is that of a very weak line, late-type subdwarf.' The spectrum of β *TrA B* is not available and the other Hyades stars in Figure 7 have not as yet been examined. In short, stars redder than $(U - V) = +0.5$ on the upper sequence do not necessarily show the abnormalities of those on the lower sequence. The course of the upper sequence in Figure 7 may be misleading and it perhaps should tend more to higher luminosities at its red end.

The whole subluminous domain is shown in Figure 8, uncluttered by conjectures as to sequence. The origin of the luminosities are as follows: crosses, members of the Hyades and Praesepe clusters; open circle, member of the Pleiades cluster; large dots, trigonometric parallaxes greater than

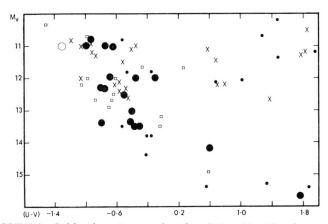

FIGURE 8 Subluminous stars in the $(M_V, \ U - V)$ plane. The luminosities were obtained from: membership in the Hyades and Praesepe clusters (crosses), membership in the Pleiades cluster (open circle), trigonometric parallaxes greater than 0″.05 with weight larger than 20 (large dots), trigonometric parallaxes greater than 0″.04 with weight less than 20 (small dots), and brighter main sequence companions with $B - V$ smaller than $+1.1$ (boxes).

0″.05 with weight larger than 20; small dots, trigonometric parallaxes greater than 0″.04 with weight less than 20; boxes, companion to brighter, main-sequence stars with $B - V$ bluer than $+1.1$. Several wide pairs with redder companions have been omitted (Eggen 1968a, Eggen and Greenstein 1965) because a later investigation of the infrared colors of these stars (Eggen 1968c) has shown a considerable dispersion in the $(M_V, R - I)$ plane. The most obvious, new feature of the $(M_V, U - V)$ diagram in Figure 8 is the growing number of stars in the upper right corner. More parallax determinations of the stars in Table III is probably the most certain method of solving the dilemma presented by these objects.

References

EGGEN, O. J. 1968a, *Ap. J. Suppl.* (in press).
EGGEN, O. J. 1968b, *Ap. J.* (in press).
EGGEN, O. J. 1968c, *Ap. J. Suppl.* (in press).
EGGEN, O. J., and GREENSTEIN, J. L. 1965, *Ap. J.*, **141**, 83.
EGGEN, O. J., and GREENSTEIN, J. L. 1968, *Ap. J.* (in press).
LUYTEN, W. L. 1957, *Catalogue of 9867 Southern Stars with Proper Motion Exceeding 0″.2 Annually* (Univ. Minn. Obs.).
LUYTEN, W. L. 1958, *Magnitudes and Colors for Southern White Dwarfs* (Univ. Minn. Obs.).
LUYTEN, W. L. 1961, *Catalogue of 7127 Stars in the Northern Hemisphere with Proper Motion Exceeding 0″.2 Annually* (Minneapol.)
SANDAGE, A. R. and EGGEN, O. J. 1959, *M.N.* **119**, 278.
WILDEY, R., BURBIDGE, E. M., SANDAGE, A. R., and BURBIDGE, G. R. 1962, *Ap. J.*, **135**, 94.

Discussion

GREENSTEIN   Would you try and re-estimate the space density of red degenerate stars now?

EGGEN   No, not until we can tell which are the sheep and which are the goats. It now appears that there are stars between the main sequence and the white dwarfs. The previous determinations of the densities of white dwarfs were based on the fact that there was nothing in there. As I understand it, you do not really have a degenerate star or phases of degeneracy in that part of the diagram. So these indeed may be white dwarf cores, stellar envelopes or something like that. That is why I prefer to call them subluminous. The density analysis still holds for subluminous stars; whether these are white dwarfs or not is something that has to be determined.

STRAND   Would Mr. Giclas comment on this remark on the colors that Mr. Eggen mentioned?

GICLAS   I will have the slides in a minute to show you these things...

Preliminary discussion of the low-luminosity stars from the Lowell proper motion survey

H. L. GICLAS

Lowell Observatory, Flagstaff, Arizona

Abstract

A proper motion program with the 13-inch photographic telescope at the Lowell Observatory has progressed to a point where it is now possible to make some preliminary analyses of the data compiled since its inception in 1957. Over 90 per cent of the northern hemisphere has been completed and some plate regions have extended into the southern hemisphere.

In this paper the number of stars found in each broad color class, broken down into two magnitude ranges (8.0 to 14.9 and 15.0 to 18), is shown in Figure 2. A short résumé of the nature of the objects comprising this list is given. All the data gathered has been stored on the disk of the IBM 1130, duplicate measures contained in 12,704 stars catalogued have been combined, and programs for utilization of the results in any of many different forms have been written. The direction of the solar motion is derived from this data and compared to the similar survey of Luyten and Erickson. Histograms of the directions of motion plotted by the computer have been examined for unique moving groups; none have yet been found. Techniques for comparing observed position angles with known convergents are given, and a list that may contain some additional Hyades members is given in Table 4. Concluding remarks call attention to some of the singular objects found.

A PROPER MOTION survey of the northern hemisphere was begun in 1957 at the Lowell Observatory. Plates taken in the early 1930's with the 13-inch photographic telescope of the Lowell Observatory in connection with the search for the planet Pluto are utilized as first epoch material. This telescope has a focal length of 66.7 inches (scale 29.4 mm/degree) and covers a net usable area in the sky of $11\frac{1}{2} \times 14$ degrees on each photographic plate (14×17 inch). The files contain at least two first epoch plates of each region and 380 different plate regions cover the sky visible at this latitude from the north pole to $-40°$ in declination. Two hundred and twenty-five plate regions have been completed to date and are discussed in this paper.

The proper motions are measured directly on a calibrated projection grid of the blink microscope used to identify the stars having perceptible motion. The working limit to which a reasonable degree of completeness can be expected has been set for motions larger than $0''.26$/year and for stars fainter than 8.0 magnitude. Several special studies of stars with motions smaller than these limits will be described later on. Each plate region is blinked and measured independently by two different individuals; the published values are the reconciled means of these. A finding chart for each star is published. For the past eight years, Messrs. R. Burnham, Jr. and N. G. Thomas have done the blink examination, measurements and reductions. A more detailed account of the procedures, including the limits of precision, estimates of completeness, and catalogs for all motion stars published to date, are contained in eleven Lowell Observatory Bulletins (Nos. 89, 102, 112, 120, 122, 124, 129, 132, 136, 138 and 140).

The Lowell program has progressed to a point where it is now possible to take a preliminary look at the nature of the data compiled during the past ten years. Ninety per cent of the sky in the northern hemisphere has been completed. Figure 1 shows the area of the sky completed at this time, and Table I contains a summary of the results.

In addition, two special studies have been made; one, an intensive survey,

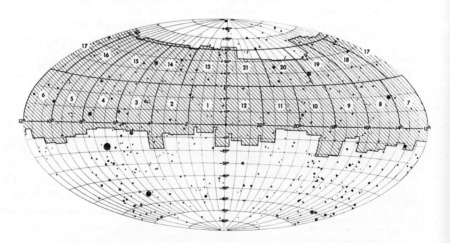

FIGURE 1 Area of Lowell proper motion program completed in 1967 showing the twenty-one areas utilized to determine the solar motion.

TABLE I   Summary of results (225 regions).

	Total
Number of motion (> 0″.26/year) stars listed	12,704
Number of different stars*	8926
Newly discovered	5086
New motion ⩾ 1″.00/year	62
New motion ⩾ 0″.50/year	799
New pairs common motion	134
New companions added to known motion stars	80
White dwarf suspects	162
White dwarf suspects < 0″.27/year	407
Hyades region	259
Blue and red binary suspects < 0″.27/year	24
Minor planets	764
Comets	4

* Combining duplicate measures from overlapping regions.

to the limit of detection of motion, of a region in the Hyades (Lowell Bulletin No. 118), and two, lists of 407 white dwarf suspects prepared from excessively blue stars with proper motion less than 0″.27/year (Lowell Bulletin Nos. 125 and 141).

The colors have been estimated in grade steps from −1 for the very bluest to +4 for the reddest. Several thousand UBV colors have been determined by Sandage and Kowal (1962, 1963), and Eggen (1968). A comparison of the $B − V$ magnitudes from all sources with our colors is shown in Figure 2 by filled dots. The crosses are similar determinations by Eggen and Greenstein (1965b) from their data alone. The two values for color class −1 are the same, where $B − V = +0^{m}.12$.

From these broad color classifications let us look into the nature of the catalogued objects in two ranges of apparent luminosity: $8^{m}.0$ to $13^{m}.9$ and from $14^{m}.0$ to the plate limit, which is not always the same, but most usually is between $16^{m}.5$ and $17^{m}.0$. In the table below Figure 2 the number of stars in each color class is given. The second figure just below is the per cent of the total number of stars in that class. The percentage distribution, the right hand vertical scale in the figure, is shown plotted on the color relationship curve. There the distribution of the bright stars is shown by a dotted curve and that of the faint ones by a solid curve. It is seen from this figure that the bulk of the brighter stars are about equally divided between color class +1 and +2. For the faint stars, about half of them are in color class +3. As expected there are many more faint red ones since only the nearer, low luminosity stars would be selectively detected by their motion. Also, as

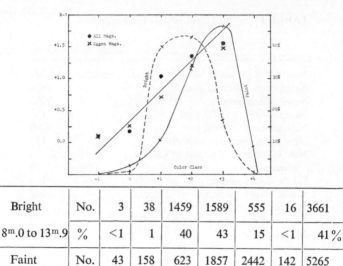

Bright	No.	3	38	1459	1589	555	16	3661
$8^m.0$ to $13^m.9$	%	<1	1	40	43	15	<1	41%
Faint	No.	43	158	623	1857	2442	142	5265
$14^m.0$ to $18^m.0$	%	1	3	12	35	46	3	59%

Total 8926

FIGURE 2 Comparison of Lowell color class with $B - V$ and the number of stars in each class

found by Eggen (1968), the space density for early type dwarfs is an order of magnitude less than for later types.

From the three papers of Eggen and Greenstein (1965a, 1965b, 1967) where they have correlated spectra with photoelectric photometry in establishing a sequential relation for white dwarfs, we can say with confidence that practically all the -1 color class are white dwarfs. This is based on over 100 confirmations. The stars of color class 0, ($B - V = +0^m.3$; $U - V = -0^m.4$), are still predominantly blue. About sixty per cent of these have been found to be white dwarfs, many others are helium-rich subdwarfs and perhaps 20 per cent horizontal branch stars. In our supplementary lists of white dwarf suspects with small proper motion (Lowell Bulletins 125 and 141) about half of the stars listed are of color class 0, the other half -1. Photometry from this list has produced further evidence that there may be as many white dwarfs present among the low-velocity population as among the high-velocity stars, and both young and old disk populations are well represented.

Sandage and Kowal (1962, 1963) made photometric studies of 1700 stars of color classes $+1$ and $+2$ to obtain data for the high-velocity subdwarfs in the solar neighborhood. Many hundreds of subdwarfs were found, among them 175 new *F* and *G* subdwarfs with ultraviolet excess values exceeding $0^m.16$. Except for the work of Eggen and Greenstein on a few red components of binaries, and those in common with Vyssotsky (1965) in the lists of late type dwarfs discovered on objective prism plates between mag. 8.0 to 11.0, very little work has been done on the $+3$ and $+4$ color classes. From these few highly selective examples, like G7–17, and G95–59, possibly the reddest dwarf known, and some of the small motion binary components, it seems safe to conclude that besides the red dwarfs there must be a large number of extreme subdwarf *M* and a few red degenerate stars in these categories.

The enormous amount of material produced by the survey has been analyzed at a reasonable cost of time and effort with the modern computer. The data for each of the 12,704 proper motion stars measured and catalogued to date were entered in the Lowell IBM 1130. A program to combine and average the duplicate measures from overlapping plates was first carried out, and the resulting list of 8926 motion stars stored permanently on the disk in order of right ascension. I am greatly indebted to Dr. Peter B. Boyce for his assistance in programming the computer for these operations.

The first analysis carried out was that of finding the stream motion from this homogeneous body of data. For this purpose the sky covered by the survey north of the equator was divided into 21 areas of 860 square degrees each. The area centers are listed in Table II and their positions in the sky are shown in Figure 1. Each area is shown outlined in heavy black lines. The number of stars moving in each ten degree interval of the observed position angle was then plotted out by the computer for each of the 21 areas in five different ways: first, for all the stars in each area, then in two magnitude ranges, bright ($8^m.0$ to $13^m.9$) and faint ($14^m.0$ to plate limit) in two color ranges, blue (-1 to $+1$) and red ($+1$ to $+4$).

An example of the plots for Area 18 is shown in Figure 3. The observed position angle from $0°$ to $360°$ is plotted as abscissa while the number of stars is shown as ordinate. Each record mark represents one star, except in the 'all' star plot where it represents two stars. The direction of streaming in each area was then determined by visual selection of the two maxima from the plotted histograms. The direction of the two Kapteyn streams are indicated by arrows in Figure 3. In some areas there were not enough stars in one or two of the categories to define stream II, which accounts for the omitted number in that column of Table II. The preliminary determinations of the direction of streaming were made by Mr. N. Thomas and then checked independently by the author. In Table II in the sixth and seventh columns the directions of motion of the two streams for all stars of each area are listed. In the succeeding columns the deviation of the direction of streaming from the 'all' category direction is given in degrees for the specific magnitude range and color

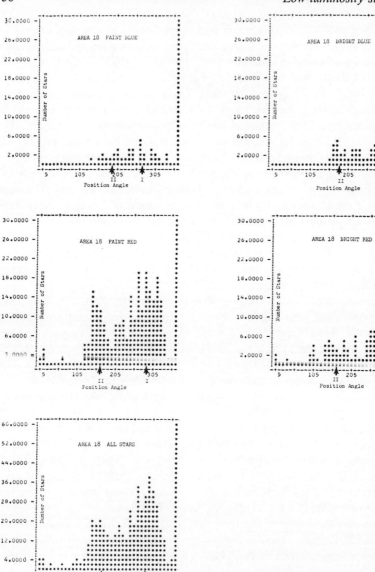

FIGURE 3 Example of the IBM 1130 plot of the directions of motion, abscissa, in 10° intervals vs number of stars, ordinate, for Area 18 centered at α 14ʰ 40ᵐ δ +45° for each color and magnitude range. Each record mark represents one star except for the all star plot where each record mark represents two stars. The arrows identify the two Kapteyn streams.

TABLE II Direction of star streaming from each area.

Area number	α	δ	l^{II}	b^{II}	All stars I	All stars II	Δ(blue—all) Bright I	Δ(blue—all) Bright II	Δ(blue—all) Faint I	Δ(blue—all) Faint II	Δ(red—all) Bright I	Δ(red—all) Bright II	Δ(red—all) Faint I	Δ(red—all) Faint II
1	$1^h\ 00^m$	+15°	127°	−48°	105°	205°	+20	−15°	0°	0°	0°	−5°	0°	0°
2	3 00	+15	164	−37	110	200		−10	+15	−10	−5	0	−5	−5
3	5 00	+15	186	−16	150		−5			0	+15		+5	+5
4	7 00	+15	201	+9	235	160		+5		0	−55	+20	−10	0
5	9 00	+15	214	+36	230	155	−5	−10	−15	−10	+15	+5	+10	+5
6	11 00	+15	234	+62	260	165	0	+20	−10	−5	+5	−5	0	0
7	13 00	+15	315	+77	260	165	+5	+10	+5	+10	+5	−10	0	0
8	15 00	+15	18	+57	270	185	+5	0	−10	+5	+5	+5	0	+5
9	17 00	+15	35	+31	315	205	−20	0	−35	−10	0	0	0	+5
10	19 00	+15	48	+4	45	205	+15	+5		−5	+10	0	0	0
11	21 00	+15	63	−20	85	210	−10	−10	0	−5	+15	0	+5	−5
12	23 00	+15	88	−40	90	210	+5	+5	−5	−10	0	−10	0	0
13	1 20	+45	129	−17	110	230	0	+25	−5	0	0	+10	0	0
14	4 00	+45	155	−6	135	—	+10	—	+5	—	+5	—	0	—
15	6 40	+45	171	+18	175	—	0	—	−15	—	+10	—	+5	—
16	9 20	+45	176	+45	225	160	−10	0	−10	−5	0	0	+5	−10
17	12 00	+45	149	+70	260	150	0	+25	−15	+20	+10	−15	−35	+5
18	14 40	+45	79	+62	285	170	0	+10	−15	+20	+5	−5	+5	−5
19	17 20	+45	71	+34	325	195	−10	+10	−20	+10	0	−15	+10	+5
20	20 00	+45	80	+8	35	200	−15	+10	−15	0	+5	+10	−15	0
21	22 40	+45	100	−12	80	215	+5	+10	+5	−10	0	+5	0	−5

indicated at the head of the column. Again, the bright range is from 8^{m}.0 to 13^{m}.9, the blue color -1 to $+1$ inclusive.

The individual directional vectors for each area in each stream for the 'all' category were then combined in the usual way to find the resultant stream motion as shown in Table III. The weighted solution is based on the number of stars in each area for each stream as its weight.

A comparison with Luyten and Erickson's (1967) analysis of the stream motions of 25,800 very faint (17^{m}.5 pg.) stars from 21 regions mostly in the northern hemisphere of the Palomar Schmidt survey is shown below. The direction of the solar motion agrees within a few degrees. The greatest difference is found in the direction of Drift II which in the Palomar data is 20° further south. This may be because the average magnitude of the Palomar data is fainter by one or more magnitudes.

Looking now at the deviation of the direction from the mean all-star group of our data for different magnitude and color groups given in the last eight columns of Table II, only the faint blue and the bright red stars show a significant deviation from the mean for Drift I. An independent solution for the direction of drift for these colors and magnitude ranges gives the values shown at the bottom of Table III.

In each case we find the position of the apex to be shifted a little east and south which indicates that these stars may be outside the main Kapteyn stream. Possibly the apex is moved south by the use of fainter stars.

The histograms of the directions of motion may also be utilized to look for anomalies in the usual distribution of stars of the two stream directions, which might indicate a unique moving group. For example, in examining the distribution of position angles for the bright red stars of area 18, Figure 3, one sees at once an excess of stars in the 295° position angle ordinate. A printout from the computer for all stars with position angles $\pm 10°$ from this point was investigated for a convergent or other singular characteristics but none were found. For our own use we have prepared a table giving the limits of position angle to six known convergents for stars within the boundaries of each of the 21 areas. This was used to look for anomalies in distribution of the position angles where the maximum resolution between a stream and a group convergent occurred.

This type of endeavour alone is not sufficient to singularly identify a unique moving group. Other parameters such as spectral type, radial velocity, quantitative color or parallax are needed to determine whether the stars selected comprise a true physical group. Since there are very little other supplementary data at this time for most of the faint stars listed, we can only try to compare the observed position angles with known convergents provided we can separate them from the normal background of motion stars.

TABLE III Direction of star streaming from Lowell proper motion data.

	Unweighted α	Unweighted δ	Weighted α	Weighted δ	Unweighted l^{II}	Unweighted b^{II}	No. of stars
Drift I	91°29'	−4°51'	107°11'	−12°54'	212°	−12°	63.8%
Drift II	244 44	−52 05	230 08	−54 25	332	−2	36.2
Solar motion	293 04	+37 34	299 57	+38 28	71	+9	All

Comparison with Luyten and Erickson

	α	δ	l^{II}	b^{II}	No. of stars
Drift I	101°	−5°	217°	−3°	67.6%
Drift II	227	−75	312	−15	32.4
Solar Motion	289	+36	68	+11	All

	Lowell faint blue Unweighted α	δ	Weighted α	δ	Lowell bright red Unweighted α	δ	Weighted α	δ
Drift I	109°34'	−13°39'	118°32'	−23°47'	95°00'	−6°21'	103°34'	−12°01'

As a trial for this technique, two known convergents were investigated: the Hyades and Bell's convergent at R.A. 14ʰ 44ᵐ Dec − 59°. All the stars with motion directed within ±30° toward the Hyades convergent, taken to be at R.A. 6ʰ 08ᵐ Dec +6°, were selected for initial comparison of observed and computed position angle. Stars approaching the Hyades convergent within ±45° of a line through its direction to nearby Drift I were omitted. The possibility of confusion with stars moving toward Drift I was minimized by normalizing the angle between Drift I and the Hyades to 100°, and adjusting the residual angles by the same factor. This procedure makes it possible to plot the Hyades residual angle near zero and in effect to remove the Drift I to 100°, thus improving the 'signal to noise' ratio of the sample. A plot of the 188 stars selected in this way by Mr. Thomas is shown in Figure 4. Sixty-one

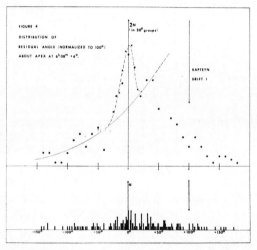

FIGURE 4   Distribution of residual angle (normalized to 100°) apex at α 6ʰ 08ᵐ δ +6°.

stars fall symmetrically about the peak toward the direction of this convergent including the stars of the Drift making up the base. Approximately sixty per cent of this number are in the normal background of stars, so that we can say that there may be an excess of around 24 stars moving in the direction of this convergent.

Most known Hyades members have proper motions less than one-half of our lower limit. For 21 of the stars the observed position angle minus the computed to the convergent is 1° or less. Four of these stars have proper motions greater than 0″.5/year which, if Hyades members, would place them at a distance of approximately 6 pc. which seems rather unlikely. It has not been possible to incorporate the smaller motions from Luyten's catalog in this discussion, but it would be interesting to do so.

The computer printout of the 61 stars is given in Table IV where convergent 1 is the direction of the Drift I stream from our data, convergent 2 the direction to the convergent at R.A. 6^h 08^m Dec $+6°$. The computed position angle for each star to the direction of the respective convergent is listed under the columns headed 1 and 2. The observed position angle minus the computed to the Hyades convergent is listed in the next column, *PA* -2. The column headed $(1-2)$ is the difference of the computed directions to the two convergents as seen from the star. This value divided into 100° gives the normalizing 'FACTOR' listed in the next to the last column of the table. This factor multiplied by (*PA* -2) gives the distance from the assumed zero of the convergent point to plot the star. This is shown in the last column of the table. The fractional angle listed under the differences comes about because the angles have been computed to several decimals and then printed out with 0°.1 accuracy by the computer program.

In the case of Bell's (1962) convergent, five of the seven known members have proper motions greater than 1".0/year, and perhaps this criterion alone might offer a means of selecting possible additional members. To investigate this possibility all stars with motions greater than 0".75/year moving within $\pm 100°$ of the convergent were plotted, as in the case of the Hyades, on a normalized angle scale to minimize the effect of Drift II. From the plot of these stars no excess of large motion stars was found to be going in the direction of the convergent. There are twenty stars with observed position angles within $\pm 5°$ of the convergent, but none of them are particularly early types which characterize the known members of the group.

While the results from these particular trial analyses are not definitive, they may suggest, with some refinements, a way to get a preliminary hold on statistically significant physical groups.

A word or two about some of the objects of special interest referred to in the notes and appendices of the eleven catalog lists published so far. Over sixty wide common proper motion pairs with motion larger than 0".30/year and separation greater than 2' of arc but less than 1° have been pointed out. Five of them are on the parallax list of the U.S. Naval Observatory; confirmation of others as physical doubles awaits further investigation. We hope to place these on our list for measurement of an accurate proper motion which will help to decide which are true binaries and worthy of more detailed observation.

Of the 24 special pairs of stars with annual proper motion less than 0".21/year, particularly chosen for the greatest contrast in color of the

Low-luminosity stars

TABLE IV The convergents are

1 = 6 hr 40 min −10°00' Drift I   2 = 6 hr 8 min +6°00' Hyades.

Number	α	δ	PM	PA	MAG	C	1	2	PA-2	1-2	Factor	N.R.AN
G160-21	3 44 4	-13 20	0.28	65	14.5	+2	91	64	0.4	26.6	3.751	1
G7-3	3 44 17	11 32	0.29	93	15.4	2	114	95	-2.7	18.8	5.318	-14
G80-29	3 49 36	4 31	0.63	82	16.0	3	109	85	-3.9	23.0	4.334	-17
G80-32	3 50 38	-1 49	0.29	78	15.7	2	104	79	-1.1	25.2	3.953	-4
G160-34	3 52 6	-8 40	0.27	63	14.3	3	96	69	-6.4	27.4	3.646	-23
G82-5	4 12 29	-5 45	0.69	72	10.9	1	101	70	1.3	30.3	3.294	4
G7-33	4 14 19	11 30	0.34	98	13.9	2	119	98	-0.6	20.6	4.843	-3
G82-13	4 18 28	5 59	0.69	91	13.2	2	114	88	2.4	25.7	3.891	9
G7-37	4 21 49	17 5	0.32	108	14.5	3	125	109	-1.9	15.9	6.274	-12
G7-40	4 26 1	16 24	0.29	110	14.8	2	126	109	0.2	16.5	6.044	1
G83-15	4 28 45	14 1	0.34	107	16.6	3	124	105	1.2	18.9	5.274	6
G82-32	4 31 40	-1 9	0.36	70	13.5	3	107	73	-3.7	33.3	2.999	-11
G8-49	4 37 23	18 17	0.58	117	15.9	3	130	116	0.7	14.3	6.951	5
G82-43	4 40 16	4 33	0.37	86	15.1	2	116	85	0.7	30.8	3.239	2
G82-51	4 49 10	1 58	0.35	78	16.0	2	113	77	0.1	36.1	2.767	0
G97-9	4 57 54	5 42	0.27	88	15.0	3	121	88	-0.1	33.4	2.987	-0
G84-36	5 3 24	4 16	0.42	76	13.0	1	120	83	-7.1	37.6	2.658	-19
G99-28	5 40 12	4 41	0.27	76	14.2	3	134	79	-3.0	55.6	1.797	-5
G99-37	5 48 46	0 11	0.27	19	15.2	0	128	39	-20.4	89.5	1.116	-22
G106-47	6 24 26	3 11	0.32	313	14.9	2	163	304	8.4	-140.9	-0.709	-5
G103-39	6 29 48	27 39	0.28	196	14.0	2	175	194	1.5	-18.5	-5.393	-8
G109-16	6 35 7	22 22	0.61	202	16.5	4	177	202	-0.7	-25.0	-3.994	3
G104-50	6 35 8	22 22	0.61	202	14.6	3	177	202	-0.8	-25.0	-3.993	3
G104-51	6 35 52	22 58	0.38	207	15.3	2	178	202	4.3	-24.5	-4.077	-17
G87-7	6 44 15	37 34	1.00	194	11.8	0	181	196	-2.8	-15.4	-6.465	18
G108-26	6 44 47	2 34	0.49	275	16.3	C	185	290	-15.7	-105.3	-0.948	14
G88-2	6 56 33	19 34	0.31	225	14.6	2	188	222	2.3	-34.3	-2.907	-6
G88-4	7 1 46	25 4	0.36	213	13.3	2	189	216	-3.2	-26.9	-3.705	12

Name	RA h	RA m	RA s	Dec °	Dec ′										
G87-24	7	4	55	29	55	0.37	211	9.9	2	189	212	-1.0	-22.5	-4.430	4
G87-26	7	6	39	38	37	1.06	209	13.5	2	188	205	3.1	-17.1	-5.822	-18
G88-19	7	14	35	19	39	0.50	239	14.5	3	196	232	6.9	-35.2	-2.834	-19
G87-36	7	19	33	30	46	0.75	217	15.3	3	194	217	-0.7	-23.0	-4.333	3
G112-12	7	19	39	-1	46	0.56	295	10.2	2	224	289	5.5	-64.5	-1.549	-8
G88-28	7	24	30	22	8	0.31	238	12.9	3	199	231	6.3	-31.7	-3.144	-19
G90-2	7	26	37	32	3	0.27	218	16.3	3	196	219	-1.3	-22.5	-4.441	5
G112-24	7	27	48	-3	34	0.27	288	12.6	3	236	293	-5.0	-56.0	-1.784	8
G88-36	7	29	11	17	25	0.30	238	13.3	3	204	242	-4.4	-37.6	-2.653	11
G88-39	7	31	26	23	22	0.28	236	15.9	3	202	232	3.5	-30.3	-3.298	-11
G90-15	7	36	2	30	1	0.29	228	13.1	3	200	225	2.8	-24.3	-4.103	-11
G90-16	7	36	22	33	34	0.28	223	13.6	2	199	221	1.4	-21.8	-4.566	-6
G50-2	7	39	23	3	55	0.28	277	16.0	3	226	276	0.8	-49.5	-2.019	-1
G50-3	7	39	24	3	55	0.28	277	16.0	2	226	276	0.8	-49.5	-2.019	-1
G112-42	7	40	12	-1	41	0.33	287	15.4	3	236	285	1.2	-49.0	-2.037	-2
G90-24	7	48	27	27	17	0.33	237	15.5	2	206	232	4.1	-26.5	-3.771	-15
G90-33	7	54	5	28	51	0.28	231	14.6	2	207	232	-1.7	-25.3	-3.952	6
G50-12	7	55	29	7	25	0.37	269	15.4	2	227	268	0.3	-41.1	-2.428	-0
G91-23	7	56	49	13	7	0.27	263	10.2	2	220	257	5.2	-37.2	-2.681	-14
G40-6	8	1	51	27	13	0.40	236	14.7	3	210	237	-1.0	-26.2	-3.814	4
G50-20	8	7	37	12	45	0.36	255	16.7	3	224	259	-4.8	-35.1	-2.842	13
G40-10	8	7	55	25	10	0.36	239	15.0	2	214	241	-2.3	-27.2	-3.665	8
G51-5	8	13	56	36	9	0.39	236	15.0	3	210	231	4.6	-21.0	-4.747	-22
G40-16	8	14	38	21	9	0.50	245	13.9	2	219	248	-3.2	-29.2	-3.419	11
G90-54	8	15	41	35	34	0.28	232	16.3	3	210	232	-0.3	-21.3	-4.685	1
G50-32	8	19	17	13	0	0.27	261	14.9	2	228	260	0.0	-32.7	-3.051	-0
G52-14	8	26	54	7	22	0.44	269	15.1	2	237	270	-1.0	-32.7	-3.051	3
G114-18	8	46	45	3	52	0.32	271	9.8	1	246	274	-3.7	-28.6	-3.485	13
G41-5	8	50	33	9	36	0.40	269	11.5	1	239	267	1.0	-28.2	-3.544	-3
G114-33	9	2	29	3	2	0.32	280	13.5	3	249	275	4.4	-26.1	-3.826	-17
G114-37	9	6	29	1	16	0.28	278	15.2	3	251	277	0.7	-25.2	-3.957	-3
G114-45	9	8	52	1	27	0.27	277	14.5	3	251	276	0.0	-25.0	-3.999	-0
G114-46	9	10	0	-6	58	0.29	282	13.8	2	260	283	-1.1	-22.9	-4.348	4

components, Eggen and Greenstein have verified white dwarf components in 13 of them; they found no certain white dwarf in 5 other pairs, and two of the remaining six that have no additional information have just been published.

The details for 21 pairs containing a white dwarf component are found in Table 3 of the first paper by Eggen and Greenstein (1965a) on 'Observations of Proper Motion Stars'. One of them, G191–B2, may contain the brightest white dwarf known with $M_V = +8^m.9$, $(B - V = -0.32; U - B = -1.20)$ which suggests it may have a parallax as great as $0''.044$. It has recently been added to the 61-inch U.S. Naval Observatory parallax program.

In the third paper on 'Observations of Proper Motion Stars,' Eggen and Greenstein mention a new type of object—dwarfs suspected of being variable. In addition to these there are two, G176–59 and G177–4, color $+3$ that are suspected of being variable, and three G40–31, G102–32 and G141–29 (LTT 15516) that are suspected flare stars.

On plate G32 there are two 16th magnitude red pairs of stars about 5° apart that show almost identical motion.

And finally, although we have paid particular attention to looking for white and red dwarf pairs, we believe that G206–17, 18 may be a white dwarf pair. The color assigned to the pair is -1 and $+1$, but the latter, because of its faint magnitude must be a late-type degenerate. Unfortunately, they are rather faint, 16.2 and 17.0 magnitude with a separation of 55 seconds of arc.

We plan now first to finish the northern hemisphere on the regular program, then we would like to extend it into the southern hemisphere at least as far south as $-30°$. Eventually we would like to go back and recheck the first 50 regions mainly to pick up the faint white dwarf suspects that may have been missed because some yellow sensitive plates instead of red sensitive plates were first used to determine color; and to see that all the smaller motions below the $0''.27$/year limit have been checked for color. Further analyses of group or singular motions will be made from time to time as need and expediency arise.

This proper motion survey has been made possible through continuing National Science Foundation grants which are herewith gratefully acknowledged.

References

BELL, R. A. 1962, *The Observatory*, **82**, 68.
EGGEN, O. J. 1968, *Ap. J. Suppl.*, **16**, No. 143.
EGGEN, O. J., and GREENSTEIN, J. L. 1965a, *Ap. J.*, **141**, 83.
EGGEN, O. J., and GREENSTEIN, J. L. 1965b, *Ap. J.*, **142**, 925.

EGGEN, O. J., and GREENSTEIN, J. L. 1967, *Ap. J.*, **150**, 927.
LUYTEN, W. J., and ERICKSON, R. 1967, *Pub. Astron. Obs.*, *University Minnesota*, II, No. 18, Minneapolis, Minn.
SANDAGE, A. R. 1962–63, *Annual Report of the Director, Mt. Wilson and Palomar Observatories*, 15.
VYSSOTSKY, A. N. 1956, *A.J.*, **61**, 201.

Discussion

EGGEN   I would like to make two points. When you are making a statistical discussion of the colors compared to what is observed, you should also try to include, somehow, variation from plate to plate. In one area all your $+2$'s have a $B - V$ around $+1$. In other areas many of them have $+1.5$. So that will make a difference when you combine everything. The other point would concern you when you are looking for the moving group, the convergent point method. I think you should use rather large tolerances because in many of these groups almost certainly one of the parameters, the motion perpendicular to the plane, will be uncoordinated with the motion in the plane itself. The so-called U and V motions will be segregated from W. So that would tend to broaden your peaks at the convergent point. There will not be a precise convergent point there, there will be a series of convergent points.

BLAAUW   I want to ask a question concerning the stars that you find in the convergent point of the Hyades. Do you find those stars especially in the region of the Hyades, or do you find them all over the sky?

GICLAS   We hope to do it all over the sky. This particular trial was made at least 90 degrees from the Hyades convergent point. We had 188 stars in this case, but we will do this on a bigger scale.

BLAAUW   Would you say whether they are evenly distributed over the sky limit area or patchy in certain directions?

GICLAS   These were over such a large area that I do not know that answer. We have not looked at it area by area; this has been combined by taking many areas and taking all the stars that would go in that direction. I do not remember whether there were more from area 21 or 6, but one could find out very easily.

Spectroscopic, spectral, and photometric parallaxes of M dwarfs

WILHELM GLIESE

Astronomisches Rechen-Institut, Heidelberg

Abstract

For application to the data of a second edition of the catalogue of stars nearer than 20 parsecs, spectroscopic, spectral, and photometric parallaxes of M dwarfs are calibrated on the basis of reliable trigonometric parallaxes. This investigation includes the absolute magnitudes of the Mount Wilson program and the McCormick program, the relations $(M_V, MK$ type$)$, $(M_V$, Kuiper type$)$, and $(M_V$, Mount Wilson type$)$, and the photometric relations $(M_V, R - I)$, $(M_V, B - V)$ and a relation $(M_V; B - V, \delta(U - B))$ which takes into account the ultraviolet excess of the M dwarf and which is derived for the first time. Table VII shows in which ranges absolute magnitude determinations are known and the different relations are calibrated. Furthermore, the expected probable errors are given. The $(M_V, R - I)$ and the $(M_V; B - V, \delta(U - B))$ relations seem to be superior to the other methods (their probable errors are ±0.3 mag and ±0.2 mag in the range down to the 16th absolute magnitude).

I. INTRODUCTION

WHEN COMPILING a second edition of the catalogue of stars nearer than 20 parsecs the question again arose which distance determinations are possible and reliable besides trigonometric measurements. Since the first edition (Gliese, 1957) only a few nearby stars previously unknown have been detected trigonometrically. The number of spectral classifications and especially the number of photometric data is growing much more rapidly.

The following investigations deal with the red main sequence stars or M dwarfs, which are the most numerous class in the solar neighborhood. Their visual absolute magnitudes are fainter than 8^m. Today we know trigonometric parallaxes only for a very small percentage. Spectroscopic, spectral, and photometric parallaxes are calibrated on the basis of reliable trigonometric measurements.

The trigonometric data are taken from the 'General Catalogue of Trigonometric Stellar Parallaxes' (Jenkins, 1952), from its 'Supplement' (1963), and from some recent lists. Strand (1963) proposes to use the relative parallaxes as published and reductions to absolute may, in general, be assumed to be $+0''.003$, $+0''.004$, or $+0''.005$ depending on galactic latitudes only. A more refined reduction on the basis of the known magnitudes of the comparison stars would be complicated and would take much time. When the proposed method in the second edition of the nearby star catalogue is used the trigonometric parallaxes are, on the average, slightly greater than the values in the Yale system. In the northern hemisphere the differences are insignificant, but in the south they grow to $+0''.005$ (Cape parallaxes) and $+0''.006$ (parallaxes from the Yale Southern Station).

The spectroscopic, spectral, and photometric parallaxes of M dwarfs are checked and partly revised on the basis of the available trigonometric parallaxes. Examination of the influence of the ultraviolet excesses on the $(M_V, B - V)$ relation leads to an $(M_V; B - V, \delta (U - B))$ relation which determines the absolute magnitudes of M dwarfs with remarkable certainty. In the following the results of the investigations and calibrations are shown and the probable errors of the absolute magnitudes determined by different methods are compared.

II. SYSTEMATIC EFFECTS DUE TO SELECTION

For calibrations only such trigonometric parallaxes can be used as are remarkably larger than their probable errors. This method prefers parallaxes with positive errors in measurement and affects the results systematically. The mean parallaxes are increased and the luminosities in the calibrated relations are too low. Several authors (e.g. Trumpler and Weaver, 1953) have shown how this effect can be eliminated from a series of measurements with a normal error distribution. But here most parallaxes used are weighted means of observations with different telescopes. Series of different accuracies and of different error distributions are combined.

To keep down this effect due to selection the material for the following calibrations is limited to stars with probable parallax errors smaller than 10 per cent. If accurate photometric data are also available the absolute magnitudes of these stars are known with probable errors smaller than 0.21. On the average the probable parallax errors of the stars used here are between $0''.005$ and $0''.006$.

Investigations of parallax series show that the errors due to selection are small in the region of nearby M dwarfs. With the limitations stated above the corrections to the absolute magnitudes are left between -0.05 and 0.00.

III. SPECTROSCOPIC ABSOLUTE MAGNITUDES

(a) The early Mount Wilson spectroscopic absolute magnitudes

The mean relation between spectroscopic absolute magnitudes determined by Adams, Joy, Humason, and Brayton (1935) and trigonometric values is in the range

$$M_V = +8.5 \text{ and } 10.5 \quad \overline{M_t} = \overline{M_{sp}} + 0.1 \ (\pm 0.05 \text{ p.e.}) \qquad 43 \text{ stars}$$

$$M_V = 10.6 \text{ and } 13 \quad \overline{M_t} = \overline{M_{sp}} + 0.35 \ (\pm 0.1 \text{ p.e.}) \qquad 11 \text{ stars}$$

For early M dwarfs the Mount Wilson system seems correct. Accurate trigonometric parallaxes being available, the Mount Wilson data of the few later types are obsolete. The probable error of a spectroscopic absolute magnitude between $M_V = 8.5$ and 10.0 is about ± 0.3.

(b) The McCormick estimations of the absolute magnitudes of M *dwarfs found spectrophotometrically*

The material of Vyssotsky and his collaborators (1943, 1946, 1952, 1956) has not lost its great value to date. A new calibration based on trigonometric parallaxes with small errors shows that the system is probably correct to 0.1, provided that 12 stars are excluded which obviously are subdwarfs (their trigonometrically determined absolute magnitudes are 1.3 to 3 fainter than the McCormick estimations). The mean relation is between

$$M_V = 8.5 \text{ and } 11.6 \quad \overline{M_t} = \overline{M_{McC}} + 0.03 \ (\pm 0.08 \text{ p.e.}) \qquad 47 \text{ stars}$$

A comparison with the photometric absolute magnitudes of the $(M_V, B-V)$ relation in the same range gives

$$\overline{M_{ph}} = \overline{M_{McC}} + 0.11 \ (\pm 0.07 \text{ p.e.}) \qquad 85 \text{ stars}$$

The probable error of a McCormick estimation is about $\pm 0.5^m$.

IV. SPECTRAL AND PHOTOMETRIC PARALLAXES

When the spectral or a photometric parallax of a star is determined its absolute magnitude is suggested on the basis of its spectral type or of a color. It is postulated as known that the star is a member of the main sequence. And it is supposed that in the range in question an obvious relation exists between the observed data and the luminosity and that the cosmic dispersion is limited. But it may not be supposed from the beginning that all dwarf stars in the solar neighborhood have the same main sequence. There are single stars and star systems, low and high velocity stars, stars without emission lines and stars with emission lines. If there is a sufficient number of stars available for a calibration, single stars of low velocity are selected for the derivation of a mean main sequence. Subsequently it will be investigated whether double star components, high velocity stars, or emission line stars deviate systematically from the mean relation. Already here it may be pointed out that M dwarf components of star systems seem to have the same main sequence as single stars.

Stars with space motions greater than 65 km/sec with respect to the mean of the M dwarfs are defined as high velocity stars (the mean velocity of M dwarfs is given by galactic solar motion components u, v, $w = -4$, $+18$, $+7$ km/sec with respect to the McCormick M star sample (Gliese, 1956)). In the Hertzsprung-Russell diagram or in a color–luminosity diagram the dispersion among them is somewhat larger than the dispersion among low velocity stars. But both groups seem to have nearly the same mean main sequence. It is known that stars of very high velocity are often subdwarfs. This luminosity class is excluded here. But also some high velocity stars classified as 'dwarfs' lie remarkably below the main sequence in the intermediate region between dwarfs and the subdwarf sequence.

Naturally, when these relations are calibrated spectroscopic binaries must be excluded. But possibly among a small group of stars the effect of some spectroscopic binaries not yet detected may be noticeable.

(a) Spectral parallaxes: The three relations (M_V, MK type), (M_V, Kuiper type), and (M_V, Mount Wilson type)

For M dwarfs the calibrations of (M_V, spectral type) relations are still uncertain; the spectral scale is contracted, and the determination of a mean luminosity for each spectral subclass is no longer as accurate as in the range of G and K dwarfs.

Only 26 stars classified as $M\,V$ were available for calibrating the (M_V, MK type) relation. But the material was supplemented by Kuiper types reduced to the MK system according to Johnson and Morgan (1953, Table 4). Among late type M dwarfs both systems agree with each other; therefore from $M2\,V$ to later types the (M_V, MK type) relation agrees with (M_V, Kuiper type).

Table I shows the mean visual absolute magnitudes for the spectral classes of the three systems. The observed data are not sufficient for defining the mean main sequence of stars intrinsically fainter than $M_V = 13$. Using these values the probable error of an absolute magnitude of an early M dwarf is about ± 0.3 in the MK or Kuiper system. It increases to ± 0.5 for the later types. Luminosities obtained from the (M_V, Mount Wilson type) relation would have a greater uncertainty; their probable errors are ± 0.6. Single stars and double star components, and apparently also high velocity stars, show the same mean main sequence, but the dispersion among the latter is somewhat greater.

TABLE I Mean absolute magnitudes for spectral subclasses of *M* dwarfs

Spectral	Visual absolute magnitude for		
type	*MK*	Kuiper	Mount Wilson
M 0	+8.7	+9.3	+8.3
0.5		9.5	
M 1	9.6	9.7	9.0
1.5		9.9	
M 2		+10.1	9.7
2.5		10.3	
M 3		10.6	10.6
3.5		10.9	
M 4		11.2	11.5
4.5		11.5	
M 5		12.0	12.4
5.5		12.6	
M 6		13.2	
M 7		(14)	
M 8		(16)	
Number of stars	101	100	93

(b) The photometric (M_V, $R - I$) relation

Eggen and Greenstein (1965) have drawn the (M_V, $R - I$) relation of main sequence stars as a straight line from $M_V = 6$ to 11.8 and continued to about $M_V = 13.4$ as another somewhat steeper line. The colors $R - I$ are in the system of Kron, Gascoigne, and White (1957). Meanwhile more data are available, mostly observed by Eggen (Wilson, 1967). Altogether there are 124 red dwarfs intrinsically fainter than $M_V = 8$ with reliable trigonometric parallaxes. After the known spectroscopic binaries, subdwarfs, and some close double stars have been excluded there are 78 low velocity stars and 23 high velocity stars with M_V between 8.00 and 16.69. Twenty-seven of these objects show emission lines.

Low-luminosity stars

In Figure 1 the low velocity stars are indicated by dots. Their mean main sequence is drawn almost as a straight line; the $(M_V, R-I)$ relation is tabulated in Table II. Between $M_V = 8.5$ and 12 the values agree well with the relation of Eggen and Greenstein. The data seem fairly reliable down to $M_V = 13$. The continuation of the straight line touches the stars near $M_V = 15$ and Wolf 359 at $(+16.69, +1.85)$. Therefore the $(M_V, R-I)$ relation seems available down to $M_V = 16$ or even 17.

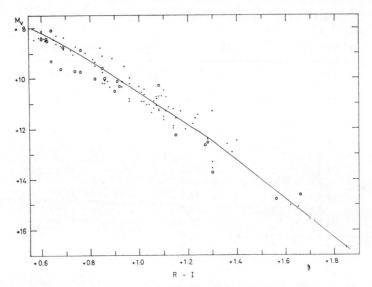

FIGURE 1 The $(M_V, R-I)$ diagram for M dwarfs with reliable trigonometric parallaxes. Stars with space velocities smaller than 65 km/sec are indicated by dots. High velocity stars are shown as open circles.

TABLE II The $(M_V, R-I)$ relation for M dwarfs of low velocities.

$R-I$	M_V	$R-I$	M_V	$R-I$	M_V	$R-I$	M_V
+0.60	+8.2	+0.90	+9.9	+1.20	+11.8	+1.50	+14.0
0.65	8.4	0.95	10.2	1.25	12.1	1.55	14.4
0.70	8.7	1.00	10.6	1.30	12.5	1.60	14.8
0.75	9.0	1.05	10.9	1.35	12.9	1.65	15.1
0.80	9.3	1.10	11.2	1.40	13.2	1.70	15.5
0.85	9.6	1.15	11.5	1.45	13.6	1.75	15.9
0.90	9.9	1.20	11.8	1.50	14.0	1.80	16.3
						1.85	16.7

The high velocity stars which are shown as open circles have a somewhat greater dispersion in the diagram. Even after exclusion of the objects classified as subdwarfs most of the high velocity stars fall below the main sequence, on the average by $+0.30 \pm 0.07$ (p.e.). While the dispersion in the absolute magnitudes of the 78 low velocity stars is ± 0.31, that of the high velocity stars is ± 0.44—the effect of the observational errors in the trigonometric parallaxes being estimated and eliminated.

The emission line stars are perhaps slightly more luminous than the *dM* stars. On the average the 21 *dMe* low velocity stars deviate from the mean main sequence by -0.13 ± 0.07 (p.e.); their dispersion is ± 0.37.

So it can be summarized that the data give some indication of possible differences between a main sequence of low velocity stars with absorption line spectra, a main sequence of high velocity stars, and another for *dMe* stars. Perhaps there are systematic differences between high velocity stars with emission lines and without emission lines too. But today only one $(M_V, R - I)$ relation for all *M* dwarfs can be derived. Possibly, the probable errors of these photometric parallaxes will be somewhat diminished if refined relations are known.

(c) The photometric relations $(M_V, B - V)$ *and* $(M_V; B - V, \delta(U - B))$

A reliable $(M_V, B - V)$ relation is known for main sequence stars of the classes *A* to *K* and also a useful $(M_V, U - B)$ relation between $M_V = 4$ and 8. For the *M* dwarfs too an $(M_V, B - V)$ relation can be determined but the dispersion in the absolute magnitudes increases remarkably. After exclusion of known and possible subdwarfs and of high velocity stars the material permits the derivation of a mean color–luminosity relation down to $M_V = 12$. The data of 48 stars available between $B - V = +1.35$ and $+1.65$ were subdivided into 7 groups. In Figure 2 their means are indicated by large open circles which define fairly well a smoothed curve. The individual objects are shown as dots.

Obviously the deviations of observed absolute magnitudes from this mean curve are correlated with the ultraviolet excesses. For stars intrinsically fainter than $M_V = 12.5$ the number of observed data is so small that probably the effects of the $\delta(U - B)$ are no longer cancelled out in the averaged absolute magnitudes.

For that reason the attempt has been made to calibrate an $(M_V; B - V, \delta(U - B))$ relation which allows a better determination of absolute magnitudes on the basis of *U*, *B*, *V* measurements. Empirically the problem is subdivided into the derivation of the three relations $(U - B, B - V)$, $(M_V, B - V)$, and

C

$(\Delta M_V,\ \delta(U-B))$ which are, in some way, connected with each other. Therefore the calibrations can be done only by successive approximations.

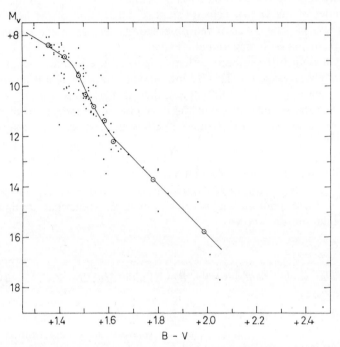

FIGURES 2 The $(M_V,\ B-V)$ diagram for M dwarfs with reliable trigonometric parallaxes. Individual objects are indicated by dots. Means of several stars are shown as large open circles. The mean curve was drawn by eye.

First step: The mean $(M_V,\ B-V)$ relation is known to about $M_V = 12$. Stars with large deviations $\Delta M_V = M_V$ (trig.) $-\ M_V(B-V)$ normally have large ultraviolet excesses. Therefore they will be unsuited for calibrating the $(U-B,\ B-V)$ relation. Stars with $\Delta M_V > 0.4^m$ are excluded. The remaining objects are subdivided into groups and their mean values are given in Table III. By these points a $(U-B,\ B-V)$ curve is fixed to about $B-V = +1.6^m$. Finally, the ΔM_V are plotted as a function of $\delta(U-B) = (U-B)_{B-V} - (U-B)_{\text{observed}}$. As a first approximation the correlation between both sets of data is expressed as $\Delta M_V \sim 8\delta(U-B)$.

Second step: The $\overline{U-B}$ are slightly corrected by reducing the $\overline{\Delta M_V}$ to

TABLE III Derivation of the mean $(U - B, B - V)$ relation for M dwarfs.

$\overline{B - V}$	$\overline{U - B}$	$\overline{\Delta M_V}$	$\overline{(U - B)}_{\text{corr}}$	Number of stars
+1.35	+1.26	−0.08	+1.25	10
1.44	1.22	−0.03	1.22	8
1.53	1.15	+0.06	1.15	10
1.62	1.19	+0.02	1.19	6
1.78	1.30			4
1.99	1.54			2

0.00^{m}. In connexion with the $(U - B, B - V)$ relation of the earlier main sequence stars a smoothed curve drawn by eye in Figure 3 gives the values of Table IV. In the $(U - B, B - V)$ diagram the (corrected) mean points are indicated by large filled circles, individual objects with small deviations ΔM_V from the mean $(M_V, B - V)$ relation are shown as small filled circles, and stars with $\Delta M_V > 0.4^{\text{m}}$ as dots. In the range to $B - V = +1.75$ the relation agrees very well with the values of H. L. Johnson (1965), but for greater $B - V$ the relation becomes very uncertain. In spite of that the curve in Figure 3 is used here as the final $(U - B, B - V)$ relation.

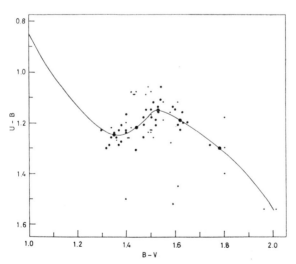

FIGURE 3 The $(U - B, B - V)$ diagram for M dwarfs. Individual stars with small deviations from the curve in Figure 2 are indicated by small filled circles, their group means by large filled circles. Stars with $\Delta M_V > 0.4^{\text{m}}$ are shown as dots.

With $\Delta M_V = 8\delta(U - B)$ the absolute magnitudes of the intrinsically faintest stars are reduced to $\delta(U - B) = 0.00$. These corrections remarkably diminish the dispersion among the M_V. In the $(M_V, B - V)$ diagram near $M_V = 12.5$ the mean curve is slightly shifted. The corrected $(M_V, B - V)$ relation seems reliable down to $B - V = +1.65$.

TABLE IV   The $(U - B, B - V)$ relation for M dwarfs.

$B - V$	$U - B$	$B - V$	$U - B$	$B - V$	$U - B$	$B - V$	$U - B$
+1.30	+1.23	+1.50	+1.16	+1.70	+1.24	+1.90	+1.41
1.35	1.25	1.55	1.16	1.75	1.28	1.95	1.47
1.40	1.24	1.60	1.18	1.80	1.32	2.00	(+1.55)
1.45	1.21	1.65	1.21	1.85	1.36		

In the range between $B - V = +1.38$ and $+1.65$ 39 stars with reliable trigonometric parallaxes and with U, B, V data are available. Their ultraviolet excesses range from -0.10 to $+0.15$ and their deviations from the mean $(M_V, B - V)$ curve are between $\Delta M_V = -1.00$ and $+1.16$. In Figure 4 the ΔM_V are plotted as a function of $\delta(U - B)$; the individual objects are shown as filled circles. A least-square solution gives a straight line which is adopted as final relation:

$$\Delta M_V = +7.5\delta(U - B)$$

It seems possible that further observational data will allow the straight line to be superseded by a non-linear curve.

Last step: With this equation the absolute magnitudes of the five intrinsically faintest stars are corrected for the effects of their ultraviolet excesses. The data of Table V provide the $(M_V, B - V)$ diagram in Figure 2 with two additional mean points at $B - V = +1.78$ and $+1.99$. Now the curve can be continued down to about $M_V = 16$ (Table VI). Some further objects of lowest luminosities cannot be used for calibration because no $U - B$ were measured.

Column $(\Delta M)_1$ in Table V gives the observed differences between trigonometrically measured absolute magnitudes and the $(M_V, B - V)$ curve, and column $(\Delta M)_2$ the differences between M_V(trig.) and the $(M_V; B - V, \delta(U - B))$ relation defined as $(M_V, B - V)$ relation $+7.5 \delta(U - B)$. Obviously most differences are remarkably diminished by eliminating the effects of the ultraviolet excesses.

The extension of the $(M_V, B - V)$ relation from $M_V = 12.5$ to 17 dependent upon 5 stars only cannot be more than a rough approximation. But the low

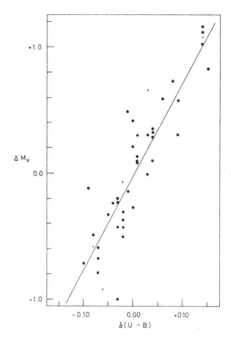

FIGURE 4   The $(\Delta M_V, \delta(U - B))$ diagram for M dwarfs. The ΔM_V are the deviations from the curve in Figure 2; the $\delta(U - B)$ are the deviations from the curve in Figure 3. Stars with $B - V$ between $+1.38^m$ and $+1.65^m$ are indicated by filled circles. The five low luminosity stars with $B - V > +1.7$ are shown as dots.

TABLE V   Five low luminosity stars with accurate data.

Star	$B - V$	$U - B$	$\delta(U - B)$	M_V(trig)	$\pm$p.e.	$(\Delta M)_1$	$(\Delta M)_2$
Barnard's star	+1.74	+1.29	−0.02	13.23	±0.02	−0.07	+0.08
+43° 44 B	1.80	1.40	−0.08	13.32	0.05	−0.58	+0.05
Wolf 424 A	1.80	1.18	+0.14	14.98	0.07	+1.08	+0.07
L 789–6	1.96	1.54	−0.06	14.60	0.06	−0.92	−0.47
Wolf 359	2.01	1.54	+0.03	16.69	0.04	+0.66	+0.47

TABLE VI   The $(M_V, B - V)$ relation for M dwarfs.

$B - V$	M_V	$B - V$	M_V	$B - V$	M_V
+1.30	+8.1	+1.55	+11.0	+1.80	+13.9
1.35	8.4	1.60	11.8	1.85	14.4
1.40	8.7	1.65	12.4	1.90	14.9
1.45	9.1	1.70	12.9	1.95	15.4
1.50	10.0	1.75	13.4	2.00	15.9

luminosities of these 5 stars are represented fairly well with the three derived relations. The 5 objects are shown as dots in Figure 4; three of them are very near to the straight line.

The 5 stars are added to the 39 other stars and after elimination of the effects of the observational errors in the trigonometric measurements the standard deviations σ of the absolute magnitudes from both relations are compared:

$$(M_V, B - V) \qquad\qquad \sigma = \pm 0.54$$

$$(M_V; B - V, \delta(U - B)) \qquad\qquad 0.20$$

These numbers justify the derivation of photometric parallaxes based on the $(M_V; B - V, \delta(U - B))$ relation:

$$(M_V; B - V, \delta(U - B)) = (M_V, B - V)_{\text{Table VI}} + 7.5\delta(U - B),$$

where

$$\delta(U - B) = (U - B)_{\text{Table IV}} - (U - B)_{\text{observed}}.$$

The advantages of these photometric parallax determinations are not only their low errors but also their availability for stars between main sequence and the subdwarfs region, at least to 1.2^{m} below the main sequence.

In Figure 5 the 10 high velocity stars are indicated by dots; the emission line stars are shown as open circles. Also these objects nearly fulfil the mean $(\Delta M_V, \delta(U - B))$ relation. But again their dispersion seems somewhat greater than among low velocity stars with absorption line spectra.

We summarize: Certainly an $(M_V; B - V, \delta(U - B))$ relation exists in the M dwarf region. When the influence of the ultraviolet excess is represented by a linear curve, reliable photometric parallaxes are determined in a wide band between $\Delta M_V = -0.8$ above and $+1.2$ below the mean main sequence.

But the derivation depends on the data of only about forty stars. Further observations are urgently needed, especially photometric data of stars with good trigonometric parallaxes. It is not sufficient to measure V and $B - V$; $U - B$ must be known too. Already a few further data may improve the mean $(M_V, B - V)$ and $(\Delta M_V, \delta(U - B))$ relations. It should be investigated how far from the main sequence the $(\Delta M_V, \delta(U - B))$ relation extends, perhaps as a non-linear curve. It is another question whether there exist different relations for *dMe* stars brighter than $\Delta M_V = -0.5$ or for high velocity stars fainter than $\Delta M_V = +1.0$.

It will be difficult and it will take a long time to measure accurate trigonometric parallaxes of stars of very low luminosity. It seems easier and definitely quicker to determine U, B, V of faint distant companions of such nearby stars for which reliable trigonometric parallaxes are known. Especially in the last years several companions suitable for such investigations have been detected. As far as we know double star components and single stars represent the same main sequence.

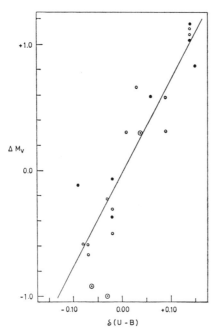

FIGURE 5   The $(\Delta M_V, \delta(U - B))$ diagram for high velocity M dwarfs (dots) and emission line M dwarfs (open circles). The straight line is the mean relation of Figure 4.

VI. COMPARISON OF SPECTROSCOPIC, SPECTRAL, AND PHOTO-METRIC PARALLAXES OF M DWARFS

The dispersions of the absolute magnitudes determined by different methods are computed by comparisons with reliable trigonometric measurements. Table VII shows in which ranges absolute magnitude determinations are known and the different relations are calibrated. Furthermore the expected probable errors are given as average values although subdivided investigations indicate that the spectral and photometric parallaxes of high velocity stars are not so certain as those of the low velocity stars. Probably the errors for the low velocity stars will be somewhat smaller than the values of Table VII while the p.e.'s of high velocity stars may be about 50 per cent greater.

Low-luminosity stars

TABLE VII Probable errors of spectroscopic, spectral, and photometric determinations of absolute magnitudes of M dwarfs.

$(M_V, \ldots)$ relation	Probable errors and their limits in M_V	Number of stars
M_{sp} (M.W.C. 511)	8.5... ±0.3...10	40
M_{sp} (McCormick)	8.5...... ±0.5......12	48
MK types	8.5.. ±0.3 increasing to ±0.5..13	91
Kuiper types	8.5.. ±0.3 increasing to ±0.5..13	98
Mt. Wilson types	8.5...... ±0.6......12	93
$R - I$	8.5...... ±0.25.....12...... ±0.35.........16	90
$B - V$	8.5................. ±0.4.................16	44
$B - V, \delta(U - B)$	8.5................. ±0.2.................16	44

The $(M_V, \ R - I)$ and the $(M_V; \ B - V, \ \delta(U - B))$ relation are certainly superior to the other methods in Table VII. Newly observed data may show whether the latter relation in particular determines the absolute magnitudes with such unusual small errors. Obviously, both relations are available to the maximum of the luminosity function and probably even beyond this maximum. And the question is whether it is possible to calibrate these relations also beyond the minimum mass on the main sequence.

References

ADAMS, W. S., JOY, A. H., HUMASON, M. L., and BRAYTON, A. M. 1935, *Ap. J.*, **81**, 187; *Contr. Mt. Wilson Obs.*, No. 511.

EGGEN, O. J., and GREENSTEIN, J. L. 1965, *Ap. J.*, **141**, 83.

GLIESE, W. 1956, *Zs. f. Ap.*, **39**, 1.

GLIESE, W. 1957, *Heidelberg Astr. Rechen-Inst.*, Mitt., Ser. A, No. 8.

JENKINS, L. F. 1952, *General Catalogue of Trigonometric Stellar Parallaxes* (New Haven, Conn.: Yale University Observatory).

JENKINS, L. F. 1963, *Suppl. to the Gen. Cat. of Trig. Stellar Parallaxes* (New Haven, Conn.: Yale University Observatory).

JOHNSON, H. L. 1965, *Ap. J.*, **141**, 170.

JOHNSON, H. L., and MORGAN, W. W. 1953, *Ap. J.*, **117**, 313, Table 4.

KRON, G. E., GASCOIGNE, S. C. B., and WHITE, H. S. 1957, *A.J.*, **62**, 205.

STRAND, K. AA. 1963, *Basic Astronomical Data*, ed. K. AA. STRAND (Chicago: University of Chicago Press), Chapt. 6.

TRUMPLER, R. J., and WEAVER, H. F. 1953, *Statistical Astronomy* (Berkeley, Calif.: University of California Press), p. 287.

VYSSOTSKY, A. N. 1943, *Ap. J.*, **97**, 381.

VYSSOTSKY, A. N. 1956, *A.J.*, **61**, 201.

VYSSOTSKY, A. N., JANSSEN, E. M., MILLER, W. J., and WALTHER, M. E. 1946, *Ap. J.*, **104**, 234.

VYSSOTSKY, A. N., and MATEER, B. A. 1952, *Ap. J.*, **116**, 117.

WILSON, O. C. 1967, *A.J.*, **72**, 905.

Discussion

EGGEN   I am very surprised that you get such a relationship for two reasons. The first reason is the observations themselves; since you are a cataloger, you are forced to take *UBV* from all the sources available. It is not widely recognized that you have to calibrate for the red stars because of the ultra-violet leak. Sometimes this is done in the ultra-violet filter. Sometimes this is done and sometimes it is not. It is on these very stars where the difficulty arises, because they are so red that you hardly get any ultra-violet from the stars themselves. Most of it could be a leak from the infrared. Now, unless this is calibrated out, it is going to enter into the $B - V$ value. The other thing that makes me surprised at the relationship is that very young stars can depart from the *UBV* diagram, for the simple reason that they are flare stars. And we know that the flare stars do show an ultraviolet excess. Yet they are the youngest stars, and, in general, would lie above the main sequence and not below it, because they are still in a state of contraction. On the other hand, the very oldest stars, the subdwarfs, are like Barnard's star and Kapteyn's star, right below the main sequence. And yet they, for a different reason, do also show ultraviolet excess. So that I think that this all means you are a victim of circumstances since you have to deal with the material that is available. It is really the fault of observers who have observed red stars in a blue system.

VON HOERNER   You mentioned a number of 3000 or 4000 low-luminosity stars; where does this estimate come from?

GLIESE   That is an estimate which comes from the data which Luyten presented at the I.A.U. General Assembly at Prague. I myself did not derive a luminosity function. He has not yet published them and, therefore, we do not know all the details.

STRAND   Well, I think I can explain this. Luyten has already surveyed the polar cap down to $+75°$ declination on the Palomar survey that he is carrying out. And I think that on the basis of information that he has from that survey, as well as from his survey in the Hyades, that he came up with this figure.

PART II

Discussion

Light variability of dMe and dM stars*

W. KRZEMINSKI†

Lick Observatory, University of California, Santa Cruz

Abstract

A sample of nine emission and non-emission line stars in the range of spectral type $dK7$ to $dM3.5$ was observed photoelectrically in the UBV system. Periodic or quasi-periodic light variations with an amplitude of a few hundredths of a magnitude and almost no change in the $(B - V)$ color index were found in some dMe but not dM stars. Periods are around a few days. Since eclipses and volume pulsation are inadequate mechanisms to explain the light variability, the most promising mode! appears to be the rotational modulation of a star with a nonuniform distribution of surface brightness.

THE DISCOVERY of periodic light variability by Evans (1959, 1964) in a single-line spectroscopic binary HD 16157 (K7Ve) and by Chugainov (1966) in HDE 234677 (K7Ve), known to be a double-line binary (Kraft 1964), raised the question of whether other late-type dwarfs with and without emission lines showed such variability and if the light variations were related to the binary characteristic. A small sample of emission and non-emission line stars in the range $dK7$ to $dM3.5$ was observed photoelectrically in the UBV system during the summers of 1966 and 1967 at the 60-inch Mount Wilson telescope; a few additional measurements were made with the 20-inch at Palomar. ADS 16557B, though of earlier spectral type, was also included in this survey. The observational data are summarized in Table I.

Four of the emission-line objects are certainly variable and for two of them the periods of light variation are fairly well established. (The period of 4.65 days suggested earlier for $AC +31°$ 70565 (Krzeminski and Kraft 1967) has not been confirmed by 1967 observations; a 2-day period seems to fit the quasi-periodic behavior of this object, however.) Light variability is accompanied by almost no change in $(B - V)$ color index. The light curve of

---

* Based on observations obtained at the Mt. Wilson and Palomar Observatories.

† On leave from the Astronomical Institute, Polish Academy of Sciences, Warsaw, Poland.

TABLE I Observational data for a sample of emission and nonemission line late-type dwarfs.

Star name	Spectrum	$\langle V \rangle$	$\langle B - V \rangle$	$\langle U - B \rangle$	Range of variation (yellow mag.)	Photometric period (days)	Spectroscopic binary; authority
+34° 106	dM0e	10.38	1.43	1.09	0.06†	2.170	Yes; Herbig
+70° 68	dM3.5e	10.00	1.43	1.05	0.03	—	—
+55° 1823	dM1.5e	9.98	1.45	1.10	0.03	—	—
HD 151288	K7 V	8.11	1.35	1.28	0.02	—	—
HDE 234677	dK 7e	8.44	1.24	1.03	0.08†	3.836	Yes; Kraft
+45° 2743	dM2	9.86	1.41	1.20	0.03	—	—
HD 193202	dM0	8.87	1.30	1.30	0.03	—	—
AC +39° 1214–608	dM3e	10.12	1.50	1.10	0.10	?	No; Kraft
AC +31° 70565	dM3.5e	11.66	1.51	1.05	0.06	?	Yes; Greenstein and Kraft
HD 218738*	dK7e	7.91	0.90	0.55	0.04	—	—

* ADS 16557 B.
† Only amplitude for 1966 observations given.

BD +34°106 is shown in Figure 1. Since *dMe* stars are flare stars as well, and during the photoelectric survey occasional eruptions have been observed, we have omitted observations during flare activity in constructing the composite light and color curves. It is of interest to note, however, that for HDE 234677 observations taken in shorter wavelengths exhibit much scatter (Figure 2); apparently small flaring activity and subsequent enhancement of

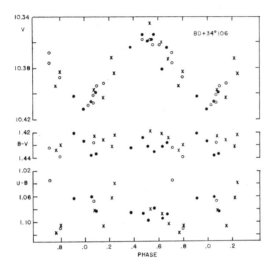

FIGURE 1   *UBV* observations of *BD* +34° 106. Phases are calculated from elements Min. $JD_{\odot}$ = 2439380.82 + 2^d.17 *E*. Crosses correspond to observations obtained in July–August 1966, open circles to those in September–October 1966 and filled circles to those taken in July–September 1967.

the emission lines is practically always present. If one compares the present observations of HDE 234677 with those of Chugainov (Figure 3) one finds a remarkable change in amplitude; there is a phase shift between the 1966 and 1967 observations though the period of variations remains constant (Figure 4). The remaining emission and non-emission line stars showed no variability in excess of 0.02 to 0.03 mag. It appears therefore that periodic or quasi-periodic light variations with periods of a few days are found in some *dMe* but probably not *dM* stars.

As quoted earlier (Krzeminski and Kraft 1967), *BD* +34°106 and HDE 234677 are spectroscopic binaries but *AC* +31°70565 shows no radial velocity variation. The spectroscopic period of 5.981 days for HDE 234677 differs markedly from its photometric period of 3.836 days. In explaining the periodic light variability, the eclipse hypothesis finds no support in view of the observed phase shifts in HD 16157 and 234677; furthermore, the sinusoidal shape of light variation in the latter star would suggest a contact system, but it is known from Kraft's orbit that the separation of the components is at least 10 times their radii. These facts strongly suggest that binary motion cannot be the cause of the light variability in the sample of *dMe* stars.

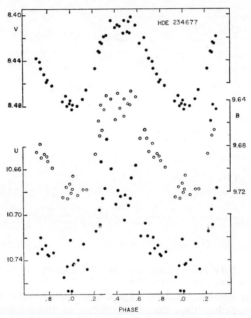

FIGURE 2 *UBV* observations of HDE 234677 in 1966. Phases are calculated from elements Min. $JD_\odot = 2439033.48 + 3^d.836\ E$.

The interpretation of light changes by radial pulsation is unlikely because the observed periods are two orders of magnitude greater than those calculated for vibrationally unstable low mass *M* dwarfs (Gabriel 1967). Hence, following Chugainov's original suggestion, the most promising interpretation is that the observed features are caused by the rotation of a star with a non-uniform brightness distribution. The presence of strong Ca II and hydrogen emission as well as flaring activity is not in contradiction with a hypothesis of existence of 'disturbed areas' on the surface of emission line late-type dwarfs. Models with bright and dark 'disturbed areas' are equally applicable. If, for simplicity, to explain the observed light variation of $BD + 34°106$ one assumes as a first approximation a model with a *single* dark spot and assigns to it a temperature 350° (i.e. two subclasses in spectral type) less than the remaining part of a star, one finds that such a spot would cover 10 per cent of the stellar surface. The presence of such a 'disturbed area' would cause the effective temperature of the stellar hemisphere on which it appears to be 30° less than the effective temperature of the spotless hemisphere. Similar values for the size of a hypothetical spot and the resulting lowering of the effective temperature of the 'disturbed' stellar hemisphere are obtained for

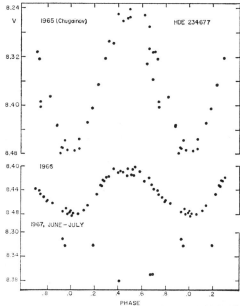

FIGURE 3 Photoelectric observations of HDE 234677 in yellow light. Chugainov's (1966) data have been transformed to the *V* band. Phases the same as in Figure 2.

HDE 234677. Such a change in the effective temperature would cause $(B - V)$ to vary by about 0.01 mag. in a photometric cycle and would be undetectable in the appearance of the integrated spectrum.

If our understanding of the observed features is correct and the observed light variations are caused by stellar rotation, then the equatorial rotational velocity would be around 10 km/sec for HDE 234677 and near 15 km/sec for $BD + 34°106$. This implies that these stars rotate faster than might be expected from their position in the H–R diagram. On the other hand, since it is thought that the *dMe* stars are rather direct descendants of the later type *T* Tauri variables, one has to account for a large loss of an angular momentum of the latter to explain the considerable difference in rotational velocities of the two groups of stars.

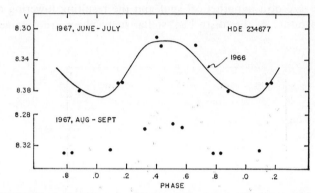

FIGURE 4   Photoelectric observations of HDE 234677 in yellow light. Phases are calculated from elements Min. $JD_\odot = 2439033.48 + 2^d.01 + 3^d.836\ E$. The solid line represents the 1966 *V* observations shifted both in ordinate and abscissa.

The author is very indebted to Dr. R. P. Kraft for illuminating discussions on late-type stars and to Dr. P. F. Chugainov for sending him his unpublished observations. The observations reported here were obtained during tenure of a Carnegie Fellowship and the author is thankful to Dr. H. W. Babcock, Director of the Mount Wilson and Palomar Observatories, for the observing facilities afforded him.

References

CHUGAINOV, P. F. 1966, *Comm.* **27** *of the IAU Bull.* No. 122 (Konkoly Obs.)
EVANS, D. S. 1959, *M.N.*, **119**, 526.
EVANS, D. S. 1964, *M.N. Astr. Soc. So. Africa*, **23**, 68.
GABRIEL, M. 1967, *Ann. d'Ap.*, **30**, 745.
KRAFT, R. P. 1964, unpublished.
KRZEMINSKI, W., and KRAFT, R. P. 1967, *A.J.*, **72**, 307.

Discussion

EGGEN   I would like to point out that Kron found the same thing some years ago, with the same explanation for YY Geminorum. The peculiar variability in 16157, which is also known as CC Eridani, could be due to the interaction between the two stars since it is a double-lined spectroscopic binary.

KRAFT   I do not know whether this is induced by the binary business or not. Certainly the spottedness is nice, for then you can explain why you do not get the phasing. It dies away secularly as it is replaced by a new set of spots to get the same period back, but you can not phase the observations like in a peculiar *A* star.

A new method for finding
faint members of the Pleiades

ROBERT P. KRAFT

Lick Observatory, University of California, Santa Cruz

JESSE L. GREENSTEIN

Mount Wilson and Palomar Observatories
Carnegie Institution of Washington, California Institute of Technology

---

Abstract

The discrepancy between the nuclear and contraction ages for several young clusters is reviewed. The contraction age in every case depends critically on the isolation of faint cluster members from the field stars. In the Pleiades, proper motions alone are inadequate for this purpose. Determination of spectral types for 49 stars fainter than $V = 13$ in the field of the Pleiades leads to the conclusion that the photographically determined colors of Johnson and Mitchell and of Ahmed, *et al.* are not sufficiently accurate to locate the faint part of the main sequence in the color-magnitude diagram.

The hypothesis advanced by Wilson that Ca II H and K emission (known as $H2$ and $K2$) of main-sequence stars dies away in a time short compared with the nuclear time is applied to the problem. It is found that in the spectral type range $K5$ V to $M3$ V, the majority of proper-motion 'members' of the Pleiades have $K2$ twice as strong as those of Hyades stars; the latter have $K2$ strengths like those of stars in the field with the most intense $K2$. Since these constitute only about 10 per cent of field stars of this spectral type, it is concluded that the presence of intense $K2$ in the spectrum of such a star in the field of the Pleiades is a powerful criterion for cluster membership. The stars with strong $K2$ on average lie on the standard Yerkes main sequence in the H–R diagram at least as far as $M1$ V. We concluded therefore that either (1) the faint Pleiades stars are more than twice as old as the brightest members or (2) the contraction ages and/or nuclear ages are incorrect. If T_K is incorrect, then the faint stars need to reach the main sequence at least twice as fast as predicted by the Iben–Hayashi theory.

I. INTRODUCTION

AS IS WELL known, the nuclear age T_N of a galactic cluster can be obtained from the luminosity of its brightest main sequence members, provided that the chemical composition is given, and provided also that in the H–R diagram, the evolutionary tracks for stars of different masses are known. A cluster contraction age T_K can also be obtained from the luminosity and corresponding mass at which stars in gravitational contraction just reach the main sequence and begin hydrogen burning; stars below this critical mass lie to the right of the main sequence. If cluster stars are, in fact, born coevally, we would expect $T_N = T_K$. This paper is concerned with observational tests of this proposition, especially in the case of the Pleiades.

Previous work on this point has focussed attention primarily on the very young clusters NGC 2264 and NGC 6530, photoelectric color-magnitude diagrams of which were first obtained and discussed by Walker (1956, 1957). From Iben's (1965, 1966a, b) post-main sequence evolutionary tracks, we estimate $T_N = 2.0 \times 10^7$ years and 1.5×10^7 years for these two clusters, respectively. However, the lower main sequence turnaway in both clusters corresponds to T_K near 1×10^6 years; further, the two clusters have so large a scatter in the luminosities of the 'contracting' members, that stars of contraction ages of order 10^5 to 10^8 years seem also to be represented (Iben and Talbot 1966, Ezer 1966). A re-evaluation in the case of NGC 2264 in the light of proper-motion criteria for cluster membership (Vasilevskis, Sanders, and Balz 1965) does not change these conclusions. The absence of stars on the main sequences of these clusters with $L/L_\odot > 10^4$ could, of course, result from a statistical fluctuation in the luminosity function with the result that T_N might be considerably shorter than 10^7 years. Further, Ezer (1966) suggested that the apparent spread in contraction ages resulted from mass-loss so that stars of apparently older than average age once had much larger masses and hence arrived at their present positions in the H–R diagram faster than the theory predicts. Qualitative support for this view follows from Kuhi's (1966) study of mass-loss in T Tauri stars. Iben and Talbot (1966), on the other hand, preferred the solution in which mass was conserved, but the hypothesis of coeval star formation was abandoned. Both pictures remain problematical, however, since they depend on the correct isolation of cluster members, and on the validity of the transformation of the observed colors and magnitudes into the theoretical effective temperatures and bolometric luminosities.

Turning to the much older Hyades in which $T_N = 4 \times 10^8$ years (cf. Iben 1965, 1966a, b), we find that the condition $T_K = T_N$ leads to a lower main-sequence turnoff near $M2\,V$, i.e. $M_V = +10.2$ (Yerkes system), or $m_V = +12\cdot3$

The nearness of the Hyades leads to the favorably large proper motion of $0''.11/yr$ at the cluster center; from proper motions, colors, and magnitudes of faint stars, van Altena (1966) concluded that the Hyades has no turn-off as far down as $m_V = +18$. Thus $T_K \approx 2 \times 10^9$ years, and $T_K > T_N$ by a significant factor.

In the younger but more distant Pleiades, the situation is less clear. Herbig (1962) was the first to point out the discrepancy between T_N and T_K and to establish that some stars having proper motions consistent with cluster membership (Hertzsprung 1947) had spectral types and absolute magnitudes which placed them on the main sequence far below the expected lower turnoff, near, in fact, $M0\ V$. From Iben's (1965, 1966a, b) tracks, we estimate $T_N = 3$ to 4×10^7 years, so that the condition $T_K = T_N$ leads to an expected lower turnoff near $G2\ V$. On the other hand, as pointed out by Herbig, the selection of members by proper motion is subject to considerable uncertainty not only because the cluster proper motion is small ($0''.046/yr$), but also because it lies in a direction similar to that from the cluster to both the solar antapex and the antapex of galactic rotation. The inclusion of non-members on the basis of such proper motion data might be expected to produce a dispersion in the lower main sequence of the color-magnitude diagram, and just such an effect has been observed by Johnson and Mitchell (1958) for the stars fainter than $m_V = +11.5$, i.e., spectral type $K0\ V$. Their data show a spread of more than 0.5 mag in $(B - V)$ at $m_V = 14$ for proper motion members from the photoelectric as well as photographic observations. In their treatment of the membership problem, Johnson and Mitchell excluded stars lying below the main sequence and suggested that the dispersion of stars from the main sequence to luminosities as much as two magnitudes above it might result from a spread in the ages of the faint cluster stars. Herbig (1962) proposed, in fact, that the existence of stars on the main sequence at these luminosities could be understood if the process of star formation was continuous for faint stars and was terminated by the formation of one or more hot stars. In that case, the nuclear age of the cluster would be shorter than its contraction age, and would represent only the time elapsed since the formation of the hottest stars. Poveda (1965), on the other hand, suggested that the spread in the lower main sequence resulted from viewing, at different aspects, stars with pre-planetary semi-opaque equatorial disks. In this way, even membership for stars below the main sequence might be admitted.

Before any of these conclusions can be accepted, however, a number of uncertainties about the observations and the theory need to be cleared up. First, we have already mentioned that pre-main sequence mass-loss may have to be accounted for in the calculation of the evolutionary tracks; this is outside the scope of the present discussion. Second, one may question how well we actually know the colors and magnitudes of Pleiades stars fainter than $m_V = +12$. Inspection of the Palomar Sky Survey prints reveals bright and irregular reflection nebulosity in the region of the Pleiades; Ahmed, Lawrence, and Reddish (1965) have criticized Johnson and Mitchell for neglecting the contribution of this nebulous fog in their treatment of the photographic photometry. Third, a comparison between the Herbig (1962) spectral types and either the Johnson-Mitchell or Ahmed, *et al.* colors indicates discrepancies much larger than can reasonably be accounted for by differential reddening or errors in the spectral types. And finally, and most important, it remains problematical that proper motions alone, independent of pre-judged considerations about the location of the main sequence, can unambiguously pick out the members.

We will now demonstrate that, so far as can reasonably be determined in the observational program about to be described, only those proper motion 'members' that populate the standard Yerkes main sequence (Keenan 1963, Vyssotsky 1963) in the visual magnitude vs. spectral type diagram have, at low dispersion, detectable emission in the H and K lines of Ca II; that this conclusion remains valid at least as far down as $M1\ V$; and thus if the presence of K emission is a valid criterion for cluster membership, and if the present stellar contraction theories are correct, there are faint Pleiades stars as old or older than 8×10^7 years, i.e., at least twice the nuclear age.

II. K-LINE EMISSION AS A CRITERION FOR CLUSTER MEMBERSHIP

In a coudé-dispersion survey of F- and G-type main sequence stars of the general field, Wilson (1963, 1966a, b) found that (1) no stars with spectral types earlier than about $F5\ V$ have components of Ca II (known as $H2$ and $K2$) in emission; (2) only about 10 per cent of the sample later than $F5\ V$ have detectable H and K emission; (3) the emission intensity statistically increases relative to the local continuum with advancing spectral type; and (4) the stars with H and K in emission lie, on the average, closer to the zero-age main sequence than those without. The last point suggested that the strength of $H2$ and $K2$ was correlated inversely with stellar nuclear age, a surmise confirmed by Wilson's work on similar stars in galactic clusters. Essentially all stars later than $F5\ V$ in galactic clusters are found to have H and K in emission, and at a fixed spectral type, $K2$ in the Pleiades ($T_N \approx 3$ or 4×10^7 years) is statistically stronger than $K2$ in the Hyades, Coma, and Praesepe ($T_N \approx 4 \times 10^8$ years for all). (Nothing can be said about still older clusters since their F- and G-type stars are too faint to be reached at coudé dispersions).

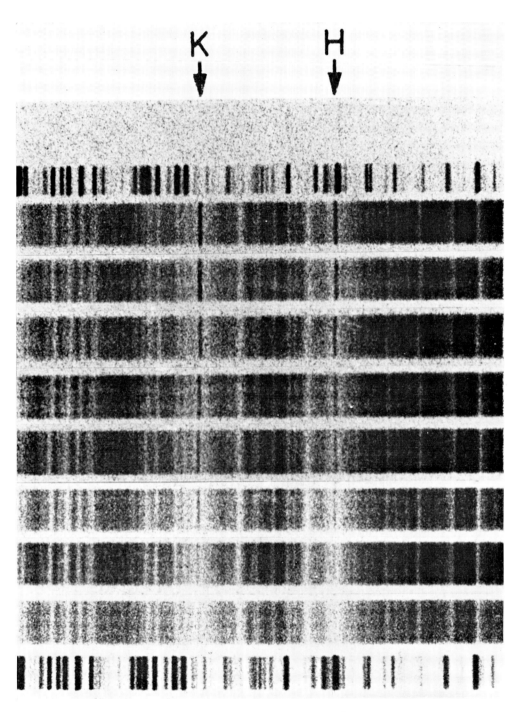

FIGURE 1 Spectrograms of *dK*5 field stars with a range of *K*2 intensity [courtesy O. C. Wilson]. The original dispersion is 38 Å/mm. The stars with the strongest *K*2 are similar to those found in the Hyades. At 200 Å/mm, emission lines of this strength cannot be detected.

Since $K2$ in the Sun, which is known to be produced in the solar chromosphere, would not be seen in the integrated sunlight at the dispersions used by Wilson, and further since the Sun is known to have a nuclear age greater than 5×10^9 years, the Wilson results can be understood if it is supposed that the stellar chromosphere, or at least that activity giving rise to $K2$, decays on a time-scale short compared with the nuclear lifetime of solar-type stars. If statistically the strength of $K2$ is inversely correlated with nuclear age, then the stars of the general field with $K2$ strengths like those of the Hyades, for example, presumably have nuclear ages similar to Hyades stars. Thus only about 10 per cent of the field solar-type stars are 'young' in this sense. However, attempts to use these conclusions to date stars, for example, must take into account the dispersion and thus resolution of the spectrograph. $K2$ in the Sun, we have already noted, would not be seen at 10 Å/mm in the integrated sunlight; dispersions of about 2 Å/mm are required to detect it. Wilson noted that some stars with weak but detectable $K2$ at 10 Å/mm showed no $K2$ at 38 Å/mm; presumably at (say) 100 or 200 Å/mm we could detect only very strong emission in H and K. But since few stars per unit volume of space would be expected to show strong $K2$, its detection is a suitable technique for selecting main sequence members of young galactic clusters.

We are concerned however not with F- and G-type stars but rather with stars later than $K0$ V; fortunately a sample of field and Hyades stars between $dK5$ and $dK7$ has been studied by Wilson (1964, 1966b) at dispersion 38 Å/mm. At this scale, it was found that (1) all Hyades stars showed $H2$ and $K2$, and (2) only about 10 per cent of field stars had $H2$ and $K2$ as strong, or stronger, than the Hyades stars. In Figure 1 is reproduced a montage of spectra of field $dK5$ stars with a range of H and K emission strength (Wilson 1966b); Hyades stars have $K2$ with strength similar to the most intense $K2$'s in the figure. We illustrate also the distribution of stars in the field sample as a function of $K2$ strength in Figure 2 (Wilson 1964). The emission line intensities are on an arbitrary scale and the age calibration is unknown, but the strength of the Hyades $K2$ in comparison with the field illustrates that detection of strong $K2$ emission would be a valuable criterion for cluster membership. In the Pleiades, however, a star at $dK5$ has $B = 14.8$, and one must work at a dispersion much lower than 38 Å/mm.

A pilot program was conducted using the Palomar nebular spectrograph with a first-order blue grating and the 1.4-inch camera; the combination gives a dispersion near 200 Å/mm. It was found that $K2$ could not be detected at $dK5$ in any of the stars illustrated in Figure 1, even those with the most intense $K2$, and similarly it could not be detected in Hyades stars of this spectral type. It could, however, easily be seen in the best-established

Low-luminosity stars

Pleiades members of this type. This suggests that detection of $K2$ at 200 Å/mm in the spectra of Pleiades stars is an extremely powerful method of isolating members. At the same time, it was deemed important to determine spectral types in order to examine the discordances between spectral type and color already noted. It was also found that $K2$ in emission could, in fact, be detected in Hyades K-type stars at dispersions near 100 Å/mm. We have

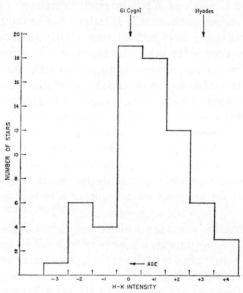

FIGURE 2   Fractional number of field stars between $dK5$ and $dK7$ as a function of $K2$ intensity on an arbitrary scale [courtesy O. C. Wilson]. There are 69 stars in the sample.

therefore studied the spectra of K- and M-type Hyades stars at Palomar with the equipment noted above, and also with the same spectrograph and grating but with the 3-inch camera (dispersion near 90 Å/mm), and we have studied also the spectra of these and other Hyades stars with the nebular spectrograph of the 120-inch Lick reflector at a dispersion of 100 Å/mm. Attempts to detect the $K2$ emission with the Palomar prime focus scanner failed owing to bad weather, but the technique was later abandoned when the importance of eliminating the nebular fog and determining the spectral types was more fully appreciated.

III. RELATIONSHIP BETWEEN SPECTRAL TYPE AND (B – V) COLOR

The relationship between spectral type and $(B - V)$ color for the standard main sequence of the Yerkes system (Johnson and Morgan 1953, Keenan

1963, Vyssotsky 1963, Johnson 1966) was first tested in the spectral type range *K2 V* to *M2 V* in the Hyades. Standards of the *MK* system were observed with each of the spectrographic combinations noted above; criteria for spectral type were established empirically based principally on line ratios and on the marginal appearance of TiO bands. These were judged likely to be least affected by the 'filling-in' of an anticipated early-type reflection nebulosity in the Pleiades. Valuable in this connection are the relative intensities of lines in the triplet centered on λ4254 Cr I, the ratios λ4254/λ4272 and λ4290/λ4325, and the strengths of TiO band-heads near λ4760 and λ4950. Secondary criteria included the strengths of Ca I λ4226 and the zero-volt Fe I lines longward of λ4383.

Since the Hyades is free from nebular fog and presumably is unreddened, we would expect a plot of spectral type vs. $(B - V)$ to have a dispersion resulting only from errors of observation. This diagram is plotted in Figure 3 for Hyades stars; the values of $(B - V)$ are taken from the photographic photometry of van Altena (1966) which in turn is based on the photoelectric sequence of Johnson, Mitchell, and Iriarte (1962). (For a few stars, among them the most discordant in the plot, photoelectric colors are available, but the amount of such material is small and its inclusion leads to some inhomogeneity.) The relationship is entirely satisfactory, and suggests that, (1) the faint Hyades stars have spectral types consistent with expectation

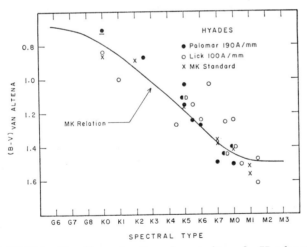

FIGURE 3 $(B - V)$ as a function of spectral type for Hyades stars. The values of $B - V$ are taken from the photographic photometry by van Altena (1966). The highly discordant star at $B - V = 1.03$, spectral type $= K6.5$ *V* is GH 7-253, for which Johnson, *et al.* (1962) give a photoelectric $B - V$ of 1.36.

on the basis of the standard main-sequence relation between spectral type and color, and (2) the criteria set up can determine a spectral type with an error no greater than about one spectral subdivision. A plot (Figure 4) of these spectral types against the absolute magnitudes determined from the

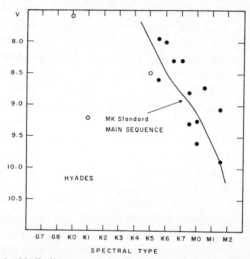

FIGURE 4   H–R diagram of the Hyades for K- and M-type dwarfs. The curve is the standard Yerkes main sequence. Filled and open circles represent stars having spectra with and without Ca II emission at 100 Å/mm dispersion, respectively.

proper motions (van Altena 1966) also gives a result entirely consistent with the standard Yerkes relation (Keenan 1963, Vyssotsky 1963). The colors, magnitudes, and adopted spectral types for these Hyades stars are listed in Table I.

Spectral types were obtained for 49 proper motion 'members' (Hertzsprung 1947) of the Pleiades with visual magnitudes between 13.0 and 15.5; a journal of observations is found in Table II. In Figure 5, these spectral types are plotted against the Johnson–Mitchell (1958) values of $(B - V)$, derived photographically from a photoelectric sequence. The systematic displacement in Figure 5 can be accounted for by several hundredths of a magnitude of reddening; the scatter, amounting to as much as 0.7 mag in $(B - V)$ at $K5\ V$, cannot however be accounted for on the basis of differential reddening derived from the brighter cluster stars or from reasonable expectation of the errors in the spectral types. These photographic colors are, however, not the only ones available, and in Figure 6 we plot the spectral types against the photographic colors of Ahmed, *et al.* (1965), which are based on the Johnson-Mitchell photoelectric sequence. Ahmed, *et al.* claim to have removed the influence of the nebular fog on the colors and magnitudes. It will be seen that, in comparing Figure 6 with Figure 5, the scatter is, if anything, worse,

TABLE I Hyades stars observed for Ca II emission.

Name GH*	V^a	$(B - V)^a$	$M_V{}^a$	Sp.[b] (MK)	EW (K2) (Å)
7–170	12.34	1.00	9.21	K1 V	<0.8
7–179	12.86	1.61	9.88	M1.5 Ve	1.3
7–182	10.12	0.71	6.67	K0 V:	<0.8
7–185	10.84	1.11	8.50	K5 V	<0.8
7–193	11.00	1.27	7.97	K5.5 Ve	1.0
7–194	12.38	1.47	9.07	M1.5 Ve	2.0
7–200	11.88	1.50	8.72	M0.5 Ve	4.0
7–202	10.55	0.87	7.38	K2.5 V	<0.8
7–205	11.16	1.15	8.59	K5.5 Ve	1.2
7–208	12.49	1.24	9.60	M0 Ve	1.7
7–215	10.64	1.49	8.31	K7 Ve	1.1
7–223	11.12	0.84	7.59	K0 V	<0.8
7–230	11.23	1.24	7.99	K5 Ve	0.7
7–238	12.33	1.40	9.25	M0 Ve	1.85
7–251	12.24	1.44	8.79	K7–M0 Ve	1.6
7–253	11.22	1.03	8.30	K6 Ve	0.9
near VB 71[c]	12.03	1.25	9.29	K7–M0 Ve	1.4

* Giclas, Burnham, and Thomas (1962).
[a] van Altena (1966).
[b] 'e' denotes emission at H and K.
[c] Incorrectly identified as VB 71 by van Altena (1966).

and the stars of latest spectral type seem to be systematically too red compared with those of earlier type. We note also that, if the stars are divided arbitrarily into two groups at H II 1653, the group with the larger running numbers is systematically redder. Thus, either the cluster's eastern section is more heavily reddened than the western, or there is an east-west systematic error in the colors.

In Figure 7 our spectral types are compared with the $(B - V)$ photoelectric colors of Iriarte (1967). Unfortunately, Iriarte did not observe very many stars; however, aside from the one star at G5 V, the scatter in Figure 7 is in reasonable accord both with the range in reddening derived from earlier-type

TABLE II Pleiades stars observed for Ca II emission.

Name H II No.*	V J&M	V A,L,&R	V I	B−V J&M	B−V A,L,&R	B−V I	Sp.† (MK)	EW(K2) (Å)
81[a]	13.59	13.63	13.56	0.90	0.64	0.88	G8 *V*	—
83	14.92	14.89	14.89	0.93	0.93	0.99	K0 *V*	—
105	13.76	13.84	13.78	0.98	0.91	0.95	G5 *V*	—
133	14.40	14.46	14.32	1.32	0.88	1.38	K5.5 *Ve*	—
134[b]	14.44	14.29	14.39	1.40	1.41	1.55	K7 *Ve*	4.9
146[b]	14.52	14.48	14.60	1.56	1.45	1.41	K7–M0 *Ve*	4.6
189	13.92	14.05	14.00	1.46	1.33	1.36	K5.5 *Ve*	1.7
191	14.38	14.43	14.39	2.15	1.21	1.46	K7–M0 *Ve*	3.8
212	14.57	14.48		1.31	1.32		K7 *Ve*	4.0
335	13.94	13.80	13.78	0.98	1.10	1.25	K5 *Ve*	2.4
347[b]	14.01	14.01		1.57	1.34		K7 *Ve*	5.3
357	13.46	13.65	13.52	1.44	0.88	1.28	K6 *Ve*	2.3
380[a]	13.19	13.14		1.39	1.08		K4 *Ve*	—
451[a]	13.25	13.44		1.43	1.08		K5 *Ve*	—
513	13.76	13.86		1.30	1.20		K7 *Ve*	2.6
554	14.09	13.82		0.92	1.32		K5 *Ve*	3.2
624[b]	15.42	15.11		1.40	1.85		M2 *Ve*	4.6
673[c]	15.46	15.44		1.47	1.39		K5 *V(e?)*	—
676	13.56	13.71		1.30	1.11		K3.5 *Ve*	2.5
740[a]	13.45	13.34	13.45	1.08	0.98	1.08	K3 *Ve*	—
890[b]	14.80	14.90		1.33	1.85		M0 *Ve*:: +B	4.1::
906[c]	15.24	15.00		1.44	1.70		K7–M0 *Ve*	—
915	13.62	13.69		1.23	1.14		K6 *Ve*	3.2
945	13.29	13.35		1.06	0.97		K2 *V*	<0.8
974	14.10	13.99		1.34	1.40		K7 *Ve*	3.1
1081[b]	14.74	14.60		1.60	1.53		K7 *Ve*	3.2
1103[b]	14.81	14.83		1.47	1.79		K7 *Ve*	4.1
1110	13.69	13.29	13.41	1.10	1.39	1.23	K6.5 *Ve*	2.1
1276	15.32	15.19		1.56	1.49		K5 *V*	<0.8
1355	13.97	14.03		1.23	1.36		K5 *Ve*	5.0
1485[b]	14.30	14.01		1.43	1.58		K5 *Ve*	4.9
1628	13.45	13.50		0.81	0.71		G0 *V*	—
1653[b]	13.31	13.48	13.69	1.45	1.19	1.28	K4.5 *Ve*	2.6
2016	13.52	13.48		1.22	1.39		K4 *Ve*	1.3:
2082	13.98	13.76		1.33	1.57		M0 *V*::	d

TABLE II—*(cont.)*

Name H II No.*	V J & M	V A,L, &R	I	B − V J & M	B − V A,L, &R	I	Sp.† (MK)	EW(K2) (Å)
2193ᵇ	14.16	14.29		1.32	1.32		K6 Ve	2.9
2199	14.31	14.39	14.38	1.13	1.10	1.08	K3 V	<0.8
2208ᵇ	14.13	14.20		1.61	1.63		K6 Ve	4.1
2209	14.38	14.35		1.47	1.61		K6.5 Ve	2.9
2402	15.15	14.92		1.16	1.28		K1 V	—
2548	13.98	14.77		1.58	0.60		K5.5 Ve	1.2
2588	13.10	13.04		1.22	1.35		K3 V(e?)	<0.8
2601ᵇ	14.99	14.83	15.09	1.55	1.78	1.56	M3 Ve	4.7
2602ᵇ	15.52	15.34		1.60	1.71		M2.5 Ve	4.6
2908ᵃ	13.02	12.90	13.41	1.38	1.41	1.15	K3 Ve	—
2927ᵇ	13.92	13.63		1.25	1.52		K4 Ve	2.8
2966ᵃ	14.72	14.71		1.67	1.82		M1.5 Ve	3.6
3069	13.85	13.82		1.12	1.14		K2 V	<0.8
3187	13.12	13.02		1.16	1.34		K4.5 Ve	1.4

* Hertzsprung (1947).
† 'e' means emission at H and K.
ᵃ Palomar plate taken at 90 Å/mm. No plate calibration available.
ᵇ Emission seen at $H\beta$ and sometimes in higher members of the Balmer series.
ᶜ Palomar plate taken at 380 Å/mm.
ᵈ Somewhat out of focus.

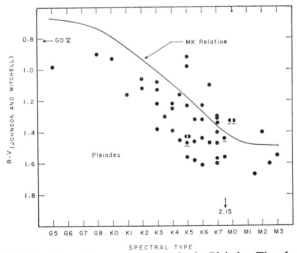

FIGURE 5 $(B − V)$ vs. spectral type in the Pleiades. The photographic colors are those of Johnson and Mitchell (1958).

Low-luminosity stars

stars (cf. Crawford 1968) and our estimate of the uncertainty in the spectral classification. From the 12 common stars that are probable members (by criteria to be discussed later), we derive from Figure 7 a mean color excess $E(B - V) = +0.075 \pm 0.025$ (m.e.) mag. This is close to the $E(B - V) = +0.05 \pm 0.01$ (m.e.) derivable from $b - y$ and Hβ photometry of the B, A, and F type stars (Crawford 1968).

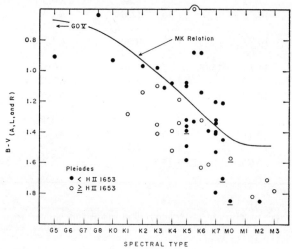

FIGURE 6 $(B - V)$ vs. spectral type in the Pleiades. The photographic colors are those of Ahmed, *et al.* (1965).

We concluded that, aside from the inclusion of non-members, the spread in the color-magnitude diagram of the Pleiades below $V = 12.0$ is a result of fairly large observational errors in the photographic colors and magnitudes of both Johnson and Mitchell and Ahmed, *et al.* There is every reason to believe that, if the non-members could be isolated and rejected, if Iriarte's list could be greatly extended, and if the stars were observed photoelectrically when flares are absent (Johnson and Mitchell 1958, Haro and Chavira 1965, Haro 1968), the lower part of the main sequence would be as narrow as the upper part.

Though the photographic colors show large random errors, the photographic visual magnitudes are good enough to be used in a classical H–R diagram in which V is plotted against spectral type. A comparison of the Johnson-Mitchell and Iriarte photometry gives in fact $V(J - M)$ *minus* $V(I) = -0.02 \pm 0.04$ (m.e.) mag. In Figure 8 this H–R diagram is shown with the ordinate taken from Johnson and Mitchell. Stars with H and K

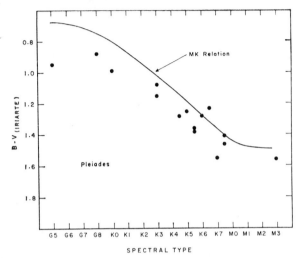

FIGURE 7 $(B - V)$ vs. spectral type in the Pleiades. The photoelectric colors are those of Iriarte (1967).

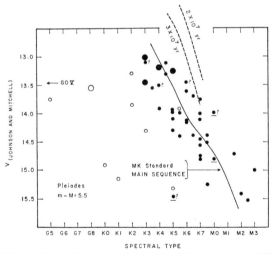

FIGURE 8 H–R diagram of the faint Pleiades stars. Filled and open circles correspond to stars with and without Ca II emission, respectively. Larger symbols correspond to plates taken at 90 Å/mm. Question marks refer to uncertainties in detection of emission and underscorings to uncertainties in the spectral types. Isochrones are those given by Iben and Talbot (1966).

in emission at 200 Å/mm are segregated. For the most part, those without Ca II emission also have spectral types too early for their magnitudes. The standard Yerkes main sequence (Keenan 1963, Vyssotsky 1963) is drawn in for a distance modulus of 5.5. The stars with detectable Ca II emission clearly lie along the standard main sequence at least as far as *M*1 *V*; not enough material is available below *M*1 *V* to state with certainty if the apparent turn-away of the main sequence to the red below this point is real.

Returning to the Johnson–Mitchell photometry, we ask if one is able to account for the stars having excessive redness or blueness. From a mean line drawn in Figure 5, we considered individually those stars in the *entire* list of Johnson-Mitchell photometry, i.e., members, probable members, probable non-members, and non-members, that deviated more than 0.2 mag. in (*B*− *V*) from the mean line. Such stars were described as 'too red' or 'too blue'. The fractional numbers of such stars in each of the one magnitude intervals centered on *V* = 13.0, 14.0, and 15.0 are illustrated in Figure 9.

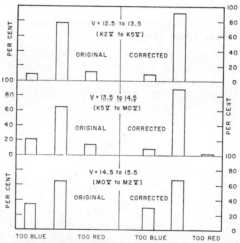

FIGURE 9 Original and corrected frequency functions per unit magnitude interval of 'normal', 'too red', and 'too blue' stars. These categories are explained in the text.

All but two of the 'too red' stars in the entire three-magnitude range had either an Iriarte color or a spectral type determined by us, and in every case, this color or spectral type moved it into the 'acceptable' range. Stars 'too blue' either could not be corrected because no other information was available, or became 'acceptable' or remained 'too blue'; none of these moved into the 'too red' category. We conclude that there are no stars in the extensive Johnson–Mitchell list in the region of the Pleiades that lie far to the right of the main sequence at least down to $M1\ V$. If, as our spectral types confirm and the Iriarte photometry suggests, the Pleiades stars narrowly define the standard main sequence to this spectral type, then $T_K \geqq 8 \times 10^7$ years, at least twice the nuclear age. Further, if the strength of $K2$ dies away with time, then the fact that there is no correlation between strength and departure in V at a fixed spectral type in Figure 8 means that all the faint Pleiades stars are coeval. Thus if $T_N < T_K$, either there is a 'jump' in formation times of faint stars compared with bright, or there is a physical factor not accounted for either in the post- or pre-main sequence evolutionary theory, or both.

IV. CORRELATION OF THE STRENGTH OF K2 EMISSION WITH AGE

The equivalent width of $K2$ has been measured in all Hyades and Pleiades stars showing it; the unit of intensity is the flux in the estimated local continuum. Hyades stars were measured mostly from Lick plates and Pleiades stars from Palomar plates; a small systematic correction was found from spectrograms of several Hyades stars taken at both telescopes. These equivalent widths are plotted as a function of spectral type in Figure 10 and are listed in Tables I and II. At a given type, $K2$ is about twice as strong in the Pleiades as in the Hyades. If this change is a consequence of the evolution of

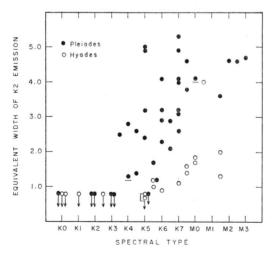

FIGURE 10 Equivalent width of $K2$ in units of the local continuum. Filled and open circles refer to Pleiades and Hyades stars, respectively.

D

stellar chromospheres, then the time for decay of the chromosphere of a Pleiades K- or M-type dwarf is about the age of the Hyades; this is several orders of magnitude shorter than the nuclear time for such stars.

One Hyades star, GH 7-200, has unusually strong $K2$, but from its position in the color-magnitude diagram (cf. van Altena 1966), it is probably a double-line spectroscopic binary. Our spectrograms have too small resolution to test the validity of this point. Wilson (1963, 1966b) has noted that a Ca II emission is usually enhanced in the components of spectroscopic binaries with periods of the order of days. Possibly some of the Pleiades stars with more intense $K2$ emission than average are spectroscopic binaries. But with this reservation, it seems possible that a study of the strength of stellar $K2$, with appropriate calibration based on stars in clusters, provides a valuable tool for studying the kinematics and space distribution of late-type field stars as a function of age.

We are indebted to Mr. W. Alschuler for making the microphotometer tracings.

References

AHMED, F., LAWRENCE, L., and REDDISH, V. C. 1965, *Pub. Royal Greenwich Obs.*, **3**, 187.
CRAWFORD, D. L. 1968, private communication.
EZER, D. 1966, *Colloquium on Late-Type Stars*, ed. M. HACK, p. 357.
GICLAS, H. L., BURNHAM, R., and THOMAS, N. G. 1962, *Lowell Obs. Bull.*, **5**, 257.
HARO, G. 1968, 'Stars and Stellar Systems', Vol. VII, ed. L. H. ALLER (Chicago: University of Chicago Press), p. 141.
HARO, G., and CHAVIRA, E. 1965, 'Vistas in Astronomy', Vol. 8, ed. A. BEER (London: Pergamon Press), p. 89.
HERBIG, G. H. 1962, *Ap. J.*, **135**, 736.
HERTZSPRUNG, E. 1947, *Ann. Sterrewacht Leiden*, **19**, part 1A.
IBEN, I. 1965, *Ap. J.* **142**, 1447.
IBEN, I. 1966a, *Ap. J.*, **143**, 483.
IBEN, I. 1966b, *Ap. J.*, **143**, 505.
IBEN, I., and TALBOT, R. J. 1966, *Ap. J.*, **144**, 968.
IRIARTE, B. 1967, *Bull. Obs. Tonantzintla y Tacubaya*, **4**, 79.
JOHNSON, H. L. 1966, *Annual Reviews of Astronomy and Astrophysics*, Vol. IV, ed. L. GOLDBERG (Palo Alto: Annual Reviews, Inc.), p. 193.
JOHNSON, H. L., and MITCHELL, R. I. 1958, *Ap. J.*, **128**, 31.
JOHNSON, H. L., MITCHELL, R. I., and IRIARTE, B. 1962, *Ap. J.*, **136**, 75.
JOHNSON, H. L., and MORGAN, W. W. 1953, *Ap. J.*, **117**, 313.
KEENAN, P. C. 1963, in 'Stars and Stellar Systems', Vol. III, ed. K. STRAND, (Chicago: University of Chicago Press), p. 78.
KUHI, L. 1966, *Ap. J.*, **143**, 991.
POVEDA, A. 1965, *Bull. Obs. Tonantzintla y Tacubaya*, **4**, 15.

VAN ALTENA, W. F. 1966, *A.J.*, **71**, 482.
VASILEVSKIS, S., SANDERS, W., and BALZ, A. 1965, *A.J.*, **70**, 797.
VYSSOTSKY, A. N. 1963, 'Stars and Stellar Systems', Vol. III, ed. K. STRAND (Chicago: University of Chicago Press), p. 192.
WALKER, M. F. 1956, *Ap. J. Suppl.*, No. 23.
WALKER, M. F. 1957, *Ap. J.* **125**, 636.
WILSON, O. C. 1963, *Ap. J.*, **138**, 832.
WILSON, O. C. 1964, *Pub. A.S.P.*, **76**, 28.
WILSON, O. C. 1966a, *Ap. J.*, **144**, 695.
WILSON, O. C. 1966b, *Science*, **151**, 1487.

Discussion

EGGEN   Unless I missed something, you do not really rule out Herbig's explanation.

KRAFT   I do not really rule it out as well as I would like. But while there is a scatter in the emission line intensities in the Pleiades, the intensities are not correlated with departure of the star from the main sequence.

EGGEN   Let us consider a cluster in which star formation is going on. In the cluster, the M dwarfs are sitting there cooking, and they could be of many different ages. Now when a bright star is forming that will blow away the dust and the star formation will cease. But, the dwarfs that are there could have a large range in age, and therefore some have emission that you detect and some that you do not.

KRAFT   That depends upon what that spread of ages is. All I can say is that the range does not result in these stars on the upper edge of the "observed" main sequence appearing younger than the ones on the lower edge.

EGGEN   The presence of the emission is not any longer the criterion of cluster membership. It is a criterion of age.

KRAFT   But the thing that I will wait for is Iriarte's colors, which he says he is going to observe.

EGGEN   Actually, I thought that it was decided that most of the trouble with Johnson's diagram was just the color error.

GREENSTEIN   I wonder if I could ask Dr. Hayashi this question: What is the plausible contraction time with the utmost acceleration that you can imagine for an $M2$ dwarf?

HAYASHI   I think that the contraction time can be calculated under the assumption that the stars are wholly convective. However, the stars may not be wholly convective when they reach the stage of quasi-hydrostatic equilibrium after the flare up. If the polytropic index of the star is larger than 1.5, then the total energy of the star would be large and the time-scales would be increased.

KRAFT   We know from Kuhi's work that there is mass-loss that occurs during contraction. So, it could be that, as Ezer suggested, the stars that we see now really start out with higher masses. So, they got down faster in the initial stage and maybe one could pick up a factor of two that way.

BLAAUW   Perhaps this is more a question for Dr. Strand. How accurate can one get proper motions for these stars?

STRAND   As far as I recall, and Mr. Riddle can correct me, we have plates going down to about 20th magnitude in the region around Pleiades.

OSTRIKER   Can you say anything about the rotation of the lowest luminosity members of Pleiades? I know it is small, but how small?

KRAFT   No, I cannot, not at 200 Å/mm, because then the rotational limitation would be about 200 km/second. I will say something more about rotations of M dwarfs in Krzeminski's paper that I will be reading later on.

Low-luminosity stars
in the Pleiades region

MARTIN F. McCARTHY, S.J.

Vatican Observatory, Vatican City State

Abstract

Twenty-two late-type stars in the Pleiades were observed with an image-tube spectrograph at dispersions of 360, 270 and 135 Å/mm with the following results: (1) The stars range in spectral type from $dK4$ to $dM4$; 12 show Hα in emission and 10 are known flare stars. (2) Membership criteria from proper motion measures, photometry, observations of flare activity emission features and dwarf characteristics are intercompared; 13 stars are considered certain members, 5 probable members, 3 as uncertain and one (LLP 73) is a known member of the Hyades. (3) Of the 13 member stars none is classified later than $M2$; later types predominate among the probable, uncertain and non-member stars. (4) Blink measures show that only 2 of the 17 stars on this list studied by Van Altena appear to be foreground stars; these two stars (LLP 73 and 111) plus one other (MT 49 not studied here) seem to share the motion of the brighter Hyades. (5) For 5 of the 12 emission stars Rubin has measured radial velocities of the Hα feature on the image tube spectra with 135 Å/mm dispersion. Four of the five stars are flare stars. All velocities are positive and average 84 km/sec for the four Pleiades stars and 116 km/sec for the Hyades member. Velocities may represent chromospheric activity in these stars. The accuracy of these measures is limited and more measures are needed. (6) Photographic photometry for 11 of the 21 Pleiades stars by Ahmed, Lawrence and Reddish shows that 9 of the 11 stars lie below the main sequence. Iriarte's photoelectric measures show significant changes in the positions of these stars on the color magnitude diagram. If such excursions from the main sequence prove to be common for low luminosity cluster members the use of color magnitude diagrams as membership criteria for these stars will require reassessment. While we find no evidence for main-sequence members in the Pleiades with spectral types later than $M2$–$M3$, the number of very faint ($B = 20$) flare stars detected by Haro and the very large number of stars blinked by Luyten which show common motion with the bright Pleiades lead one to believe that at this time we cannot be sure that we have reached the termination of the main sequence of the Pleiades cluster.

I. INTRODUCTION

THE LENGTH and breadth of the main sequence of the Pleiades has been a matter of interest and concern for the past forty years, since the pioneering researches of Hertzsprung, Voûte and Trumpler. Current theories of stellar evolution have intensified this concern and today we are still asking: 'Where does the main sequence terminate?'

To put the problem in perspective we can recall the very early color-magnitude diagram presented by Hertzsprung (1929) in his George Darwin Lecture. This diagram he drew first for the stars whose membership he had established by his proper motion measures at Leiden and after that for the field stars in the area. The remarkable length of the Pleiades main sequence can be seen clearly by comparing it with the corresponding sequences in other nearby clusters.

In addition to the length of the main sequence, the Pleiades cluster was unusual in its lack of yellow giants and white dwarfs and in the presence of reflection nebulosity. These features were all pointed out by Hertzsprung. Johnson and Mitchell (1958) using a combination of photoelectric and photographic photometry derived a new color-magnitude diagram for the Pleiades which once again showed a long main sequence, no evidence of a contracting sequence but instead a rather striking broadening near $M_v = +6.5$ with stars lying above and below the main sequence. They also observed the first flare star in this cluster HZ 1306. A year earlier at the Vatican Conference on Stellar Populations, Herbig (1958) had discussed the hypothesis that dMe stars might be younger members of open clusters while the non-emission dM stars were more ancient. However, the complete absence of dMe stars in the Pleiades cluster stood as a puzzling obstacle to his hypothesis. The present paper reports evidence for 11 new Hα emission stars in the Pleiades region plus one other discovered independently by Haro and the author.

Within the past decade several studies of low-luminosity stars in this cluster have been carried out, each bearing on the fundamental problem, yet none arriving at a full and final solution. We summarize these researches in Table I. At this conference Wilson, Haro, Rosino, Greenstein and Kraft have discussed their results. We summarize our own research under the following headings:

Part One: Vatican Observatory Objective Prism Survey of M Stars in the Pleiades. McCarthy and Treanor (1964).

Part Two: Finding List of Late-Type Members of the Pleiades Cluster. McCarthy and Treanor (1968).

Part Three: Image Tube Spectra of 22 Dwarf Stars in the Pleiades Region (this paper).

TABLE I Summary of recent research on low-luminosity stars in the Pleiades.

Astrometric Studies
Luyten (1967). Blink Studies; Palomar Schmidt; Baseline: 1946–64
Van Altena (1968). Blink Studies; Lick 20″ Astrograph; Baseline: 1949–1965

Photometric Studies
Johnson and Mitchell (1958). Color Magnitude Study; UBV; p.e. and p.g.
Ahmed, Lawrence and Reddish (1965). Color Magnitude Study: UBV; p.g. only
Iriarte (1967). Color Magnitude Study; UBV; p.e.
Johnson and Mitchell (1958). Flare Star Study
Haro and Chavira (1966). Flare Star Study
Haro (1968). Flare Star Study
Rosino (1966). Flare Star Study

Spectroscopic Studies
Wilson, O.C. (1963, 1964). High Dispersion Spectra of Early K Stars
Pesch (1961). Objective Prism Study; M Stars
Herbig (1962). Crossley Nebular Spectrograph; K and M stars
Kraft and Greenstein (Pres. Conf.) 200″ Coudé Spectra of Late Type Spectra
McCarthy and Treanor (1964). Objective Prism Study; M type stars
McCarthy, Treanor and Ford (1967). Spectra with Carnegie Image Tube; 3 M Stars
McCarthy (Present Conference). Image Tube Spectra of 22 dwarf K and M Stars

---

II. OBJECTIVE PRISM SURVEY

The objective prism survey was made with hypersensitized IN plates and RG 2 Filter, the Vatican Schmidt and its 2°.5 objective prism. This combination gives a dispersion of 3600 Å/mm at 7600A. A region of 4°.6 square centered on the Pleiades was searched for possible M dwarf members of the Pleiades cluster. We detected 125 stars and divided these into three natural groups of possible members in the spectral ranges of $M2$–$M3$ (Group A), $M0$–$M1$ (Group B) and $K7$–$K9$ (Group C). A fourth Group D, we considered to contain non-member field stars, mostly giants. Because of the low dispersion employed, no luminosity classification was possible. From our analysis of the surface distribution, magnitudes and motions of these stars, we inferred that the three groups A, B and C contained many probable cluster members. A select group of stars, fainter than $B = 16$, was chosen for further study. It was pointed out that a well populated main sequence of late-type stars was present in the Pleiades down to about $B = 17$.

III. FINDING LIST OF 144 PROBABLE PLEIADES K-M TYPE STARS

Recently the stars of our selected list (some 44 of the original 125 red stars) were combined by Treanor and the author (1968) with other stars whose astrometric, photometric or spectroscopic data indicated that they might be members of the faint end of the main sequence of the Pleiades cluster. These stars are assembled together with identification maps into a Finding List of 144 likely candidates for cluster membership. This list is with the printer at Vatican City now and contains the following information:

1. Designation of the star in this and other catalogues.
2. Blue magnitudes and spectral classifications where avaiable.
3. Characteristic features indicative of Membership status:
   (a) proper motion data from p.m. measures and blink surveys
   (b) dwarf characteristics in spectra
   (c) presence of Hα in emission
   (d) presence of flare activity
   (e) photometric characteristics.

This information is given for each of the stars; obviously the fainter stars have been less thoroughly observed. The stars are considered in four groups:

Part One: 88 stars listed in Hertzsprung's General Catalogue (1946).

Part Two: 21 stars *not* in Hertzsprung's Catalogue and brighter than $B = 17$.

Part Three: 23 stars *not* in Hertzsprung's Catalogue and fainter than $B = 17$.

Part Four: 12 stars from McCarthy–Treanor Catalogue (1964) which require further study.

We note that while a convergence of probabilities exists for many of the 144 stars on our Finding List, there can be no certainty until improved photometry, accurate proper motion studies and especially radial velocity measures are obtained.

IV. IMAGE TUBE SPECTRA OF 22 LATE-TYPE STARS IN THE PLEIADES

We shall now concentrate our attention on the 22 late-type stars in the Pleiades for which we have obtained image tube spectra during the past three observing seasons. All observations were made with Carnegie image tubes in the DTM spectrograph at the 72-inch (1.83 meter) Perkins reflector of the Ohio State University and the Ohio Wesleyan University at the Lowell Observatory in Flagstaff, Arizona. W. K. Ford, Jr., V. C. Rubin and the author made the observations with occasional assistance from W. A. Baum and P. B. Boyce of the Lowell Observatory. Observations of one star were repeated during an observing session at the University of Virginia when the 31-inch (78 cm) Fan Mountain reflector was used. Details of the observations are summarized in Table II.

TABLE II Instrumental details: Image-tube spectra of 22 stars in the Pleiades.

	Set one	Set two	Set three
Camera	178 mm Perkins	225 mm Bowen	225 mm Bowen
Relay lens	1:1 Burke James	1:1 Burke James	1:1 Burke James
Grating	150 lines/mm	150 lines/mm	300 lines/mm
Dispersion	360 Å/mm	270 Å/mm	135 Å/mm
Range	5500–8500 A	4500–8000 A	4500–7500 A
Comparison source	Same for all: Fe and Ne Hollow-Cathode Lamp.		
Voltage	20.5 kV	20.0 kV	18.5 kV
Exposures	40^m to 60^m	30^m to 120^m	20^m to 108^m
Emulsion	IIa 0	IIa 0 (Baked)	IIa 0 (Baked)
Date	1965 3–4 Dec.	1966 15–21 Dec.	1966 22–24 Dec.
			1967 10 Jan.
			1968 4–10 Jan.

(a) Spectral classification

Several criteria have proven useful in classifying spectra in the green–yellow–red region at the dispersions listed above in Table II; we summarize them here.

For temperature classification the strength of the TiO bands is a most useful criterium beginning at *K*4 or *K*5. This feature increases throughout the range of the *M* dwarfs; for our present material this reaches to *M*4. As a check on temperature classification the Mg triplet can be used as well as the Na *D* lines and the Ca feature at 6162 A, though these are more useful as luminosity criteria once the temperature classification has been made from the TiO bands. The bands of TiO most evident in the spectral range covered by the image tube plates are those at 7589 A, 7054 A, 6651 A, 6158 A, 5847 A, 5809 A, 5589 A, 5167 A, 4955 A and 4761 A. On the plates discussed here the VO bands, which are so prominent in extremely cool *M* stars, do not show any significant strength.

Luminosity classification on objective prism plates employs as its main criterion for dwarf stars, the presence and strength of the Na *D* lines, a criterion discovered by Luyten (1923). On slit spectra obtained with image tubes, long exposures (which are required to reach the faint stars in the Pleiades cluster) are affected by the presence in emission of the night sky lines of Na. At low dispersions a blending with TiO at 5847 further complicates luminosity classification. Still, the best single feature for eye estimates of luminosity on low dispersion image-tube plates is the intensity of the *D* line feature.

Other atomic lines of use in luminosity classification are the Mg 'b' lines 5167–5184 A. The use of this criterium was first described by Fitch and

Morgan (1951). The strong Ca I line at 6162 A is useful especially in the early *K* stars and it was first employed by Burwell (1930). One should note that the Mg 'b' feature can be complicated by TiO at 5167 A, by ZrO at 5185 A and by MgH bands at 5211 A; the Ca I line coalesces with the nearby band of TiO at 6158 A near *K*9 or *M*0. All of these features show themselves to be strong in dwarfs and weak in normal giants of the same spectral subtype. In the far blue at 4227 A the Ca I line is visible often in great strength on the image-tube plates, but as it is at the limit of the effective spectral range on most of the present plates, it has not been used in luminosity classifications.

The hydride bands of magnesium and calcium described first by Öhman (1936) find special application for luminosity classification in the yellow–red region, a region ideally suited for image-tube studies. One image-tube spectrum covers the same spectral range which would be covered by a combination of the O, D, E, F, U and N emulsions. To illustrate this point the reader is invited to compare an objective prism spectrogram tracing made at a dispersion of 316 Å/mm (at Hα) with the 12° prism attached to the Vatican Schmidt (Treanor and McCarthy 1967) with the tracings shown below in Figure 1 of the present paper. On the tracing of the objective prism spectrum of a dwarf *M*2 star one sees the strength of the Na *D* lines and the presence of CaH at 6385 A and 6885 A. Note that the

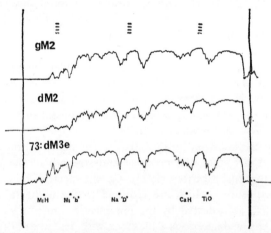

FIGURE 1 Tracings of two control spectra of spectral type *M*2 and one faint star in the Pleiades (LLP 73 = Hz 2411 = MT 102) the Hyades member in the Pleiades region. See Figure 3. Top spectrum shows *gM*2 star BD + 36° 2139, *V* = 6.1. Middle spectrum shows *dM*2 star BD + 36° 2147, *V* = 7.2. Bottom spectrum: LLP 73 = Hz 2411 = MT 102, *dM*3*e*, *dM*4*e* V = 14.2.

latter feature occurs on the descending branch of the emulsion sensitivity curve. In the image-tube spectrum tracing of a dwarf *M*2 star shown in Figure 1, the features of CaH are more prominent and there is no drop off of emulsion sensitivity. At later types the strength of TiO at 6158 Å and 6651 Å can mask the CaH features but if allowance is made for this and for the presence of the water vapour *B* band, there is no more useful discriminant for dwarf *K* and *M* stars than the CaH band near 6900 Å.

At shorter wavelengths in the green-yellow region one observes the MgH bands first discussed by Öhman; the bands at 5211 Å and 5622 Å and especially the one at 4788 Å are strong even in *K* stars. Vardya (1967) pointed out that these bands should be even more effective as luminosity discriminants than the CaH bands. It is only in our most recent set of observations described under Set 3 of Table II that this spectral region has been sufficiently well exposed for these features to be useful in classification of image-tube spectra.

In addition to the spectra of the 22 dwarf stars in the Pleiades described below, spectra of two stars (MT 51 and MT 71 both listed as non-member field giants by McCarthy and Treanor (1964)) were obtained with the image-tube spectrograph; both stars proved to be giants as judged by the criteria outlined above. Thirty-three stars with well known spectral types and luminosities were also photographed to control the spectral classification.

In Table III we present the designation and catalogue data for the 22 stars for which we have obtained image-tube spectra. The first column, designated LLP for Low-luminosity Pleiades Finding List gives the number in the new Finding List by McCarthy and Treanor described above in Part Two. In subsequent columns we list the number from the Hertzsprung II General Catalogue (1946), then the entry in the McCarthy–Treanor *M* Star Catalogue (1964) and the corresponding entries from the lists of Haro and Chavira (1966) and Rosino (1966), which refer to Flare Stars, and from Van Maanen's list of proper motion stars (1945). Positions for 1900 for all of these stars are given by Hertzsprung (1946) and deGraeve (1967). Identification charts are given in Hertzsprung (1946) and McCarthy–Treanor (1964, 1968).

The spectral types and luminosities estimated according to the criteria outlined above are listed in Table IV. The number of plates used in establishing the classification is given in Column 3; for the most part only one plate was available; however, for stars LLP 9 and 73, both flare stars, several plates were obtained at dispersions of both 270 Å/mm and 135 Å/mm; no marked changes were observed in either star on these plates. The spectrum of LLP 73 = MT 102 = Hz 2411 is reproduced in Figure 2. In Table IV the photographic magnitudes are listed in Column 4; these are all taken from

TABLE III Identification data for the 22 stars of late type studied with the image tube.

LLP	Hz II	MT	R	HC	VM
5	191			7	
8	347	52			
9	357	54	10	8	
21	673	59			
23	686	59A		13	
26	793			*	
31	906	65		14	
41	1113	68			
50	1355	73			25
56	1531			19	
60	1653			21	
66	2016	95			
73	2411	102	11	1 (Hya)	
89		72	13	15	16
99		19			
100		21			
105		61			
106		78			
107		81			
109		91			
111		92			
121		79		18	

the Finding List (McCarthy and Treanor 1968) where full references to individual values are presented. Data for the dispersions used for each of the image-tube spectral plates are listed among the remarks to Table IV.

Both eye estimates and tracings made with the Junkes high-speed micro-photometer were used in classifying the spectra. Sample tracings of three stars are reproduced in Figure 1; these stars are (1) *BD* $+36°2139$, a giant *M*2 star; (2) *BD* $+36°2147$, a dwarf *M*2 star and the faint star LLP 73, described above and shown in Figure 2. These three spectra were all obtained at a dispersion of 135 Å/mm. Intercomparisons of spectral features for classification purposes especially of the dwarf features are no more difficult for stars observed at 360 Å/mm and 270 Å/mm than for those at our highest dispersion, 135 Å/mm. What is improved at the higher dispersion is the ability to detect the Hα feature in emission as was pointed out by McCarthy and Treanor (1965). Uncertainties indicated by colons in Column 2 of Table IV stem from exposure effects and from the lack of sufficient standards among the late-type *K* dwarf stars.

The main purpose of the image-tube survey was the detection and study

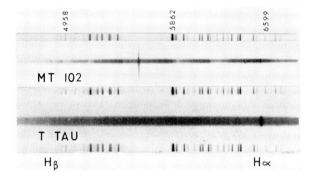

FIGURE 2 Image tube spectrum of LLP 73 = Hz 2411 = MT 102, a member of the Hyades cluster located in the region of the Pleiades. Exposure 105 minutes with 1.83 meter Perkins reflector at Flagstaff, Ariz. The spectrum shows Hα and Hβ in emission plus strong night sky (OI) emission at 5577 Å; the strongest TiO absorption features plus the prominent Na *D* lines, Mg '*b*' lines are evident. A tracing of part of this spectrum is given in Figure 1. The spectrum of T Tauri is included to indicate the position of the Hα emission feature in the spectrum of the Hyades member.

of dwarf characteristics in the spectra of late-type stars in the cluster. The spectral classification is basically that of the stars observed as reference and control stars; for the most part these have temperature classifications given in the Mt. Wilson system but transferred from the blue region to the green–yellow–red region described above. Precision calibration of spectral types

TABLE IV Spectral classification for 22 late-type stars observed with. the image tube.

LLP	Spectral type	n	m_{pg}	Remarks
5	dK7	1	15.85	3
8	dM1	1	15.35	2
9	dK4e	4	14.80	2, 3
21	dM1e	1	16.83	3
23	dK7e	1	14.40	2
26	dM0, dM1	1	15.65	3
31	dM2e	1	16.70	2
41	dK5	1	15.30	3
50	dK7e, dM0e	2	15.39	3
56	dK7e, dM0e	2	14.63	3
60	dK7	2	14.97	3
66	dK7::	1	14.87	3
73	dM3e, dM4e	3	15.77	2, 3, 4
89	dM1e, dM2e	1	Var	2, 5
99	dM2:	1	16.3	1
100	dM4::	1	16.3	1
105	dM3e	1	16.6	2
106	dM3e	1	16.5	2
107	dK5e	1	16.3	2
109	dM2:	1	16.1	1
111	dM0	1	17.0	2
121	dM3e	1	17.2	2, 6

Remarks
1. Spectra obtained at dispersion of 360 Å/mm.
2. Spectra obtained at dispersion of 270 Å/mm.
3. Spectra obtained at dispersion of 135 Å/mm.
4. Hyades member: Hz 2411.
5. Range: 14 to $17 + m_{pg}$.
6. Also observed independently by Haro as Hα em, $M3 - M4$.

in this region will require many more observations of standard stars already classified in the *MK* system. The formation of this collection of standards for image tube work is among the first projects planned for the new image–tube spectrograph at Castel Gandolfo.

Five stars classified on slitless Crossley spectrograms by Herbig (1962) and one reported by Haro (see Iriarte, 1967) can be compared with our classifications as given in Table V. We note that three of Herbig's Pleiades stars are non-emission stars and we confirm the absence of emission on our image-tube spectra. We find emission at Hα in LLP 9.

TABLE V   Spectral types of McCarthy, Herbig, and Haro compared.

Star	McCarthy	Herbig	Haro
5	*dK*7	*dK*7	—
8	*dM*1	*dK*5	—
9	*dK*4e	*dK*5	—
60	*dK*7	*dK*7	—
73 Hyades	*dM*3e, *dM*4e	*dM*4e	—
121	*dM*3e	—	*dM*3e, *dM*4e

It is much more difficult to compare the spectra listed in Table IV with the results of objective prism surveys; the difference in dispersion employed (3600 Å/mm as compared with3 60, 270 or 135 Å/mm), spectral regions (the near infrared from 7000 Å to 8900 Å for the objective prism work as compared with 4500 Å to 7500 Å for the image-tube spectra) and the classification systems followed (Case system for the objective prism classifications and Mt. Wilson system for the Image-tube spectra) plus the known variations of late-type stars all point to the need for caution in comparison. Seventeen of the 22 stars studied with the image tube are listed in the McCarthy–Treanor catalogue (1964). Table VI shows the assignments of the *MT* stars into rough

TABLE VI   Comparison of image-tube spectral types with prism survey.

Obj. prism classification	Slit classification		
	*M*1–*M*4	*K*7–*M*0	*K*4–*K*5
Group A: *M*2–*M*3	9	3	1
Group B: *M*0–*M*1	1	1	0
Group C: *K*7–*K*9	0	0	2

natural groups A (*M*2–*M*3), B (*M*0–*M*1) and C (*K*7–*K*9) appear to have been basically correct as indicated in the later survey made with the image tube. These results are somewhat influenced by observational selection in the sense that the stars of the *MT* Group A were chosen precisely as the most likely candidates by reason of astrometric and photometric properties as well as those of spectral type. What is most gratifying is that none of the stars chosen as dwarfs have turned out to be giants and that the two stars chosen from Group D (*MT* Nos. 51 and 71) when studied with the image–tube spectrograph proved to be as predicted 'stars with pronounced spectral characteristics of giants'.

Table VII combines the available astrometric, photometric and spectroscopic data which are presently available for the 22 stars studied in this image–tube survey. There are large gaps in our information; the most outstanding one is, of course, the lack of radial velocity measures for faint Pleiades stars.

TABLE VII Membership criteria for 22 late-type stars observed with image tube.

LLP	p.m. status	Photometric status				Spectra
	(1, 2, 3, 4)	ALR (5)	I (6)	Flare (7, 8)	*H*α Em	Sp.
5	Yes (1)	Yes	Yes	Yes (7)	—	*dK*7
8	Yes (1, 3)	Prob	—	—	—	*dM*1
9	Yes (1)	No	Yes	Yes (7, 8)	Yes	*dK*4*e*
21	Yes (1, 3)	Prob	—	—	Yes	*dM*1*e*
23	Yes (1)	No	—	Yes (7)	Yes	*dK*7*e*
26	Yes (1)	Yes	Yes	Yes (7)	—	*dM*0, *dM*1
31	Yes (3), Fg (1)	Yes	—	Yes (7)	Yes	*dM*2*e*
41	Yes (1)	—	—	—	—	*dK*5
50	Yes (1) (2)	Yes	—	—	Yes	*dK*7*e*, *dM*0*e*
56	Yes (1)	Prob	Yes	Yes (7)	Yes	*dK*7*e*, *dM*0*e*
60	Yes (1)	Prob	Yes	Yes (7)	—	*dK*7
66	Yes (1)	Yes	—	—	—	*dK*7::
73	Hyades (1)	—	—	Yes (7) (8)	Yes	*dM*3*e*, *dM*4*e*
89	Yes (2)	—	—	Yes (7) (8)	Yes	*dM*1*e*, *dM*2*e*
99	—	—	—	—	—	*dM*2
100	—	—	—	—	—	*dM*4::
105	Yes (3)	—	—	—	Yes	*dM*3*e*
106	—	—	—	—	Yes	*dM*3*e*
107	Yes (3)	—	—	—	Yes	*dK*5*e*
109	Yes (3)	—	—	—	?	*dM*2
111	Yes (3), Fg (4)	—	—	—	—	*dM*0
121	—	—	—	Yes (7)	Yes	*dM*3*e*

References

1. Hertzsprung (1946)
2. Van Maanen (1945).
3. Luyten (1968).
4. Van Altena (1968).

5. Ahmed, Lawrence, Reddish (1965).
6. Iriarte (1967).
7. Haro and Chavira (1966).
8. Rosino (1966).

The first two columns in Table VII list the designation and the proper motion status of each star. Here the assignments given by Van Maanen (1945), Hertzsprung (1946), Luyten (1967) and Van Altena (1968) are listed according to the references given at the end of the table. The next three columns present the photometric data on cluster membership. Column 3, designated ALR, presents the membership status assigned by the Edinburgh astronomers on the basis of their color-magnitude and color-color diagrams. Column 4, designated Ir, presents the qualifications of candidates observed by Iriarte (1967) who used the Zero Age Main Sequence (see Johnson and Iriarte, 1958) as the basis for assigning stars to cluster membership. In Column 5, we list the stars which have proven to be flare stars according to the studies of Haro and Chavira (1966) and Rosino (1966). Column 6 calls attention to the emission characteristics of 12 of the 22 stars. The last column lists the spectral types adopted as a result of the image-tube study.

In Table VIII, we summarize the membership status of the 22 stars listed in Table VII according to the different criteria discussed above.

TABLE VIII

Membership		Characteristics				
	P M	Photom. ALR	Ir	Flare	H Em	Dwarf spectrum
Yes	15	5	5	10	11	21
Uncertain	2	4	0	0	1	0
No	0	2	0	0	0	0
Not observed	4	10	16	11	9	0
Hyades member	1	—	—	1	1	1

The following are the main conclusions from an examination of the data presented in Tables VII and VIII.

(1) Of the 22 stars selected for study with the image tube all show pronounced dwarf characteristics discussed above. The hydride features vary from star to star within the same spectral class, but all 22 stars are clearly dwarfs. Lacking standards for subdwarfs, we cannot make any further refinements of spectral classification at this time.

(2) Of the 22 stars 12 show $H\alpha$ emission; 11 of these are reported here for the first time; one (LLP No. 121) was discovered independently by Haro and by the author. We note that 7 of the 12 emission stars have shown flare activity while 5 have not.

(3) For 5 of the 22 stars studied here V. C. Rubin has measured the

Hα emission feature to determine radial velocities. The measures concern four of the Pleiades stars (LLP Nos. 9, 23, 50 and 56), the Hyades member, LLP 73 and a control star of spectral type AO, HD 23632. The measures were made on image–tube spectra with a dispersion of 135 Å/mm. The Mann two coordinate measuring engine of the DTM laboratory was used; the measurements and reduction methods are described fully elsewhere (Rubin 1968a, 1968b). Only the Hα emission feature was measured in the late-type spectra; lines from a hollow cathode Fe + Ne spectrum were used for comparison with the stellar spectrum. The results of these measurements are presented in Table IX. We emphasize the reserve which accompanies the

TABLE IX   Radial velocity measures of the Hα emission features in five Pleiades stars.

Star	Sp	m_{pg}	$V_{reduced\ to\ Sun}$	Remarks
9	K4	14.8	+173 km/sec	Flare; Plate 1232
23	K7	14.4	+15	Flare; Plate 1403
50	K7, M0	15.4	+83	Plate 1412
56	K7	14.6	+77	Flare; Plate 1400
73	M3	15.8	+116	Flare; Plates 1225, 1226 Hyades
HD 23632	A1	6.8	−2	GCRV: =2.4 km/sec Pl. 1420

publication of these data. The plates were not taken for radial velocity determinations but for the classification of these faint Pleiades stars. They were measured on plates of the highest available dispersion to see if meaningful radial velocities could be obtained. Only the Hα emission feature was measured as it was not practicable to attempt to measure any other features in a spectrum dominated by molecular absorption bands. An estimate of the relative accuracy, based on measures of the only star (LLP No. 73) for which we have more than two plates of suitable dispersion for measurements, yields a probable error near ±20 km/sec. The velocities reduced to the sun are given in Table IX. The striking features of these measures are the positive sense of all of the velocities measured and the very high value for the mean velocity of the Pleiades stars: $V = +87$ km/sec. This value differs significantly from the average value for the brighter Pleiades stars where $V = +5$ to $+15$ km/sec. Perhaps we are observing here the infall of chromospheric material in these late-type dwarf stars. Wilson has studied chromospheric activity as indicated by the intensity of H and K emission in main-sequence stars of types $G0$–$K2$ and finds that this is appreciably higher for the Pleiades than for the Hyades, Praesepe, or Coma and much higher than for similar local field stars (Wilson,

1963, 1964). We are dealing here with stars of later type and the region near the *H* and *K* lines is not available to us. Clearly more observations (of these stars) are required at much higher dispersions.

(4) Proper motion data have proven easier to obtain than radial velocity measures. The studies of Van Maanen (1945), Hertzsprung (1946) and of Luyten (1967) point to membership for 15 of the 22 stars in Table VII. There is some disagreement for two of the stars on the list. LLP No. 31, a 16.7 m_{pg} star is considered a foreground star by Hertzsprung and as a candidate for membership by Luyten. The same star should be considered a member according to the photometric criteria of the Edinburgh astronomers; it has been observed as a flare star by Haro and we find that the spectrum shows Hα in emission. The other star LLP No. 111 whose blue magnitude is 17.0, is considered a likely candidate for membership by Luyten but is assigned to the foreground by Van Altena (1968). It is interesting that this star and one other not studied here (MT 49) show motions which in magnitude and direction resemble very much the motion of the well known Hyades member, LLP No. 73. The data are assembled in Table X. The blink estimates

TABLE X Van Altena's motion estimates of three foreground stars.

Star	Total p m	Position angle	Remarks
LLP 73	0″.13	108°	Hyades member
111	0″.11	111°	
MT 49	0″.10	113°	

of Van Altena differ from those of Luyten in two ways. Van Altena used the Lick 20-inch (50 cm) astrograph plates with a 20 year baseline whereas Luyten used 1945 Palomar Schmidt plates and compared these with plates he took with the same instrument in 1965. Van Altena compared the motions of 125 stars in the McCarthy–Treanor catalogue in an effort to detect foreground stars. Luyten first of all marked all stars in the area which showed indication of motion and he recorded the discovery of 756 faint objects of this nature. Then after the author had checked the identification of these stars against the stars listed in the McCarthy–Treanor catalogue, Luyten re-examined his plates to select those stars on this list which showed motions similar to those of the bright Pleiades stars. The results of these estimates are presented in the Finding List of Late-type Stars in the Pleiades (discussed in Part Two above; McCarthy and Treanor 1968); they have also been incorporated into the data of Table VII. A full scale measuring program to determine proper motions for these stars would appear to be required and rewarding. For example, such measures will reveal whether LLP No. 111 which is a dwarf $M0$ star, but shows no other photometric or spectroscopic peculiarities really belongs to the Pleiades, Hyades or neither.

(5) The photometric status symbol for membership in a cluster is the place the star occupies on the color-magnitude and the color–color diagrams of the cluster. As described above, these diagrams in the case of the Pleiades have provided problems for theoreticians and observers alike. Theoretically, the length of the Pleiades main sequence seems too great in comparison with similar diagrams for other clusters. Observationally, the determination of blue magnitudes has been extremely difficult because of the reflection nebulosity over a large part of the cluster. The best prospects for remedying this situation would seem to lie in the application of VRI photometry after the manner of Greenstein and Eggen (1967) or perhaps in the electronographic photometry perfected recently by Walker and Kron (see, for example, Walker 1967).

Among the recent contributions to color-magnitude studies in the Pleiades are the photographic studies of Ahmed, Lawrence and Reddish (1965) and the photoelectric photometry of Iriarte (1967). Of the 22 stars listed in Table VII, 11 are contained in the Edinburgh work. Membership is accorded to 5 stars, 4 are considered probable members and 2 are excluded from membership. In Table VII, we have considered the Edinburgh classification CC as equivalent to recommendation for membership and we have entered 'Yes' in Table VII; their CF or FC designation we regard as 'Probable' and their FF qualification as non-membership or 'No.'

In Figure 3 we present the color-magnitude diagram for 11 of the 22 stars for which we have image–tube spectra. We have drawn the main sequence as given by the Edinburgh astronomers, but to avoid confusion, we have not included the other 89 stars measured by them. The stars which are known flare stars have been circled and those which are $H\alpha$ emission stars are marked with a bar at the end of the arrow. Of the 11 stars plotted 9 are below the main sequence as drawn, one is above it and one is located on it though it is both a flare star and an emission star. Note that the stars Nos. 9 and 23, which have a $(B - V)$ value of $+0.8$ mag. according to the Edinburgh photometry, are both flare stars and both emission stars and according to the listed proper motion data both are cluster members.

Iriarte has observed only five of the 22 stars under discussion here and he considers all five as members. Interestingly enough he considers LLP No. 9, discussed above, to be a member and he elevates two of the candidates considered 'Probable' by Ahmed, Reddish and Lawrence to full membership. In Figure 4, we see Iriarte's color-magnitude diagram for the faint end of the

Pleiades sequence. The solid line represents the Zero Age Main Sequence and the dotted line represents the mean Hyades sequence. Large circles represent the known flare stars on the diagram; circles indicated by arrows (also all flare stars) are those listed in Table VII. Note that these are all photoelectric observations and as plotted have been corrected for +0.04 mag. of reddening and ' or +0.16 mag. of absorption. We see that the star LLP

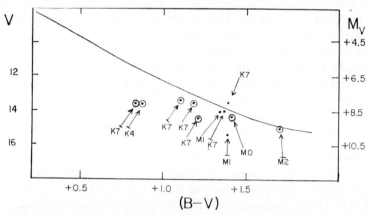

FIGURE 3   Color-magnitude diagram for 11 late-type stars studied with image tube as plotted from the photographic observations of Ahmed, Lawrence, and Reddish (1956). They have corrected observed data for the effects of nebulosity but not for the variable absorption over the cluster. Stars circled have been detected as flare stars by Haro and Chavira or by Rosino. Hα emission stars are indicated by bars at the end of the arrows. Nine stars are found to lie below the main sequence.

No. 9, which was rejected by Ahmed, Lawrence and Reddish as too far off, the main sequence, has in this Figure returned from its eccentric position at $(B - V) = +0.8$; this position is marked with an 'X' on Figure 4. In Table XI. we compare the *UBV* data for 5 stars according to the published photometry of Ahmed, Lawrence and Reddish (1965), of Iriarte (1967) and of Johnson and Mitchell (1958). The best agreement is found for the *V* magnitudes This table points up the photometric problems connected with intrinsic variations in the stars and with the effects of uneven nebulosity in the region of the Pleiades.

If future studies show that such excursions in the color-magnitude plane are not uncommon for flare stars, then it may become necessary for us to revise our thinking about the significance of the place of these stars on or near the main sequence. It is important to test this observationally; it will also be difficult to do this.

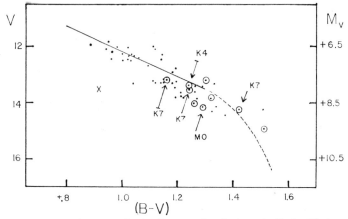

FIGURE 4 Color–magnitude diagram for 5 stars studied with image tube which have *BV* data from the photoelectric photometry of Iriarte (1967). Circled stars are flare stars. Bars at the end of the arrows indicate stars observed to show Hα in emission. 'X' marks the locus of star LLP 9 on color magnitude diagram by Ahmed, Lawrence and Reddish; note the change in $B - V$ for this star from $+0.8$ to $+1.3$.

TABLE XI Comparison of photometric measures of 5 flare stars in the Pleiades.

Star	V magnitudes		
	ALR	I	JM
5	14.43	14.39	14.38
9	13.65	13.52	13.46
26	14.26	14.32	14.34
56	13.36	13.41	13.30
60	13.48	13.69	13.31
	$(B - V)$ colors		
5	$+1.21$	$+1.46$	$+2.15$
9	0.88	1.28	1.44
26	1.41	1.33	1.04
56	1.11	1.22	1.15
60	1.19	1.28	1.45
	$(U - B)$ colors		
5	$+1.55$	$+1.12$	—
9	1.10	1.15	0.69
26	1.44	1.12	—
56	0.87	1.05	—
60	1.20	1.05	—

V. CONCLUSION

The following list summarizes the results of our study of the 22 Pleiades stars for which we have image-tube spectra.

Members: (13) LLP 5*, 8, *9*, *21, 23*, 26*, 31*, 41, *50, 56*, 60*, 66, *89*.
Probable members: (5) LLP *105, 106, 107*, 109, *121*.
Uncertain: (3) LLP 99, 100, 111.
Non-member: (1) LLP *73*.

Stars with asterisks superposed are flare stars; those in italic are ones which have shown Hα in emission. We find a high percentage of stars studied are flare stars and more than half show Hα in emission. Eight of the 13 member stars are flare stars and 6 of these same 13 are Hα emission stars.

The distribution of the 13 stars listed above as members into the various spectral classes shows (Table XII) that the latest spectral type in this group is *dM*2. Among the 'probable' members the latest type is *dM*3 and among the

TABLE XII   Spectral distribution of stars of different membership class.

Class	Spectral type						
	K4–K5	K7	M0	M1	M2	M3	M4
Members	2	6	1	3	1	0	0
Probable	1	0	0	0	1	3	0
Uncertain	0	0	1	0	1	0	1
Nonmember	0	0	0	0	0	0	0

stars listed as 'uncertain' the latest is *dM*4. Of course this distribution may be influenced by the fact that the stars most easily observed and the ones most likely to show sufficient evidence for membership are the brighter stars, whereas those which are late in spectral type can be detected easily and studied for dwarf characteristics in their spectra but are more difficult objects for photometry and for motion studies.

The high velocities derived from the Hα emission measures for 5 of the stars observed with the image-tube spectrograph require further investigation; the accuracy of the present plate material is limited.

In summary, it would appear that without further observations it seems unwise to write 'finis' as yet to the main sequence of the Pleiades cluster. In support of this, we point out that Haro and colleagues have detected flare stars in this region at $B = 20.9$ and that Luyten has found several hundred faint stars in the area which may share the proper motion of the Pleiades.

Acknowledgements

The work described in the first two portions of the present paper was done largely with the collaboration of Dr. P. J. Treanor. The work reported in the final portion is based on observations made with Dr. W. Kent Ford, Jr. and Dr. Vera C. Rubin of the Department of Terrestrial Magnetism, Carnegie Institution of Washington: Dr. Rubin also measured the radial velocities of the Hα features on the image–tube spectra. I thank them and also Dr. John S. Hall, Director of the Lowell Observatory for the use of the facilities of the Lowell Observatory, Dr. Arne Slettebak, Director of the Perkins Observatory for the use of the Perkins telescope of the Ohio State and Ohio Wesleyan Universities, and Dr. W. Luyten and Dr. W. Van Altena for communicating the results of their blink estimates in advance of publication. The hospitality and assistance of the Directors and Staff of the Department of Terrestrial Magnetism of the Carnegie Institution of Washington during visits to Washington is gratefully acknowledged.

References

AHMED, F., LAWRENCE, L. C., and REDDISH, V. C. 1965 *Royal Obs. Edinburgh*, **1**, No. 7.

BURWELL, C. G. 1930 *Publ. A.S.P.*, **42**, 351.

FITCH, W. S. and MORGAN, W. W. 1951, *Ap. J.*, **114**, 548.

DE GRAEVE, E. 1967, *Ric. Astron. Spec. Vat.* **7**, No. 9.

GREENSTEIN, J. L., and KRAFT, R. 1968 (Private Communications).

GREENSTEIN, J. L., and EGGEN O. 1967, *Vistas in Astronomy* (Ed. A. BEER and K. AA. STRAND, Oxford, Pergamon Press) Vol. 8, 63.

HARO, G., and CHAVIRA, E. *Ibid*, p. 96.

HARO, G. (1968). In Press. See Iriarte B. (1967) below.

HERBIG, G. 1958 *Ric. Astr. Spec. Vat.* **5**, 127.

HERBIG, G. 1962, *Ap. J.*, 135, 736.

HERTZSPRUNG, E. 1929, *M.N.R.A.S.*, **89**, 660.

HERTZSPRUNG, E. 1946, *Leiden Ann.* 19, Part 1.

IRIARTE, B. 1967, *Bol. Tonantzintla y Tacubya*, **4**, 28.

JOHNSON, H. L., and IRIARTE, B. 1958, *Lowell Obs. Bull.* no. 81, 47.

JOHNSON, H. L., and MITCHELL, R. I. 1958, *Ap. J.*, **128**, 31.

LUYTEN, W. J. 1923, *Publ. A.S.P.*, **35**, 175.

LUYTEN, W. J. 1967, (Private Communication).

McCARTHY, M. F., and TREANOR, P. J. 1964, *Ric. Astr. Spec. Vat.* **6**, 535.

McCARTHY, M. F., and TREANOR, P. J. 1965, *Ibid.*, **7**, No. 3.

McCARTHY, M. F., and TREANOR, P. J. 1968, *Ibid.* In Press.

McCARTHY, M. F., TREANOR, P. J., and FORD, W. K. 1967, Trieste Colloquium on Late Type Stars (Ed. M. HACK, Trieste: Osservatorio Astronomico di Trieste), p. 100.

ÖHMAN, Y. 1936, *Stockholm Ann.*, **12**, Nos. 3 and 8.

PESCH, P. 1961, *Ap. J.*, **133**, 1085.

ROSINO, L. 1966, *Mem. Soc. Astr. Ital.*, **37**, 717.
TREANOR, P. J., and MCCARTHY, M. F. 1967, Trieste Colloquium on Late-Type
    Stars (Ed. M. HACK, Trieste: Osservatorio Astronomico di Trieste), p. 114.
RUBIN, V. C. (1966). Paper presented at the Toronto Symposium No. 30 IAU on
    Radial Velocity Determinations (In Press).
RUBIN, V. C. and Ford Jr., W. K. (1968) *Ap. J.* (In Press).
VAN ALTENA, W. 1968, *Publ. A.S.P.* (In Press).
VAN MAANEN, A. 1945, *Ap. J.*, **102**, 26, 1945.
VARDYA, M. S. See McCarthy, Treanor and Ford (1967) above, p. 115.
VARDYA, M. S. 1963, *Ap. J.*, **68**, 247.
WALKER, M. F. 1967, XIII Gen. Assembly IAU, Prague, Draft Report, p. 114.
WILSON, O. C. (1963) *Ap. J.*, **138**, 832.
WILSON, O. C. (1964) *Ibid.*, **140**, 1401.

Discussion

KRAFT   Would you say that Hα is, in your opinion, a good criterion for
membership?

MCCARTHY   I think that, because of its correlation with the other criteria,
especially flare star activity and from what we know from the proper motion
measures from Hertzsprung and others, Hα would seem to be a very good
criterion for membership.

KRAFT   Did you see Hα emission in the $M4$ star that was not in the cluster?

MCCARTHY   No. It does not show Hα emission.

Chromospheric variations in main-sequence stars

O. C. WILSON

Mount Wilson and Palomar Observatories
Carnegie Institution of Washington, California Institute of Technology

Abstract

During the past two years extensive measurements of the flux at the centers of stellar H and K lines have been made, using the coudé scanner of the 100-inch telescope as a two-channel photoelectric photometer. All observed stars are on, or close to, the main sequence; a majority are from the catalogue of Strömgren and Perry. More recently a selection of later-type main-sequence stars has been included. During the period June–December 1967, flux variations in both components of 61 Cygni have been found. It is not yet possible to state whether these variations are indicative of cycles analogous to the solar-activity cycle. Observations are being continued.

THIS REPORT is based in part on material which is in press, and in part on work which is in a very preliminary state. Therefore it will not be given here in great detail.

It is reasonable to suppose that the 22-year solar-activity cycle is not unique and that stars in the appropriate parts of the $H - R$ plane behave in a similar fashion. In fact, if such cycles could be detected, it is reasonable to assume further that the dependence of their amplitudes and periods upon such stellar parameters as mass, age, etc., would provide valuable clues to aid in comprehensive theoretical understanding of the cycles.

With this primary goal in view, observations were begun about two years ago. The coudé scanner of the 100-inch telescope is used as a two-channel photometer to measure the fluxes in 1 Å bands at the centers of the stellar H and K lines. The fluxes are measured with respect to the combined radiation in two 25 Å bands on either side of H and K. These monitor bands are the same for both lines in a given star, and virtually the same for all stars except

for small Doppler shifts which are mostly less than 1 Å. Thus the measured
$H - K$ fluxes are composed of the residual photospheric light in the absorption lines, plus any chromospheric components which may be present.
Since Ca II spectroheliograms show very pronounced differences in the
chromospheric H and K emission over the solar surface between sunspot
maximum and minimum, the stellar observations should in principle, be
capable of revealing similar changes in chromospheric activity in the stars.

During the first year observations were restricted to about 140 main-
sequence stars from the Strömgren–Perry catalogue with values between
0.24 and 0.44, i.e., with spectral types from about F3 to G3. Spectrograms
at 10 Å/mm were available for all these objects, so that those selected included
the stars with known chromospheric emission, and stars with appreciable
rotation could be avoided. The Strömgren–Perry photometry provided
accurate measures of surface temperature, luminosity, and relative age,
and the chromospheric emission strengths were studied within this framework. Details of this work are given in a forthcoming publication. In
particular, while the slow decrease in chromospheric activity with age is well
illustrated by the results, no short-term trends of undoubted reality have yet
been demonstrated for the emission-line stars. This may, of course, be due
entirely to the relatively short period covered by the observations and further
flux measurements will be needed to see whether the methods used can
uncover cyclical variations in this part of the H R plane. One further
important result from this first phase of investigation is that the probable
error of one measure of one line is 1.8 per cent.

Beginning in the summer of 1967 the program of observation was somewhat
changed. In the first place, further observation of the Strömgren–Perry stars
was restricted to those with $H - K$ fluxes above a certain limit set to include
all objects known to have observable emission from 10 Å/mm spectrograms,
plus a few others in which the emission is probably only slightly less. In
addition, a few 'standard stars' from the $S - P$ list are observed each night,
in order to insure instrumental constancy by providing small corrections to
the directly observed data. These standard stars are those which have minimum measured $H - K$ fluxes and in which, therefore, it may be assumed that
chromospheric contributions are very small and unlikely to produce any
intrinsic variations.

Secondly, the observations have been extended to later-type main-
sequence stars, chiefly those in which H and K emission is known to be
present from existing spectrograms. There are a fair number of such objects
within range of the equipment, but owing to time limitations a reasonably
good series of measures has thus far been obtained only for 61 Cygni *A* and *B*.
Results for these two stars are shown in Figure 1 where each point represents

the measures of one night. Each of the larger points, the majority, is the mean of four observations, two on *H* and two on *K*; for the smaller points only one observation of *H* and *K* is involved. For comparison, the fluxes for two of the standard stars over the same period of time are shown in Figure 2.

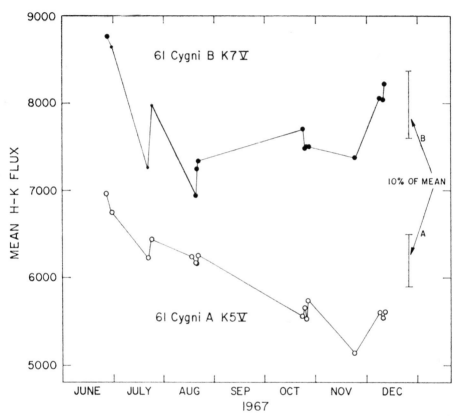

FIGURE 1 Mean *H − K* flux (in instrumental units) for 61 Cygni *A* and *B* during 1967.

It appears that both of the late-type stars have undergone variations of *H − K* flux of the order of 15 per cent within a period of six months, much larger than any variation of the standard stars, although all observations were made and reduced in the same way. For this reason I feel that the changes in *H − K* flux in 61 Cyg *A* and *B* must be largely real and that they reveal genuine variations in chromospheric activity. This conclusion is strengthened because, while the fluxes in both stars declined at first, this parallelism has not been maintained throughout the period of observation. In fact, after the initial decline, the flux in 61 Cyg *B* has increased while that in *A* has continued to drop, or, at most, to begin levelling off.

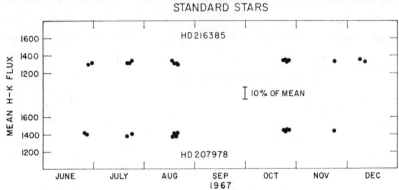

FIGURE 2   Mean $H - K$ flux for two standard stars during 1967.

Are these observations to be interpreted in terms of stellar analogs of the solar cycle? It is obviously too soon to do so and this question can be answered only by further observation. All one can say at present is this: if the variations do turn out to be cyclical in nature, then the periods are probably functions of the mass such that the smaller the mass the shorter the period.

Discussion

EGGEN   Are your standards giants or dwarfs?

WILSON   They are all main-sequence stars. I have not had time to observe a single giant.

WRUBEL   I missed the identification of the lower boundary in the Strömgren's sequence. Is that an observational effect or is it physical?

WILSON   The lower boundary is simply drawn in through the points, and you notice that it comes down and there is no straggling. It is a sharp boundary and then they straggle upwards from there. No, it is truly observational.

On the scatter in the blanketing-free color-magnitude diagram for solar-neighborhood K dwarfs

BENJAMIN J. TAYLOR

Astronomical Department, University of California, Berkeley

As PART of my thesis investigation, which consists of scanner abundance studies of late G and K dwarfs, I am obtaining red magnitudes and yellow–red colors for some 280 of these stars. These observations are made with the Wampler scanner at the prime foci of the Crossley and 120-inch telescopes at Lick Observatory, those at the latter telescope being secured for me by Dr. Hyron Spinrad. The bandpasses used are 15 Å shortward and 30 Å longward of $\lambda 5360$. After the observations are corrected for extinction using standard coefficients and transformed to a standard system set up by Dr. Spinrad and myself, magnitudes and temperature indices are derived, the former at $\lambda 7000$, the latter from the relation

$$T = \frac{I(7000) + I(7400)}{I(5360)} \times 1000, \qquad (1)$$

where I is the intensity at a specified wavelength. A preliminary reduction for these data for stars so far observed with trigonometric parallaxes equal to or greater than $0''.50$ forms the basis for this investigation.

A plot of T versus the $R - I$ of Kron, Gascoigne, and White (1957) as measured by Eggen (1968) is given in Figure 1 for stars common to the two sets of observations. In view of the *rms* error bars (± 1 per cent in T, unquoted but assumed to be $\pm 0^m.015$ in $R - I$), the correlation appears to be satisfactory. The open circle is Groombridge 1830; its closeness to the mean relation defined by the other stars indicates that T does not suffer significantly from differential blanketing effects. The Hyades, if plotted, would be systematically displaced from the relation defined by the field stars; if $R - I$'s are assigned to the Hyades which I have observed on the basis of their $B - V$'s and the $B - V$ versus $R - I$ correlation for Eggen's

Hyads*, the resulting $R - I$'s are redder than those for field stars of the same
T's by about $0^m.05$. Since my observations yield the same T's for field stars
on nights when the Hyades were observed as on nights when the Hyades
were not observed, I can offer no satisfactory explanation for this discrepancy.

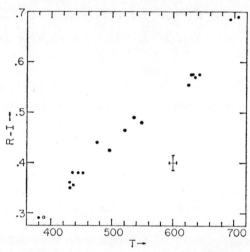

FIGURE 1   Relationship between T and $R - I$ as measured by Eggen
(1968) for field stars in common. The open circle is Groombridge 1830.

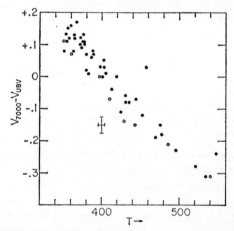

FIGURE 2   Relationship between $V_{7000} - V_{UBV}$ and T, using averages
of published V_{UBV} magnitudes. Open circles are Hyades; filled circles
are field stars.

* The term 'Hyads' refers to several stars in the Hyades, while 'Hyad' refers to one star
  in the Hyades.

In Figure 2 is shown a plot of my magnitude at $\lambda 7000$, V_{7000}, minus the V of *UBV* photometry versus T, using published V_{UBV} magnitudes (Johnson, Morgan, and Harris 1953; Johnson and Knuckles 1955; Gliese 1957; Arp 1958; Sandage and Eggen 1959; Argue 1963; Johnson 1963; Hoffleit 1964; Argue 1966; Eggen 1968). When more than one determination has been listed for a given star, a straightforward average has been taken. Again, in view of the *rms* error bars (about $\pm 0^m.01$ in V_{7000}, possibly $\pm 0^m.02$ or $0^m.03$ in V_{UBV}, due to the variety of sources used), the correlation appears to be satisfactory. The displacement of the Hyads (open circles) below the field stars (filled circles) may not be statistically significant; in any case, it is much smaller than the parallax-error-induced absolute magnitude errors in the field stars, and hence may be ignored.

The plot of absolute red magnitude, M_{7000}, versus T for all the stars observed from spectral types $G8$ to about $K5$ (plus some earlier Hyads) is given in Figure 3. The Hyades (open circles) are plotted using the individual

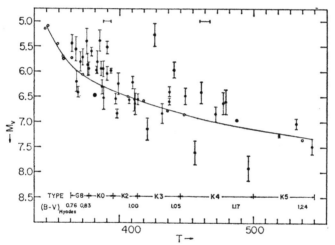

FIGURE 3   The blanketing-free, color-magnitude diagram. Filled circles are field stars; open circles are Hyads; the solid line is the eye-estimated mean Hyades line. The circled cross is the 'equivalent sun', whose placement and definition are explained in the text.

distance moduli of Heckmann and Johnson (1956). The field stars satisfy the following conditions: (1) they are not known spectroscopic, astrometric, or otherwise unresolved binaries, unless the components are known to be nearly equal (in which case, their average is plotted; see Gliese 1957); (2) they have trigonometric parallaxes equalling or exceeding $0''.050$ (Jenkins 1952, 1963); (3) their parallaxes have been determined at two or more observatories; (4) the probable errors of their parallaxes, assigned in the manner used by Gliese (and usually those quoted by Gliese himself) give rise to absolute magnitude errors equal to or less than $0^m.25$. The error bars actually exhibited are the Pythagorean sums of the parallax error bars and the magnitude equivalents of the error bars in T, computed for each star from the slope of the Hyades relation at its T. (The error bars for the Hyades relation have been neglected.) Note the presence of both considerably overluminous and considerably underluminous stars; this resembles the results of Eggen (1968) and, to a lesser degree, those of Johnson (1965), Greenstein (1965), and others, for M stars. Note also that the Hyades are displaced below the field star mean, in disagreement with Eggen's results; the reason for this is the T versus $R - I$ discrepancy mentioned previously.

The circled cross represents the approximate position of a zero-age $K0$ dwarf of solar composition; note that it (and, equivalently, the sun) is a fairly decisive subdwarf. The placement of this 'equivalent sun' was accomplished as follows: Allen (1963) gives $M_V = 4.79$ and $B - V = 0.62$ for the present sun; corrected for solar age brightening according to computations made by Ulrich (1968) and the θ versus $B - V$ relation of Oke and Conti (1966), these become $M_V = 5.25$ and $B - V = 0.66$ or 0.67. This places the sun about $0^m.2$ below the Hyades in the M_V, $B - V$ plane; one may note for comparison that the Hyad HD 28099 has $M_V = 4.95$ and $B - V = 0.657$, according to the photometry of Johnson and Knuckles and the distance modulus of Heckmann and Johnson. Since the sun is slightly redder for its θ in $B - V$ than the Oke–Conti relation, differential blanketing in $B - V$ may probably be neglected; if differential blanketing in V is also neglected and similar shapes for the zero-age solar composition and Hyades main sequences are assumed, the placement of the 'equivalent sun' follows.

Let us consider first the observed field star scatter. The statistical significance of this scatter may be judged from Table I. Of the three possible mean relations tested, the second is the best, but one can see immediately that there are too many stars more than three probable errors away from any of the relations; this impression is confirmed by chi square tests. Eight of the twelve

TABLE I Statistical distribution of observed stars.

δ/p	$\leqslant 1.0$	1.01–2.0	2.01–3.0	> 3.0
Hyades	10	11	12	11
Hyades $-0^m.3$	20	5	7	12
Hyades $-0^m.6$	7	11	5	21
Expected	22	14.2	5.9	1.9

of these stars listed for the (Hyades $-0^{m}.3$) relation (which will be used hereafter as the field mean) have velocities less than 60 km/sec; hence the scatter is not the property of the high-velocity stars alone.

The reliability of the parallaxes of the widely deviant stars may be judged from Table II. The first and third columns are self-explanatory; the second

TABLE II   Parallaxes of deviant stars.

HD	ΔM_{7000}	$\pi(".001)$				HD	ΔM_{7000}	$\pi(".001)$			
10436	+0.5	73 ± 5	78	M	10	149661	−0.8	87 ± 5	83	A	16
			66	Yk	8				49	M	10
			40	D	4*				104	Y	12
			77	A	28				122	C	7
25329	+0.7	54 ± 5	45	M	8	160346	−0.5	81 ± 6	80	A	20
			39	Yk	6				77	M	10
			72	S	10						
			54	V	12	171314	+1.1	70 ± 8	69	M	10
47752	+1.0	83 ± 9	78	M	8				70	D	4*
			96	D	4*	188088	−1.1	67 ± 7	63	Y	12
4628	+0.4	143 ± 5							82	C	7
						103095	+0.8	116 ± 5			
10700	+0.7	275 ± 8									
						201091	+0.3	293 ± 4			
17925	+0.4	123 ± 6									

column lists the deviations in M_{7000} from the adopted mean, while the fourth lists for those stars with mean parallaxes less than 0″.1 the relative parallaxes, abbreviations for the observatories at which they were obtained, and the assigned weights, as given by Jenkins. The Dearborn parallaxes have been corrected by the apparent magnitude precept given by Jenkins; these are starred. The assigned error bars appear provisionally satisfactory in all cases, including the three weakest: HD 47752, HD 171314, and HD 188088.

It is of interest to examine the compositional properties of the deviant stars, in order to discover some possible common property or correlation with the amount of deviation. To this end, assessments of Z/X may be made from the remainder of the same scans from which the colors and magnitudes were obtained. In these scans, intensities are obtained in passbands containing strong atomic or molecular features and also in one or two side bands per feature; after these have been corrected for extinction and transformed to
E

the standard system, ratios (called 'indices') are then formed according to the equations

$$r = \frac{I_a + I_b}{2I_f} \quad \text{or} \quad r = \frac{I_a}{I_f}, \tag{2}$$

where I_a and I_b are side band intensities and I_f is the feature band intensity. This is basically the same procedure as that used by the Cambridge observers (cf. Griffin and Redman 1960). The atomic and molecular species measured and the feature and side bands for each are listed in Table III. Background

TABLE III   Features scanned.

Feature	Feature band	Side band(s)
CN	$\lambda4200$	$\lambda\lambda4040, 4500$
Ca	$\lambda4227$	$\lambda\lambda4040, 4500$
CH	$\lambda4300$	$\lambda\lambda4040, 4500$
Mg	$\lambda5175$	$\lambda\lambda5000, 5300$
Na	$\lambda5892$	$\lambda5864$

absorption is of course present in all the blue passbands, but its effect is cancelled to first order by its presence in both feature and side bands. Once these indices have been calculated, abundances for each species in each field star may be assessed by comparing the field star index with the relation formed by the Hyades in the index versus T plane at the field star's measured T. (See Spinrad and Taylor 1967; Taylor 1967.) Possible gravity effects will be considered subsequently.

The results of these observations and reductions, together with the M_{7000} deviations, are shown in Table IV for each of the deviant stars. The values tabulated are the equivalent width ratios between the stars and the Hyades, calculated from the equation

$$\frac{W_1}{W_2} = \frac{1 - r_0/r_1}{1 - r_0/r_2}, \tag{3}$$

which is derived from equation (3) of Price (1966); r_0 is the no-feature value of the index. The probable errors are estimated to be usually five to 10 per cent, except for measurements of weak features, where they may be considerably higher. The colons denote values which are less than usually accurate because of observational scatter or, in the case of calcium, because the wings of the line extend beyond the feature band limits; neither of these sources of error is judged to be serious for this purpose. Inspection of the tabulated ratios yields the following conclusions. First, the run of the ratios reveals no trends which can consistently be interpreted as being due to gravity effects; hence these may probably be neglected. Second, the ratios tabulated for the well-known metal-poor stars HD 10700 (τ Ceti), HD 25329, and HD 103095 (Groombridge 1830), indicate that those for cyanogen and sodium are probably best correlated with overall metal abundance. Third, there appears to be no strict one-to-one correlation between these two ratios (or, for that

TABLE IV   Equivalent width ratios for deviant stars.

HD	ΔM_{7000}	W/W (Hyades): CN	Ca	CH	Mg	Na(D)
201091	+0.3	0.9	1.0:	1.1	1.05	0.8
4628	+0.4	0.7	1.0	1.0	1.2	0.8
17925	+0.4	0.75	0.95	1.0	0.95	0.8
10436	+0.5	0.8	0.9:	1.0	1.0	0.7
10700	+0.7	0.5	1.0	1.2	1.1	0.6
25329	+0.7	0.45	0.7	1.0	0.75	0.2
103095	+0.8	0.4	0.8	1.0	0.8	0.0
47752	+1.0	0.9	0.9	1.15	1.1	0.9
171314	+1.1	0.9	0.9:	1.0	1.1	1.0
160346	−0.5	0.8	0.8	0.9	1.0	0.9
149661	−0.8	0.7	0.8	1.0	1.0:	1.0
188088	−1.1	1.2	1.0	1.1	1.1	1.4

matter, the other three) and amount of over- or under-luminosity; note particularly HD 47752 and HD 171314. Previous discussions of under-luminosity, such as that of Dennis (1968) for the metal-poor stars of Eggen and Sandage (1962) and that of Alexander and Pepper (1967) for the par-ticular case of Groombridge 1830, have tended either explicitly or implicitly to attribute these luminosity effects to metallicity variations alone. Clearly, such an interpretation will not do here; variation in either helium abundance or the convection parameter l/H or both is indicated. If l/H is assumed to be constant, one may guess the range of variation of helium abundances from the observed magnitude residuals, guesses of Z/X for each star from the tabulated equivalent width ratios, and the appropriate quasi-homology relation given by Faulkner (1967); these are no better than guesses, since some of the stars involved are redder than the red limit of validity given by that author. Assuming with Bodenheimer (1965) $X = 0.66$ and $Z = 0.0264$ for the sun, one obtains in this manner values of Y of about 0.5 for the faintest stars, zero or less for the brightest ones, and 0.1 for a star on the mean line with solar Z/X. In the absence of the requisite convection theory, all that can be said at present is that these values represent a *possible* range of helium abundances in the solar neighborhood.

Two comments should be made about the stars above the mean. First, since the hottest (HD 149661 = 12 Oph) is a low-velocity $K0$ and the other two are $K3$'s, it appears unlikely that evolution is significant in their cases. Second, the scans reveal that none are emission-line stars, thus ruling out the possibility that they are in the act of contracting to the main sequence.

The presence of the Hyades below the field star mean requires further comment. The observations indicate that the displacement is about $0^m.3 \pm 0^m.1$ (p.e.). If they are correct (as seems likely from their consistency) and this displacement is real, its existence most emphatically does *not* demand a revision of the distance modulus of the Hyades by this amount. Once it is admitted that zero-age main-sequence stars of a given temperature can have different luminosities, the reason for demanding that the Hyades coincide with the field mean is nullified. Because of the importance of the displacement, it will be necessary to check its existence by photometry of the late K and M stars of the cluster, particularly those observed by Eggen.

The major conclusions of this paper may be summarized as follows. First, the stars in the solar neighborhood do not all fall on any given zero-age main sequence, either before or after correction for metallicity effects; a range of helium abundances by weight of zero to 0.5 may be present. Second, the Hyades appear to fall below the field star mean, pending further photometry. Third, the sun is a subdwarf with respect to the field star mean (by an amount depending on the position of the Hyades), and may have more helium than the field average.

I wish to thank Dr. Spinrad for suggesting this topic, and for his continued advice and interest. I also wish to thank Dr. I. R. King for his advice on statistical procedures, and Drs. O. J. Eggen, A. R. Sandage, and L. G. Henyey for interesting and helpful conversations. This research was supported in part by the George Davidson Fund.

References

ALEXANDER, J. B., and PEPPER, S. D. 1967, *Observatory*, **87**, 267.
ALLEN, C. W. 1963, *Astrophysical Quantities* (2nd ed.: London: Athlone Press).
ARP, H. C. 1958, *Hdb. d. Phys.*, ed. S. FLÜGGE (Berlin: Springer-Verlag), **51**, 75.
ARGUE, A. N. 1963, *M.N.*, **125**, 557.
ARGUE, A. N. 1966, *M.N.*, **133**, 475.
BODENHEIMER, P. 1965, *Ap. J.*, **142**, 451.
DENNIS, R. T. 1968, *Ap. J.* (Letters), **151**, L47.
EGGEN, O. J. 1968, *Ap. J. Suppl.*, **16**, No. 142.
EGGEN, O. J., and SANDAGE, A. R. 1962, *Ap. J.*, **136**, 735.
FAULKNER, J. 1967, *Ap. J.*, **147**, 617.

GLIESE, W. 1957, *Astr. Rechen-Inst. Heidelberg, Mitt. Ser. A*, No. 8.

GREENSTEIN, J. L. 1965, *Galactic Structure*, eds. A. BLAAUW and M. SCHMIDT (Chicago: University of Chicago Press), chap. xvii.

GRIFFIN, R. F., and REDMAN, R. O. 1960, *M.N.*, **120**, 287.

HOFFLEIT, D. 1964, *Catalogue of Bright Stars* (New Haven, Conn.: Yale University Observatory).

HECKMANN, O., and JOHNSON, H. L. 1956, *Ap. J.*, **124**, 477.

JENKINS, L. F. 1952, *General Catalogue of Trigonometric Stellar Parallaxes* (New Haven, Conn.: Yale University Observatory).

JENKINS, L. F. 1963, *Supplement to the General Catalogue of Trigonometric Stellar Parallaxes* (New Haven, Conn.: Yale University Observatory).

JOHNSON, H. L. 1963, *Basic Astronomical Data*, ed. K. AA. STRAND (Chicago: University of Chicago Press), chap. xi.

JOHNSON, H. L., and KNUCKLES, C. F. 1955, *Ap. J.*, **122**, 209.

JOHNSON, H. L., MORGAN, W. W., and HARRIS, D. L. 1953, *Ap. J.*, **118**, 92.

KRON, G. E., GASCOIGNE, S. C. B., and WHITE, H. S. 1957, *A.J.*, **62**, 205.

OKE, J. B., and CONTI, P. S. 1966, *Ap. J.*, **143**, 134.

PRICE, M. J. 1966, *M.N.*, **134**, 135.

SANDAGE, A. R., and EGGEN, O. J. 1959, *M.N.*, **119**, 278.

SPINRAD, H., and TAYLOR, B. J. 1967, *A.J.*, **72**, 320.

TAYLOR, B. J. 1967, *P.A.S.P.*, **79**, 441.

ULRICH, R. 1968, unpublished.

Duplicity characteristics of low-luminosity stars

CHARLES E. WORLEY

U.S. Naval Observatory, Washington, D.C.

Abstract

This paper summarizes current knowledge of the duplicity characteristics of low-luminosity stars, with comments on the serious problems
of observational selection. Groups of stars discussed include: stars
within 10 parsecs, M dwarfs, white dwarfs, and high-velocity subdwarfs.
In the case of the first three groups, binaries are found to be common.
The situation with regard to the subdwarfs is less clear, but it is shown
that binaries, including close pairs, are by no means absent from this
group.

I. INTRODUCTION

ONE ONLY has to glance at a list of nearby stars to be struck immediately by
two salient features: first, most of these stars are considerably less luminous
than the sun; and second, many of them do not occur as single objects, but
rather are grouped into double or multiple systems. As we shall see, these
systems probably embrace a majority of the stars in the solar neighborhood.

In principle, binaries may be detected by visual, photographic, spectroscopic, and photometric techniques, but in practice only the first three
methods are of importance for the nearby stars discussed here. We will consider first the observational techniques and the resulting surveys which have
revealed new binary stars, and then discuss four groups of nearby stars which
provide data on the question of stellar duplicity. Since observational selection
has an important effect on the statistics of duplicity, we will emphasize its role
in the following discussion.

II. DISCOVERY SURVEYS FOR LOW LUMINOSITY BINARIES

(*a*) *Visual surveys*

The comprehensive double star discovery surveys carried out early in this century (van den Bos 1963) provided little data concerning the duplicity of low-luminosity stars, since these surveys were restricted to Durchmusterung stars brighter than apparent magnitude 9.0 or 9.1 in the northern zones, and 9.5 in the southern zones. Consequently, existing double star catalogues give an entirely false picture of the true frequency of binaries of various luminosities, since the majority of pairs listed in them are more luminous, while the majority of binaries are less luminous, than the sun.

Special surveys are required to detect binaries among the faint nearby stars. The first such survey was that carried out by Kuiper (1934) some thirty years ago. He looked at all accessible stars then known to have either large parallaxes or proper motions greater than $0''.3$/year. This initial survey yielded a rich harvest of close binaries in rapid orbital motion; many of these pairs have since provided us with good orbits and masses. More recently, the writer (Worley 1962) surveyed 700 M dwarfs and found 27 new pairs. Nine of these systems had separations less than $1''$, and some of these have proved to be in rapid motion. Couteau (1966) has also found a number of companions to proper motion stars in his visual survey of stars contained in the Paris Zone Catalogue. New pairs have been found occasionally in the course of other observing programs, but it is significant that the vast majority of close pairs have been detected in programs aimed deliberately at their discovery. In such a program, the probability of visual detection of a companion is a complicated and subjective function of the apparent magnitude, magnitude difference, separation, sky and seeing, observer, and telescope. Even for the brighter stars, an appreciable fraction of potentially 'discoverable' binaries is missed in each survey, as has been illustrated by the success of the repeat surveys carried out by Kuiper and Couteau. Of course, some binaries also will be missed at any given epoch because they are then too close. Therefore, we can say only that the fraction of missed, but discoverable binaries must be higher for the faint stars of low luminosity than it is for the brighter stars.

(*b*) *Photographic surveys*

Most of the known binaries among low luminosity stars have been discovered as by-products of photographic proper motion surveys. For a limited number of nearby stars, intensive astrometric observations (van de Kamp 1965) have revealed new, 'unseen' companions.

Early photographic proper motion studies of limited portions of the sky conducted by Ross (1925–39) and Wolf (1919, 1920–29) yielded a number of wide and intrinsically faint pairs, but the discovery of large numbers of these systems awaited the much more comprehensive surveys conducted by Luyten and Giclas. Luyten's work comprises two distinct surveys. The Bruce Proper Motion Survey (Luyten 1960), principally encompassing southern hemisphere objects, resulted in the discovery of many hundreds of new common proper motion pairs; a portion of these have been catalogued by Luyten (1941). Important discoveries by Luyten include the double white dwarf LDS 275 and the nearby red dwarf pair L 726–8. More recently Luyten has begun a survey using Palomar 48-inch Schmidt plates and reports (Luyten 1967) discovery of 285 faint double and multiple systems in the first 700 square degrees surveyed. He therefore estimates that there are potentially 10,000 pairs detectable over the entire sky accessible to the instrument. Giclas and his collaborators (Giclas 1965) have been conducting a survey with the Lowell 13-inch astrograph to a brighter limiting magnitude than Luyten's work, but have now nearly completed their survey. Their lists of proper motion stars include about 200 new double stars. Probably the most interesting double noted in their survey is the faint pair Stein 2051 (= G 175–34), which, while discovered by Stein long ago, had not been recognized previously for the important pair that it is. It now appears that this nearby binary consists of a red dwarf together with a red degenerate companion showing appreciable orbital motion, so that we may eventually hope to obtain masses for the components. In addition to the major proper motion surveys cited above, less comprehensive surveys have yielded a number of new proper motion pairs. References to these additional surveys have been given by Luyten (1963).

In all photographic proper motion studies, detection of common proper motion pairs involves severe selection effects, the most serious of which is undoubtedly the limitation to wide pairs imposed by the combined resolving power of the plate and telescope. While statistical studies of the semi-major axes of visual binaries (Heintz 1967) show a pronounced frequency maximum around 50 A.U., the average pair discovered in the photographic studies has a semi-major axis of several thousand A.U. Therefore, as Luyten points out, the photographic discoveries represent only the tail of the true frequency distribution. Such pairs, while very useful for photometric and spectroscopic studies of low luminosity objects, will not yield information on masses in the foreseeable future.

In a statistical discussion of common proper motion pairs, Luyten (1939, 1967 loc. cit.) develops formulae relating the observed separation, s, and proper motion, μ, to the linear semimajor axis, a:

$$a = 12d, \quad \text{where} \quad d = s/\mu.$$

The writer has prepared a list of the Lowell pairs, and their observed separations are presented in Table I. Table II lists the distribution of their semi-major axes according to Luyten's formulae, along with his values for the Palomar, Bruce, and Greenwich surveys. Note that the Lowell discoveries appear quite comparable with the other surveys: the slight deficiency of very wide pairs simply reflects the writer's hesitation to include pairs where there seems some doubt of physical connection.

TABLE I   Observed separations of Lowell common proper motion pairs.

Observed separation	No.	%	Cum. %
< 10″	16	8.8	
10″.0–19″.9	48	26.6	35.4
20.0–29.9	24	13.3	48.7
30.0–59.9	32	17.7	66.4
1′.0– 1′.9	26	14.4	80.8
2.0– 4.9	16	8.8	89.6
5.0– 9.9	10	5.5	95.1
10.0–59.9	8	4.4	99.5
>1°	1	0.5	100.0
	181	100.0	

TABLE II   Distribution of binary semimajor axes for proper motion pairs.

d ($=a/12$)	Palomar	Bruce	Greenwich	Lowell	(No.)
4	1.4	0.3	0.6	0.5	(1)
10	7.0	2.8	3.0	6.1	(11)
25	22.1	15.2	11.0	28.2	(51)
63	26.7	29.6	24.1	31.5	(57)
158	20.0	26.4	31.9	14.9	(27)
400	15.5	15.2	20.5	8.3	(15)
1000	6.6	7.8	7.7	6.1	(11)
2500	0.7	2.7	1.2	4.4	(7)

One group of photographically discovered objects, the astrometric binaries, is at least potentially capable of providing good stellar masses as well as statistical information on other parameters. Among the stars nearer than ten parsecs, perturbations in the proper motion of the visible star are reported definitely for ten systems and tentatively for a number of others. As van de

Kamp has stressed, short-period perturbations are difficult to discover because of their small amplitude, while long-period perturbations are revealed only by extended observations over many years. As time passes and the astrometric data accumulate, we may anticipate the discovery of more of these interesting systems.

(c) Spectroscopic surveys

Spectroscopic observations also reveal the presence of binaries among the nearby stars. Since most of these stars are so faint, radial velocity data can only be obtained with large telescopes and the expenditure of sizeable amounts of observing time. Consequently, very little or no work has been done on many of the nearby stars. Often, even when a few spectrograms have been secured, they do not suffice to show any but the most obvious cases of velocity variation.

The large reflectors of the Mt. Wilson and Palomar Observatories have provided most of our radial velocity data on intrinsically faint stars. Early observations were discussed by Joy (1947) and Joy and Mitchell (1948). These astronomers found a considerable number of spectroscopic binaries among the stars observed. Dyer (1954) published radial velocities for 166 red dwarfs, but unfortunately was able to secure only single spectra for many of his stars, so that he reports only one new spectroscopic binary. Recently, Wilson (1967) has published new data secured with the Palomar reflector for more than 300 *dK* and *dM* stars. He finds a number of spectroscopic binaries and very tentatively concludes that their frequency among the late-type main sequence stars is not greatly different from that found among stars of earlier type.

III. THE DUPLICITY CHARACTERISTICS OF LOW-LUMINOSITY STARS

The observational techniques and surveys described above have resulted in the discovery of many binary and multiple systems among the nearby stars. We now examine in detail the duplicity characteristics of four groups of stars; these groups include stars within ten parsecs, the *M* dwarf stars found in the McCormick survey, white dwarfs, and high-velocity subdwarfs.

(a) Stars within ten parsecs

New trigonometric parallaxes were collected from the literature and combined, according to the Yale precepts, with the values previously available in the Yale parallax catalogue (Jenkins 1952) and its supplement (Jenkins 1963). No preliminary parallax values were used. The 173 objects found to be nearer than ten parsecs yielded the 66 binary or multiple systems which are listed in Table III. An attempt has been made to limit this list to definite, physical

TABLE III Double and multiple stars nearer than ten parsecs.

R.A. (1900)	Dec.	Yale	Name	Mag. A	Mag. B	Mag. C	Sp A	Sp B	Sp C	Notes
00^h 12^m.7	+43°27'	49	+43°44	8.07	11.04		M1V	M6V		A is a SpB
00 43.0	+57 17	155	η Cas	3.45	7.24		G0V	dM0		P = 480^y
00 55.3	+71 09	202*	+70°68	10.01	10.42	12.3	dM3	dM4	M4+	
00 57.0	+61 50	205	+61°195	9.57	13.66		dM2	dM7		
01 01.6	+54 26	219	μ Cas	5.16			G5VI			Astrom. binary
01 34.0	−18 28	343.1	L 726-8	12.4	12.9		dM6e	dM6e		P = 55^y
01 36.0	−56 42	352	−56°329	5.76	5.26		K2V	K5V		P = 500^y
02 30.6	+06 25	520	+6°398	5.82	11.65		K3V	M4		
03 16.0	−62 53	701	ζ Ret	5.24	5.52		G1V	G2V		
04 10.7	−07 49	945	o² Eri	4.48	9.50	11.0	K1V	DA	dM4e	P = 250^y
04 55.9	−05 52	1129	−5°1123	6.20			dK5			SpB
04 58.2	−21 24	1135	−21°1051	8.4	10.4		dM1			P = 48^y
05 33.4	+53 27	1291	+53°934,5	6.22	9.71		dK2	M1		
05 40.3	−22 29	1316	γ Lep	3.63	6.-7	15:	F6V	K2V	M4±	
06 05.4	+10 22	1426	+10°1032	10.6	12.6		dM3			
06 24.3	−02 44	1509	R 614	11.1	14.3		dM4e			P = 16^y
06 40.7	−16 35	1577	α CMa	−1.47	8.5		A1V	DA		P = 50^y
06 47.4	−05 03	1606	−5°1844	6.57	10.4		dK6	M2		
06 54.9	−44 09	1641.1	−44°3045	11.5	11.7		M4+	M4+		
07 03.3	+38 43	1668	R 986	11.67			dM5e			SpB

RA (h)	RA (m)	Dec	No.	Name	m₁	m₂	m₃	Sp 1	Sp 2	Sp 3	Notes
07	34.1	+05 29	1805	α CMi	0.3	10.8		F5IV-V	DF		$P = 41^{\mathrm{y}}$
07	35.8	−17 10	1813*	L 745-46	13.0	18		DF	m		
09	07.6	+53 07	2198	+53°1320, 1	7.60	7.70		MOV	MOV		
09	26.4	−13 03	2267	−12°2918	10.8	10·8		dM4			
09	29.7	+36 16	2280	+36°1979	5.41	13		G8IV-V			
10	45.8	+07 22	2524	W 358	11.63			dM5			SpB
10	57.9	+36 38	2576	+36°2147	7.47			M2V			Astrom. binary
11	00.5	+44 02	2582	+44°2051	8.77	14.8		M2V	dM5.5e		
11	12.9	+32 06	2625*	ξ UMa	4.29	4.75		GOV	GOV		$P = 60^{\mathrm{y}}$
12	22.8	−70 34	2873	L 68-27, 28	14.7:	16.7:		k − m	k − m		
12	28.4	+09 34	2890	W 424	12.7	12.7		M7	M7		
13	23.2	−01 50	3074	R 486	11.35	14.2		dM4	M5		
13	49.9	+18 54	3175	η Boo	2.69			GOIV			SpB, $P = 492^{\mathrm{d}}$
14	32.8	−60 25	3309	α Cen	0.0	1.4	10.68	G2V	dK	dMe	$P = 80^{*}$
14	46.8	+19 31	3360	ξ Boo	4.70	6.85		G8V	K4V		$P = 152^{\mathrm{y}}$
14	49.5	+23 58	3369	R 52	11.6	12.2		M5			
14	51.6	−20 58	3375	−20°4123, 5	5.82	8.11		dK5	dM2		
16	13.5	−37 19	3701	−37°10765	10.6	15:		M4			
16	24.7	−12 25	3746	−12°4523	10·12			dM5			
16	50.1	−08 08	3844*	W 629, 30	9.8	9.8	11.95	dM3e	dM4	sdM4	SpB; $P = 1^{\mathrm{y}}.7$ (for W 630)
17	09.2	+45 50	3907	+45°2505	9.98	10.35		M3	M3+		$P = 13^{\mathrm{y}}$
17	09.2	−26 27	3908	−26°12026	5.07	5.10	6.31	KOV	dK1	K5V	
17	11.5	−46 32	3919	−46°11370	5.54	8.66		G8V	MOV		$P = 700^{\mathrm{y}}$
17	12.1	−34 53	3924	−34°11626	6.13	7.6	10.16	K3V	K5V	M2	$P = 42^{\mathrm{y}}$
17	25.5	+05 38	3975.1	+5°3409	9.2	14		M2			

table continues

TABLE III—(cont.)

R.A.	Dec.	Yale	Name	Mag	Sp	Notes
17 33.4	+18 37	4009	CC 2347	9.62	M1	Astrom. binary
17 37.0	+68 26	4029	+68°946	9.15	M3.5V	SpB & astrom. binary
17 42.5	+27 47	4060	μ Her	3.42 10.5 10.7	G5IV M4 M4	$P = 43^y$
17 52.9	+04 25	4098	Barnard's star	9.54	M5V	Astrom. binary
18 00.4	+02 31	4137	70 Oph	4.21 5.94	K0V K5V	$P = 88^y$
18 22.9	+72 41	4245	χ Dra	3.58	F7V	SpB & astrom. binary
18 41.7	+59 29	4330	+59°1915	8.90 9.59	dM4 dM5	
19 02.8	+20 44	4451	R 730, 31	10.8 10.3	sdM2 sdM2	
19 03.6	+32 21	4459	17 Lyr C	11.8 12.1	dM4	
19 12.1	+05 03	4494	+4°4048	9.1 18.7	dM3.5	
19 15.2	−07 51	4519.1	LDS 678	12.5 13:	DC M3	
20 04.6	−36 21	4782	−36°13940	5.32 11.5	K3V M5	SpB
20 09.1	−27 20	4804	−27°14659	5.74	K0V	
20 35.6	−32 47	4929	−32°16135	8.64 10.7 10.9	M0Ve dM4e dM4e	CPM with −31°1781
21 02.4	+38 15	5077	61 Cyg	5.19 6.02	K5V K7V	$P = 650^y$
22 12.1	−09 18	5397	W 1561	13.5 14.5	dM4e dM4e	
22 24.4	+57 12	5438	Krüger 60	9.82 11.4	dM3 dM4e	$P = 44^y$
22 42.5	+43 49	5520	+43°4305	10.04	dM5e	Astrom. binary
22 50.8	−32 06	5562	α PsA	1.15 6.48	A3V K5V	
22 51.8	+16 02	5563	+15°4733	8.66	dM2	SpB
23 26.8	+19 21	5694	+19°5116	10.47 12.52	dM4e dM6e	

Notes to Table III

Yale

202   CPM with the close pair, which is Yale 86 (ADS 433).

1813   The star 1810.1 in the supplement to the parallax catalogue is evidently the same as Yale 1813.

2625   Both visual components are themselves SpB.

3844   Quintuple. Wolf 630 is the close visual pair. Wolf 629 is a SpB. The fifth component is 18 mag. at 221″ separation.

systems. Thus, some stars reported to be astrometric binaries at one epoch, but not subsequently confirmed, such as $BD + 20°2465$, are not listed. However, Barnard's star is included, since the perturbation seems real and the companion may be a degenerate star of small mass. Note that while some astrometric and spectroscopic binaries are listed in Table III, the vast majority of companions are visible stars at comparatively large separations from their primaries. Table IV summarizes our knowledge of the multiplicity of these nearby stars. Most striking is the fact that more stars occur in the binary and multiple systems (58 per cent) than as single objects. Also, as has been pointed out previously for field binaries with visual orbits (Worley 1967), multiple systems of three or more stars are quite common.

TABLE IV Multiplicity of stars within ten parsecs of the sun.

No. of components	No. of systems	No. of stars	% of stars
1	107	107	42.3
2	55	110	43.4
3	9	27	10.7
4	1	4	1.6
5	1	5	2.0
Totals	173	253	100.0

At the present time, construction of an accurate Hertzsprung–Russell or color-magnitude diagram for the nearby stars is impossible. Aside from the close binaries whose components cannot be observed separately with present techniques, spectral types do not exist in a consistent system, bolometric corrections are very uncertain for late-type stars (Harris 1963), and the photometric *UBV* system provides little color resolution for those objects which can be observed (Kron, Gascoigne and White 1957). Examination of a preliminary plot (which is not given here) reveals, besides the expected preponderance of late-type dwarfs down to $M_v \sim 14$, that the components of the binary and multiple systems populate the entire main sequence, and show no lack of numbers at the fainter absolute magnitudes. Thus the apparent dearth of late-type binaries reported by Aitken (1935) is not substantiated for these nearby dwarfs, and it is probable that Aitken's sample is an undifferentiated collection of giants and dwarfs, with giants predominating. Observational selection alone can account easily for the lack of objects fainter than $M_v = 14$. Such objects may be numerous, particularly if the 'turnover' in the luminosity

TABLE V Additional *M* dwarf binaries contained in Vyssotsky's lists.

Vyssotsky no.	DM or AC catalog	α (1900)	δ	Mag (V)	Sp	Notes
6	BD +39° 710	03ʰ 00ᵐ.4	+39° 59′	9.67	M0	ADS 2343, triple; component A is a SpB
20	BD −13° 2439	08 08.4	−13 36	9.38	M2	ADS 6664, triple; component A is a SpB
49	BD +36° 2393	13 33.0	+36 16	9.07	M2	SpB
199	BD +61° 2068	20 51.3	+61 48	8.54	M0	SpB
321	BD +5° 3409	17 25.5	+05 38	9.5	M	VB, sep. 10″
322	BD +68° 946	17 37.0	+68 26	9.15	M5	Both a SpB and an astrometric binary
330	BD −20° 5833	20 04.6	−20 45	8.94	M0p	SpB
340	BD −26° 16420	22 58.0	−26 51	9.5	M0	VB, RST 1154; one component is also a SpB, hence triple
538	BD −8° 2582	09 01.9	−08 25	9.51	M0	SpB

function reported by Luyten (1939 loc. cit.) does not in fact occur, as presently suspected (van Rhijn 1965). Finally, despite the fact that there must still be undiscovered companions among the stars nearer than ten parsecs, it is already evident that the ordinary star exists in a binary or multiple system, and that the single star is very much in a minority.

(b) M dwarf stars

The spectrophotometric surveys of *M* dwarf stars carried out by Vyssotsky and his collaborators (Vyssotsky 1943; Vyssotsky *et al.* 1946; Vyssotsky and Mateer 1952; Vyssotsky 1956) at the Leander McCormick Observatory provide a unique sample of the more luminous red dwarf stars in the solar neighborhood, since this sample is unbiased by the usual selection of nearby stars on the basis of proper motion alone. The incidence of multiplicity in this sample has been reported upon previously by the writer (1963 loc. cit.), and a catalogue of *M* dwarf binaries presented therein. To this list now should be added the objects contained in Table V. Most of these additions are newly discovered spectroscopic binaries identified by Wilson, who also lists Vys 142 (= ADS 8862) as a spectroscopic binary. However, examination of the predicted velocity curve (Dommanget 1967) reveals that orbital motion in the visual pair could account for the observed variation, so that Vys 142 is not listed in the table. Table VI summarizes the multiplicity characteristics of the 561 *M* dwarfs contained in the Vyssotsky lists. The data in Table VI is largely, but not entirely, independent from the data concerning the nearby stars (Table IV), since 22 of the Vyssotsky stars are also included there. While the *M* dwarf sample has not been nearly so thoroughly investigated for binaries as the nearby stars have, it still reveals a very high incidence of multiplicity.

TABLE VI   Multiplicity of 561 *M* dwarfs.

No. of components	No. of systems	No. of stars	% of stars
1	436	436	61.2
2	104	208	29.2
3	17	51	7.2
4	3	12	1.7
5	1	5	0.7
Totals	561	712	100.0

(c) White dwarf stars

Among the stars nearer than ten parsecs are 11 degenerate stars, of which five belong to multiple systems. Of these five, Sirius B and Procyon B have primaries approximately ten magnitudes brighter, while the other three, o^2 Eri B, L 745–46 A, and LDS 678 A, are of roughly the same visual apparent magnitude as their red dwarf companions. Luyten and Giclas both have listed a considerable number of wide binaries consisting of red–white dwarf pairs, and detailed photometric and spectroscopic work on these objects has been done by Eggen and Greenstein (1965a, 1965b, 1967). The resulting photometric parallaxes indicate that a number of these systems are also within ten parsecs and only await trigonometric parallaxes to be included in Table III. It is a difficult observational problem to find more Sirius or Procyon-like binaries, but the situation is more promising for systems resembling o^2 Eri. Hopefully, examination of the many new proper motion stars discovered recently will reveal more close binaries among the white dwarfs.

It is obvious from the very meager data available at present that our information concerning the duplicity characteristics of white dwarfs is in an unsatisfactory state. We can only say that there are numerous wide red–white dwarf pairs, and that those white dwarfs among the nearby stars appear to exhibit an incidence of duplicity roughly comparable to that found for other stars.

(d) High-velocity subdwarfs

There appear to be difficulties both in the definition of and identification of low-luminosity subdwarfs. Indeed, Greenstein and Eggen (1965) have remarked that 'the existence of subluminous M dwarfs is still in doubt', and Greenstein (1965b) also has stated that 'it is generally assumed that close binary systems are rare in Population II'. The genesis of this latter notion apparently dates back to a study of high-velocity stars made by Oort (1926), who found only about a quarter as many binary stars among the high-velocity stars as among a sample of nearby stars of low velocity. Reinforcing the notion of the lack of close binaries in Population II have been remarks such as Kopal's (1959) that eclipsing binaries seem totally absent from globular clusters. However, Miss Roman (1965) mentions that the low frequency of binaries among high-velocity stars found by Oort is not confirmed in her own catalogue of these objects, where she finds a normal frequency compared to stars listed in the Bright Star Catalogue (Hoffleit 1964), while Partridge (1967) finds a similar normal incidence of binaries among the high-velocity stars contained in Gliese's (1957) catalogue of stars nearer than 20 parsecs. Furthermore, Walker (1964) has remarked that the lack of eclipsing binaries in globular clusters may be simply a problem of observational selection, since stars that are close doubles may not appear on the horizontal branch just because they are double and therefore evolve differently from single stars. Recent theoretical work supports this hypothesis (Plavec 1967).

TABLE VII Subdwarf binaries among 127 high-velocity stars.

Pos. (1900) R.A.	Dec.	Name	Sp	m vis (V)	Notes
00h 06m.1	−21°17′	LDS 1	sdF4, sdG	11.8–12	8″, VB
01 01.9	+54 25	μ Cas	G5VI	5.16	astrometric binary
02 29.4	−12 48	−13° 482	sdF4	9.75	var. vel.
03 07.5	+18 29	W 134	sdM0	14.32	SB
03 21.0	+23 27	R 374	sdG5	11.12	SB?
03 40.1	+41 10	ADS 2757	sdK1, sdK2, sdMp	8.14–8.74	visual triple
05 57.2	+19 22	+19° 1185	sdG8, K3	9.32–13.25	VB
08 42.0	−13 00	−12° 2669	sdF	10.25	var. vel.
09 35.6	+01 29	R 889	sdF	10.50	SB
09 43.3	+44 46	+44° 1910	sdF	10.92	var. vel. ?
10 41.9	+28 57	R 626	sdG5	10.24	var. vel. ?
12 00.3	−26 02	−25° 9024	sdF–G	10.39–14	VB, 12″
14 38.4	+06 14	W 540	sdG2	10.43–13	VB, 4″
17 53.0	−13 04	ADS 10938	sdF8	9.62–11	VB, <1″
18 31.4	+28 38	R 711	sdG0	11.30	var. vel.
21 38.7	+43 07	SS Cyg	sdB + dG	7.7	eclipsing binary
23 11.9	−14 22	ADS 16644	sdG2	8.8–9.0	VB, <1″
23 56.7	−26 21	ADS 17173	sdF7	9.5–9.9	VB, <1″

Eggen (1964) has published a catalogue of 656 high-velocity stars which includes only the more extreme objects having space motions greater than 100 km/sec. The writer has examined this material and finds 53 binary systems. This is a smaller incidence of binaries (8 per cent) than found by Oort (13 per cent) or Miss Roman (14 per cent) in their own catalogues. But if we restrict ourselves to those 127 systems identified positively as subdwarfs by spectral type (including a few stars listed as 'weak-lined' or 'very weak-lined'), we find that no fewer than 15 of these objects are binaries; some of them close visual or spectroscopic pairs. In addition, there are three more stars suspected of having variable velocities. All of these 18 objects are listed in Table VII.

Most of the subdwarfs just discussed are of spectral types *F* or *G*. It would be very informative to examine for duplicity a substantial sample of late-type subdwarfs, provided that they could be so identified. Eggen and Greenstein have listed a number of red–white dwarf pairs where the red dwarf is called a subdwarf, and among the nearby binary stars of late type, Barnard's star, BD +43°44 B, and Ross 614 A are sometimes called subdwarfs. Thus, without being able to make a very precise statement, it is apparent that there do exist appreciable numbers of binary, low-luminosity subdwarfs as well as those of higher luminosity.

The problem of the general frequency of subdwarf binaries is once again bound up with the question of observational selection. The eight visual pairs among the 127 subdwarfs in Eggen's catalogue give a frequency which is exactly the same as found in the general double star surveys made early in this century. While it is obvious that the observed incidence of binaries among the subdwarfs is lower than among the nearby stars, it is not so obvious what role observational selection has played in these statistics. As an illustration of one such selection effect, Table VIII gives the frequency distribution of the

TABLE VIII   Magnitude distribution of subdwarf binaries.

m vis	No. of stars	No. of binaries
≤5.99	1	1
6.0–6.99	0	0
7.0–7.99	6	1
8.0–8.99	15	3
9.0–9.99	35	3
10.0–10.99	37	4 + (3)
11.0–11.99	26	2
12.0–12.99	3	0
≥13.0	4	1
Total 127		15 + (3)

127 subdwarfs according to apparent visual magnitude. More than half of these stars are fainter than $m_v = 10$ and therefore would not have been included in the systematic surveys for double stars, so that there is every likelihood that more pairs remain to be discovered. Astrometric observations to detect unseen companions are not practicable for the majority of these stars because they are too distant, but more intensive spectroscopic observations would also certainly reveal additional binaries.

IV. CONCLUSION

The question of stellar duplicity is a profound one, since it is intimately connected with the processes of star formation and evolution in our galaxy. Observational evidence increasingly reveals that stars are born in groups which often become stable binary and multiple systems. Sampling of the stellar population to delineate the frequency of binaries is therefore required. For low-luminosity stars, present techniques are inadequate to detect all the binary stars existing in the solar neighborhood. Nevertheless, their systematic application will reveal a statistically significant fraction of the binaries. For those stars known to be within ten parsecs, we are undoubtedly at the point of diminishing returns for visual surveys, but intensive astrometric and spectroscopic work should still reveal new binaries in this sample, particularly for the neglected southern objects. We have every reason to think that the true percentage of nearby stars in binary systems will be of the order of 70 per cent. For the unique sample of *M* dwarfs found by Vyssotsky, the percentage of stars known to be in binary systems is already 40 per cent, and there is no reason to think that the true frequency for this sample will prove to be any less than that for the nearby stars. Knowledge of the binary frequency among the white dwarfs is unsatisfactory because of lack of observation, but there still appear to be many pairs in the small sample available. Finally, the writer believes that it has not been proved that binaries are any less frequent among the Population II stars than among younger objects, since it is both possible and probable that present statistics are beset by severe selection effects.

Deficiencies in our observations of nearby low-luminosity binaries are many, and the astronomical problems needing solution are important. Discovery alone provides information on the true frequency of binaries, but only subsequent observation can yield other parameters, the most significant of which is undoubtedly the stellar mass.

References

Aitken, R. G. 1935, *The Binary Stars* (McGraw-Hill Book Co.), p. 268.
Couteau, P. 1966, *J. Observateurs*, **49**, 221.
Dommanget, J. 1967, *Comms. Obs. Royal de Belgique*, Ser. B, No. 15.

DYER, E. R., Jr. 1954, *A. J.*, **59**, 218.

EGGEN, O. J. 1964, *Royal Obs. Bull.* No. 84.

EGGEN, O. J., and GREENSTEIN, J. L. 1965 a, *Ap. J.*, **141**, 83.

EGGEN, O. J., and GREENSTEIN, J. L. 1965 b, *Ap. J.*, **142**, 925.

EGGEN, O. J., and GREENSTEIN, J. L. 1967, *Ap. J.*, **150**, 927.

GICLAS, H. L. 1965, *Vistas in Astronomy*, eds. A. BEER, and K. A. STRAND (Pergamon Press Ltd., London), **8**, p. 215.

GLIESE, W. 1957, *Mitt. Astron. Rechen-Inst. Heidelberg*, Ser. A, 8.

GREENSTEIN, J. L. 1965, *Galactic Structure*, eds. A. BLAAUW, and M. SCHMIDT (Univ. of Chicago Press, Chicago), p. 361.

GREENSTEIN, J. L., and EGGEN, O. J. 1965, *Vistas in Astronomy*, eds. A. BEER, and K. A. STRAND (Pergamon Press Ltd., London), **8**, p. 69.

HARRIS, D. L., III, 1963, *Basic Astronomical Data*, ed. K. A. STRAND (Univ. of Chicago Press, Chicago), p. 263.

HEINTZ, W. D. 1967, *Comms. Obs. Royal de Belgique*, Ser. B, No. 17, p. 49.

HOFFLEIT, D. 1964, *Catalogue of Bright Stars* (Yale Univ. Obs., New Haven, Conn.), 3rd. rev. ed.

JENKINS, L. F. 1952, *General Catalogue of Trigonometric Stellar Parallaxes* (Yale Univ. Obs., New Haven, Conn.).

JENKINS, L. F. 1963, *Suppl. to the General Catalogue of Trigonometric Stellar Parallaxes* (Yale Univ. Obs., New Haven, Conn.).

JOY, A. H. 1947, *Ap. J.*, **105**, 96.

JOY, A. H., and MITCHELL, S. A. 1948, *Ap. J.* **108**, 234.

KOPAL, Z. 1959, *Close Binary Systems* (John Wiley and Sons, New York), p. 7.

KRON, G. E., GASCOIGNE, S. C. B., and WHITE, H. S. 1957, *A.J.*, **62**, 205.

KUIPER, G. P. 1934, *Publs. A.S.P.*, **46**, 235.

LUYTEN, W. J. 1939, *Publs. Univ. of Minn. Obs.*, **2**, No. 1.

LUYTEN, W. J. 1941, *Publs. Univ. of Minn. Obs.*, **3**, No. 3.

LUYTEN, W. J. 1960, *Publs. Univ. of Minn. Obs.* **2**, No. 15.

LUYTEN, W. J. 1967, *Comms. Obs. Royal de Belgique*, Ser. B., No. 17, p. 45.

LUYTEN, W. J. 1963, *Basic Astronomical Data*, ed. K. A. STRAND (Univ. of Chicago Press, Chicago), p. 46.

OORT, J. H. 1926, *Groningen Obs., Publs.*, No. 40.

PARTRIDGE, R. B. 1967, *A.J.*, **72**, 713.

PLAVEC, M. 1967, *Comms. Obs. Royal de Belgique*, Ser. B., No. 17, p. 83.

ROMAN, N. G. 1965, *Galactic Structure*, eds. A. BLAAUW and M. SCHMIDT (Univ. of Chicago Press, Chicago), p. 345.

ROSS, F. E. 1925–39, *A.J.*, Nos. 835, 856, 862, 871, 886, 900, 926, 935, 957, 1073, 1101, 1118.

VAN DEN BOS, W. H. 1963, *Basic Astronomical Data*, ed. K. A. STRAND (Univ. of Chicago Press, Chicago), p. 320.

VAN DE KAMP, P. 1965, *Vistas in Astronomy*, eds. A. BEER and K. A. STRAND (Pergamon Press Ltd., London), **8**, 215.

VAN RHIJN, P. J. 1965, *Galactic Structure*, eds, A. BLAAUW and M. SCHMIDT (Univ. of Chicago Press, Chicago), p. 27.

VYSSOTSKY, A. N. 1943, *Ap. J.*, **97**, 381.

VYSSOTSKY, A. N., JANSSEN, E. M., MILLER, W. J., and WALTHER, M. E. 1946, *Ap. J.*, **104**, 234.

VYSSOTSKY, A. N., and MATEER, B. A. 1952, *Ap. J.*, **116**, 117.

VYSSOTSKY, A. N. 1956, *A.J.*, **61**, 201.
WALKER, M. F. 1964, *Royal Obs. Bull*, No. 82.
WILSON, O. C. 1967, *A.J.*, **72**, 905.
WOLF, M. 1919, *Veröff, Sternw. Heidelberg*, **7**, 195.
WOLF, M. 1920–29, *Astr. Nachr.*, Nos. 5033, 5035, 5039, 5050, 5074, 5079, 5084, 5089, 5090, 5120, 5128, 5133, 5214, 5243, 5262, 5281, 5289, 5293, 5305, 5307, 5319, 5374, 5388, 5391, 5396, 5422, 5451, 5470, 5495, 5516, 5658.
WORLEY, C. E. 1962, *A.J.*, **67**, 396.
WORLEY, C. E. 1967, *Comms. Obs. Royal de Belgique*, Ser. B., No. 17, p. 221.

Discussion

HAYASHI   What is the mean distance between the binaries?

WORLEY   The statistical studies show that the average separation is about 50 astronomical units for the visual binaries. Dr. Heintz has studied this problem recently, and he can give you more detailed information.

KUMAR   Would you give an estimate of the fraction of all stars that are members of multiple systems?

WORLEY   I expect at the moment 60 per cent of the stars in the solar neighborhood that have trigonometric parallaxes are members of binary systems. The percentage can only increase and I expect that it will be somewhere around 70 or 80 per cent. In fact, maybe there are no single stars. That is the situation I am working towards!

STRAND   What is the companion to the Sun?

WORLEY   Some people have thought they have found it before, but I do not know whether it is real or not.

The Herstmonceux parallax program

D. V. THOMAS

Royal Greenwich Observatory

Abstract

The Herstmonceux parallax program is described, and a new method for the determination of the trigonometric parallaxes of nearby open clusters is presented.

TRIGONOMETRIC PARALLAXES have been determined at the Royal Greenwich Observatory for more than 50 years. Three parallax programs were completed at Greenwich. The first two volumes of results have been published (Dyson 1925, 1934), and the results of the third program have been distributed. In all, some 760 parallaxes were determined at Greenwich.

The telescope used for this work, the Thompson 26-inch refractor (focal length 6.9 m, giving a plate scale of 30″/mm) was re-erected at Herstmonceux in 1958. Observing began the following year on a further program of parallaxes consisting of more than 300 stars. About 200 of these were stars begun, but not completed, at Greenwich. In 1965 this observing program was drastically pruned, as it was realised that a large proportion of the stars carried over from the old Greenwich programs were unlikely to have parallaxes greater than the standard errors of the determinations. The revised program now contains only 100 stars. All are stars lacking trigonometric parallaxes, or for which a confirmatory parallax is desirable. Eighty per cent are stars in a catalogue by Gliese (1957), and fifty per cent are known M-dwarfs. The remaining twenty per cent consists of stars in the Yale Catalogues, Jenkins (1952) with discordant parallaxes, recently discovered stars with large proper motion, and a few binaries and stars of astrophysical interest (including one quasar). When completed, this program should add significantly to the observational data on low luminosity stars.

The observing program has not made very rapid progress in the last few years, because of conflicting demands on telescope time. However, a number of stars now have series of more than 30 plates, and those of six M-dwarfs have been measured. The reductions await the completion of computer programs, which were only recently begun. To some extent the reduction methods are experimental at this stage. The computer will enable us to obtain a great deal more information than can be obtained by a simple dependence-reduction of measures in right ascension alone. The plates are being measured on a two-coordinate digitised measuring machine, and will be reduced by the method of differential plate constants to a standard reference frame defined by the mean positions of the six reference stars on all the measured plates. Each plate contains one direct and one reversed exposure, and is measured by two people. By making separate parallax determinations for the individual exposures and measurers and for various combinations of them, we hope to isolate the principal components of the standard errors, viz. errors due to refraction anomalies, random and systematic emulsion distortions, and measuring errors. This should lead to more accurate parallaxes, both by indicating the weights which should be assigned to individual plates, and also by indicating where changes in observing, plate processing, or measuring techniques might be beneficial.

One of the biggest problems of trigonometric parallax work is the existence of systematic differences between the parallaxes determined by different instruments. A significant part of these differences probably arises from the use of different sets of reference stars. We are paying particular attention to this aspect at Herstmonceux. The Greenwich (and Herstmonceux) reference stars have a mean photographic magnitude of about 11. As nothing is known at present about their colors nor their proper motions, they can be considered to constitute a random sample of 11th magnitude stars. Their mean parallax should therefore be about .″002 at low galactic latitudes, increasing by a factor of about 2 or 3 towards the galactic pole. The proportion of nearby dwarfs among the reference stars should be almost negligible at low latitudes, but is likely to be significant at higher latitudes, where their inclusion would impair the reduction from relative to absolute parallax. As we have six reference stars on each plate, it will be possible to determine both the proper motion and the parallax of each star relative to the mean of the others. Combining this information with their colors should enable us to identify most of the intruders. We hope eventually to improve the published Greenwich parallaxes by extending this study to the reference stars used in the old Greenwich programs.

Breaking away from traditional parallax work, we are considering a project which amounts to the addition of the further dimension of the annual parallax to the combined proper motion and photometric analysis of cluster fields which has been successfully developed at Herstmonceux during the last few years by Murray, Corben, and Allchorn (1965); Murray, Jones, and Candy

(1965); Clube (1966); Murray and Clements (1968); Murray (1968). There are not many clusters which would be suitable—only those nearer than a few hundred parsecs, having an apparent diameter of the order of 1°, and containing at least fifty stars brighter than 16th magnitude. Praesepe is the most obvious candidate. According to Klein Wassink (1925), there are about 150 cluster members and 600 field stars brighter than 16th magnitude in the useable field of a 26-inch plate (approximately 1° square). Suppose a series of about fifty plates were to be taken at times of favorable parallax factors, and reduced using all known cluster members to define the reference system. With such a large number of reference stars, it would be possible to determine second-order plate constants, including magnitude and color terms. The reduced measures would yield the relative parallax of each field star with respect to the cluster. The accuracy would probably be of the same order as that attainable in conventional parallax work.

The proper motion analysis and photometric data should enable one to select with some confidence the field stars which would be expected to be giants at a distance of more than 500 parsecs. The mean photometric parallax of these stars would be of the order of $.''001$ and would be determined with relatively high precision. The combination of the mean measured parallaxes of these stars relative to the cluster with their mean photometric parallaxes then gives the absolute parallax of the cluster. This result could be strengthened by including also any groups of stars sufficiently well identified, on the basis of their proper motions and photometry, for their photometric parallaxes to be determined with adequate precision. One would hope ultimately to obtain the parallax of the cluster from perhaps 400 field stars, with a standard error of about one-twentieth that of the individual stars, i.e. somewhere in the region of $\pm .''0005$. This would be a standard error of about ten per cent of the parallax of Praesepe, and would correspond to the determination of the distance modulus with a standard error of ± 0.2 mag. An independent kinematic parallax would be obtained, of course, from the proper motions alone.

The Hyades is probably the only cluster for which one could expect to determine a reliable annual parallax by conventional methods. The formal accuracy, taking the mean of the trigonometric parallaxes of individual members, is only a little better than that estimated above. However, the trigonometric parallax of the Hyades may be subject to systematic error due to inadequate data on the reference stars and to the significant depth of the cluster in relation to its distance. Combining a number of different approaches, Alexander (1967), has recently concluded that the distance of the Hyades could be in error by up to five per cent (± 0.1 mag. in the distance modulus), but Hodge and Wallerstein (1966), have suggested that the adopted distance of the Hyades is as much as twenty per cent too small (± 0.4 mag.). In this context, the determination of the distance modulus of Praesepe with a standard error of ± 0.2 mag. would be quite valuable.

A project such as this would be fairly extravagant as regards telescope time, though not more so than is conventional parallax work. It could hardly be contemplated without the aid of an automatic measuring machine and an electronic computer. Even then, success will depend quite critically on the accuracy ultimately achieved. A standard error of ± 0.1 in the distance modulus would be a very valuable result, but ± 0.5 would hardly be worth the effort. For this reason, a very thorough appraisal will be necessary before this project is started.

References

ALEXANDER, J. B. 1967, *The Observatory*, **87**, 213.
CLUBE, S. V. M. 1966, *R. Obs. Bull.*, No. 111.
DYSON, F. 1925, *Observations of Stellar Parallax at the R. Obs.*, Greenwich, Vol. I.
DYSON, F. 1934, *Observations of Stellar Parallax at the R. Obs.*, Greenwich, Vol. II.
GLIESE, W. 1957, *Astr. Rechen. – Inst. Heidelberg. Mitt.*, *A*, No. 8.
HODGE, P. W., and WALLERSTEIN, G. 1966, *Publ. A.S.P.*, **78**, 411.
JENKINS, L. F. 1952, *Gen. Cat. of Trig. Stellar Parallaxes*, *Suppl.* to above 1963.
KLEIN WASSINK, W. J. 1925, *Publ. Kapteyn Astron. Lab.* No. 39.
MURRAY, C. A., CORBEN, P. M., ALLCHORN, M. R. 1965, *R. Obs. Bull.*, No. 91
MURRAY, C. A., JONES, D. H. P., CANDY, M. P. 1965, *R. Obs. Bull*, No 100.
MURRAY, C. A., and CLEMENTS, E. D., 1968, *R. Obs. Bull.*, No. 139.
MURRAY, C. A., 1968, *R. Obs. Bull.*, No. 141.

Intermediate band photometry of white dwarfs*

J. A. GRAHAM

Kitt Peak National Observatory†

Abstract

Four-color photometry on Strömgren's u, v, b, y photometric system is given for 45 white dwarfs which have been previously observed by Eggen and Greenstein. Three of the four pass bands of Strömgren's system are essentially free of strong hydrogen line blocking while the v band, centered near $\lambda = 4100$ Å, is very sensitive to the strength of the Hδ line in the observed star. Strömgren's m_1 index, which measures the depression of the v band relative to the b and y bands, is therefore an excellent index of hydrogen line strength in this context.

In the $(u - b) - (b - y)$ diagram, which is analagous to the two-color diagram of the U, B, V, system, the advantage of the line-free colors are clear. The DA-type stars lie along a very narrow sequence which is close to the main-sequence relation for $(b - y)$ negative but which becomes horizontal for $(b - y)$ positive. The helium DB stars and the continuous spectrum DC stars lie close to the black-body line. The white dwarfs with peculiar spectra tend to lie close to or above the black-body line. It is suggested that the higher spectral resolution of the u, v, b, y transmission bands should make it easier to fit theoretical white dwarf model atmospheres to the observed colors.

I. INTRODUCTION

UNTIL RECENTLY very little reliable data on the magnitudes and colors of white dwarfs was available at all. This deficiency has now been remedied by Eggen and Greenstein (1965, 1967) who now have obtained new photometric and spectroscopic data for more than 200 stars of this type. They used the Johnson U, B, V system for their observations. There are some advantages

* Contribution No. 367 from the Kitt Peak National Observatory.

† Operated by the Association of Universities for Research in Astronomy, Inc., under contract with the National Science Foundation.

in using this system. The wide transmission bands of the filters make it possible to observe stars to faint limiting magnitudes and still to obtain accurate colors. On the other hand, the appreciable blocking of the continuous spectrum by the extremely strong hydrogen lines in the DA-type stars seriously complicates the interpretation of the observed U, B, V colors. Observations with intermediate band photometric systems, which use transmission band widths of the order of 200 Å, should provide the extra special resolution to enable observations of hydrogen-line free colors to be made. There is the disadvantage of having a limiting magnitude some two magnitudes brighter than that allowed by the U, B, V system, but this is offset by the more accurate knowledge of the effective wavelength of the measurement. Photometric measurements of line strength may also be made by centering a filter on lines of particular interest.

At first sight, the 4-color u, v, b, y system introduced by Strömgren (1963) seems particularly useful for this application. The b and y bands of this system are centered at 4670 Å and 5470 Å respectively and can be combined to give a nearly line-free color gradient. The u band, at 3500 Å, can be used to measure the size of the Balmer discontinuity. The v band, centered at 4100 Å, includes the $H\delta$ line and consequently the index $m_1 = (v - b) - (b - y)$ is a sensitive measure of hydrogen line intensity provided that there are no other prominent lines in this spectral region. The m_1 index is known in other contexts as a metallic line index, but here there is no such implication. The m_1 index, in this instance, is more closely analogous to the $H\beta$ index which has been used by Crawford and others (see Crawford 1966).

II. OBSERVATIONS

Four-color observations of 45 white dwarfs from the Eggen-Greenstein lists are given in Table I. These were made with the original set of 4-color filters which Strömgren used to define his system. Transmission curves for these filters have been published by Crawford (1966). The 84-inch telescope at Kitt Peak National Observatory was used with a standard photometer with an R.C.A. 1P21 photomultiplier for nearly all the observations. Table I contains the following information.

1. Number in Eggen–Greenstein Catalogue.

2. Other identification. These come from a number of sources and it is suggested that Eggen and Greenstein's papers be consulted for details.

3. Yellow magnitude, y. This has been transformed to that of the V magnitude of the U, B, V system by using observations of bright standard stars which have been assigned V magnitudes by Johnson and his co-workers.

TABLE I White dwarf photometry.

E.G. no.	Other catalog no.	y	$b-y$	m_1	$u-b$	n	V	$B-V$	Sp	W_γ
9	W 1516	13.82	+.132	+.015	+.269	2	13.82	+.11	DC	—
11	L 870-2	12.79	+.264	+.053	+.651	2	12.84	+.34	DA s	6.0
19	F 22	12.79	-.067	+.254	+.205	3	12.65	-.06	DA	27.4
20	F 24	12.39	-.069	+.031	-.340	3	12.25	-.23	DA wke	4.5
26	HZ 4	14.47	-.005	+.364	+.471	2	14.47	+.15	DA	38.4
30	HZ 10	14.15	+.031	+.358	+.583	2	14.14	+.07	DA s	37.4
31	HZ 2	13.89	-.078	+.260	+.153	2	13.86	-.05	DA wk	22.7
33	40 Eri B	9.52	-.047	+.310	+.360	4	9.52	+.03	DA	31.3
36	VR 7	14.27	-.058	+.275	+.201	2	14.29	-.02	DA	32
37	VR 16	14.05	-.096	+.245	+.011	2	14.02	-.09	DA	25
38	HZ 9	13.90	+.220	+.135	+.411	2	13.95	+.33	DAe comp.	30.2
39	HZ 7	14.11	-.097	+.276	+.117	2	14.18	-.03	DA	23.7
42	HZ 14	13.86	-.116	+.192	-.111	2	13.83	-.15	DA	20.7
50	He 3	12.05	-.082	+.261	+.092	2	12.10	-.08	DA n	28.8
182	GD 47-18	15.14	+.226	-.339	-.058	1	15.18	.00	λ 4670	—
64	GD 116-16	15.35	+.153	+.171	+.529	3	15.32	+.24	DA s	19.5
67	SA 29-130	13.29	-.005	+.317	+.549	3	13.32	+.07	DA	39.3
70	L 825-14	13.00	-.092	+.190	-.040	2	12.97	-.15	DA n	17.1
71	F 34	11.20	-.159	+.054	-.407	2	11.12	-.30	DO	3.3
74	L 898-25	14.27	+.231	+.022	+.529	1	14.28	+.32	DA	—
76	L 970-30	13.09	-.031	+.330	+.458	2	12.92	+.09	DA	35.1
79	R 627	14.20	+.287	+.038	+.614	1	14.24	+.31	DF	—
80	F 43	14.97	-.114	+.209	-.071	2	14.89	-.13	DA	21.4
184	GD 140	12.46	-.045	+.236	+.087	4	12.50	-.06	DA wk	—
81	F 46	13.30	-.129	+.074	-.312	2	13.24	-.30	DO-B	5.6
86	HZ 21	14.71	-.129	+.008	-.458	3	14.22	-.36	DO	0.8
91	HZ 29	14.11	-.061	+.166	-.132	4	14.18	-.23	DBp	—

table continues

TABLE I—(*cont.*)

E.G. no.	Other catalog no.	y	$b-y$	m_1	$u-b$	n	V	$B-V$	Sp	$W\gamma$
93	HZ 34	15.77	−.147	+.047	−.432	2	15.66	−.28	DO	4.1
187	GD 153	13.40	−.152	+.123	−.310	2	13.42	−.25	DA wk	—
98	HZ 43	12.67	+.010	+.007	−.280	4	12.86	−.10	DA wk comp.	9.2
99	W 485 A	12.34	−.031	+.369	+.489	4	12.30	+.08	DA	31.9
102	Gw 70° 5824	12.80	−.104	+.291	+.152	2	12.79	−.09	DA	35.2
107	F 93	15.37	−.137	+.151	−.210	1	15.33	−.23	DA wk	10.0
193	GD 190	14.72	−.068	+.154	−.063	1	14.72	−.10	DB	—
113	+1° 3129 B	15.19	−.118	+.069	−.349	2	15.27	−.32	DA wk	9.6
118	L 770-3	13.47	−.135	+.165	−.201	1	12.40	−.25	DA wk	18.7
197	G 169-34	14.14	+.134	+.176	+.556	1	14.08	+.24	DA ss	—
129	Gw 70° 8247	13.26	+.073	+.069	+.124	2	13.19	+.05	D pec	—
131	LDS 678 B	12.33	+.117	+.007	+.154	1	12.24	+.07	DA wk	2.3
133	L 1573-31	14.51	+.025	+.051	−.060	2	14.51	−.09	DB	—
139	W 1346	11.54	−.081	+.252	+.119	2	11.54	−.07	DA	27.0
147	Gw 82° 3818	12.94	−.050	+.297	+.321	2	13.02	−.02	DA	24.4
148	L 1363-3	13.27	+.206	−.007	+.343	2	13.23	+.17	DC	—
157	F 108	12.94	−.119	+.104	−.135	2	12.90	−.28	DA s	7.4
158	F 110	11.81	−.131	+.058	−.327	3	11.50	−.30	DO s	4.7

Both y and V measure the intensity of the continuum in a comparatively line free part of the spectrum and no difficulty is expected in effecting this transformation.

4. $(b - y)$ color index.

5. m_1 index, defined as being the color difference $(v - b) - (b - y)$.

6. $(u - b)$ color index.

7. Number of observations.

8, 9. V and $B - V$ magnitudes and color index according to Eggen and Greenstein.

10, 11. Spectral type and $H\gamma$ equivalent widths from the same source.

It is possible to estimate roughly the standard deviation of a single observation from the range of the repeated measurements of a particular star. Using all the stars in Table I which were observed more than once, the following standard deviations were computed: y:.03, $(b - y)$:.013, m_1:.018, c_1:.029. The values given in Table I are somewhat more precise since they are usually the mean of 2 or more observations.

In several cases, the 4-color measurements of visual magnitudes differ radically enough from those of Eggen and Greenstein to cause comment. These disagreements are indicated in Table II.

In cases E.G. 20 and 98 the disagreement may be due to the composite nature of the star. For E.G. 86 and E.G. 118, the error is evidently clerical. The discrepancy in the case of E.G. 118 has already been pointed out by

TABLE II* Disagreements between photometry of Graham and that of Eggen and Greenstein.

Star E.G. number	$y_{\text{Kitt Peak}}$	$V_{\text{Eggen–Greenstein}}$
19	12.79	12.65
20	12.39	12.25
76	13.09	12.92
86	14.71	14.22
93	15.77	15.66
98	12.67	12.86
118	13.47	12.40
158	11.81	11.50

* Dr. Eggen has kindly sent me some additional information on the white dwarf magnitudes. He writes: The (previously published) magnitudes of E.G. 98, 86, and 118 are misprints for 12.68, 14.72 and 13.40. New observations of E.G. 76 confirm the published (Eggen–Greenstein) value. Harris's result for E.G. 93 (see GREENSTEIN, *Ap. J.* **144**, 496) is 15.63.

F

Hardie (1967). The remaining stars deserve the attention of observers working with the U, B, V system since there is a possibility that some of the variations might be due to the stars themselves.

(*a*) *The reproducibility of the 4-color system for white dwarfs*

Although all the observations listed in Table I were made with the original set of 4-color filters, there are in use at Kitt Peak a number of other sets of filters which have been found to reproduce this system well for main-sequence and giant stars with spectral types between O and G. One of these additional sets (filter numbers 281, 271, 264, 350) has been used to make some observations of white dwarfs. As usual, the transformation equations relating the natural observed system to the Strömgren system were defined by the observations of a set of main-sequence and giant standard stars. For the hydrogen line DA-type white dwarfs it has proved very difficult to reproduce the system in this way. Systematic differences of the order of .02 to .04 magnitudes are found in each photometric index when a comparison is made with the observations carried out with Strömgren's filter set. In order to check the influence of the hydrogen line blocking on the $(b-y)$ index measured with each set, the white dwarf hydrogen line profiles given by Greenstein (1960) were superposed on the transmission curves of the b and y filters of each set. The total effect of the line blocking was very small and it does not seem at all possible to explain this discrepancy in this way unless the hydrogen lines are far broader than Greenstein's data indicate. Because of these difficulties, it seems that the techniques of photoelectric scanning may be more suitable for studying the radiation of white dwarfs since one can more easily define passbands of arbitrary position and width.

(*b*) *The m_1 index as a measure of hydrogen line intensity*

In the introduction, it was pointed out that, for white dwarfs, the Strömgren m_1 index is expected to be very sensitive to the strength of the $H\delta$ line. This can be verified by examining the correlation between m_1 and the $H\gamma$ equivalent widths given by Eggen and Greenstein, since it is likely that the $H\gamma$ and $H\delta$ strengths are very closely related. This has been done in Figure 1 for all DA- and DO-type stars with the data available. A well defined relation between the two quantities evidently exists. The scatter is no more than that expected from observational uncertainties. The two composite objects E.G. 38 and E.G. 98 deviate from the mean relation. The two DC stars in the sample denoted by crosses lie close to the origin as one would expect.

For most white dwarfs of either DO- or DA-types, the m_1 index is an excellent measure of hydrogen line intensity. In Figure 2, m_1 is plotted against the $(b-y)$ color index. At the time of writing only a few white dwarfs have

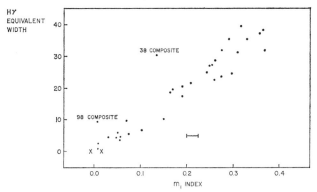

FIGURE 1 $H\gamma$ equivalent width in ångströms is plotted against m_1 index for the DA- and DO-type white dwarfs. The $H\gamma$ equivalent widths are taken from Eggen and Greenstein's papers. Two composite objects, E.G. 38 and E.G. 98 are indiacted. The crosses represent the two DC-type stars in the sample. The expected standard error of each m_1 value is shown by the bar.

been observed with positive $(b - y)$ but even so it is clear that m_1 reaches a maximum near $(b - y) = 0$. If the effective temperatures, given by Weidemann (1963) for stars in common, are used to calibrate the $(b - y)$ index in terms of effective temperature, it is found that this maximum occurs close to $11,500°$K

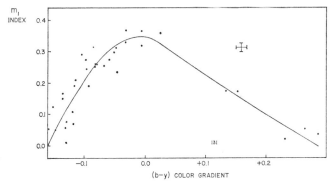

FIGURE 2 The hydrogen line index m_1 is plotted against the $(b - y)$ color gradient for the DA- and DO-type white dwarfs. The expected standard error of each point is shown by the bars. The peculiar star LDS 678B is shown by its E.G. number.

(c) *The two-color* $(u - b) - (b - y)$ *diagram*

The $(u - b)$ index, a measure of the Balmer discontinuity is plotted in Figure 3 against the $(b - y)$ color index. The white dwarfs of various spectral types are plotted with different notation. The positions of several peculiar objects are indicated by the appropriate Eggen–Greenstein catalogue numbers. An estimate of the relation applicable black bodies is shown by the upper straight line while the curve represents the main-sequence star relation given by Slettebak, Wright, and Graham (1968). The two continuous spectrum DC white dwarfs lie very close to the black-body relation while the peculiar stars tend to lie close to or above this line.

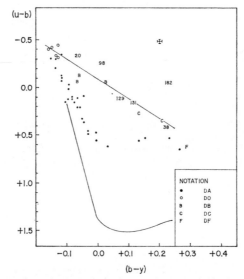

FIGURE 3 The two color $(u - b) - (b - y)$ diagram for all white dwarfs observed is shown. Different spectral classes are plotted with different notations while spectroscopically peculiar stars are represented by their Eggen–Greenstein numbers. The main-sequence curve and black-body line are also drawn. The bars again represent the expected standard error of each point.

There are a number of distinct differences between this diagram and the analogous one of the *U, B, V* photometric system (see, for example, Eggen and Greenstein (1965) Figure 6, and (1967) Figure 3). The DA-type stars, for example, lie along a very narrow sequence which is close to the main-sequence relation for $(b - y)$ negative but which becomes nearly horizontal, at $(u - b) = .550$ for positive $(b - y)$ to at least $(b - y) = +.200$. Some of the dispersion along this sequence is real. Stars E.G. 102 and E.G. 184, for example, seem certainly to have unusual colors. The scatter among the DO-type stars and the bluest DA-type stars is also significant and may be a result of including a few hot subdwarfs in the list. There is no certain indication that the 4-color system can distinguish between the bluest white dwarfs and the more luminous O-type subdwarfs.

Figure 3 suggests one potentially important application of intermediate band photometry to the study of white dwarfs. Because of the increased spectral resolution of the transmission bands, one can obtain more precise knowledge of the stellar energy distribution. With *uvby* photometry for example, models could possibly be fitted to the observed $(u - b)$ and $(b - y)$ colors for some classes of white dwarfs. The corresponding matching of U, B, V photometric indices should be much more difficult because of the large influence of the strong hydrogen lines on the observed colors. However, it should be remembered that the *uvby* system may not be entirely successful in this respect and some residual contamination of the transmission bands by the very broad hydrogen lines probably remains.

References

CRAWFORD, D. L. 1966, *Spectral Classification and Multicolor Photometry*. Eds. K. LODÉN, L. O. LODÉN, and U. SINNERSTAD. (New York, Academic Press), 170.

EGGEN, O. J., and GREENSTEIN, J. L. 1965, *Ap. J.*, **141**, 83.

EGGEN, O. J., and GREENSTEIN, J. L. 1967, *Ap. J.*, **150**, 927.

GREENSTEIN, J. L. 1960, *Stellar Atmospheres*. Ed. J. L. GREENSTEIN. (Chicago, University of Chicago Press), 676.

HARDIE, R. H. 1967, *Pub. A.S.P.*, **79**, 173.

SLETTEBAK, A., WRIGHT, R. R., and GRAHAM, J. A. 1968, *A.J.* **73**, 152.

STRÖMGREN, B. 1963, *Basic Astronomical Data*. Ed. K. AA. STRAND. (Chicago, University of Chicago Press), 123.

WEIDEMANN, V. 1963, *Zs. f. Ap.*, **57**, 87.

Discussion

WOOD   Did the faintness of the star prevent you from measuring the hydrogen lines in the white dwarfs with standard Hβ type photometry? What aperture telescope did you use for these observations?

GRAHAM   No, the filter is much too narrow. I used mostly the 84" telescope which takes you down to magnitude 16.

OSTRIKER   Would you be able to look for the possibly broad shallow hydrogen lines in a DC star to any better accuracy than can be done by the standard photoelectric techniques?

GRAHAM   I think one could probably do almost as well with scanning.

The continuous energy distribution of T Tauri stars

LEONARD V. KUHI

Department of Astronomy
University of California, Berkeley

T TAURI stars have now been accepted as stars still in the pre-main sequence gravitational contraction phase of stellar evolution. Their existence in the very young galactic cluster NGC 2264 where they clearly lie above the zero-age main sequence is perhaps the best confirmation of this view and of our ideas concerning the contraction phase. However, because their spectra have very strong emission lines of hydrogen and Ca II, and in addition have several unidentified sources of continuous emission (most notably the so-called ultraviolet excess), it has been rather difficult to locate these objects very precisely on the H – R diagram. The chief reason for this difficulty is the contamination of the *UBV* colors usually used to determine the intrinsic colors. In order to alleviate this situation an observing program was started at Lick Observatory in 1965 to measure the continuous energy distribution of T Tauri stars using a photoelectric scanner. A range of wavelengths from 3200 to 11,000 Å was measured using exit slits of 48 Å in the blue and 64 Å in the red. The wavelengths were chosen to be as free of emission lines as possible except where it was actually desired to measure the emission-line intensities themselves. These measurements were in turn tied in to standard stars and absolute fluxes per unit frequency via the new absolute calibration of Vega by D. Hayes.

Typical results are shown in Figures 1 to 3 (the abcissa is in reciprocal wavelengths in inverse microns, the ordinate is in mag. per unit frequency). Figure 1 shows T Tauri, *dK1e*, with no strong ultraviolet emission but still fairly red; Figure 2 is BP Tauri, *dK5e*, with moderate ultraviolet emission and quite red; and Figure 3 shows DF Tauri, *dM0e*, with very strong ultraviolet emission and also very red. Thus, we note that there is a general increase in the strength of the ultraviolet emission with later spectral type and that T Tauri stars are already quite red at wavelengths less than 1 micron. This

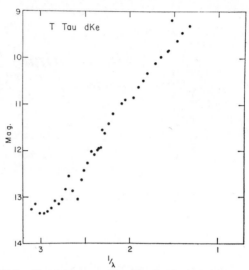

FIGURE 1   Continuous energy distribution of T Tauri (*dK1e*).

of course agrees very nicely with the recent work of Mendoza in which he finds large excesses of radiation in the 1 to 5 micron region and smaller excesses at shorter wavelengths. He also found the same correlation of ultraviolet excess with spectral type as determined from his wide-band measurements.

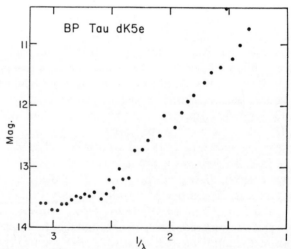

FIGURE 2   Continuous energy distribution of BP Tauri (*dK5e*).

What is this large red excess due to? Mendoza ruled out interstellar reddening as the sole cause because of the unreasonable absolute magnitudes needed (e.g. $M_v = -7$ for R Mon). Low and Smith then suggested that in the case of R Mon, at least, the excess may be the thermal emission from a circumstellar cloud which absorbs and reradiates the energy produced by the central star at a temperature characteristic of the cloud. The intrinsic variability of T Tauri stars provides a way in which to check the circumstellar origin of the red excess, whether it be due to reddening or thermal emission.

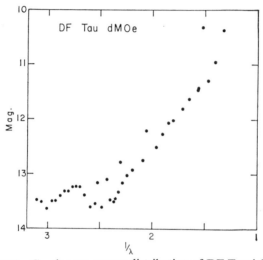

FIGURE 3 Continuous energy distribution of DF Tauri (*dM0e*).

For the stars showing very strong emission lines, we can determine the relative line strengths with some confidence since the underlying stellar absorption will be quite small as will any narrow shell component. Thus, we can plot the ratios of Hα to Hβ and Hβ to Hγ following Osterbrock *et al.* to separate the effects of reddening and optical thickness in the Balmer lines.

Figure 4 shows the observed intensity ratios for those T Tauri stars having an emission line index of 4 or 5 on Herbig's scale. The most interesting thing is the behavior of these ratios when the brightness of the star changes. If all the red excess were due to interstellar reddening alone then there should be no component of motion of the plotted point parallel to the reddening line. That is to say, changing the physical characteristics of the emitting region should have no effect on the portion of the line-intensity ratios contributed by interstellar reddening. However, this is clearly not the case: individual stars wander in all directions in the diagram as the emission line intensities change. In fact, this behavior is quite peculiar: for both AS 205 and AS 209 the star is brightest (at 5556 Å) when the Hα to Hβ ratio is largest, but for DF Tauri just the opposite is true. In either case, it is clear that the redness must be partly circumstellar in origin, since it is extremely unlikely that these changes can be attributed to the interstellar medium at any great distance from the star.

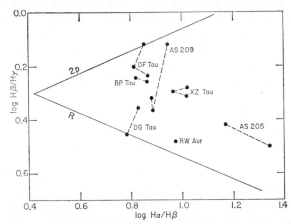

FIGURE 4   Observed intensity ratios for emission-line index of
4 or 5 on Herbig's scale.

We must also conclude that the mechanism for the production of the emission lines and the red excess are intimately connected. What this connection or mechanism is, however, will have to await further study.

Observations of rapidly pulsating radio sources

A. HEWISH

Cavendish Laboratory, University of Cambridge

I wish to present a brief summary of recent observations of pulsating sources which have been carried out at Cambridge. The initial work on the discovery of sources was made in conjunction with Dr. Pilkington and Miss Bell. Detailed investigation of the pulse period and structure has been carried out by Dr. Scott, Mr. Collins and Mr. Cole. Accurate positional measurements were made by Sir Martin Ryle, Miss Bailey and Mr. Mackay. First, it is worth pointing out that the discovery of these interesting objects depended upon the use of a high sensitivity, a very short time constant and repeated observations. We covered the survey area at least once per week, and even then it was several months before the first pulsating source was recognized with confidence.

Following the discovery of the pulsed nature of the radio emission in 1967, December, a search for additional sources was made and three others were found. The characteristics of all the sources are presented and they are, indeed, remarkably similar. We call the sources CP 0834, 0950, 1133 and 1919—a more satisfactory nomenclature than L.G.M. Of particular interest is CP 0950 which has the short period of 0.25 sec. All the sources exhibit a similar random variability and we are now sure that the most rapid variations are genuine source effects which cannot be ascribed to interplanetary scintillation. Frequency dispersion of the pulses indicates that the sources are distributed in a distance range 30–100 parsec if a mean electron density of 0.1 is assumed.

Accurate timing of the pulses has shown, in all cases, that Doppler variations of period can be accounted for by the Earth's motion alone. Further investigation indicates that there is no systematic difference between alternate pulses as might be expected on the binary neutron star model with gravitational focusing. Similarly, for a rotating white dwarf model, only one radiating source can be present on the star and this must have a lifetime of several months. Some observations of the radio frequency structure of the pulses indicates that CP 0950 radiates pulses in which the energy of a single pulse varies considerably within a bandwidth of 300 kHz, while the detailed structure within the pulse may change appreciably in a short time. CP 0950 appears to be more variable in this respect than the other three sources.

Positions of high accuracy have been derived for CP 1919 and CP 0950. The position of CP 1919 agrees well with that of a yellow star of about 18.5 mag.

Discussion

SAVEDOFF   Mr. Meltzer informs me that the neutron star models that have been picked out are very unlikely to be obtained physically, for the reason that these models are unstable. They are positive energy models. They are reached by a collection of neucleons that are compressed to a certain density and you have to do work on dispersed matter to bring them into this particular region in the diagram. They are stable for small perturbations, because if you decrease the density or increase the density you need more energy. So they are a local minimum and not an absolutely negative energy stage. We found by telephone inquiry that this was also known to Thorne.

The second point that I would like to make is that a electron density of 0.1 is very optimistic. A decade or more ago we were using figures like 10^{-3} for a typical H I region and perhaps 1 or .1 for a typical H II region. And I think this would move this object to a distance of about 6500 parsecs or 6100 parsecs. We have a star model which is near the white dwarf stage, and I think the periods would be much too long, and it would be much too bright. It has nuclear energy generation in it, its absolute visual magnitude would be about -5, and that would put the thing back to about 100,000 parsecs, which is ridiculous.

GREENSTEIN   I would like to make a comment concerning the electron density. I looked carefully at the field, and there is not a trace of emission. You see, this is not an ionized region. You may have noticed on your photograph that coming in from the top there is a great big blank. This is in a very heavily reddened region and therefore, it is unlikely to be at 6500 parsecs mainly because of this enormous absorpiton that would be there. There is a lot of dust and gas, but there is no H II region in the area. So, Dr. Savedoff's suggestion that one would have only residual electrons that come from either cosmic ray bombardment or carbon ionization would give a much better density. When you try to do that, however, for these others which are at high galactic latitudes, you run out of the galactic plane in about 100 parsecs and though I have not looked at these, I would be willing to bet that there are no H II regions in this area. So, you may have from these a maximum number of electrons indicated with considerable accuracy, unless they are out in the halo where we do not know the electron density.

SALPETER   I would like to report on a couple of pieces of information obtained by Drake and his group at Arecibo. One of the points is about the structure of the pulse. What seems fairly clear is that the pulses seem to have a width of about 37 ms or so, pretty closely the same at various frequencies. Drake and his co-workers have looked at four different frequencies ranging from 75 to 430 MHz, and the width is the same in all of them. There seems to be a tendency for a typical pulse to have a triple structure. Some time-scales are as short as about 4 ms which would indicate that not necessarily the whole source, but at least some part of the source is even smaller than the 5000 kilometers Dr. Hewish has just mentioned. The other point has to do with the dispersion which Dr. Hewish has mentioned. They have obtained a precision measurement for the number of electrons per square centimeter assuming the dispersion to be slight, and the data agree very well at all the frequencies. This indicates that not only is it the electron density, that makes up this number of electrons per square centimeter, less than the critical density for the lowest frequency, but it is much less. I do not know the exact numbers, but I hope that an upper limit to the electron density of something like 10^3 electrons per cc will eventually come out of this. Now, of course, that is much higher than what the density is in the interstellar medium; it is rather low compared to what you would expect for a corona surrounding a neutron star or a white dwarf. So these are the two observations I wanted to mention. On the theoretical side, there exists another possibility, which many theorists are grasping, and that is that it is the rotation rather than the pulsation of a neutron star that has the extra degree of freedom. A star can rotate at any period longer than one that would make the neutron star unstable, so therefore, having neutron stars with pulsation periods of milliseconds and rotation periods of the order of a second is certainly no great difficulty. Of course, the difficulty will be in inventing some clever mechanism of having a searchlight pointing from the neutron star and nicely collimated at the medium around in order to give these very sharp pulses.

EGGEN   Obviously, we have to wait for the optical data to become available, but I think you talked yourself out of the white dwarf too easily, for the only stated observed fact is that the proper motion is no larger than 0.05. And from that diagram that I showed you this morning, it could still be within 100 parsecs and have a transverse motion of 25 kilometers per second. I think that you are a little premature in throwing that away.

KRAFT   I would just like to emphasize once again that the star picked out by Professor Ryle is not very blue. In fact, if you just look at the Palomar prints you would say it is red, it is simply not as red as a lot of the stars in the field of similar magnitude which are extremely red. I would say that unless it is very heavily reddened it is not a particularly blue object.

OSTRIKER   One question for Dr. Salpeter and one statement on rotation. Models of Meltzer and Thorne are carried through to the end point of thermonuclear evolution, which is perhaps not a realistic interior state for white dwarfs. And they find a minimum period of about 8 seconds. Now the models that Salpeter and Hamada calculated have higher densities near the mass limit; and one would guess correspondingly shorter periods for them. Has anyone calculated the normal modes of oscillation, particularly the first radial mode, for those stars?

SALPETER   At least I have not; I do not know of anybody else who has done that.

OSTRIKER   One further point on the question of rotation. The rotation periods for neutron stars can be anything greater than milliseconds, but there would be no particular reason or fact to expect them to be in the range of seconds. The natural periods of rotation for the more massive ones, white dwarfs, on the contrary, fall into the range of seconds. In other words, if you take a massive white dwarf and more or less have an equilibrium of the kinetic energy and gravitational energy, then the period of rotation is of the order of one or two seconds.

CAMERON   In reply to this question about shorter periods, there are some recent models calculated by Thorne and Ipser, in which they looked at the vibrations of helium white dwarfs. In that case the fundamental period can come down to about 2.7 seconds which is shorter than what even the carbon can do, but still not short enough for this. But no investigation has been made of the kind of complex structure that you can have with variation in composition part way through, and I think that this sort of thing is needed before the last word has been said on the subject.

KIPPENHAHN   I just want to say that we have investigated, a white dwarf model which has a helium core and shell burning at the interface. The fundamental period that we get is something like 60 seconds, and Thorne and Ipser can get small periods only by assuming high overtones. And these high overtones are not at all excited by the normal mechanisms. We tried to find instability in our models, but we never succeeded. So, I think that the estimate in this paper by Thorne and Ipser are due to the wrong idea that the damping is inside and the excitation is outside, but actually it is just the other way around. The damping is in the atmosphere and the excitation is below where the amplitudes are low (see paper by Kippenhahn and Weigert in this volume).

Observation of a rapidly pulsating radio source*

A. HEWISH, S. J. BELL, J. D. H. PILKINGTON,
P. F. SCOTT, and R. A. COLLINS

Mullard Radio Astronomy Observatory
Cavendish Laboratory, University of Cambridge

Abstract

Unusual signals from pulsating radio sources have been recorded at the Mullard Radio Astronomy Observatory. The radiation seems to come from local objects within the galaxy, and may be associated with oscillations of white dwarf or neutron stars.

IN JULY 1967, a large radio telescope operating at a frequency of 81.5 MHz was brought into use at the Mullard Radio Astronomy Observatory. This instrument was designed to investigate the angular structure of compact radio sources by observing the scintillation caused by the irregular structure of the interplanetary medium.[1] The initial survey includes the whole sky in the declination range $-08° < \delta < 44°$ and this area is scanned once a week. A large fraction of the sky is thus under regular surveillance. Soon after the instrument was brought into operation it was noticed that signals which appeared at first to be weak sporadic interference were repeatedly observed at a fixed declination and right ascension; this result showed that the source could not be terrestrial in origin.

Systematic investigations were started in November and high speed records showed that the signals, when present, consisted of a series of pulses each lasting ~ 0.3 s and with a repetition period of about 1.337 s which was soon found to be maintained with extreme accuracy. Further observations have shown that the true period is constant to better than 1 part in 10^7 although there is a systematic variation which can be ascribed to the orbital motion of

* Reprinted from *Nature*, **217**, 709–713 (1968), by permission of the publishers.

the Earth. The impulsive nature of the recorded signals is caused by the periodic passage of a signal of descending frequency through the 1 MHz pass band of the receiver.

The remarkable nature of these signals at first suggested an origin in terms of man-made transmissions which might arise from deep space probes, planetary radar or the reflexion of terrestrial signals from the Moon. None of these interpretations can, however, be accepted because the absence of any parallax shows that the source lies far outside the solar system. A preliminary search for further pulsating sources has already revealed the presence of three others having remarkably similar properties which suggests that this type of source may be relatively common at a low flux density. A tentative explanation of these unusual sources in terms of the stable oscillations of white dwarf or neutron stars is proposed.

POSITION AND FLUX DENSITY

The aerial consists of a rectangular array containing 2048 full-wave dipoles arranged in sixteen rows of 128 elements. Each row is 470 m long in an E–W direction and the N–S extent of the array is 45 m. Phase-scanning is employed to direct the reception pattern in declination and four receivers are used so that four different declinations may be observed simultaneously. Phase-switching receivers are employed and the two halves of the aerial are combined as an E–W interferometer. Each row of dipole elements is backed by a tilted reflecting screen so that maximum sensitivity is obtained at a declination of approximatly $+30°$, the overall sensitivity being reduced by more than one-half when the beam is scanned to declinations above $+90°$ and below $-5°$. The beamwidth of the array to half intensity is about $\pm\frac{1}{2}°$ in right ascension and $\pm3°$ in declination; the phasing arrangement is designed to produce beams at roughly $3°$ intervals in declination. The receivers have a bandwidth of 1 MHz centered at a frequency of 81.5 MHz and routine recordings are made with a time constant of 0.1 s; the r.m.s. noise fluctuations correspond to a flux density of 0.5×10^{-26} W m^{-2} Hz^{-1}. For detailed studies of the pulsating source a time constant of 0.05 s was usually employed and the signals were displayed on a multi-channel 'Rapidgraph' pen recorder with a time constant of 0.03 s. Accurate timing of the pulses was achieved by recording second pips derived from the *MSF* Rugby time transmissions.

A record obtained when the pulsating source was unusually strong is shown in Figure 1a. This clearly displays the regular periodicity and also the characteristic irregular variation of pulse amplitude. On this occasion the largest pulses approached a peak flux density (averaged over the 1 MHz pass band) of 20×10^{-26} W m^{-2} Hz^{-1}, although the mean flux density integrated over one minute only amounted to approximately 1.0×10^{-26} W m^{-2} Hz^{-1}. On a more typical occasion the integrated flux density would be several times smaller than this value. It is therefore not surprising that the source has not been detected in the past, for the integrated flux density falls well below the limit of previous surveys at meter wavelengths.

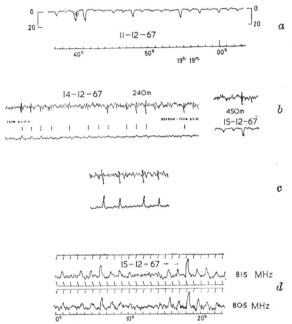

FIGURE 1 (a) A record of the pulsating radio source in strong signal conditions (receiver time constant 0.1 s). Full scale deflexion corresponds to 20×10^{-26} W m^{-2} Hz^{-1}. (b) Upper trace: records obtained with additional paths (240 m and 450 m) in one side of the interferometer. Lower trace: normal interferometer records. (The pulses are small for $l = 240$ m because they occurred near a null in the interference pattern; this modifies the phase but not the amplitude of the oscillatory response on the upper trace.) (c) Simulated pulses obtained using a signal generator. (d) Simultaneous reception of pulses using identical receivers tuned to different frequencies. Pulses at the lower frequency are delayed by about 0.2 s.

The position of the source in right ascension is readily obtained from an accurate measurement of the 'crossover' points of the interference pattern on those occasions when the pulses were strong throughout an interval embracing such a point. The collimation error of the instrument was determined from a similar measurement on the neighbouring source 3C 409 which transits about 52 min later. On the routine recordings which first revealed the source the reading accuracy was only ± 10 s and the earliest record suitable for position measurement was obtained on August 13, 1967. This and all subsequent measurements agree within the error limits. The position in declination is not so well determined and relies on the relative amplitudes of the signals obtained when the reception pattern is centered on declinations of 20°, 23° and 26°. Combining the measurements yields a position

$$\alpha_{1950} = 19\text{h } 19\text{m } 38\text{s} \pm 3\text{s}$$
$$\delta_{1950} = 22° \ 00' \pm 30'$$

As discussed here, the measurement of the Doppler shift in the observed frequency of the pulses due to the Earth's orbital motion provides an alternative estimate of the declination. Observations throughout one year should yield an accuracy of $\pm 1'$. The value currently attained from observations during December–January is $\delta = 21° \ 58' \pm 30'$, a figure consistent with the previous measurement.

TIME VARIATIONS

It was mentioned earlier that the signals vary considerably in strength from day to day and, typically, they are only present for about 1 min, which may occur quite randomly within the 4 min interval permitted by the reception pattern. In addition, as shown in Figure 1a, the pulse amplitude may vary considerably on a time-scale of seconds. The pulse to pulse variations may possibly be explained in terms of interplanetary scintillation,[1] but this cannot account for the minute to minute variation of mean pulse amplitude. Continuous observations over periods of 30 min have been made by tracking the source with an E–W phased array in a 470 m × 20 m reflector normally used for a lunar occultation programme. The peak pulse amplitude averaged over ten successive pulses for a period of 30 min is shown in Figure 2a. This plot suggests the possibility of periodicities of a few minutes' duration, but a correlation analysis yields no significant result. If the signals were linearly polarized, Faraday rotation in the ionosphere might cause the random variations, but the form of the curve does not seem compatible with this mechanism. The day to day variations since the source was first detected are shown in Figure 2b. In this analysis the daily value plotted is the peak flux density of the greatest pulse. Again the variation from day to day is irregular and no systematic changes are clearly evident, although there is a suggestion that the source was significantly weaker during October to November. It therefore appears that, despite the regular occurrence of the pulses, the magnitude of the power emitted exhibits variations over long and short periods.

INSTANTANEOUS BANDWIDTH AND FREQUENCY DRIFT

Two different experiments have shown that the pulses are caused by a narrow-band signal of descending frequency sweeping through the 1 MHz band of the receiver. In the first, two identical receivers were used, tuned to frequencies of 80.5 MHz and 81.5 MHz. Figure 1d, which illustrates a record made with this system, shows that the lower frequency pulses are delayed by about 0.2 s. This corresponds to a frequency drift of ~ -5 MHz s^{-1}. In the second method a time delay was introduced into the signals reaching the receiver from one-half of the aerial by incorporating an extra cable of known length l. This cable introduces a phase shift proportional to frequency so that, for a signal the coherence length of which exceeds l, the output of the receiver will oscillate with period

$$t_0 = \frac{c}{l}\left(\frac{dv}{dt}\right)^{-1}$$

where dv/dt is the rate of change of signal frequency. Records obtained with $l = 240$ m and 450 m are shown in Figure 1b together with a simultaneous record of the pulses derived from a separate phase-switching receiver operating with equal cables in the usual fashion. Also shown, in Figure 1c, is a simulated record obtained with exactly the same arrangement but using a

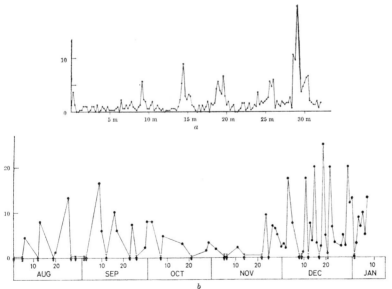

FIGURE 2 (a) The time variation of the smoothed (over ten pulses) pulse amplitude. (b) Daily variation of peak pulse amplitude. (Ordinates are in units of W m^{-2} Hz^{-1} $\times 10^{-26}$.)

signal generator, instead of the source, to provide the swept frequency. For observation with $l > 450$ m the periodic oscillations were slowed down to a low frequency by an additional phase-shifting device in order to prevent severe attenuation of the output signal by the time constant of the receiver. The rate of change of signal frequency has been deduced from the additional phase shift required and is $dv/dt = -4.9 \pm 0.5$ MHz s^{-1}. The direction of the frequency drift can be obtained from the phase of the oscillation on the record and is found to be from high to low frequency in agreement with the first result.

The instantaneous bandwidth of the signal may also be obtained from records of the type shown in Figure 1b because the oscillatory response as a function of delay is a measure of the autocorrelation function, and hence of the Fourier transform, of the power spectrum of the radiation. The results of the measurements are displayed in Figure 3 from which the instantaneous bandwidth of the signal to exp (-1), assuming a Gaussian energy spectrum, is estimated to be 80 ± 20 kHz.

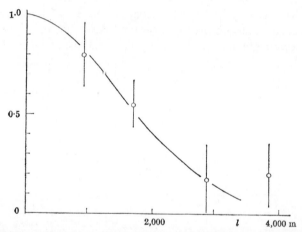

FIGURE 3   The response as a function of added path in one side of the interferometer.

PULSE RECURRENCE FREQUENCY AND DOPPLER SHIFT

By displaying the pulses and time pips from *MSF* Rugby on the same record the leading edge of a pulse of reasonable size may be timed to an accuracy of about 0.1 s. Observations over a period of 6 h taken with the tracking system mentioned earlier gave the period between pulses as $P_{obs} = 1.33733 \pm 0.00001$ s. This represents a mean value centered on December 18, 1967, at 14 h 18 m UT. A study of the systematic shift in the frequency of the pulses was obtained from daily measurements of the time interval T between a standard time and the pulse immediately following it as shown in Figure 4. The standard time was chosen to be 14 h 01 m 00 s UT on December 11 (corresponding to the centre of the reception pattern) and subsequent standard times were at intervals of 23 h 56 m 04 s (approximately one sidereal day). A plot of the variation of T from day to day is shown in Figure 4. A constant pulse recurrence

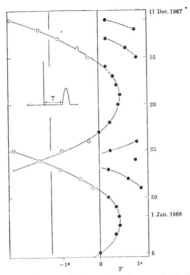

FIGURE 4   The day to day variation of pulse arrival time.

frequency would show a linear increase or decrease in T if care was taken to add or subtract one period where necessary. The observations, however, show a marked curvature in the sense of a steadily increasing frequency. If we assume a Doppler shift due to the Earth alone, then the number of pulses received per day is given by

$$N = N_0 \left(1 + \frac{v}{c} \cos \phi \sin \frac{2\pi n}{366.25}\right)$$

where N_0 is the number of pulses emitted per day at the source, v the orbital velocity of the Earth, ϕ the ecliptic latitude of the source and n an arbitrary day number obtained by putting $n = 0$ on January 17, 1968, when the Earth has zero velocity along the line of sight to the source. This relation is approximate since it assumes a circular orbit for the Earth and the origin $n = 0$ is not exact, but it serves to show that the increase of N observed can be explained by the Earth's motion alone within the accuracy currently attainable. For this purpose it is convenient to estimate the values of n for which $\delta T/\delta n = 0$, corresponding to an exactly integral value of N. These occur at $n_1 = 15.8 \pm 0.1$ and $n_2 = 28.7 \pm 0.1$, and since N is increased by exactly one pulse between these dates we have

$$ 1 = \frac{N_0 v}{c} \cos \phi \left[\sin \frac{2\pi n_2}{366.25} - \sin \frac{2\pi n_1}{366.25} \right] $$

This yields $\phi = 43° \, 36' \pm 30'$ which corresponds to a declination of $21°$ $58' \pm 30'$, a value consistent with the declination obtained directly. The true periodicity of the source, making allowance for the Doppler shift and using the integral condition to refine the calculation, is then

$$ P_0 = 1.3372795 \pm 0.0000020 \text{ s} $$

By continuing observations of the time of occurrence of the pulses for a year it should be possible to establish the constancy of N_0 to about 1 part in 3×10^8. If N_0 is indeed constant, then the declination of the source may be estimated to an accuracy of $\pm 1'$; this result will not be affected by ionospheric refraction.

It is also interesting to note the possibility of detecting a variable Doppler shift caused by the motion of the source itself. Such an effect might arise if the source formed one component of a binary system, or if the signals were associated with a planet in orbit about some parent star. For the present, the systematic increase of N is regular to about 1 part in 2×10^7 so that there is no evidence for an additional orbital motion comparable with that of the Earth.

THE NATURE OF THE RADIO SOURCE

The lack of any parallax greater than about $2'$ places the source at a distance exceeding 10^3 A.U. The energy emitted by the source during a single pulse, integrated over 1 MHz at 81.5 MHz, therefore reaches a value which must exceed 10^{17} erg if the source radiates isotropically. It is also possible to derive an upper limit to the physical dimension of the source. The small instantaneous bandwidth of the signal (80 kHz) and the rate of sweep (-4.9 MHz s^{-1}) show that the duration of the emission at any given frequency does not exceed 0.016 s. The source size therefore cannot exceed 4.8×10^3 km.

An upper limit to the distance of the source may be derived from the observed rate of frequency sweep since impulsive radiation, whatever its origin, will be dispersed during its passage through the ionized hydrogen in interstellar space. For a uniform plasma the frequency drift caused by dispersion is given by

$$\frac{d\nu}{dt} = -\frac{c}{L}\frac{\nu^3}{\nu_p^2}$$

where L is the path and ν_p the plasma frequency. Assuming a mean density of 0.2 electron cm^{-3} the observed frequency drift (-4.9 MHz s^{-1}) corresponds to $L \sim 65$ parsec. Some frequency dispersion may, of course, arise in the source itself; in this case the dispersion in the interstellar medium must be smaller so that the value of L is an upper limit. While the interstellar electron density in the vicinity of the Sun is not well known, this result is important in showing that the pulsating radio sources so far detected must be local objects on a galactic distance scale.

The positional accuracy so far obtained does not permit any serious attempt at optical identification. The search area, which lies close to the galactic plane, includes two twelfth magnitude stars and a large number of weaker objects. In the absence of further data, only the most tentative suggestion to account for these remarkable sources can be made.

The most significant feature to be accounted for is the extreme regularity of the pulses. This suggests an origin in terms of the pulsation of an entire star, rather than some more localized disturbance in a stellar atmosphere. In this connexion it is interesting to note that it has already been suggested[2,3] that the radial pulsation of neutron stars may play an important part in the history of supernovae and supernova remnants.

A discussion of the normal modes of radial pulsation of compact stars has recently been given by Meltzer and Thorne,[4] who calculated the periods for stars with central densities in the range 10^5 to 10^{19} g cm^{-3}. Figure 4 of their paper indicates two possibilities which might account for the observed periods of the order 1 s. At a density of 10^7 g cm^{-3}, corresponding to a white dwarf star, the fundamental mode reaches a minimum period of about 8 s; at a slightly higher density the period increases again as the system tends towards gravitational collapse to a neutron star. While the fundamental period is not small enough to account for the observations the higher order modes have periods of the correct order of magnitude. If this model is adopted it is difficult to understand why the fundamental period is not dominant; such a period would have readily been detected in the present observations and its absence cannot be ascribed to observational effects. The alternative possibility occurs at a density of 10^{13} g cm^{-3}, corresponding to a neutron star; at this density the fundamental has a period of about 1 s, while for densities in excess of 10^{13} g cm^{-3} the period rapidly decreases to about 10^{-3} s.

If the radiation is to be associated with the radial pulsation of a white dwarf or neutron star there seem to be several mechanisms which could account for the radio emission. It has been suggested that radial pulsation would generate hydromagnetic shock fronts at the stellar surface which might be accompanied by bursts of X-rays and energetic electrons.[2,3] The radiation might then be likened to radio bursts from a solar flare occurring over the entire star during each cycle of the oscillation. Such a model would be in fair agreement with the upper limit of $\sim 5 \times 10^3$ km for the dimension of the source, which compares with the mean value of 9×10^3 km quoted for white dwarf stars by Greenstein.[5] The energy requirement for this model may be roughly estimated by noting that the total energy emitted in a 1 MHz band by a type III solar burst would produce a radio flux of the right order if the source were at a distance of $\sim 10^3$ A.U. If it is assumed that the radio energy may be related to the total flare energy ($\sim 10^{32}$ erg)[6] in the same manner as for a solar flare and supposing that each pulse corresponds to one flare, the required energy would be $\sim 10^{39}$ erg yr^{-1}; at a distance of 65 pc the corresponding value would be $\sim 10^{47}$ erg yr^{-1}. It has been estimated that a neutron star may contain $\sim 10^{51}$ erg in vibrational modes so the energy requirement does not appear unreasonable, although other damping mechanisms are likely to be important when considering the lifetime of the source.[4]

The swept frequency characteristic of the radiation is reminiscent of type II and type III solar bursts, but it seems unlikely that it is caused in the same way. For a white dwarf or neutron star the scale height of any atmosphere is small and a travelling disturbance would be expected to produce a much faster frequency drift than is actually observed. As has been mentioned, a more likely possibility is that the impulsive radiation suffers dispersion during its passage through the interstellar medium.

More observational evidence is clearly needed in order to gain a better understanding of this strange new class of radio source. If the suggested origin of the radiation is confirmed further study may be expected to throw valuable light on the behaviour of compact stars and also on the properties of matter at high density.

We thank Professor Sir Martin Ryle, Dr. J. E. Baldwin, Dr. P. A. G. Scheuer and Dr. J. R. Shakeshaft for helpful discussions and the Science Research Council who financed this work. One of us (S. J. B.) thanks the Ministry of Education of Northern Ireland and another (R. A. C.) the SRC for a maintenance award; J. D. H. P. thanks ICI for a research fellowship.

References

1. HEWISH, A., SCOTT, P. F., and WILLS, D. *Nature*, **203**, 1214 (1964)
2. CAMERON, A. G. W. 1965, *Nature*, **205**, 787.

3. FINZI, A. 1965, *Phys. Rev. Lett.*, **15**, 599.
4. MELTZER, D. W., and THORNE, K. S. 1966, *Ap. J.*, **145**, 514.
5. GREENSTEIN, J. L. 1958, in *Handbuch der Physik. L.*, 161.
6. FICHTEL, C. E., and McDONALD, F. B. 1967, in *Annual Review of Astronomy and Astrophysics*, **5**, 351.

Observations of some further pulsed radio sources*

J. D. H. PILKINGTON, A. HEWISH,
S. J. BELL, and T. W. COLE

Mullard Radio Astronomy Observatory
Cavendish Laboratory, University of Cambridge

Abstract

Details are now given of three of the four pulsating radio sources discovered at Cambridge.

IN A RECENT communication[1] an account was given of the discovery of a new class of radio source characterized by the emission of short pulses of radiation having an extremely constant repetition frequency. The records on which the source was first detected were taken during a survey for the investigation of compact radio sources using the method of interplanetary scintillation. Following the recognition of the first pulsed source the survey records, which covered the region $-08° < \delta < 44°$, were examined for evidence of further similar sources. Where these records indicated that the intensity fluctuations of a particular source were more impulsive than those caused by interplanetary scintillation, further observations were made. These led to the discovery of three additional pulsed sources. Even though each area of sky was observed on about twenty separate occasions during this survey, the large day-to-day variations of flux density from the known sources indicate that this program should not be regarded as an exhaustive search of the entire region, and observations are continuing.

The three sources and the original one have been given numbers of the form *CP*.1919 to indicate the Cambridge pulsed source at $\alpha = 19^h 19^m$.

The three additional sources emit pulses which are remarkably similar to those from the first source, and their characteristics have been obtained by

* Reprinted from *Nature*, **218**, 126–129 (1968), by permission of the publishers.

similar methods. Examples of the observed pulses are shown in Figures 1 and 2. All the measurements were made at a frequency of 81.5 MHz with a bandwidth of 1 MHz, using the 470 m × 45 m north–south phased array and the 470 m × 20 m east–west phased array at the Mullard Radio Astronomy Observatory.

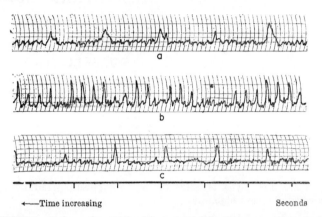

←——Time increasing   Seconds

FIGURE 1   Pulses observed with a recording time constant of about 0.03 s on March 21, 1968, (a) *CP.0834.* (b) *CP.0950,* during a period of intense activity. (c) *CP.1133.*

The approximate positions of the sources were derived from the response of the aerial systems. The position of *CP.0950* has been determined with greater precision from observations at 408 MHz with the one-mile telescope (following communication) using the same method as previously employed[2] for locating *CP.1919.* The best available positions for all four sources are given in Table 1.

The periodicity of the pulses was determined in the same way as previously reported.[1] The time of occurrence of a pulse at approximately the same sidereal time each day was determined, and the incremental time interval between this pulse and a standard time differing by successive units of 23 h 56 m 04 s was plotted as shown in Figure 3.

In this way the variation of periodicity caused by the motion of the Earth has been compared with the expected variation, and found to be consistent in each case. The true periods P_0 in solar seconds are given in Table 1.

The rate of change of frequency during the pulse and the intrinsic pulse duration have been derived, as before, from observations in which the signal from one aerial of the interferometer was delayed relative to the other by passage through an additional length of cable. In the case of *CP.0950* it was

TABLE I Characteristics of the four pulsed radio sources.

	CP.0834	CP.0950	CP.1133	CP.1919
α(1950.0)	$09^h\,34^m\,07^s \pm 15^s$	$09^h\,50^m\,28^s.95 \pm 0^s.7$	$11^h\,33^m\,32^s \pm 20^s$	$19^h\,19^m\,37^s.0 \pm 0^s.2$
δ(1950.0)	$07°\,00' \pm 45'$	$08°\,10' \pm 1'$	$17°\,00' \pm 45'$	$21°\,47'\,02'' \pm 10''$
P_0 (s)	1.27379 ± 0.00008	0.253071 ± 0.000008	1.1880 ± 0.0004	1.3372795 ± 0.0000020
$-(d\nu/dt)$ at 81.5 M Hz (M Hz s^{-1})	5.3 ± 0.5	20 ± 5	11 ± 3	5.15 ± 0.03 (ref.6)
Integrated electron density Nl (cm^{-3} pc)	12 ± 1	3.2 ± 0.8	6 ± 2	12.55 ± 0.06 (ref.6)
Emitted pulse duration (Gaussian) (ms)	35 ± 10	<10	12 ± 4	16 ± 4
Mean flux density at 81.5 M Hz (10^{-26} W m^{-2} Hz^{-1})	0.3	0.8	0.3	0.4
ℓ"	220°	230°	240°	56°
b"	26°	44°	70°	4°

found that the results were often inconsistent with the emission of a broad-band pulse; individual pulses sometimes occurred in which nearly all the energy received within a receiver bandwidth of 1 MHz was confined to a band of less than 0.3 MHz. This feature will be discussed in more detail later.[3]

The frequency sweep can be interpreted in terms of dispersion in the intervening medium, and the integrated electron density Nl is given in Table 1. The smaller dispersion of $CP.0950$ suggests that for an assumed value of $N \sim 0.1$–0.2 cm^{-3} the distance is likely to be only 15–30 pc.

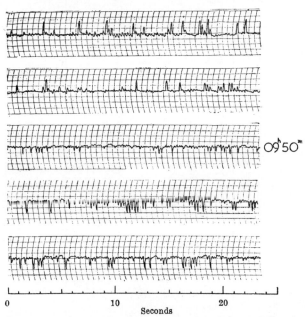

FIGURE 2   A continuous set of observations of $CP.0950$ at a time of strong activity, showing the irregular variations of intensity from pulse to pulse. The deflexions change sign as the source moves through the interference pattern of the aerial. Recording time constant 0.1 s.

The short pulse lengths of all the sources indicate physical dimensions in the range 3000–10,000 km. The variations in the peak pulse amplitude observed from day to day are shown in Figure 4.

The similarity in the quantities given in Table I shows that pulsating sources are, indeed, a new class of object in which the intrinsic powers, pulse widths, variability and periodicities are similar. Although $CP.0950$ has a periodicity somewhat shorter than the other three, and it would be difficult to detect still shorter pulse periods with the present system, there are no observational selection effects which would reduce the probability of detecting sources having periodicities as long as 10 s. The limited range in periodicity clearly has great significance in relation to the nature of the sources, and the shorter period of *CP.0950* may make it more difficult to account for them without invoking the very high densities of neutron stars.

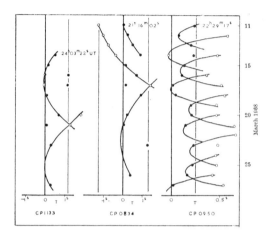

FIGURE 3 The day-to-day variation of pulse arrival time.

All the sources are characterized by a period of emission which is very much shorter than the repetition period and which occurs at a precisely defined phase of the cycle. This places limitations on the eccentricity of the orbits in a binary neutron star model[4] because of precession of the perihelion; in the case of *CP.0950* the eccentricity must be less than about 0.1.

All the sources show an extremely variable flux density both from pulse to pulse and on a longer time scale. In the first communication[1] it was suggested that the rapid variation from pulse to pulse might be caused by interplanetary scintillation. The recent observations were, however, carried out during the night when interplanetary scintillation is known to be small, and it seems that the rapid variations in flux density must be interpreted in terms of the source; the pulse to pulse variation cannot be attributed to interstellar scintillation (unpublished results of P. A. G. Scheuer).

Using the approximate distances given by the measured dispersion, we may conclude that the local density of pulsating sources is $\sim 10^{-5}$ pc^{-3}. The lifetime during which radio pulses are emitted from each source will depend on the supply of energy and on the mechanism by which it is converted into radio pulses; there may be many other 'dead' sources which are no longer

G

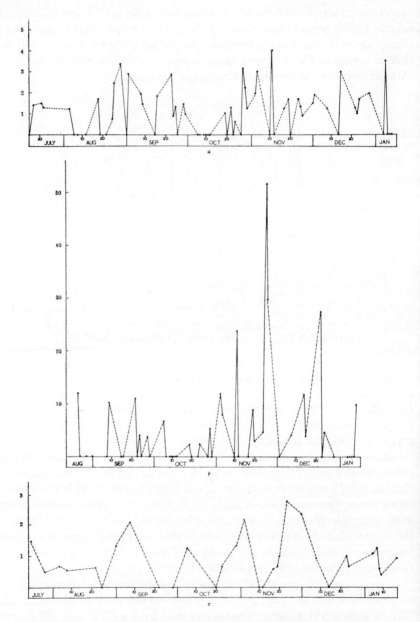

FIGURE 4 The day-to-day variation of peak pulse energy. The ordinates are in units of 10^{-26} J m^{-3} Hz^{-1}. (a) *CP*.0834, (b) *CP*.0950 (c) *CP*.1133.

observable. If the sources are associated with superdense stars of mass $\sim M_\odot$ and if they emit radio pulses for T years, the total number which have occurred over a period of $\sim 10^{10}$ years will give a local density of matter of

$$\sim \frac{10^{10}}{T}.10^{-5} \ M_\odot \ pc^{-3}$$

Oort[5] has derived an upper limit of $0.14 \ M_\odot \ pc^{-3}$ for all matter in the solar neighbourhood, and we must therefore conclude that the radio pulses must be emitted for a period of at least 6×10^5 years.

If the source of energy is the vibrational modes of a white dwarf or neutron star, a conversion efficiency much higher than those occurring in solar flares must occur. Alternatively some other source model must be found in which a larger supply of energy is available.

We thank Professor Sir Martin Ryle for his encouragement and Dr. P. A. G. Scheuer for helpful discussions. One of us (J. D. H. P.) thanks ICI for a research fellowship; S. J. B. and T. W. C. acknowledge maintenance awards from the Northern Ireland Ministry of Education and CSIRO, respectively.

References
1. HEWISH, A., BELL, S. J., PILKINGTON, J. D. H., SCOTT, P. F., and COLLINS, R. A., Paper 1–14 in these proceedings (p. 159); *Nature*, **217**, 709 (1968).
2. RYLE, M., and BAILEY, J. A., *Nature*, **217**, 907 (1968).
3. SCOTT, P. F., and COLLINS, R. A., *Nature* (in the press).
4. SASLAW, W. C., FAULKNER, J., and STRITTMATTER, P. A., *Nature*, **217**, 1222 (1968).
5. OORT, J. H., *Bull. Astro. Inst. Netherlands*, **15**, 45 (1960).
6. DAVIES, J. G., HORTON, P. W., LYNE, A. G., RICKETT, B. J., and SMITH, F. G., *Nature*, **217**, 910 (1968).

Section II

Observations of nebular variables and flare stars in young clusters

L. ROSINO

Astrophysical Observatory of Asiago
University of Padua, Italy

Abstract

An extended survey of nebular variables and flare stars in young clusters has been carried out during the last years at the Astrophysical Observatory of Asiago, using a 122 cm parabolic telescope and a 67–92–215 cm Schmidt telescope. The fields under observations are centered in: (a) Orion Trapezium; (b) ξ Orionis; (c) Pleiades and the Praesepe; (d) h and χ Persei; (e) NGC 7635 and NGC 6530. Many new nebular variables and flare stars have been found. Preliminary results of the survey are presented here.

I. INTRODUCTION

A SHORT ACCOUNT is given of the researches on irregular variable stars of low luminosity (henceforward called RW Aurigae or nebular variables) associated with young clusters involved in nebulosities, which have been carried out and are still in progress at the Astrophysical Observatory of Asiago of the University of Padua. The researches are directed to ascertain: (1) the relative number of nebular variables in stellar aggregates of different age and distance, the character of their light curves in blue and infrared, their distribution and extension over the field; (2) the presence of flare stars in the aggregates, their physical properties and relation with the RW Aurigae variables.

For the study of distant aggregates or for the central compact parts of clusters and nebulae, the 122 cm parabolic telescope was employed. Large field surveys were made at first with a 40 cm, f:2.5 Schmidt telescope; later on, however, the whole program was carried out by a new Schmidt of 67 cm, f:3.3 covering a field of 25 square degrees, which proved to be the best instru-

181

ment for this kind of work. 103a–0 plates and films were used for the blue, 1N + RG5 for the infrared. The infrared plates were hyper-sensitized by pre-flash in order to avoid the formation of white spots which are so frequent with ammonia.

The Asiago survey for nebular variables began with a systematic search on the Orion aggregate, which was later extended to an increasing number of young clusters and associations embedded in dark and bright nebulosities. The principal result of this survey was that emission-line stars and nebular variables (including flare stars) are always found in young aggregates when proper means and techniques are used.

Table I gives a list of fields extensively observed by the writer and his associates at Asiago and summarizes some of the results obtained. Other fields have been examined by P. Maffei (1963, 1965, 1966) and some more are now under observation, but they will not be discussed here.

In Table I the successive columns give for each object: position, variable stars recently found at Asiago, total number of variables (flare stars excluded), uncorrected distance modulus, mean photographic magnitude of variables at the top of the frequency distribution curve, total number of flares known to 1965 (Haro 1965), new flares found by the writer and associates after 1965. We pass on now to a more detailed discussion.

II. ORION AGGREGATE IN THE TRAPEZIUM AND NEAR ξ ORIONIS

The Trapezium area observed at Asiago covers 25 sq. degr. including the nebulae NGC 1976, 1977, 1982 and 1999. Maps and an extensive catalogue of stars of this region are found in Parenago's classical work on Orion Nebula (1954). The survey work on the nebular variables made at Loiano and Asiago prior to 1962 has been published already (Rosino 1946, 1956, 1962; Maffei 1963).

After 1962, forty-two new variable stars have been found by the writer all over the field. Twenty-two are concentrated around the Trapezium (Zone E); 25 are visible only in infrared ($m_{pg} > 18.5$). The total number of variables (flare stars not included) in the field is more than 456. Positions and charts of the new variables will be given in a forthcoming paper. Figure 1 shows the distribution of the variables included in the Catalogues of Kukarkin and Parenago and Suppl.: the density is particularly high near the Trapezium cluster, where 70 per cent of the stars are found to be variable, and along the line going from NGC 1999 to NGC 1977, north of the Trapezium, and continuing towards the Horsehead nebula. As observed elsewhere (Rosino 1962), in heavily obscured regions the variables are mostly found near the fringes of the dark nebulosity.

TABLE I Nebular fields observed at Asiago.

Field	R.A. 1950	D.	New neb. var.	Total number var.	Dist. mod.	m_p	Flares (1965)	New-Asiago flares
Orion Trapezium	5ʰ 33ᵐ	−5°25'	42	456	8.1	16.5	183	39
ζ Orionis	5 40	−2 20	30	86	8.0	16.5	13	—
NGC 2264	6 38	+9 57	—	143	10.0	17.0	*	—
NGC 7023	21 1	+67 58	13	18	7.5	17.5	—	—
NGC 7635	23 18	+60 54	5	13	7.5?	18	—	—
NGC 6530	18 2	−24 20	33	102	11.5	(18.5)	—	—
h and χ Per	2 15	+56 55	4	8	11.8	18	—	—
Pleiades	3 44	+23 58	—	—	5.5	—	33	13
Praesepe	8 37	+19 52	—	—	6.0	—	7	2

* Some flares in members of the association near NGC 7203 have recently been discovered at Bjurakan Observatory (E. PARSIMIAN, *Budapest Intern. Colloq. Variable Stars*, 1968).

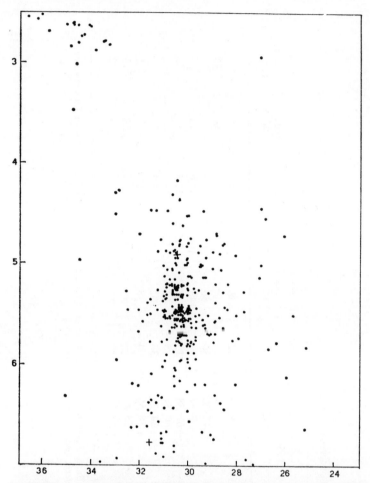

FIGURE 1 Distribution of RW Aur variables in the field of the Orion nebula. (Vertical axis is negative dec.; central number (30) on horizontal axis is 5^h30^m for R.A.)

Considering the light curves, a rough subdivision can be made between variables which are most of the time at minimum with sudden U-Gem like outbursts (not flares) and variables which are most of the time near maximum or at an intermediate magnitude, with fluctuations of brightness and a tendency to show irregular deep minima somewhat similar to those of eclipsing variables, but without any definite periodicity. Stars with erratic variations without preference for any level are also observed, so that all four Parenago classes are confirmed.

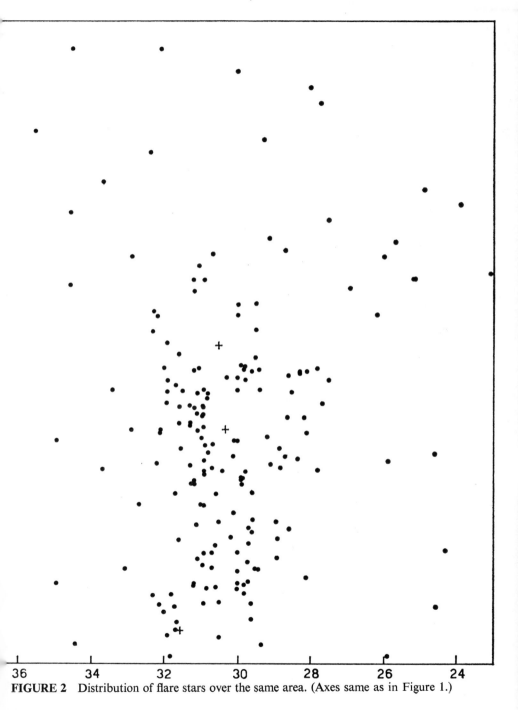

FIGURE 2 Distribution of flare stars over the same area. (Axes same as in Figure 1.)

The Asiago observations after 1962 have shown another interesting property of the nebular variables (particularly those with ultraviolet excess): they have at times sudden changes of brightness in the duration of one or two hours or less (YY, HS, SY, XX, YZ Ori, etc., Figures 3, 4). These rapid fluctuations, sometimes with an amplitude of one magnitude or more, have nothing in common with flares, although they may simulate a flaring when the brightness is rising. Photoelectric monitoring of faint Orion variables should be of extreme interest to detect quick flickering of light in these stars. Also radio observations should be worthwhile.

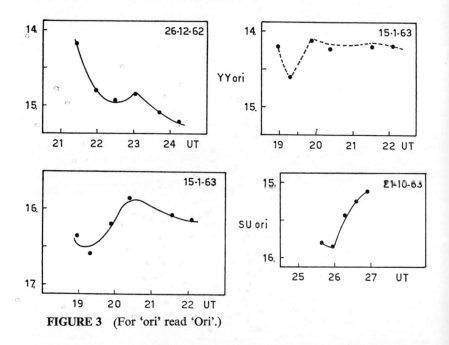

FIGURE 3   (For 'ori' read 'Ori'.)

The mean amplitude of the nebular variables in Orion is 1.60 pg, but some of the stars reach much higher amplitude. The frequency distribution of magnitudes shows a maximum near 16.5 pg, corresponding to absolute magnitude +8.5 (Figure 5). It is likely, however, that the number of variables may be still higher at lower luminosity. An infrared survey of the Trapezium area with a larger instrument than that used at Asiago would be of great importance in this connection. Since many of the faintest variables are infrared, photographs in the far infrared (from 8000 to 11,000) will be made at Asiago as soon as possible with the new infrared RCA-Carnegie intensifier.

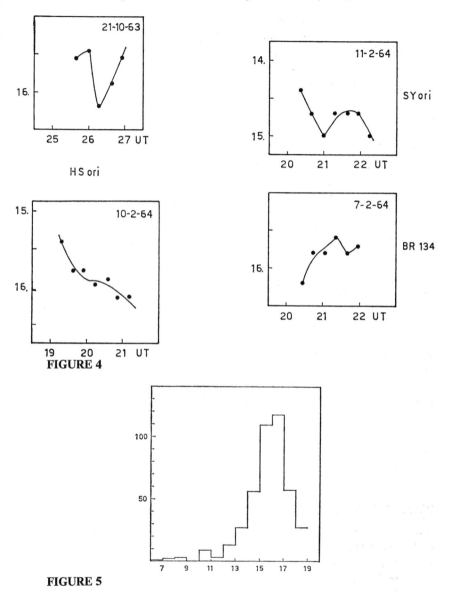

FIGURE 4

FIGURE 5

III. THE REGION OF THE HORSEHEAD NEBULA

The region of the Horsehead Nebula, including IC 434, NGC 2024, etc., has also been surveyed in the last few years at Asiago with the 122 cm telescope and the Schmidts. Thirty new variable stars have been discovered. Positions and charts will be reported elsewhere. The light curves of these variables are not different from those observed in the Trapezium region, with the same variety of forms. A sort of periodicity has been observed sometimes in the variables V 505 Ori (Haro–Moreno No. 10) and No. 39 ($5^h 32^m 55^s$; $-1°09'8$, 1900.0); but this semiregular activity disappeared after a few months. The possibility that eclipsing binaries may be present among nebular variables, should not be disregarded. However, the erratic variation would mask the eclipses unless a detailed analysis of the light curves over a long period of time could be made.

In conclusion, the two regions, near the Horsehead Nebula and the Trapezium, do not sensibly differ from the point of view of the variables, if not for the relative frequency which is much higher near and south of the Trapezium than in the Horsehead Nebula.

The other regions under investigation are shown in Table I. Near and around NGC 6530, 33 new variable stars, most of which are visible only in infrared, have been discovered on Asiago plates (charts and positions will soon be given in the Asiago contributions). Considering also the variables discovered by Walker, Lampland and Herbig, the number of nebular variables found in this cluster exceeds one hundred and many others are certainly undiscovered. Good light curves have not been obtained, since the cluster is difficult to observe at the latitude of Asiago and the variables are too faint to be followed at minimum. In the few cases in which light curves have been drawn, they appear to be similar to those of Orion.

Very faint nebular variables have been found, as shown in Table I, in h and χ Persei, and in NGC 7635, a region rich of obscuring matter, where the variable stars appear to be strongly reddened. The material obtained in the last years is still under reduction. No flares have been found, but this, as in NGC 6530, is an obvious consequence of the long exposure necessary to reach the faintest stars.

IV. FLARE STARS IN ORION, PLEIADES, AND PRAESEPE

Flare stars in Orion (Trapezium and ξ Orionis) have been occasionally searched at Asiago in the past years with the 122 cm telescope and the 40cm Schmidt. A number of flares (23) were found, but not as high as expected, considering the time dedicated to the survey. The situation, however, greatly improved when the new Schmidt of 67 cm, f:3.3, entered into the research. We have found 29 flares in 2880 minutes of observation over a field of 25 square degrees around the Trapezium (1 flare every 100 minutes of effective observation) which is about the same frequency as observed by Haro (62 flares in 6770 minutes or 1 flare every 112 minutes). In a field of the same area centered

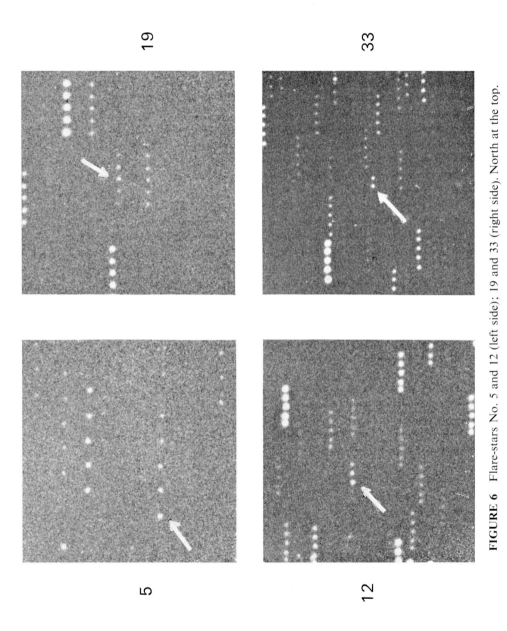

FIGURE 6 Flare-stars No. 5 and 12 (left side); 19 and 33 (right side). North at the top.

in ξ Orionis, we found 5 flares in 1185 minutes (1 flare every 4 hours). In total the flares discovered (mostly by Miss L. Pigatto) in the Orion aggregate (Trapezium + ξ Orionis) from 1962 to now have been 39. Positions and other data on these flares are reported in Table II. Charts and light curves will be given in a forthcoming paper. Some of the flares are reproduced in Figure 6. All of the stars in which flares were found are fainter than m_{pg} 16.5. The mean m_{pg} is 17.2. The corresponding mean value for 121 flare stars in the list of Haro and Chavira in *Vistas in Astronomy*, Vol. 8, p. 89, 1965, is $m_{pv} =$ 15.8, so giving a mean color index of the order of $+1.4$, with an absolute visual magnitude 7.7 and pg magnitude 9.1. On the average, the Orion flare stars are therefore slightly to the right of the main sequence. Although we don't have very precise photometric data, comparing blue and infrared plates we agree with Haro that some of the flare stars lie on the main sequence and even to the left of the main sequence, while the majority are to the right.

It may be interesting to look on a possible dependence of the number of flare stars in different parts of the Orion aggregate from the number of RW Aurigae variables. Such dependence undoubtedly exists, but it is not so strong as we believed. Figures 1 and 2 illustrate this point. Normal RW Aurigae variables in Orion are strongly concentrated around the Trapezium, in a narrow strip from -7 to -4 and from $5^h\,28^m$ to $5^h\,32^m$. They reappear north of -3, between $5^h\,33^m$ and $5^h\,37^m$. To the contrary the flare stars are not so strongly concentrated in the Trapezium (but one has to consider that this may be partly due to various selection effects, for example the presence of the bright background of the nebula) and, although they follow the general pattern of dark and bright nebulosities, they are found spread all over the field, with a tendency to avoid the regions where most of the RW Aur variables are found.

Flares are sometimes observed in typical RW variable stars of Orion,* but they are rather unusual (10 in 456 variable stars) and the amplitude of the flares is, on the average, smaller than in normal flare stars.

In general, the amplitude of the Orion flares increases with decreasing brightness. This is true for the flares of Table II (A <1, m_{pg} 16.0; 1–2, m_{pg} 17.0; 2, 3, m_{pg} 17.8) as well as for the 121 flare stars listed by Haro and Chavira (A <1, m_v 14.6; 1–2, 15.2; 2–3, 15.8; 3–4, 16.5; 4, 18.5). However, this fact has only a rough statistical meaning, since it is well known that flares of different amplitude can be observed in the same star.

By looking at the 222 flare stars of Orion (121 in Haro–Chavira list, 62 in their Supplement list (1965), 39 in Table II) it can be observed that only

* The following RW variables of Orion have shown flares: HO? (1^m); XX (0.8); IZ (2.0); YZ (0.8); BW (1.0); OT (2.0; 1.5; 0.6; 0.9); OX (0.8); II (0.9); NS (1.5); V 365 (0.6).

TABLE II Flare stars in Orion recently discovered at Asiago.

N.	P.	1900 R.A.	1900 D.	m	Date	Dur.	Notes
1		5h 22m 00	−6° 04.0	15.2–17.4	11 Dec. 1966	40m	—
2		5 23 00	−4 24.7	15.8–18.5	8 Dec. 1967	30	—
3		5 24 34	−6 37.6	14.5–(18	2 Dec. 1967	—	
4		5 25 00	−7 09.0	16.6–17.8	30 Jan. 1968	—	Haro 38
5		5 25 09	−4 27.6	15.3–17.5	23 Jan. 1966	>20	
6	981	5 27 47	−5 03.9	16.0–16.8	30 Jan. 1968	10	—
7		5 28 13	−5 23.8	16.2–17.5	4 Jan. 1968	20	—
8	1333	5 29 08	−5 40.5	16.2–(17.1	30 Jan. 1968	15	II Ori
9		5 29 09	−4 11.7	15.8–(18.5	18 Jan. 1966	10?	—
10		5 29 27	−6 22.8	16.2–(18.5	12 Jan. 1967	37	—
11		5 29 35	−6 03.3	16.4–17.2	4 Jan. 1968	20?	—
12		5 29 38	−0 31.0	15.0–17.5	7 Dec. 1967	>30	—
13		5 29 42	−6 12.3	15.0–(18.5	1 Feb. 1968	>30	—
14		5 29 55	−1 50.0	15.6–17.1	27 Feb. 1968	30	—
15	1625	5 29 55	−5 50.0	15.5–16.8	19 Jan. 1964	>60	—
16		5 30 29	−6 51.2	15.5–17.2	23 Jan. 1966	25	—
17	2039	5 30 32	−6 05.4	15.0–16.5	25 Feb. 1963	—	NS Ori
18	2112	5 30 40	−5 33.2	15.0–(16.8	1 Feb. 1964	35	—
19		5 30 42	−7 06.1	15.8–17.3	30 Jan. 1968	30?	—
20		5 30 52	−5 34.5	15.2–(17	19 Jan. 1966	—	—

No.	No.	RA	Dec	Mag	Date		Desig.
21	2210	5 30 52	−5 44.9	15.1–16.8	8 Jan. 1968	25?	V 378 Ori
22	2235	5 30 55	−5 39.3	15.8–16.8	27 Dec. 1967	10	—
23	2245	5 30 56	−5 19.0	16.4–(17	12 Dec. 1966	10	V 379 Ori
24	2246	5 30 56	−5 20.4	16.0–16.9	30 Jan. 1968	>20	OT Ori
25		5 30 56	−6 21.8	15.8–18	1 Feb. 1968	>20	—
26		5 31 04	−4 22.3	16.4–(17	19 Jan. 1966	—	—
27	2295	5 31 06	−5 27.3	16.6–(17.2	21 Jan. 1968	—	V 365 Ori
28		5 31 12	−6 29.2	15.4–16.8	8 Dec. 1967	—	—
29		5 31 32	−5 34.4	16.0–17.5	27 Feb. 1965	40	—
30		5 31 40	−6 44.3	16.6–(17.2	15 Jan. 1966	15	—
31		5 31 41	−6 46.0	15.5–16.9	19 Jan. 1963	10	—
32		5 31 45	−6 38.0	16.2–17.2	6 Jan. 1968	10	—
33		5 32 10	−2 55.0	15.0–(18	9 Dec. 1966	—	—
34		5 32 18	−6 49.3	16.2–(17.2	19 Jan. 1966	—	—
35		5 32 52	−0 49.5	15.5–17.2	31 Jan. 1968	20	—
36		5 33 40	−3 50.0	15.9–16.7	27 Jan. 1968	20	—
37		5 34 33	−2 55.0	16.2–18.2	24 Jan. 1966	—	—
38		5 36 31	−1 45.0	15.9–(17.2	27 Jan. 1968	20	—
39		5 40 02	−1 11.9	14.8–16.7	8 Feb. 1959	—	—

23 stars (10 per cent) have been caught in flare activity more than once. Of these only three have had a third flare and only one (besides OT Ori) has had more than three flares. This may signify that the mean interval between two successive flares in a star of Orion must be higher than 10–15 days, because the 90 per cent of the flare stars have not shown a second flare in about 300^h of effective observation.

Finally, the duration of a flare varies from 10^m to some hours. The mean value is around 30 to 40 minutes, from minimum to minimum. Of course, shorter flares or flares of small amplitude may escape detection; so that the inferior limit of duration, as well as the number of flares, are partly determined by the techniques of observation.

Flare stars have also been searched at Asiago on NGC 2264, the Pleiades and Praesepe. The results are contained in Table I. In NGC 2264, thirteen flares have been found (1957); this field contains many nebular variables, and flare stars will be searched systematically with the 67 cm Schmidt in the future. In the Pleiades, 13 flares have been found at Asiago (1966) in 12 stars. In the Praesepe only two flares have been observed, but only 21 hours of effective observations were made with the 122 cm telescope. Also these fields will be included in our future programs.

As for the characters of the flare stars of these aggregates, Dr. Haro has recently presented a complete review. The Asiago observations confirm his conclusions in every respect.

References

HARO, G., and CHAVIRA, E. 1965, *Vistas in Astronomy*, Vol. 8, p. 89.
HARO, G. 1965, private communication.
MAFFEI, P. 1963, Asiago Contr. No. 136, 140, 141, 146.
MAFFEI, P. 1965, *Mem. SAI*, **36**, 487; **36**, 493.
MAFFEI, P. 1966, *Mem. SAI*, **37**, 459.
PARENAGO, P. P. 1954, *Trud. Astr. Inst. Sternberg* XXV.
ROSINO, L. 1946, Publ. Bologna, **5**, no. 1.
ROSINO, L. 1956, *Asiago Contr.* No. 69, *Mem. SAI*, **27**, 335.
ROSINO, L. 1960, *Asiago Contr.* No. 109.
ROSINO, L. 1962a, *Rend. Scuola Int. Fisica Varenna*, XXVIII.
ROSINO, L. 1962b, *Asiago Contr.* 125, *Mem. SAI*, **32**, 297.
ROSINO, L. 1966, *Asiago Contr.*, 189, *Mem. SAI*, **37**, 717.
ROSINO, L. GRUBISSICH, C., MAFFEI P. 1957, *Asiago Contr.*, No. 82.

Discussion

EGGEN   I would like to make two points. One is on the RW Aurigae stars. When you assign them to the RW Aurigae class, is that based entirely on the discovery observations? The reason I ask this is that some of them look as if they are eclipsing. Can you eliminate the eclipsing from the RW Aur stars?

ROSINO   In general, the observations are sufficient to decide whether a variable belongs to the RW Aur type or not.

EGGEN   And the other question is: Does the distribution of the RW Aurigae stars reflect the distribution of the association members?

ROSINO   Yes, that's right. The flare stars also follow the distribution of the aggregate.

McCARTHY   Would you comment on the criterion of membership of flare stars in clusters? I mean especially the faint flare stars which Professor Haro has detected in the Pleiades. He has a group of five stars whose blue magnitude is 21. And I wonder if you think that the occurrence of flares is a good criterion for cluster membership.

ROSINO   I think that it is certainly a good criterion, although not absolute,

VARDYA   I would like to know what the frequency distribution of flare stars is with respect to spectral class.

ROSINO   I am afraid I cannot answer this question because we do not have sufficient data.

AUMAN   Has anybody ever gotten a spectrum of a flare? What spectral features show up?

ROSINO   Dr. Haro has taken some spectra, but let me discuss the photographic observations. In a flare star the continuum increases strongly in the blue ultraviolet. There is also strong increase in the emission lines of hydrogen, particularly Hα. So you can find a flare in Hα, or in the ultraviolet.

KRAFT   I have two remarks. The first spectrum of a flare that was taken, aside from the sun, was Dr. Joy's spectrum of *UV* Ceti, taken about 20 years ago. The other thing that I would comment on is the stars in Pleiades that show flares that are below the main sequence. I know the stars individually, but at least one of them is a fake and the reported position of the star below the main sequence is wrong. That star is not as blue as the photometry indicates. It is considerably redder and the spectral type is redder. The proper motion list of Van Altena accepts it now as a member, and it shows strong calcium emission. I think that the color is incorrect.

A luminosity-dependent phenomenon in flares of dMe stars*

WILLIAM E. KUNKEL

Cerro Tololo Interamerican Observatory

Abstract

Flare decay rates of five solar neighborhood *dMe* stars are compared, and a correlation with luminosity is confirmed, in the sense indicated by Haro and Chavira. A calibration of this relation with binaries of known mass, and Johnson's infrared photometric data indicates that the average flare decay rate varies as the inverse first power of the stellar surface gravity.

SOME YEARS AGO Haro and Chavira (1955) reported a correlation between the duration of stellar flares and the spectral type of the flare star, in the sense that flares of shorter duration were encountered on stars of later type. In the work, stars of earlier type, *dK*5, were in the Orion Association while stars of later type, *dM*0 to *dM*5, were in the Taurus dark clouds and the solar neighborhood. Thus, the sample consisted of three distinct groups of stars according to the region in the sky. Within any single group, data were insufficient to establish the reported correlation, so that one may not conclude that the phenomenon is characteristic of all flare stars, or that it suffices only to establish differences between stellar aggregates that contain flare stars.

The purpose of this paper is to present some pertinent data from work on solar neighborhood flare stars begun at the McDonald Observatory, and currently in progress at Cerro Tololo. Continuous U-band photoelectric photometry is now available for five stars. Three of these, AD Leo, YZ CMi, and Wolf 359, have been described earlier (Kunkel 1967) and were shown to have flare decay rates that correlate well with stellar luminosity, in the sense indicated by Haro and Chavira. Since flares are surface phenomena, it appears of great interest to learn what correlation might exist between flare decay

---

* Contributions from the Cerro Tololo Interamerican Observatory, No. 32.

rate and surface gravity. Therefore, UV Ceti and 40 Eridani C were added to the observing program. UV Ceti is a well-known visual binary with masses of 0.040 and 0.046 $M_\odot$ (Luyten 1961). (These stars are too close to one another to permit the exclusion of one star from the photometer diaphragm—though the nearly equal masses make this a relatively unimportant requirement.) 40 Eridani C with a mass of 0.20 $M_\odot$ (Baize 1966) is the fainter companion of the white dwarf 40 Eridani B, from which it is sufficiently distant to permit conventional photoelectric monitoring of just one star.

The data on the five stars are summarized in Table I. The star name appears in Column 1, and the absolute visual magnitude appears in Column 2. Column 3 is the flare time scale parameter. This is defined as the logarithm (base 10) of the decline in magnitudes per minute, τ_3, measured at a point in time when the flare light (quiescent photosphere subtracted) is three magnitudes below maximum light. Column 4 indicates the dispersion in the log of τ_3. The dispersion is in chromospheric structure over the stellar surface. The number of flares used to construct the average (Column 3) is given in Column 5.

TABLE I   Flare decay rates for five solar neighborhood *dMe* stars.

Star	M_v	$(\log_{10} \tau_3)$	σ	N
AD Leo	11.07	$-.87 \pm .17$	0.33	4
YZ CMi	12.44	$-.68 \pm .09$	0.19	5
40 Eri C	12.5	$-.55 \pm .14$	0.42	6
UV Ceti	15.88*	$-.28 \pm .09$	0.34	15
Wolf 359	16.82	$-.24 \pm .10$	0.37	6

* Value of the fainter component.

The choice of the time scale variable, τ_3, is based on the observation that the dispersion in flare decay rate decreases as the decay proceeds. Another way to do it is to divide a flare event into two components: A fast component comprising light fluctuations of higher frequency is associated with the rise to maximum light—though it generally persists somewhat beyond (giving rise to phenomena like Roques 'stillstand' (Roques 1961)). The slow component dominating the lower frequencies is the decay—a monotonic decrease of light. τ_3 is a measure of the slow component, and the delay between maximum light and the point three magnitudes down is simply a device to minimize the influence of the fast component.

By using Johnson's (1965) photometry of M dwarfs to transform the absolute visual magnitudes of Table I to bolometric magnitudes, one can represent the relation between the flare decay rate and the bolometric magnitude (Figure 1) by the empirical relation

$$M_{bol} = 5.16 \log_{10} \tau_3 + 13.25 \tag{1}$$

If we postulate that the decay rate τ_3 varies as the power η of the surface gravity g

$$\tau_3 = \text{const.}_1 \, g^{\eta} \tag{2}$$

then from

$$g = \text{const.}_2 \, Mr^{-2} \tag{3}$$

and

$$r^2 = \text{const.}_3 \, L \, T^{-4} \tag{4}$$

one obtains a value for η,

$$\eta = -1.00 \pm 0.15 \tag{5}$$

In these equations, M is the mass, T is the effective temperature, and L is the luminosity.

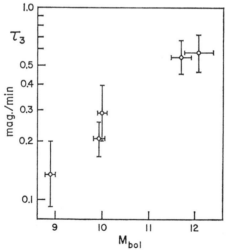

FIGURE 1 Relation between flare decay rate τ_3 in magnitudes per minute and bolometric magnitude. In order of increasing magnitude, the stars are AD Leo, YZ CMi, 40 Eri C, UV Cet, and Wolf 359.

In other words, the decay time of stellar flares varies approximately as the inverse first power of the stellar surface gravity. The uncertainty quoted in equation (5) is a somewhat subjective estimate, since the value of the error estimate depends primarily on the accuracy with which bolometric correction and the stellar mass are known, and, for dwarfs as faint intrinsically as UV Ceti, bolometric corrections are uncertain, and few mass estimates exist. However, the very small spread observed in the relation (Figure 1) together with the long main sequence phase of *M* dwarf evolution, implies that the luminosity dependence of flare time scales is relatively free of evolutionary effects that might be expected to appear among stellar aggregates of different ages. So the possibility exists that relations (1) and (2) may be useful for determining the surface gravity and luminosity of single red dwarfs on which flare activity is observed.

References

BAIZE, P., 1966, *J.O.* **49**, 1.
HARO, G., and CHAVIRA, E., 1955, *Bol. Ton. y Tac.*, No. 12, 3.
JOHNSON, H. L. 1965, *Ap. J.*, **141**, 170.
KUNKEL, W. E., 1967, unpublished *Ph.D. Dissertation*, Univ. of Texas.
LUYTEN, W. J., 1961, *Publ. Obs. Univ. of Minnesota*, **2**, No. 16.
ROQUES, P. E., 1961, *Ap. J.*, **133**, 914.

Discussion

ALLER   Does the sun fit on this relationship that you just described?

KUNKEL   I would be very eager to put the sun on the relation, but the spectroscopy gives problems. One of the reasons for picking the U-band is that most of the light that we see in the U-band happens to be optically thin, while the B and V bands are optically thick. Another difficulty with the sun is that the solar flare observers work with Balmer lines, and it is just a little bit difficult to make the transition, because the observational techniques differ so vastly. Most of the solar flare work is done on Balmer lines and I am doing broad-band photometry.

EGGEN   It may be that the sun does not belong on the plot, because I believe all of them are binaries.

KUNKEL   Yes, in fact AD Leonis is supposed to be a binary; Wolf 359 is the only one I have not heard of as a binary.

EGGEN   That is the shortest period binary known: Wolf 359 and 630.

KUNKEL   Wolf 630 is the star that I have paid some attention to because it is an old one. And it does show a good deal of flare activity and from the model calculations that were made a year and a half ago, one gets the impression that the star radiates about 0.1 per cent of its total energy for flare activity.

A new flare star?

PETER VAN DE KAMP

Sproul Observatory, Swarthmore College

THE HIGH VELOCITY subdwarf Groombridge 1830 = BD + 38° 2285 (R.A. $11^h 50^m .1$, Decl. $+38° 5'$, 1950) has a proper motion of $7''.04$ in position angle 145°, radial velocity -98.3 km/sec., parallax $+''.116 \pm .005$, vis. mag. 6.46, spectrum G8 VI. A close companion appears on the eight exposures of $2\frac{1}{2}$ minutes each, taken by Worth on four consecutive plates with the Sproul 24-inch refractor in the morning of February 27, 1968, over the interval $1^h 18^m - 1^h 39^m$ E.S.T. A slight increase of brightness with time is indicated. The companion appears about two magnitudes fainter than Groombridge 1830, i.e., of apparent magnitude 8.5 or absolute magnitude 8.8.

The relative position of B with respect to A was obtained from the best four exposures and is found to be

$$\rho = 1''.7 \qquad \theta = 166°$$

There is no sign of a companion on any of the other Sproul plates covering the interval 1937–1968, including a few nights in March 1968. There is no star in or near the present position of Groombridge 1830 on any of the plates, including some long exposure plates.

It is difficult to avoid the obvious explanation that the observations reveal a flare up of a companion of Groombridge 1830, which normally is several magnitudes fainter than the observed estimated value of 8.5, i.e., a companion perhaps some 4 or more magnitudes fainter than the primary.

The parallax of Groombridge 1830, from the Yale Catalogue is $+''.116 \pm ''.005$ (Jenkins 1952), while the recent Sproul value is $+''.084 \pm ''.012$ (Harrington 1964). The latter determination covering 25 years did not and could hardly reveal any perturbation because of the limited accuracy of the material. For an adopted combined mass equalling the sun's mass, a period of revolution of the order of 50 or 100 years may be expected.

It is to be hoped that other flare-ups may occur and attempts be made elsewhere to detect the companion.

References:

JENKINS, L. F. 1952, *General Catalogue of Trigonometric Stellar Parallaxes* (Yale University Observatory, New Haven, Conn.
HARRINGTON, R. S. 1964, *A. J.*, **69**, 804.

Discussion

GREENSTEIN   Since Groombridge 1830 is such a bright star, I am curious as to why its companion had not been seen by visual observers. It is within the magnitude limit of the surveys, and, therefore, the flare must be much bigger than that.

VAN DE KAMP   Well, it is up to 8.5. But let us say that the normal magnitude for this star is about 12 magnitude. To separate the 6.5 and 12 magnitude stars at a separation of 1.7 seconds is probably not very easy. But we will see what we have before long.

GREENSTEIN   Well, I can try. But the other thing that I want to point out is that Groombridge 1830 is a member of the Groombridge 1830 group, and is one of the very old metal-poor subdwarfs. So if you have flare activities in a star which is approximately 10 billion years old, it is fun—no reason why not, but just fun.

EGGEN   I think that this would discomfort Dr. Wilson more than it does me. Another example of this that may not be known is that Wolf 630, which has been mentioned several times in the last few days, consists of a very close pair with a period of about a year and a half. A third component is quite distant and a 4th component is extremely faint. And the close pair has shown flares on two occasions on being observed photoelectrically. Yet it is as old as this system, whereas at least one of the other components shows no *H* and *K* emission whatsoever. So we may be building up evidence that there's discontinuity in the age-chromospheric activity relation.

WILSON   May I make a comment here? Groombridge 1830 is the only subdwarf that I know which shows *H–K* emission. The luminosity obtained from the width of the emission line agrees very well with the luminosity from parallax measurements. This, I fear, is a very serious thing which has been bothering me now for some time because all the other criteria here indicate age. But the presence of chromospheric activity indicates youth. Accordingly, I have begun observing a few other subdwarfs to see if I can find these effects in any of them. But, the program has been hampered by poor weather.

GLIESE   The star is the fundamental star 1307 in the FK_4 and therefore, there should be a lot of meridian circle observations. Would it be possible to detect astrometric perturbations by meridian circle observations?

VAN DE KAMP   I have not looked into that. Of course, the meridian circle observations are not very accurate.

WORLEY   Last Friday, I received a letter from Dr. van de Kamp reporting this important discovery and since that time I have been attempting to see the companion with the 26-inch refractor in Washington. I have actually been able to observe on three nights, two of those nights had quite bad seeing, the third had fair seeing. On the best of the three nights, I did see the companion. The magnitude difference is five or six magnitudes. The position angle which I measured is 173° and the separation is about 2 seconds. Now this is a very weak measure and I hope that a better job will be done. I certainly intend to keep observing this very important star. So far about an hour and 30 minutes have been spent observing and I have not yet been fortunate enough to see the companion flare; that would be a much easier object to measure.

GREENSTEIN   I would like to ask a question of my colleague Olin Wilson. It seems to me rather interesting that all the peculiar objects that have been flaring or have H and K emission are binaries. That is in the old population, even in very wide binaries and I admit many of them are extremely wide binaries, there does seem to be some kind of linkage, maybe by the stellar wind back through magnetic fields through a perturbation which produces every now and then flare-like activity. Do you have any comments about duplicity?

WILSON   The only cases that I know of where you find exceptionally strong and indeed broadened calcium emission are close binaries. But these wider ones, I have always supposed would be too far apart for any interaction to take place.

Section III

Equations of state and stellar evolution*

E. E. SALPETER

*Newman Laboratory of Nuclear Studies, Physics Department, and
Center for Radiophysics and Space Research
Cornell University, Ithaca, New York*

IN LATER PAPERS we shall hear detailed reports on how the equations of state affect the various evolutionary stages of low-mass stars. I want to give a qualitative introduction to these talks and start with a short review on the ranges of average temperature T and average density ρ to be expected in stellar interiors under different conditions. Let us work with 'natural units' (with $\hbar = m = c = 1$ where m is the electron mass) for energy ϵ and for temperature T and re-express density ρ in terms of a dimensionless parameter x,

$$mc^2 = 0.511 \text{ MeV}, \quad mc^2/k = 5.93 \times 10^{9\circ}\text{K},$$

$$\rho = x^3 \times 9.74 \times 10^5 x^3 \text{ gm/cm}^3 \tag{1}$$

For simplicity I consider at first a star consisting only of ionized hydrogen (helium is qualitatively similar). The parameter x is then the Fermi momentum of the electrons (in units of mc) and the radius (in units of $\hbar/mc$) of a sphere containing one electron is $(4/9\pi)^{1/3}x^{-1}$. According to the Virial Theorem for any star in hydrostatic equilibrium the average kinetic energy (thermal or Fermi) ϵ_K per electron is comparable to the gravitational energy ϵ_G per nucleon. Expressing the mass M of the star in units of the 'Chandrasekhar limiting mass' M_{Ch} of about 5.6 solar masses, one finds

$$\epsilon_K = Tf(T, x) \sim -\epsilon_G \sim xM^{2/3} \tag{2}$$

where f is unity for non-degenerate electrons (and large for degenerate ones). If the electrons are non-relativistic ($x \ll 1$) the Fermi-energy E_F per electron at zero temperature is about x^2 in our units. At finite temperature T, the electrons are non-degenerate (degenerate) when the ratio T/x^2 is large (small).

* Supported in part by the Office of Naval Research and by the National Science Foundation Grant GP-6928.

We are considering low-mass stars, i.e. $M \ll 1$ in our units, and the electrons are then always non-relativistic. One then finds that the electrons are non-degenerate ($f \sim 1$ in eqn. (2)) for x well below a critical value x_{cr} but become highly degenerate as the density-parameter x approaches x_{cr}. The temperature T thus rises linearly with x for $x \ll x_{cr}$, reaches a maximum value T_m (at x slightly less than x_{cr}) and then decreases towards zero at the maximum density $x = x_{cr}$ (when, in eqn. (2), $\epsilon_K \sim E_F \sim x^2_{cr}$). The evolution of T with increasing density-parameter x is shown schematically in Figure 1. The

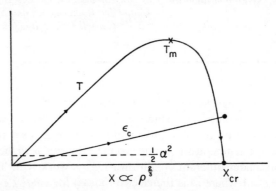

FIGURE 1

decreasing part of this curve, of course, represents the cooling off of a white dwarf. The dependence on mass M (in units of M_{Ch}) of the maximum temperature and density is given by

$$T_m \sim x^2_{cr} \sim M^{4/3} \quad \text{for} \quad M \ll 1 \tag{3}$$

When we come to Coulomb effects, the Sommerfeld fine structure constant α plays an important role and an analogous gravitational 'fine structure' constant α_G can also be defined.

$$\alpha^{-1} = \frac{\hbar c}{e^2} = 137.04, \qquad \alpha_G^{-1} = \frac{\hbar c}{GH^2} \sim 10^{38}, \tag{4}$$

where $H = 1837 \, m$ is the mass of a hydrogen atom. The Chandrasekhar limiting mass M_{Ch} is of the order of $\alpha_G^{-3/2}H$. We shall find another mass-value M_{pl} of interest when considering Coulomb effects, in our units given by

$$M_{pl} \equiv \alpha^{3/2} \sim (\alpha/\alpha_G)^{3/2}H \sim 3 \times 10^{-3} \, M_\odot, \tag{5}$$

a little larger than the mass of Jupiter.

The importance of Coulomb effects can be estimated as follows. Let ϵ_c be the average Coulomb interaction energy between a nucleus (or an electron) and the nearest nucleus. The typical nuclear separation is x^{-1} in our units and

$$\epsilon_c \sim ax \sim (a/M^{2/3})\epsilon_K \sim \left(\frac{M_{pl}}{M}\right)^{2/3} \epsilon_K \qquad (6)$$

From now on we shall restrict ourselves to stellar masses M which are large compared with M_{pl}. With this inequality, $M \gg M_{pl}$, eqn. (6) gives $\epsilon_c \ll \epsilon_K$ and Coulomb effects are a relatively small perturbation on the kinetic energy of the electrons whether they are degenerate or not. For $M \sim M_{pl}$ about half of the electrons could be unionized and for $M \ll M_{pl}$ Coulomb effects would be all important. M_{pl} (a few times the mass of Jupiter) is thus the dividing line between planets or rocks on the one hand, where the electronic structure of atoms is important and gravitation is a perturbation, and stars on the other hand where gravitation is important and Coulomb effects perturb the electrons little.

Having restricted ourselves to $M \gg M_{pl}$ the electrons form an almost perfect gas under all circumstances and the next question is how the nuclei behave. As long as the electrons are non-degenerate, the thermal energy of a nucleus equals the energy ϵ_K per electron and the Coulomb effects ϵ_c are then also unimportant for the nuclei. However, because of the large mass-ratio H/m the nuclei remain non-degenerate when the temperature T starts decreasing and the importance of Coulomb effects for the nuclei is measured by the numerical value of the ratio Γ of ϵ_c to T. As can be seen from Figure 1, the maximum value of ϵ_c is much smaller than the maximum value T_m of T (although much larger than the ionization potential, $I = (\frac{1}{2})a^2$ in our units) whenever $M \gg M_{pl}$. As x increases, the ratio Γ is first small and almost constant while the electrons are non-degenerate, but Γ increases indefinitely when the electrons become degenerate and T decreases.

Since $M \gg M_{pl}$, our ratio Γ can exceed unity only when the electrons are highly degenerate (but the protons are not) and the electrons provide the largest part of the pressure. When $\Gamma > 1$, the proton contribution to the pressure is affected percentage-wise by the Coulomb effects but is still small compared with the Fermi pressure of the electrons. The Coulomb effects on the total pressure are thus small at all stages of evolution of a (hydrogen) star whose mass is very large compared with $M_{pl} \sim a^{3/2}M_{Ch} \sim 10^{-3}\ M_\odot$. On the other hand, the specific heat of the electrons decreases drastically when they become more degenerate and the nuclei provide most of the heat capacity of the star during its cooling-off period through the white dwarf stage towards the eventual 'black-dwarf' stage. For cool white dwarfs the Coulomb effects can thus dominate questions about the heat content of the star even though $M \gg M_{pl}$.

As far as the nuclei are concerned, the electrons merely provide a uniform

H

neutralizing charge density and the nuclei represent a 'one component Coulomb plasma'. When Γ exceeds some critical value Γ_m, the nuclei undergo a first-order phase transition and solidify to form a rigid lattice (of the body-centered cubic type). Γ_m is a dimensionless number which should be of order unity, but is actually somewhere between 10 and 100. The numerical value of Γ_m as well as the latent heat of solidification will be discussed in a later talk. The specific heat of a lattice is $3k$ compared with $1.5k$ for a gas. For $1 \ll \Gamma \lesssim \Gamma_m$ the nuclei form a liquid with short-range order similar to that of the lattice and the specific heat is still close to $3k$. At the appropriate stages in the cooling curve of a white dwarf, the cooling times are thus lengthened due to the release of latent heat and the increase in specific heat. At an even later stage of cooling the opposite is true: the Coulomb lattice formed by the nuclei has, like any other solid, a 'Debye-temperature' T_D and the specific heat of the lattice decreases drastically as T becomes much less than T_D (very crudely $kT_D \sim \hbar\omega_0$ where ω_0 is the vibration frequency of a nucleus about its equilibrium position in the lattice). This drastic increase in cooling speed in the final stages of cooling of a white dwarf will also be discussed in later talks.

So far I have discussed only stars made of the two common elements of low atomic charge Z, hydrogen and helium. It is at least of academic interest to see how the Coulomb effects would be altered if a star consisted of an element with large atomic charge Z. With x^{-1} the typical separation between nearest pairs of electrons, the typical separation between nuclei is $Z^{1/3}x^{-1}$ and the Coulomb energy per electron in eqn. (6) becomes $(Z^{2/3}\alpha)x$. The critical mass M_{pl}, below which Coulomb forces dominate the equation of state, is then generalized to

$$M_{pl} \equiv (Z^{2/3}\alpha)^{3/2} M_{Ch} \sim Z \times 3 \times 10^{-3} M_\odot. \qquad (7)$$

In the Thomas–Fermi theory for a complex atom the ionization potential of a 'typical' atomic electron is of order $(Z^{2/3}\alpha)^2$ in our units. With α replaced by $Z^{2/3}\alpha$ in eqn. (7), Coulomb effects on the equation of state are then again small if $M \gg M_{pl}$ and the atoms will be 'mostly' ionized. Note, however, that the K-shell ionization potential is $\frac{1}{2}Z^2\alpha^2 \gg (Z^{2/3}\alpha^2)$ and the innermost atomic electrons may not be fully ionized.

For ordinary stellar composition $M_{pl} \sim 0.01\ M_\odot$. The low-mass stars we are considering in this conference are more massive than M_{pl}, but not necessarily by a very large factor. For accurate work on the equation of state, one then has to consider also the Coulomb deviations from a perfect gas by the electrons. This difficult quantum mechanical problem will also be discussed in two later talks.

Discussion

DeWITT   Can you say something about the temperature at which lattice formation occurs?

SALPETER   Well, I was hoping you or somebody else in the later talks would discuss that. But let me just say something about the Monte Carlo calculations for this classical coulomb lattice problem which will probably be discussed in detail by Hugh Van Horn. There are numerical factors which ought to be of order unity, but are about 30, meaning that the ratio of the coulomb energy to thermal energy at the place where the melting actually occurs is about 20 or 50 or 80 rather than about unity. What you have then is an in-between region where you have a sort of viscous fluid, where in fact you do not have any really long, long-range order, but as far as interaction with the nearest few nuclei is concerned you could not tell the difference between it and the lattice.

Statistical mechanics of dense ionized gases*

HUGH E. DeWITT

Lawrence Radiation Laboratory, University of California, Livermore

Abstract

This report is a summary of present day knowledge of quantum statistical mechanics of dense ionized gases, particularly the calculation of the Coulombic interaction contributions to the equation of state of stellar interiors. The quantum perturbation expansion of the partition function including wave mechanics and Fermi statistics is described. Results are given for the quantum ring sum contribution to the free energy which generalize the Debye–Huckel results, and the next higher order results beyond the ring sum are given. It is shown that the importance of Coulomb corrections is measured by the plasma parameter, $\Lambda_s = 1/4\pi n\lambda_s^3$ where n is the charged particle number density and λ_s is the screening length as modified by Fermi statistics for the electrons. Asymptotic results are given for the free energy for $\Lambda_s \ll 1$. For numerous small stars the value of Λ_s is roughly 1, and it is shown that the pressure can be numerically calculated to within a few percent even when the Coulomb interactions contribute a large fraction of the ideal gas pressure.

I. INTRODUCTION

THE CALCULATION of the equation of state of the dense, partially ionized gases that form stellar interiors may be roughly divided into two aspects: (1) evaluation of Coulombic interactions among all charged particles; and (2) ionization equilibrium with accompanying problems of atomic bound states. The second aspect, ionization equilibrium, is generally the more significant as well as the more complex for numerical calculations; basically this is because the pressure depends so much on how many free particles are present. The numerous problems, conceptual and calculational, concerning

* Work performed under the auspices of the U.S. Atomic Energy Commission.

ionization equilibrium will be discussed at this conference by H. Graboske. This talk will deal exclusively with the first aspect, evaluation of Coulombic interactions, and will describe the results of quantum mechanical perturbation expansions in statistical mechanics of ionized gases. It is assumed that one knows the temperature of the gas and the number density of each charged species. For temperatures so high that one has essentially complete ionization, the methods of statistical mechanics give quite good results for the Coulombic deviations from the ideal gas pressure. When these results are combined with a good procedure for calculating ionization equilibrium for partially ionized gases, it is still possible to get numerical results of satisfactory accuracy.

In the interior of most large hot stars, the temperature is high enough that the equation of state is given fairly well by the ideal gas law when the partial degeneracy of the electrons is taken into account. In this situation it is commonly assumed that the Coulombic corrections to the ideal gas pressure are given by the Debye–Huckel theory. Thus, a simple result for the equation of state for temperature $\beta = 1/kT$ and total number density of $n = n_e + \Sigma n_i$ $(n_e = N_e/V)$ is

$$\beta P/n = \left(f_e \frac{\mathscr{I}_{3/2}(a_e)}{\zeta_e} + \Sigma f_i \right) - \frac{1}{6} \Lambda \tag{1}$$

where $f_e = n_e/n$ and $f_i = n_i/n$ are the fractions electrons and various ionic species. The quantum ideal gas pressure of the electrons is determined by the usual Fermi integrals:

$$\mathscr{I}_{3/2}(a_e) = \frac{1}{\Gamma(5/2)} \int_0^\infty \frac{dz\, z^{3/2}}{e^{-a_e + z} + 1} = e^{a_e} - \frac{e^{2a_e}}{2^{5/2}} + \cdots \tag{2}$$

$$\zeta_e = \frac{(2\pi\hbar)^3 n_e}{2(2\pi m_e kT)^{3/2}} = \mathscr{I}_{1/2}(a_e) = \frac{1}{\Gamma(3/2)} \int_0^\infty \frac{dz\, z^{1/2}}{e^{-a_e + z} + 1} = e^{a_e} - \frac{e^{2a_e}}{2^{3/2}} + \cdots.$$

The parameter a_e is the chemical potential divided by kT, $a_e = \beta\mu_e$. The classical Debye–Huckel parameter Λ is

$$\Lambda = \frac{1}{4\pi n \lambda_D^3} = \frac{\beta\langle z^2 \rangle e^2}{\lambda_D} = 2\pi^{1/2}\langle z^2 \rangle^{3/2} e^3 \beta^{3/2} n^{1/2} \tag{3}$$

$$\lambda_D = (kT/4\pi e^2 \langle z^2 \rangle n)^{1/2}, \quad \langle z^2 \rangle = \frac{n_e + \Sigma z_i^2 n_i}{n} \tag{4}$$

and λ_D is the classical multi-component Debye screening length. As long as the Debye term in (1), $-\Lambda/6$, is not too large (i.e. only a small fraction of the ideal gas term) and the electrons are not too degenerate, this formula is quite satisfactory. The Debye result quoted here is completely classical and the electrons and ions are treated on an equal footing. The dimensionless parameter Λ is of the form $(\langle z^2 \rangle e^2 / \lambda_D)/kT$ and is evidently a measure of the ratio of the average Coulomb interaction energy to the kinetic energy.

As one goes to the temperature and density regimes found in low-mass stars, Eq. (1) begins to fail for several reasons. First, even for classical particles the Debye theory is quite inadequate when Λ is of the order of 1 or 2, the region of intermediate coupling. The Debye term is only the first term in a rather complicated cluster expansion (Abe, 1959). The form of this cluster expansion has a considerable resemblance to the Mayer cluster expansion for ordinary non-ideal gases; a major difference is that the screened Coulomb potential $u_s = (e^2/r) \exp(-r/\lambda_D)$ replaces the Coulomb potential, $u = e^2/r$ at least in the classical theory. Secondly, quantum effects become quite significant. The electron–electron and electron–ion interactions are very much affected by the uncertainty principle, and this gives rise to wave-mechanical diffraction effects. Also, Fermi statistics for the electrons change the classical Debye screening length, and ultimately when there is a considerable degree of electron degeneracy, the form of the Debye term is also quite different. Both effects of quantum mechanics enter the calculation of the next term beyond the Debye term. In addition, there are electron–electron exchange interactions which ultimately become numerically significant. Thus the first necessity is a fully quantum generalization of the Debye result, i.e. including both wave mechanics and quantum statistics (Montroll and Ward, 1961). Beyond this is needed the quantum theory of the higher order terms (DeWitt, 1966). Theoretical results for these problems have existed for some time, but they are of sufficient complexity that they are only recently being used for numerical evaluation of equation-of-state calculations.

In the previous paragraph we have envisaged low-mass stars in which both electrons and ions are not too strongly coupled gases, i.e. the Coulombic effects do not cause strong correlations among charges. The ions because of their large mass are essentially classical, but the electrons are definitely quantum mechanical and thus quantum effects enter both electron–electron and electron–ion interactions. In white dwarf stars, another situation prevails. The ions, still classical, are very strongly coupled; so strongly correlated in fact that many white dwarf stars may have a lattice structure, as discussed by Van Horn at this conference. The electrons on the other hand are completely degenerate but only weakly coupled. Most of the pressure comes from the electron quantum ideal gas formula for zero temperature since kT is much less than the electron Fermi energy. Deviations from the quantum ideal gas law are discussed at length by Salpeter using the Thomas–Fermi method.

Thus, in white dwarf stars one has the interesting situation in which the ionic pressure contribution is completely dominated by the Coulombic interaction, but the electrons are little affected by Coulomb interactions with each other because their average kinetic energy, $3\epsilon_F/5$, is so much greater than the average Coulombic interaction energy.

In this report it is only possible to give a descriptive survey of the results of quantum statistical mechanics of multi-component plasmas. Accurate calculations are possible for fully ionized gases such as hydrogen and helium, and these results should tie on to Thomas–Fermi results for plasmas with a significant amount of high Z nuclei. The method used is standard statistical mechanics with a perturbation expansion of the logarithm of the partition function,

$$Z(\beta, n_e, n_i) = Tr\ e^{-\beta(H_0+H_I)} = e^{-\beta(F_0+F_I)} \qquad (5)$$

where H_0 is the sum of all particle kinetic energies,

$$H_I = \sum_{\alpha<\beta} z_\alpha z_\beta e^2/r_{\alpha\beta}$$

is the Coulomb interaction energy, and $F = E - TS$ is the Helmholtz free energy, and F_I is the portion due to the Coulomb interactions. The evaluation is subject to the restriction of electrical neutrality,

$$n_e = \sum_i z_i n_i.$$

The pressure is found from

$$\beta P = \frac{\partial}{\partial V} \log Z = -\frac{\partial}{\partial V}\beta(F_0 + F_I). \qquad (6)$$

II. FUNDAMENTAL LENGTHS AND DIMENSIONLESS PARAMETERS

Before discussing the various terms which come into F_I, it will be useful to mention the fundamental lengths in the theory of multi-component plasmas and the dimensionless parameters that may be formed from them. The magnitude of the dimensionless parameters determines the importance of various contributions to F_I, and also something about the difficulty of calculation of F_I for various values of temperature and density. The fundamental lengths ordered as they would be for most stellar temperature and densities are

$$\begin{array}{cccc} \lambda_\iota & < \quad r_0 & < \quad \lambda_e & < \qquad\qquad \lambda_s \\ \| & \| & \| & \| \\ \dfrac{\hbar}{(2m_ikT)^{1/2}} & \dfrac{e^2}{kT} & \dfrac{\hbar}{(2m_ekT)^{1/2}} & \left[\dfrac{kT}{4\pi e^2 n(\theta_e f_e + \Sigma z_i^2 f_i)}\right]^{1/2} \end{array} \qquad (7)$$

where r_0 is the classical distance of closest approach of two repulsive charges in a thermal distribution, and λ_i and λ_e are the thermal de Broglie wavelengths of ions and electrons. Since m_i/m_e is at least 2000, we have $\lambda_i \sim \lambda_e/40$. Consequently, λ_i is generally less than r_0 and hence the ions are essentially classical, whereas typically $\lambda_e > r_0$ and the electrons are quantum mechanical. One may think of the electrons no longer as points but as wave packets with a radius of roughly λ_e, and these finite size wave packets overlap considerably when two electrons collide. The term λ_s is a generalized screening length which takes into account the fact that the electrons at the bottom of the Fermi momentum distribution are frozen and cannot participate in screening. This effect gives rise to a degeneracy parameter for electrons in the screening length

$$\theta_e = \frac{\mathscr{I} - 1/2(a_e)}{\zeta_e}, \quad \mathscr{I}_{-1/2}(a) = \frac{1}{\Gamma(1/2)}$$

$$\int_0^\infty \frac{dz\, z^{-1/2}}{e^{-a+z} + 1} = e^a - \frac{e^{2a}}{2^{1/2}} + \ldots \text{ for } e^a < 1, \quad (8)$$

where ζ_e is given by (2). For high temperature where the electrons are $\mathscr{I}_{-1/2}$ non-degenerate θ_e is 1, and the screening length is the same as the classical Debye length. As the temperature is lowered (or density increased), θ_e decreases and finally goes to $\theta_e = (3/2)kT/\epsilon_F$. Note that $\zeta_e \sim n_e \lambda_e^3 = (\lambda_e/a)^3$ with $n_e = 1/a^3$. For high T and small n one has $\zeta_e \ll 1$, or λ_e much less than the interelectron distance a; in this case the electrons are non-degenerate and described quite well by Maxwell–Boltzmann statistics. The electrons become partially degenerate at higher densities when $\lambda_e \sim a$ and reaches extreme degeneracy $\lambda_e \gg a$, at low temperature or extremely high density. To summarize, we can talk about three degeneracy regions:

I. $\zeta_e \ll 1$, little degeneracy, $a < -4,\ \theta_e \to 1$

II. $\zeta_e \sim 1$, partial degeneracy, $-4 < a < 4,\ \theta_e \sim 1/2$

III. $\zeta_e \gg 1$, extreme degeneracy, $a > 4,\ \theta_e \to 0$.

For a moment now let us consider a one-component gas, an electron gas in which the ions are considered to be completely smeared out and serve only to keep the system electrically neutral. This one component gas may sound unrealistic in a meeting on real stars, but it is a system that statistical mechanicians work on first before going to the additional complexities of the rea multi-component plasmas. For the one-component gas, there will be only two dimensionless parameters,

$$\Lambda_e = \frac{r_0}{\lambda_s} = \frac{1}{4\pi n_e \lambda_s{}^3} = 2\pi^{1/2} e^3 \theta_e{}^{3/2} \beta^{3/2} n_e{}^{1/2} \sim \frac{\langle PE \rangle}{\langle KE \rangle} \qquad (9)$$

and

$$\gamma = \frac{\lambda_e}{\lambda_s}. \qquad (10)$$

In Region I ($\zeta_e \ll 1$) the Coulomb interaction parameter defined by (9) is just the classical Debye parameter for a one-component system as given by (3). On a density-temperature plot, a line of constant Λ_e would have a slope of 3, whereas a line of constant ζ_e has a slope of 3/2. In Region II the line of constant Λ_e begins to curve upwards and backwards because of the effect of the degeneracy parameter θ_e. In Region III for zero temperature this Λ_e becomes

$$\Lambda_e = \left(\frac{6\pi n_e e^6}{\epsilon_F{}^3} \right)^{1/2} \sim \frac{1}{n_e{}^{1/2}}. \qquad (11)$$

Λ_e is a measure of Coulombic effects for electrons at least in the weak coupling region, i.e., that is for values of Λ_e up to about 1. The other parameter γ_e is the quantum diffraction parameter. Generally, for values of density and temperature in any of the degeneracy regions such that Λ_e is very small, the Coulombic corrections to the quantum ideal gas are also very small, and these corrections are given by the quantum generalization of the Debye–Huckel term. For larger values of Λ_e, say approaching 1, higher order terms of the cluster expansion must be added to the Debye term to get sufficient accuracy. The behavior of Λ_e on a log n versus log T plot is shown in Figure 1. The number density of electrons has been translated into mass density to make the plot more useful for astronomers. The value of Λ_e was calculated for a number of low-mass stars, and the location of these low-mass stars on this plot is indicated. It is apparent that the electrons in numerous low-mass stars are not strongly coupled, but the values of Λ_e are such that Coulombic corrections to the pressure can be quite significant. Some stars are in Region I where calculations are relatively easy, and some are in Region II where electron degeneracy effects and wave mechanics get mixed up and calculations are quite tedious. The electrons in white dwarf stars are in Region III on this plot and since Λ_e is generally fairly small for them, the Coulomb interactions do not contribute much to the total pressure.

For some purposes it is useful to discuss the coupling of the ions with each

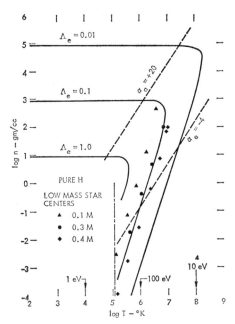

FIGURE 1 The Coulomb interaction parameter, Λ_e, as a function of temperature and density.

other in terms of the Debye parameter. Since the mass is so large we have usually $\theta_i = 1$, and hence

$$\Lambda_i = \frac{z_i^2 \beta e^2}{\lambda_i} = 2\pi^{1/2} z_i^3 e^3 \beta^{3/2} n_i^{1/2}. \tag{12}$$

In some parts of a star the electrons may be weakly coupled ($\Lambda_e < 1$), but the ions can be fairly strongly coupled ($\Lambda_i \sim 1$, or $\Lambda_i \gg 1$). The existing quantum perturbation methods in statistical mechanics cannot handle this situation very accurately yet. If the ionic density is such that Λ_i is very large, it is no longer a suitable parameter for indicating the importance of Coulombic effects. In this case the 2/3 power of Λ_i is more appropriate. Brush, Sahlin, and Teller (1966) in their very extensive Monte Carlo calculations of the thermodynamic properties of a one component gas of classical ions use

$$\Gamma = \frac{\beta e^2}{a}, \quad n_i^{-1} = \frac{4\pi}{3} a^3 \tag{13}$$

(With the average interionic distance defined as above, one finds $\Lambda_i = \sqrt{3}\Gamma^{3/2}$.) When $\Gamma = 125$ they find that the one component gas of classical charges goes into a lattice. This rather high value of Γ at which the transition from gas to lattice takes place is quite interesting for Van Horn's study of white dwarf stars in which the nuclei are the classical charges and the electrons in a completely degenerate Fermi distribution form the neutralizing background.

Finally for multi-component plasmas in which both the electrons and ions are weakly coupled and hence treated more or less on an equal basis, the Coulomb interaction parameter is

$$\Lambda_s = \frac{\beta e^2 \langle z^2 \rangle}{\lambda_s} = \frac{1}{4\pi n \lambda_s^3} \tag{14}$$

$$\langle z^2 \rangle = \theta_e f_e + \sum_i z_i^2 f_i.$$

For low mass stars, other than white dwarfs, in which Λ_s is not too large, $\Lambda_s \lesssim 2$, perturbation methods in statistical mechanics yield reasonably accurate results. For small cool stars and planetary interiors the value of Λ_s is so large that exact statistical mechanical methods are not of much use at the present time.

III. SUMMARY OF QUANTUM CLUSTER EXPANSION RESULTS FOR WEAKLY COUPLED PLASMAS

Since the numerical calculations for the equation of state of various assumed stellar compositions now being done by Graboske begin with the Helmholtz free energy, it is appropriate to quote the free-energy results here, rather than expressions for the pressure. Numerical values of the pressure will be obtained generally by numerical differentiation of the free energy. We use the free energy in the form

$$F = (F_{oe} + \sum F_{oi}) + F_{ex} + F_{\text{ring}} + F_{2,ee} + 2F_{2,ei} + F_{2,ii} + \dots \tag{15}$$

The first term on the right-hand side gives the classical and quantum ideal gas pressure for ions and electrons written down in Eq. (1). The exchange free energy, F_{ex}, is due to electrons alone; it is generally small but can contribute significantly when Λ_e is large (~ 1) and the electrons are partially degenerate (Region II). The term F_{ring} is the quantum generalization of the Debye–Huckel theory. The subscript ring comes from the ring diagrams that are summed to get the Debye–Huckel theory rigorously. The F_2 terms for electron–electron, electron–ion, and ion–ion interactions are sums of the so-called ladder diagrams in the full quantum theory. Let us take these in turn. In Region I the results for the exchange free energy is

$$\beta F_{ex} = -N_e \frac{1}{2} \left\{ \frac{\gamma_e^2}{2} - \frac{\pi^{1/2} \log 2}{2^{5/2}} \Lambda_e \gamma_e + \dots \right\}, \gamma_e = \frac{\lambda_e}{\lambda_s} = \frac{\hbar (2\pi e^2 n_e)^{1/2}}{kT m_e^{1/2}}$$

$$\tag{16}$$

In Region II there is no simple answer for even the first-order exchange; as an integral the result is

$$\beta F_{ex} = -N_e \frac{1}{2} \cdot \frac{\gamma e^2}{\zeta e^2} \int_{-\infty}^{a} da' \, \mathscr{I}_{-1/2}{}^2(a').$$ (17)

The general form of the ring sum-term is

$$\beta F_{ring} = -\frac{V}{2} \sum_{\nu=0,\pm1,\dots\pm\infty} \int_0^\infty \frac{4\pi k^2 \, dk}{(2\pi\hbar)^3} \{X(k,\nu) - \log[1 + X(k,\nu)]\}$$ (18)

$$X(k,\nu) = \frac{4\pi\beta e^2}{k^2} \sum_j z_j{}^2 f_j L(x_j{}^2, 2\pi i\nu), \quad x_j = \lambda_j k$$

$$L(x_j{}^2, 2\pi i\nu) = \int \frac{d^3y}{\pi^{3/2}\zeta} \frac{f^-(|\bar{y} + \bar{x}|) - f^-(y)}{(|\bar{y} + \bar{x}|^2 - y^2) - 2\pi i\nu}$$

$$f^-(y) = \frac{1}{e^{-\alpha+y2} + 1}.$$

This rather opaque formula makes use of the dynamic screened potential rather than the static screened potential $u_s = (e^2/r) \exp(-r/\lambda_s)$, that gives the classical Debye result. The dynamic screened potential can only be expressed as a Fourier transform,

$$u_s(k,\omega) = \frac{4\pi e^2}{k^2 + X(k,\beta\hbar\omega)} \xrightarrow[\omega=0]{} \frac{4\pi e^2}{k^2 + 1/\lambda_s{}^2}$$ (19)

$$= \frac{u(k)}{\epsilon(k,\omega)}, \quad \epsilon(k,\omega) = 1 + X,$$

where $\epsilon(k,\omega)$ is the dielectric function for the plasma. Besides inclusion of wave mechanics, quantum statistics and proper screening, the ring sum formula (18) has contributions from another physical effect, i.e. plasma oscillations. These oscillations occur when the real part of the dielectric function vanishes. In Region I F_{ring} may be evaluated analytically with the result (DeWitt, 1962),

$$\beta F_{ring} = -N \frac{\Lambda_s}{3} \left(1 - \frac{3\pi^{1/2}}{2^{3/2}} \gamma_e \cdots\right)$$

where the first term is the Debye term and the term with γ_e is due to wave mechanics. In Region II there seems to be no hope for getting a useful analytic answer, and one must resort to numerical evaluation. The general ring sum has been programmed, though unfortunately numerical results were not available at the time of this conference.

The next terms in Eq. (14) describe three and more scatterings of two particles via the dynamic screened potential. These terms are described by the ladder diagrams in field theory; in the classical limit they have a formal resemblance to the second virial coefficient of ordinary gases. The classical form appropriate for the ion–ion interaction is

$$\beta F_{2,ii} = -\frac{N_i^2}{2V} \int_0^\infty 4\pi r^2 \, dr \left\{ e^{-q_{ii}} - 1 + q_{ii} - \frac{1}{2} q_{ii}^2 \right\}, \quad q_{ii}$$

$$= \frac{z_i^2 \beta e^2}{r} e^{-r/\lambda_s} \tag{20}$$

$$= -N_i f_i^2 \frac{(z_i^2 \beta e^2 / \lambda_s)^3}{12 \Lambda_s} \left[\log \frac{z_i^2 \beta e^2}{\lambda_s} + D_C \right]$$

$$D_c = 2C + \log 3 - 11/6, \quad C = 0.5772.$$

The evaluation of the integral in (20) is useful when $z_i^2 \beta e^2 / \lambda_s < 1$; otherwise the integral must be done numerically. One notes that in this result the logarithm term involves the two lengths λ_s and $z_i^2 \beta e^2$, and the logarithm itself arises from the fact that the first term in the bracket of the integrand in (20) when expanded in powers of the screened potential goes as $(1/r)^3 \exp(-3r/\lambda_s)$. Thus λ_s is a long-distance cutoff that would appear if the pure Coulomb potential had been used, while the classical distance of closest approach for two charges $z_i^2 \beta e^2$ provides the short distance cutoff.

For the electron–electron and electron–ion for which the de Broglie wavelength is greater than the classical distance of closest approach, one would provide the short-distance cutoff. An exact evaluation bears out this expectation. A somewhat imprecise expression for $F_{2,ei}$ which is suitable for numerical work is

$$\beta F_{2,ei} = +\frac{N_e N_i}{2V} \int_{\lambda_e^a{}_i}^\infty 4\pi r^2 \, dr \left\{ e^{|q_{ei}|} - 1 - |q_{ei}| - \frac{1}{2}|q_{ei}|^2 \right\} \tag{21}$$

$$q_{ei} = \frac{z_e z_i \beta e^2}{r} e^{-r/\lambda_s},$$

where the electronic charge number is $z_e = -1$. Without the short-range cutoff provided by wave mechanics this expression would diverge. The exact result for high-temperature ($\lambda_{ei} > z_i \beta e^2$) with λ_{ei} as the thermal de Broglie wavelength of the electron ion system $\lambda_{ei} = \hbar/(2\mu_{ei}k\lambda)^{1/2}$ and $\mu_{ei} = m_e m_i/(m_e + m_i)$ is

$$\beta F_{2,ei} = +N f_e f_i \frac{(z_i \beta e^2/\lambda_s)^3}{12\Lambda_s}\left[\log\frac{\lambda_{ei}}{\lambda_s} + D_Q\right] \qquad (22)$$

$$D_Q = \log 3 + \tfrac{1}{2}C - \tfrac{1}{2}.$$

The very involved calculation of the integrals that give the constant D_Q have only recently been carried out by Hoffman and Ebeling (1968). Again this result (22) is useable when $z_i\beta e^2/\lambda_s < 1$ and $\mu_{ei} > z_i\beta e^2$; otherwise the integral form (21) must be evaluated numerically. The lower cutoff in (21), $a\lambda_{ei}$, is chosen so that at high temperature one recovers (22). The expression for $F_{2,ee}$ is the same as $F_{2,ei}$ when one takes into account the fact that electrons repel each other while the electrons and ions attract,

$$\beta F_{2,ee} = -\frac{Ne^2}{2V}\int\limits_{a\lambda_{ee}}^{\infty} 4\pi r^2\, dr\left\{e^{-q_{ee}} - 1 + q_{ee} - \tfrac{1}{2}q_{ee}^2\right\}, q_{ee} = \frac{z_e^2\beta e^2}{r}e^{-r/\lambda_s}$$

$$(23)$$

$$= -N f_e^2 \frac{(\beta e^2/\lambda_s)^3}{12\Lambda_s}\left[\log\frac{\lambda_{ee}}{\lambda_s} + D_Q\right].$$

For temperature and density regions such that the electrons are only slightly degenerate and the coupling is not too strong, a very suitable expression for the equation of state of the multi-component system is

$$\beta(P - P_{\text{ideal}})/n =$$

$$-\frac{1}{6}\Lambda_s\left(1 - \frac{3\sqrt{\pi}}{2^{7/2}}\gamma_e\right) + \frac{4\pi n(\beta e^2)^3}{12}\left[f_e^3\left(\log\frac{\lambda_{ee}}{\lambda_s} + D_Q + \frac{1}{2}\right)\right. \qquad (24)$$

$$\left. - 2f_e f_i z_i^3\left(\log\frac{\lambda_{ei}}{\lambda_s} + D_Q + \frac{1}{2}\right) + f_i^3 z_i^6\left(\log\frac{z_i^2\beta e^2}{\lambda_s} + D_Q + \frac{1}{2}\right)\right].$$

The expressions for the free energy given so far have been for electrons and point ions (protons, helium nuclei, etc.). The situation becomes considerably more complicated when there are an appreciable number of nuclei with bound electrons, thus finite size ions. Graboske discusses a finite size modification of the Debye–Huckel term. For the $F_{2,ei}$ and $F_{2,ii}$ the previous discussion of a short-range cutoff at the thermal de Broglie wavelength becomes somewhat academic. The easiest procedure that is still reasonable is to treat the finite size ions as charged hard spheres and in the integral definitions (21) and (22) to take as the lower limit of the integration the radii of the finite size ions.

Numerical calculations of the equation of state using Eq. (14) are now being carried out. More results from the quantum perturbation theory of the partition function will be added as needed. In general, it seems that as long as the Coulomb interaction effects are not more than roughly 40 per cent of the ideal gas pressure it should be possible to calculate the equation of state with existing theory to within 2 or 3 per cent.

References

ABE, R., 1959, *Prog. Theor. Phys.* **22**, 213.
MONTROLL, E. W., and WARD, J. C. 1958, *Phys. Fluids* **1**, 55.
DE WITT, H. E. 1961, *J. Nucl. Energy, Part C: Plasma Phys.* **2**, 27.
DE WITT, H. E. 1966, *J. Math. Phys.* **7**, 616.
BRUSH, S. G., SAHLIN, H. L., and TELLER, E. 1966, *J. Chem. Phys.* **45**, 2102.
DE WITT, H. E. 1962, *J. Math. Phys.* **3**, 1216.
HOFFMAN, H. J., and EBELING, W. 1968, *Beitr. Plasmaphysik* **8**, 43.

Discussion

SAVEDOFF   Dr. Salpeter mentioned the crystallization region. Have you integrated into that region yet?

DEWITT   I could incorporate crystallization into my treatment.

Ionization equilibrium and equations of state for low-mass stars*

HAROLD C. GRABOSKE, Jr

Lawrence Radiation Laboratory, University of California, Livermore

Abstract

A general statistical mechanical method has been utilized to compute the equilibrium composition and thermodynamic properties of a dense, partially ionized gas. Several free-energy models have been applied to study the non-ideal behavior of such a system, taking into account the Coulombic interactions of the charged particles, the excluded volume effect of the extended atoms and molecules, and the density perturbations of the bound states. It is found that the bound state perturbation is the key factor in a realistic model for pressure ionization and dissociation. Thermodynamic properties of a weakly ionized dense gas depend sensitively on the bound state perturbation and excluded volume correction. The highly ionized gas requires a high-order plasma theory to obtain accurate thermodynamic properties. A specific non-ideal model is applied to low mass stars (0.3 and 0.1 $M_\odot$), and it is found that all three non-ideal effects are important in the interiors of stars in the mass range $M \leqslant 0.3\ M_\odot$.

I. INTRODUCTION

THE STUDY of the structure and evolution of low-mass stars requires an accurate knowledge of the constitutive physics which describe the properties of matter in their interiors. Thermodynamic descriptions of the astrophysical mixtures in the interiors of low-mass stars ($M \leqslant M_\odot$) have been generally formulated as ideal or nearly ideal gases (Kumar, 1963; Hayashi and Nakano, 1963). In dense gases such as the interiors of low-mass stars, significant deviations from ideal behavior occur, requiring much more complex thermo-dynamic models to describe the non-ideal gas.

* Work performed under the auspices of the U.S. Atomic Energy Commission.

Non-ideal behavior can be separated into three distinct parts. First, one must account for the Coulombic interactions of the charged components of the gas (ions, bare nuclei, electrons). This is the problem of the quantum-mechanical, fully-ionized equilibrium plasma, the many-body problem discussed by DeWitt (1969). Second, the perturbations of the bound eigenstates of an atom or molecule by its neighboring atoms and ions render invalid the use of isolated atom internal partition functions. Instead of energy eigenvalues taken from experimental measurements or theoretical calculations for the free or isolated atom or molecule, complicated theoretical calculations of the eigenstates of perturbed systems must be made. Third, at high density the volume occupied by the atoms and ions becomes a significant fraction of the total volume, and the effects of this excluded volume must be incorporated into the theory.

Each of these non-ideal effects—the plasma interaction, the bound-state perturbation, and the excluded volume correction—are formidable problems in themselves and admit of no rigorous statistical mechanical solutions at the present time. This study is concerned with utilizing recent theoretical models for these various effects in an attempt to determine their importance for low-mass star interiors.

II. THERMODYNAMIC FORMALISM

The approach chosen for this problem is a numerical application of standard statistical mechanical methods and is correct in the limits of low density, low temperature, and high temperature. It presents an alternative, low-density approach to the statistical model of the atom and its developments (Thomas–Fermi models), and emphasizes the use of exact thermodynamic relations for a real gas.

The description of a system is given by a total partition function, separable into three components: the translational, the configurational, and the internal partition functions. The astrophysical independent variables ρ and T translate directly into the set ($\{N_i\}$, V, T) which are the basic variables for the canonical partition function and the associated thermodynamic state function F, the Helmholtz free energy. Thermodynamic equilibrium requires that the free energy be minimized with respect to the particle numbers at constant V and T, subject to the stoichiometric constraints. For given values of V and T, this set of minimization equations determines the equilibrium values of the free energy and the corresponding particle numbers. The equilibrium having been determined, the equations of state can then be computed. For the canonical ensemble, the equations of state are the pressure P, the entropy S, and the chemical potentials μ_i. Other thermodynamic potentials such as internal energy U and enthalpy H are now readily computed from the known parameters. Last, one can compute thermodynamic properties such as specific heats C_p and C_v and their ratio γ.

The thermodynamic formalism is:

$Z = Z(\{N_i\}, V, T) \equiv Z(A_j)$ Total partion function for independent variables A_j

$F = -NkT \ln Z$ State function

$dF = 0 \rightarrow \left(\dfrac{\partial F}{\partial N_i}\right)_{N_j, V, T} = 0$ Equilibrium condition for all species i, subject to chemical constraints

$\text{EOS} = \left(\dfrac{\partial F}{\partial A_j}\right)_{A_{k \neq j}}$ Equations of state

$\text{TDP} = \dfrac{\partial^2 F}{\partial A_j \partial A_k}$ Thermodynamic properties

For a complicated, non-ideal, partially ionized gas, the free energy is a highly involved function of $\{N_i\}$, V, and T, which is not solvable in analytic form (primarily due to the internal partition function). To solve this problem, the free energy minimization is performed on a computer. The method described here is part of a program carried out by H. Graboske and D. J. Harwood at the Lawrence Radiation Laboratory, and was originally developed by G. M. Harris (1959, 1961). It is, in essence, a numerical solution of the thermodynamic equilibrium equations. There are three steps:

1. The free energy is written for the entire system as a function of the independent variables:

$$F = F(\{N_i\}, V, T)$$

This free-energy expression involves various approximate forms or theoretical models for the non-ideal terms, and constitutes the fundamental approximation of the method.

2. The free energy of a system of specified chemical composition at fixed V and T is minimized with respect to the particle numbers, subject to the stoichiometric constraints. This is done by a gradient search method in composition space on a large computer. The result is an equilibrium-free energy and corresponding equilibrium concentrations of all the species.

$$F_{\text{equil}}, \{x_i\}_{\text{equil}} \text{ at } V, T$$

where x_i are the mole numbers of species i, N_i/N_0 (Avogadro's number)

3. The equations of state and thermal properties as functions of ($\{N_i\}$, V, T) are computed.

Two models were investigated with the above method. The first, which shall be called the isolated atom (IA) model, is defined by the following forms:

$$F = \sum_{i=1}^{6} F_i$$

F_1 = translational free energy of heavy particles (assumed Maxwellian here)

$$= - \sum_{\substack{\text{all} \\ \text{heavy particles}}} N_i kT \ln \left(\frac{V e \zeta_i}{N_i} \right)$$

where

$$\zeta_i = (2s_i + 1) \left(\frac{2\pi m_i kT}{h^2} \right)^{3/2};$$

$$s_i = \text{spin of particle of type } i;$$

$$m_i = \text{mass of particle of type } i.$$

F_2 = internal free energy of all species with bound states

$$= - \sum_{\substack{\text{all} \\ \text{bound} \\ \text{species}}} N_i kT \ln (Z^0{}_{\text{int}})_i$$

where

$$(Z^0{}_{\text{int}})_i$$

is the internal partition function for the isolated (free) atom of species i, a function of T only.

F_3 = translational free energy of electrons

$$= N_e kT \left[a_e - \frac{I_{3/2}(a_e)}{I_{1/2}(a_e)} \right]$$

where a_e is the degeneracy parameter for electrons, defined by

$$\frac{N_e}{V \zeta_e} = I_{1/2}(a_e); \quad I_n(a_e) \equiv \frac{1}{\Gamma(n+1)} \int_0^\infty \frac{x^n dx}{e^{x-a} + 1}.$$

F_4 = configurational free energy due to Coulomb interaction of free charges, given as a modified ring term (De Witt, 1961);

$$= - \sum_{\substack{\text{all} \\ \text{charged} \\ \text{species}}} N_i kT \left(\frac{-\Lambda}{3} \right) \tau(x)$$

where Λ is the generalized plasma parameter

$$\Lambda = \frac{\beta e^2 \langle Z^2 \rangle}{\lambda} \quad \text{and} \quad \beta = \frac{1}{kT}$$

$$\langle Z^2 \rangle = \frac{\sum Z_i^2 N_i \theta(a_i)}{\sum N_i}$$

$$\lambda = [4\pi \beta e^2 \sum_i Z_i^2 n_i \theta(a_i)]^{-1/2}$$

$$\theta(a_i) \equiv \frac{I_{-1/2}(a_i)}{I_{1/2}(a_i)}$$

and $\tau(x)$ is the correction for finite ion size,

$$\tau(x) = \frac{3}{x^3}\left[\ln(1 + x) - x + \frac{x^2}{2}\right]$$

with

$$x = r_{e/\lambda}, \quad r_e = \frac{e^2 \langle z \rangle}{E_F}$$

$$E_F = kT \frac{I_{3/2}(a_e)}{I_{1/2}(a_e)}.$$

The short-range cutoff distance r_e is a minimum interaction distance, computed by assigning as free electrons only those with positive energy. The total energy of an electron is computed as the sum of the mean kinetic energy (E_F) plus its Coulomb interaction with neighboring screened nuclei of effective charge $\langle Z \rangle$. This short-range cutoff for the Coulomb interaction is a density- and temperature-dependent function which includes the quantum statistical effects of the fermions.

F_5 = configurational free energy due to excluded volume of extended (non-point) species

$$= \frac{kTV}{\pi}\left\{6\xi_0 \ln\left(\frac{1}{1 - \xi_3}\right) + \frac{\xi_2}{1 - \xi_3}\left(18\xi_1 + \frac{9\xi_2^2}{(1 - \xi_3)^2}\right)\right\}$$

$$\xi_l = \frac{\pi}{6}\sum_{i=1}^{m}\frac{N_i}{V}(2R_i)^l$$

where m = number of extended species, and R_i is the hard-sphere radius for particles of species i.

The excluded volume correction is derived by Lebowitz *et al.* (1965) for the configurational free energy of a mixture of hard spheres. This analytical form agrees well with values computed for a hard-sphere gas by Monte Carlo and molecular dynamics methods, and is valid for $v/V \rightarrow 1$ (where v

is the excluded volume), as opposed to the more usual linear hard-sphere approximation, valid only for $v/V \ll 1$.

F_6 = configurational free energy due to weak (attractive) interaction of molecular species

$$= \frac{-\sum N_i^2 a_i}{V}$$

where a_i is a constant for molecular species i.

The weak molecular interaction is of importance only near the liquid region, but is included for completeness, in the form of a Van der Waals attractive interaction.

The IA model is close in form to the model of Gabriel (1966), which gives the best previously available low-mass star equation of state. Differences between these models are threefold. The internal partition function is computed here for all states, rather than using ground-state statistical weights alone. This is necessary since the contribution of the excited states, generally small at normal densities and temperatures, becomes important at high densities. In the IA model, the short-range cutoff for the plasma interaction is density- temperature-dependent parameter which is related to the interacting Fermi–Dirac characteristics of the particles. Last, the IA model's excluded volume correction is a high-order theory, accurate in the region where it is an important factor.

The second model to be considered is identical to the IA model in all terms except F_2. The confined atom (CA) model is a development of the work of Harris (1961). It introduces a volume-dependent internal partition function, which includes a perturbation of the bound states by the neighboring atomic and molecular systems. Each atom is assumed to be confined in a spherical box of radius r_0, by the presence of the other atoms. The Schrödinger equation for hydrogenic atoms and ions in a hard spherical box is identical to that of the isolated atom. The boundary condition at the edge is no longer

$$\psi(r) \rightarrow 0 \quad \text{at} \quad r = \infty$$

but instead becomes

$$\psi(r) \rightarrow 0 \quad \text{at} \quad r = r_0$$

where r_0 is the radius of the atomic cell.

This change still leaves a radial wave equation with analytic solutions, the confluent hypergeometric function (De Groot and Ten Seldam, 1946). These solutions to the Schrödinger equation for a confined atom have two significant characteristics. The energy eigenvalues for state nl are dependent on the radius of the confining volume, and are shifted toward higher (less negative) values as r_0 decreases. Also, at some critical value of r_0, the energy for state nl reaches zero, and for smaller r_0, is positive. From these characteristics of the confined atom model, we see a system whose energies are dependent on the volume into which it is compressed, and a system which at some radius has electron energy states which are changed from negative to positive, interpretable as a change from a bound to a free state.

For non-hydrogenic species, the confined atom hydrogenic eigenstates are combined in a simple atomic orbital calculation to create a scaled confined atom model for all species with bound states. The confined atom orbitals for all species are normalized to the observed energies and radii for the isolated atoms and molecules. In this way, the correct free atom values are recovered in the limit of high volume or low density.

The volume- or density-dependent energy eigenvalues for the confined atom models of H and H_2 are illustrated in Figure 1. Using the volume-dependent eigenvalues, $E_j(r)$, and maximum bound quantum numbers, $n_{max}(r)$, for all bound species, one can construct an internal partition function

$$(Z_{int}^{CA})_i = \sum_{j=1}^{n_{max}(r)} g_j\, e^{-\beta E_j(r)} \sum_{k=0}^{k_{max}(j)} e^{-\beta(kV_j^0 + k^2 W_j^0)} \sum_{l=0}^{l_{max}(k)} (2l+1)\, e^{-\beta B_j^0 l(l+1)}$$

where g_j is the statistical weight of level j, V_j^0 and W_j^0 the vibrational energies, B_j^0 the rotational energy.

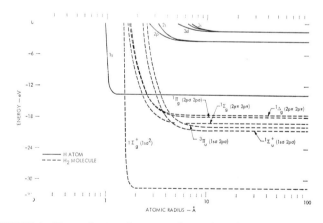

FIGURE 1 Dependence of energy eigenvalues for H and H_2 on the radius of the confined atom.

The free energy of the CA model is then

$$F = \sum_{i=1}^{6} F_i$$

with $F_2 = - \sum_{\substack{\text{all} \\ \text{bound} \\ \text{species}}} N_i kT \ln (Z_{\text{int}}^{CA})_i$

III. THERMODYNAMIC EQUILIBRIUM CALCULATIONS

The thermodynamic properties of a dense, partially ionized hydrogen gas were computed using the two non-ideal models described above. The free-energy minimization method was used to calculate the equilibrium compositions given in Figure 2. At low densities (ρ $1 \leqslant 10^{-4}$ g/cm^3) both models give similar results, essentially those of the ideal classical gas. At moderate densities differences appear, but the major effect is recombination (at 2 and 10 eV) and association (at 0.5 eV), which for both models are decreased over the ideal gas values by the plasma interaction (F_4) and excluded volume (F_5) effects. At high density, the IA model shows the gas compressed into its largest cluster systems, the neutral atom (10 eV) and molecule (2 and 0.5 eV). The CA model, however, predicts the opposite, reaching nearly complete ionization at densities of approximately 100 g/cm^3.

The most important fact of this comparison is that while the plasma interaction (F_4) and excluded volume correction (F_5) do cause enhanced ionization and dissociation over the ideal gas model, the correct limiting form of ionization equilibrium at high density can only be achieved consistently by a realistic treatment of the perturbed bound states (F_2). The internal partition function and its component perturbed energy eigenvalues are the key to a correct high-density ionization equilibrium. The IA model, with free atom, unperturbed bound-state energy levels, is reasonable at low and intermediate densities, and unquestionably is a great improvement over simple Debye–Huckel (DH) theory. It should be useful for the interiors of stars in the mass range between $M_\odot \gtrsim M \gtrsim 0.5\,M_\odot$. But no matter how sophisticated a treatment is used for the other non-ideal effects, the most important effect is the density-perturbation of the bound states.

The characteristic features of the equilibrium concentrations for the CA model in Figure 2 are the near-ideal behavior at low densities, enhanced ionization (over the ideal and the IA models) at intermediate densities, and rapid and continuous rise to complete ionization at high density, for all temperatures. The CA model indicates that for a pure hydrogen gas, the presence of bound states is negligible for $T > 20$ Ve, at all densities. The onset of ionization (or dissociation) due to density effects occurs at somewhat higher density, the lower the temperature. Also, for $T \leqslant 2$ eV and $\rho > 0.05$ g/cm³, it is found that the excited states ($n \geqslant 2$) of the atom contribute significantly to Z_{int}. Therefore, the accurate tabulation of the temperature- and volume-dependent upper levels should not be neglected by using the statistical weights of the ground state or other simplifications.

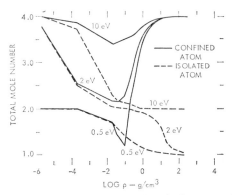

FIGURE 2 Equilibrium composition for hydrogen using isolated atom and confined atom internal partition functions. Both models with plasma interaction and excluded volume contributions. Initial composition is 1 mole H_2.

While the thermodynamic equilibrium is essentially dependent on the form of F_2, the other non-ideal effects are significant. A comparison of the contributions of the F_4 and F_5 terms is illustrated in Figure 3. A series of equilibrium concentrations are shown as a function of density, using various modifications of the CA model. The basic model including only the three terms F_1, F_2, and F_3 produces the equilibrium shown as curve A, again demonstrating the effects of the confined atom model, this time with F_4 and F_5 set equal to zero. The results of the next model, given by curve B, show the effects of including the plasma interaction F_4 (with F_5 still zero), in the form of the modified ring term. The increased ionization due to plasma effects is naturally most important at 10 eV where free charges are abundant, producing equilibrium concentrations 5 to 10 per cent higher than the basic model. At 1 eV, the plasma effects on the equilibrium concentrations are only a few per cent, and at 0.5 eV, where the system is molecular, the plasma term is negligible. Curve C depicts the basic model plus the excluded volume term F_5, with F_4 set to zero. Again the non-ideal term produces enhanced ionization, at 10 eV and at 1 Ve. Even at low temperature the molecular gas has 5 to 10 per cent higher dissociation when the excluded volume is considered.

The effect of the excluded volume is quite large in this model, due specifically to the relation between the bound-state model and the excluded volume term. Because the bound states are destroyed very close to the hard sphere radius, the volume occupied by the atoms can approach close to the total available volume. This accentuates the importance of F_5. In a model with long-range perturbation of the bound states (such as the screened Coulomb potential) the excluded volume correction will be less important than here. In Figure 3, curve D shows the equilibrium concentration with all corrections combined (F_2, F_4, F_5), which is the CA model as previously defined.

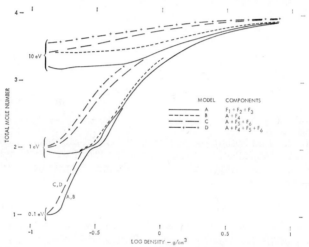

FIGURE 3 Comparison of various free-energy models for hydrogen.

Having determined the equilibrium values for the system, it is simple to compute the equations of state. For these models, there are six pressure terms corresponding to the volume derivatives of the six free-energy terms. The results for this equation of state are shown for hydrogen at 0.5 eV (Figure 4) and 10 eV (Figure 5). At 0.5 eV, the dominant contribution at low density is the pressure of the classical molecules (P_1). At intermediate density, at the onset of dissociation and ionization, the hard-sphere term P_5 and the bound-state term P_2 rise sharply and dominate, causing a significant 'stiffening' of the gas over the ideal values. As ionization becomes significant, the electron pressure and the plasma term (a negative pressure) increase rapidly, and at high density, the dominant term is that of the strongly

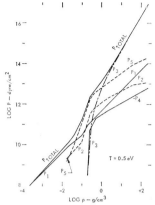

FIGURE 4   Equation of state for hydrogen at 0.5 eV. P_1 = ion pressure; P_2 = pressure due to volume dependence of bound states; P_3 = electron pressure; P_4 = pressure due to plasma Coulomb interaction; P_5 = excluded volume contribution.

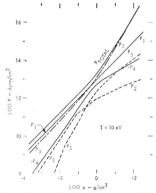

FIGURE 5   Equation of state for hydrogen at 10 eV. P_1 = ion pressure; P_2 = pressure due to volume dependence of bound states; P_3 = electron pressure; P_4 = pressure due to plasma Coulomb interaction; P_5 = excluded volume contribution.

degenerate electron gas. The significance here lies in the importance of all the non-ideal terms, two of which are dominant at intermediate density. It should be noted that the strong increase of P_5 results from the presence of a large excluded volume just prior to pressure disruption of the atoms, and the sharp increase of P_2 is due to the very strong dependence of the confined atom energy levels on the cell radius (see Figure 1). In a model with longer-range perturbations, these two terms would be smaller, although still important.

In Figure 5, the pressure is shown for hydrogen at 10 eV. The situation here is simpler because there are much fewer bound systems. At low density, the pressure is chiefly from the classical electron-ion gas. At intermediate density, the pressure due to the plasma interaction P_4 increases relative to P_1 and P_3 (being negative it causes a decrease in the total pressure below the ideal gas pressure), while the other non-ideal terms contribute less. At high density, the dominant term is again the pressure of the degenerate electron gas.

To summarize the foregoing results, it is clear that all three of the non-ideal effects are important in determining the thermal properties of dense gases. The bound-state perturbation is the key element in high-density ionization equilibria, while plasma Coulomb interaction and excluded volume effects are also important. The equations of state for dense gases at low temperature (many bound systems) depend strongly on the F_2 and F_5 models, while the high temperature (plasma) equations of state are influenced mainly by F_4.

It should also be noted that sophisticated, high-order theoretical models are needed for accurate results. The use of a DH model, or a generalized (degenerate) DH model, for the plasma interaction results in the production of negative total pressures at high density, before the F_4 term can produce significant pressure ionization. A much more powerful method, such as the quantum cluster expansion must be utilized for an accurate treatment. Similarly, a simple linear hard-sphere term for the excluded volume correction is seriously in error long before F_5 becomes a dominant factor in the equilibrium.

The most difficult part of the problem is the construction of a proper model of the perturbed internal partition function. The confined atom model represents an atom perturbed by the effects of neighbouring hard-sphere atoms, a short-range, 'mechanical' model. Additional perturbations of the bound states can be caused by long-range Coulombic interactions with the charged components of the gas, or perturbation by a quasi-continuous distribution of charge within the atomic cell itself, as described by the Thomas–Fermi potential. These perturbation effects are currently being investigated (Rouse, 1967; Zink, 1968).

Before continuing, it is worth re-emphasizing the importance of the equilibrium calculation in computing the equation of state. The specific non-ideal pressure terms are only part of the complete model—they must be evaluated at the given V and T, for the correct values of $\{N_i\}$ as determined in a consistent manner in the equilibrium calculation.

IV. APPLICATION TO LOW-MASS STARS

Using the thermodynamic models developed here, the specific behavior of dense gases in the interiors of low-mass stars can be evaluated. The thermo-

TABLE I Equilibrium concentrations at the center of an evolving 0.1 $M_\odot$ model.

ρ(g/cm³)	T (°K)	CA model					IA model				
		$X(H)$	$X(H^+)$	$X(He)$	$X(He^+)$	$X(He^{++})$	$X(H)$	$X(H^+)$	$X(He)$	$X(He^+)$	$X(He^{++})$
2.94 − 3	1.12 + 5	0.145	0.776	0.007	0.050	0.108	0.461	0.460	0.018	0.029	0.028
7.68 − 3	1.55 + 5	0.125	0.796	0.004	0.036	0.035	0.498	0.423	0.011	0.030	0.034
7.40 − 2	3.41 + 5	0.142	0.779	0.002	0.021	0.053	0.562	0.359	0.006	0.036	0.033
7.88 − 1	7.55 + 5	0.044	0.877	0.002	0.012	0.061	0.794	0.127	0.008	0.039	0.028
3.05	1.13 + 6	0.021	0.900	0	0.012	0.063	0.839	0.082	0.011	0.044	0.020
1.22 + 1	1.71 + 6	0.004	0.917	0	0.015	0.060	0.869	0.052	0.016	0.046	0.013
5.14 + 1	2.60 + 6	0	0.921	0	0.011	0.064	0.889	0.032	0.024	0.044	0.009
2.44 + 2	3.72 + 6	0	0.921	0	0.004	0.071	0.905	0.016	0.036	0.037	0.002
4.71 + 2	3.99 + 6	0	0.921	0	0.002	0.073	0.909	0.012	0.043	0.031	0.001
8.49 + 2	3.80 + 6	0	0.921	0	0.001	0.074	0.915	0.006	0.046	0.029	0
1.44 + 3	2.86 + 6	0	0.921	0	0	0.075	0.921	0.0	0.048	0.027	0

dynamic properties of these star models are computed for a Population I
chemical composition of hydrogen, helium and heavier elements with a mass
distribution of (0.739, 0.24, 0.021). Using the variation of central density
and temperature for an evolving 0.1 $M_\odot$ model computed by Ezer and
Cameron (1967), the equilibrium compositions are computed with both
thermodynamic models. Although different stellar models may have somewhat different values, these densities and temperatures can be taken as
reasonable estimates for the interior of a real 0.1 $M_\odot$ star.

The equilibrium concentration calculations for these models show that
pressure ionization is a basic feature of low-mass star interiors. In Table I,
the equilibrium concentrations are tabulated for hydrogen and helium at
densities and temperatures corresponding to the center of the evolving
0.1 $M_\odot$ model. For a population I mixture ($X = 0.739$, $Y = 0.24$, $Z = 0.021$)
the total mole numbers for hydrogen and helium are:

$$X(\text{H}) = 0.9209$$
$$X(\text{He}) = 0.0753$$

when $X(\text{H}) \rightarrow 0.92$, hydrogen is in the form of neutral atoms; when $X(\text{H}) \rightarrow 0$
and $X(\text{H}^+) \rightarrow 0.92$, the hydrogen is fully ionized. Similarly, when any helium
species approaches 0.075, that is the dominant species. The results of the CA
model show that even at the center of the star near the top of the Hayashi
track, there is still some neutral He and H. As the star evolves, the amount
of neutral H rapidly falls to zero, as do the concentrations of the He and
He$^+$. The mechanism for this ionization is the density-dependent F_2 term
in the free energy.

The IA model, even with the detailed plasma interaction and excluded
volume corrections, predicts that initially half of the hydrogen is neutral,
and as the star evolves (the density increases), the gas recombines to almost
complete neutral hydrogen at the center and throughout the interior. The
helium equilibrium likewise proceeds toward larger clusters (He$^+$ and then
He). The failure of the classical, ideal gas (Saha) equilibrium is still
present, even in this sophisticated model. This demonstrates that for low-
mass star interiors, no theory is valid unless it treats the bound-state perturbation in an accurate manner. Pressure ionization can be achieved with a
model of the IA type, by inclusion of various scaling factors or 'empirical'
models. But a consistent use of such models for both the equilibrium calculations and equation-of-state calculations will yield totally erroneous values
for one or the other of these thermodynamic properties.

The degeneracy and Coulomb interaction effects in the equations of state
for the interior of evolving low-mass stars are shown in Figures 6 and 7,
computed with the CA model. In Figure 6, the centers of evolving 0.1 $M_\odot$

and 0.3 $M_\odot$ models computed by Ezer and Cameron (1967), are plotted in a_e, Λ space. The ordinate, a_e, gives the value of the electron degeneracy, and the abscissa, Λ, denotes the value of the plasma parameter, an estimate of the strength of the plasma Coulomb interaction, where

$a_e \leqslant -4$ Maxwell Boltzmann statistics

$-4 < a_e < +20$ partially degenerate Fermi Dirac

$a_e \geqslant +20$ completely degenerate Fermi Dirac

$0 \leqslant \Lambda \leqslant 0.1$ weakly interacting system

$0.1 < \Lambda < 1$ moderately interacting system

$\Lambda > 1$ strongly interacting system

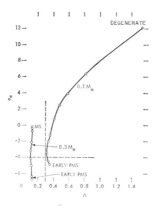

FIGURE 6   Degeneracy and Coulomb effects at central densities and temperatures of evolving low-mass stars. MS = main sequence; early PMS = high on Hayashi track. Low-mass models of Ezer and Cameron (1967) $-X = 0.739$; $Y = 0.24$; $Z = 0.021$.

The gas at the center of the 0.3 $M_\odot$ model, high on its Hayashi (pre-main sequence) track, is a weakly interacting classical gas, well described by a classical ring (DH) term for F_4. As the star evolves down the Hayashi track onto the main sequence, the electron gas becomes moderately degenerate, requiring a generalized (degeneracy-modified) ring term for adequate descirption of the plasma. The center of the 0.1 $M_\odot$ model is much more difficult, being classical only at its start on the Hayashi track. Even here the plasma interaction is moderate (~ 0.35), requiring at least second-order perturbation terms in the perturbation expansion. By the time the model has passed the lower end of the main sequence, the gas at its center is in the region of strong degeneracy and strong Coulomb interactions. Obviously, for stars of masses $\lesssim 0.1\ M_\odot$, a sophisticated plasma theory, such as that discussed by DeeWitt (1968), is required.

In Figure 7, the variations of α_e and Λ throughout the interior of 0.1 $M_\odot$ and 0.3 $M_\odot$ models are plotted, illustrating the behaviour of the gas throughout the entire interior of these low-mass stars, at one point in their evolution. In the 0.3 $M_\odot$ model (Ezer and Cameron, 1967) near its main-sequence position, one sees the expected constant electron

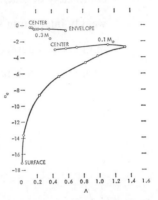

FIGURE 7   Degeneracy and Coulomb effects throughout the interior of low-mass stars. 0.3 $M_\odot$ model near main sequence; 0.1 $M_\odot$ model near start of Hayashi track.

degeneracy in the convective region of the interior. A significant feature, also characteristic of low-mass interiors, is the strong increase in the Coulomb interaction toward the surface, reaching moderate values (0.6) at $M/M_{\mathrm{tot}} = 0.99$. This means that Coulomb interaction is even more important in the envelope than at the center of the model, and the treatment of the plasma interaction is crucial to studies of stability which depend strongly on the thermodynamic behavior of the gas throughout the hydrogen ionization zone.

A more complete curve is given for a 0.1 $M_\odot$ model of Graboske, which shows similar behavior to the more massive model, and further, shows the rapid fall of both α_e and Λ as the hydrogen ionization and dissociation zone is reached. This rapid decrease of degeneracy and Coulomb interaction is due to the disappearance of free charges in the recombination zone. At the photosphere the gas is molecular, highly classical, and almost non-interacting ($\Lambda \to 0$). Both star models demonstrate that the plasma interaction is strong throughout the bulk of the interior of low-mass stars, and in regions where electron degeneracy is significant.

The results of this study have led to the conclusion that a detailed and consistent model is required to describe the constitutive physics of low-mass stars. One such model (CA) has been presented and its effects discussed. A continuing effort is under way to include in the plasma interaction term the highest-order quantum-mechanical cluster expansion results, to give a complete description of the fully ionized plasma at all values of a_e for $\Lambda \lesssim 2$. The solutions for bound-state perturbations produced by both static and dynamic screened Coulomb potentials (SCP) are to be included in the model, and the SCP solutions for each species are to be computed via self-consistent methods. Eventually a high-order, reasonably accurate thermodynamic description of astrophysical mixtures should be available for low-mass star models over the entire range $M_\odot < M_\odot 0.01 \lesssim M_\odot$.

References

DE GROOT, S. R., and TEN SELDAM, C. A. 1946, *Physica* **12**, 669.

DEWITT, H. E. 1961, *J. Nucl. Energy, Part C: Plasma Phys.* **2**, 224.

DEWITT, H. E. 1969, *Low-luminosity Stars*, Ed. S. S. KUMAR (these proceedings), p. 211.

EZER, D. W., and CAMERON, A. G. W. 1967, *Canad. J. Phys.* **45**, 3461.

GABRIEL, M. 1966, *Colloquium on Late-Type Stars*, Ed. M. HACK, Osservatorio Astronimico di Trieste, Italy.

HARRIS, G. M. 1959, *J. Chem. Phys.* **31**, 1211.

HARRIS, G. M., and TRULIO, J. 1961, *Plasma Phys.* **2**, 224.

HAYASHI, C., and NAKANO, T. 1963, *Prog. Theor. Phys.* **30**, 460.

KUMAR, S. S. 1963, *Ap. J.* **137**, 1121.

LEBOWITZ, J. L., HELFAND, E., and PRAESTGAARD, E. 1965, *J. Chem. Phys.* **43**, 774.

ROUSE, C. A. 1967, *Phys. Rev.* **159**, 41.

ZINK, W. 1968, *Phys. Rev.* **176**, 279.

Discussion

VARDYA Can you tell me as to how you have taken into account the molecules?

GRABOSKE We are not really claiming this to be a non-ideal molecular model; we are putting it in only for completeness. We hope to do an exact calculation for molecular hydrogen in the near future.

HUBBARD Is this method limited to small values of the ion-coupling parameter as well?

GRABOSKE Yes, it is. Originally, we thought the best we could do was to discuss a star of mass 0.1 $M_\odot$. You can see that Λ exceeds 1 for a 0.1 $M_\odot$

I

star, and the electron gas is becoming very degenerate—essentially becoming a weak coupler. So what we are hoping to do is to use the Monte Carlo results for the classical ion gas, which are valid here.

HUBBARD   In the limits of low temperature would this approach be in any way equivalent to the work of Wigner, for example, on metallic hydrogen?

GRABOSKE   No.

The effect of some physical parameters on the lower mass limit

W. C. STRAKA

Department of Astronomy
University of Calfornia, Los Angeles

Abstract

Several of the parameters entering into the calculation of the lower-mass limit to the main sequence are, at present, poorly determined. The qualitative effect of errors in nuclear reaction rates, equation of state, and opacities are considered.

ONE ASPECT of the problem of the lower mass limit to the main sequence that is often overlooked is the fact that several of the most important physical parameters are only known qualitatively. The actual importance of these parameters tends to be forgotten. For example, we know roughly what happens in pressure ionization, but it is not clear how close we are to the real values. Although work on water vapor opacities and their importance has been carried out, the full treatment has not yet been done. And since the structure of a fully convective star is critically dependent on the outer boundary conditions, the numbers we obtain will be only as good as the atmosphere we use. Further we find that one of the major reference articles on nuclear reactions has internal inconsistencies, whether due to misprints or for other reasons, and the various sources often disagree by factors of 2 or more. In addition, it is of interest to examine the effect of composition on the value obtained for the lower mass limit.

The first problem we encounter is the definition of the lower-mass limit, and indeed what we mean by the main sequence. A working definition used in the present work is that a main-sequence configuration is a hydrostatic model in which the energy generation at each mass point is due only to the chain of nuclear reactions commonly known as the proton–proton chain, including the CNO cycle, the emergent flux at the surface thus being due entirely to nuclear sources, with no contribution from gravitational sources. We, of course, want the first step in the proton–proton chain to dominate. That is, we are not interested in any deuterium, helium, lithium, or other exotic 'main sequences'. These are interesting problems in themselves, but are not relevant here.

There are several ways to define a lower-mass limit. The traditional way, used by Kumar, Masani, and others, is to examine a series of models of various masses. In each series a maximum central temperature is found for each mass. Then, a minimum temperature for which a sufficient amount of energy would be generated by nuclear sources is chosen.

More recently, the approach has been to examine a series of masses in terms of their pre-main sequence evolution. Each mass would be started back on its Hayashi track and allowed to gradually settle onto the main sequence. There is a minimum mass found by this which can form a main-sequence configuration.

A third method, which has not been used to any great extent, is to take a model which is apparently on the main sequence, then perturb it and see if it returns to the original configuration. One perturbation would be to artificially shut off the nuclear reactions for, say, one Kelvin time, then turn them on again.

Each of these methods will probably yield a different mass limit, although, in principle, they should agree. In the present work the second approach is being used.

Let us consider then the effects of the various parameters. First let us examine the effects of errors in the nuclear reaction rates. If at a given temperature and density, the energy generation rate is systematically off, the effect is obvious. Less energy means that a larger mass is required for the minimum main-sequence object.

On the other hand, if the computed reaction cross-sections are off, in such a way that the temperature dependence is, say, stronger, while the generation rate is correct at, for example, the conditions obtaining at the center of Castor C, the effect is to concentrate the energy generation region into a smaller and smaller fraction of the model as we go to smaller masses. Hence, again, the minimum mass we obtain is larger.

One might conclude from this that the effect of changing composition would be as follows: since at low temperatures, only the first two steps in the proton–proton chain go, a lower percentage of hydrogen means less energy yield per gram at a given density and temperature, and thus a larger mass would be required for high metal, low hydrogen composition than for low metal, high hydrogen one. However, the change in composition also effects the opacity in the outer layers significantly, so that the change in energy yield is more than compensated for. We shall return to this point in a moment.

The next effect we might look at is the question of the equation of state, specifically pressure ionization. Much recent work uses a correction to the ionization equation of the form given by Rouse (1967), that is:

$$\frac{C_{p-1}}{C_p} = P_e K_p(T)\phi_{p-1}(\rho) \tag{1}$$

where C_p, C_{p-1} are the concentrations of species p and $p-1$, P_e is the electron pressure, K_p is the normal ionization function, and

$$\phi = e^{\,(\gamma a_n/r_0)^m}$$

r_0 being the average separation of the ions, a_n being the classical Bohr orbit for the state n, γ and m being adjustable factors. While the value of m is probably close to 3, the value of γ is rather undecided, and certainly differs from one species of ion to another. Although the interior is completely ionized, the outer part of the envelope will be quite profoundly affected. In effect, we are changing the mean adiabatic gradient, by changing the rate of ionization, from $(\mathrm{d} \log T)/(\mathrm{d} \log P) = 0.40$ for the perfect gas case to about 0.30 for the normal ionization equation to perhaps 0.25 or less when including a pressure ionization correction of this form.

Since this gradient is less steep when including pressure ionization than when neglecting it, we will find that the temperature at the photosphere will be higher for the atmosphere fitted to a given interior model (Figure 1).

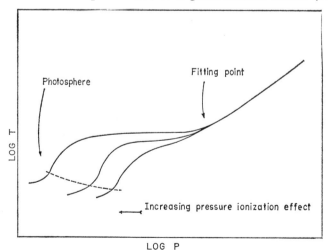

FIGURE 1   The effect of including pressure ionization.

Thus the main-sequence model for a given mass must be smaller in radius (to raise the central temperature), hotter, and more luminous, and requires more nuclear energy generation. Hence a larger mass will be required for the minimum. The models in the current work indicate that this change in luminosity is, for example, about a factor of 2 at 1/4 solar mass, going from a mean adiabatic gradient of 0.33 to 0.25.

The last parameter to be considered here is the effect of opacity. Recently some numbers on the size of water vapor opacities have become available. Model atmospheres done by Auman (1966) and by Gingerich, Latham, Linsky, and Kumar (1967) indicate the effect is quite sizeable. The difference in pressure at a given depth for 2500°K models was found to be between a factor of 7 and 10, depending on whether the opacities were approximated as black lines, or as continuous opacities. We may readily see the direction of the effect by noting that a larger opacity means the optical depth increases more rapidly with physical depth, and hence we are looking at a cooler layer as the photosphere. Thus, we will have, for a given mass, a less luminous star, and one which is larger in radius (to drop the central temperature and thus the energy generation rate). Hence, the minimum mass will be found to be smaller by inclusion of the water vapor opacities or, indeed, anything which increases the opacity.

Also, we may consider further the effect of composition. In general, the atmospheric opacities for a low hydrogen, high metal mixture will be higher than for high hydrogen, low metal. This is due, among other things, to the greater abundance of electron contributors in high metal mixtures, hence more H^- and H_2^-, as well as the greater abundance of oxygen and hence water vapor in the outer, cooler layers. Hence, less flux will pass out of the star as we saw earlier, requiring less energy generation for a given mass. This effect, in fact, will more than compensate for the lesser abundance of hydrogen to participate in nuclear reactions. Although the current calculations have not yet reached the stage where this may be confirmed, the comparison of values obtained by various workers as summarized by Ezer and Cameron (1967) bears this out.

In general, then, these are the effects of some of the unknown physical parameters. There are others which should be considered and will affect the numbers we finally obtain as a definitive value for the minimum mass main-sequence object.

I wish to thank Dr. E. K. L. Upton for helpful discussions in preparing this paper.

References

AUMAN, J. R. 1966, A. J., **71**, 154.
EZER, D., and CAMERON, A. G. W. 1967, *Evolution of Stars of Low*, Institute for Space Studies, Goddard Space Flight Center, NASA, New York.
GINGERICH, O. J., LATHAM, D., LINSKY, J., and KUMAR, S. S. 1967, *Colloquium on Late-type Stars* ed., M. HACK, p. 291.
ROUSE, C. A. 1967, *A Finite Electronic Partition from Screened Coulomb Interactions*, NRL Report 6556, Naval Research Laboratory, Washington, D. C.

Discussion

SALPETER   In reply to the question raised about the numerical values of the rates of the proton–proton chain, I just want to point out that John Bahcall at Cal Tech has recalculated that rate recently. I think there is a paper in press, but the rates have not changed by more than 10 per cent from previous values.

The surface boundary condition and approximate equation of state for low-mass stars*

ALLEN S. GROSSMAN

Indiana University

Abstract

A method by which model stellar atmospheres, in the temperature range 2500° to 3500° and log g range 1 to 5, can be constructed for the surface boundary condition in stellar evolution calculations is discussed. The ionization equilibrium equations used in the calculation of the gas pressure for the interior and atmosphere models have been modified to include pressure ionization and dissociation effects. Finally a portion of evolutionary track for a 0.1 $M_\odot$ star in the pre-main-sequence contraction phase is computed for the composition $X = 0.68$, $Y = 0.29$ and $Z = 0.03$.

I. INTRODUCTION

It has been shown by Ezer and Cameron (1966) and Hayashi (1963) that for a star of mass less than 0.2 $M_\odot$ the transport of energy during the pre-main sequence contraction phase will be by convection, except for a thin surface layer in the region of the photosphere which will be in radiative equilibrium. Since the temperature gradient throughout approximately 97 per cent of the mass of one of these objects will be very close to the adiabatic gradient, the luminosity and effective temperature will be to a great extent determined by this outer radiative layer. In this radiative layer it will be the opacity which controls the amount of energy that can be radiated by the star. For a given radius the higher the opacity, the smaller the amount of energy radiated, and

* Reprint of the Goethe Link Observatory, Indiana University, No. 44.

consequently the lower the effective temperature and the slower the contraction rate. The studies referred to indicate that the evolutionary paths of these objects in the HR diagram will be nearly vertical and cover an effective temperature range of 2500° to 3500°. This investigation will concern itself with the construction of a set of approximate model atmospheres which can be used for the surface boundary condition in the calculation of a pre-main sequence evolutionary track for a 0.1 $M_\odot$ object and an approximate treatment of pressure effects in the ionization equilibrium calculations in these atmospheres. A preliminary evolutionary track for a 0.1 $M_\odot$ star will also be given.

The composition for these calculations is $X = 0.68$, $Y = 0.29$, and $Z = 0.03$

II. ATMOSPHERE

The atmosphere solution is separated into two parts. The first part considers the atmosphere above the point $T = T_{\text{eff}}$. In this region it will be assumed that the local acceleration of gravity (g) and the luminosity will not vary and will be equal to GM/R^2 and the total luminosity, respectively. A T vs τ (Rosseland mean) distribution is assumed (Vardya 1966), approximately valid in radiative equilibrium:

$$T^4 = \tfrac{3}{4} T_{\text{eff}}^4(\tau + 1.4 - 0.825\, e^{-2.54\tau} - 0.025\, e^{-30\tau}). \tag{1}$$

The equation of hydrostatic equilibrium with g constant

$$\frac{dP}{d\tau} = \frac{g}{\kappa} \tag{2}$$

is integrated inward and the run of gas pressure, temperature and density *vs* τ is obtained. At each step in the atmosphere $\nabla_{\text{ad}} = (d\log T/d\log P)_{\text{ad}}$ is computed and compared with $\nabla_{\text{rad}} = (d\log T/d\log P)_{\text{rad}}$. As long as ∇_{ad} is greater than ∇_{rad} the integration of Eq. (2) continues. If ∇_{rad} becomes greater than ∇_{ad} then the convective flux and the average gradient of the surroundings $\nabla = (d\log T/d\log P)_{\text{act}}$ are evaluated by the mixing length theory as outlined by Henyey *et al.* (1965). In these calculations it has been assumed that the mixing length to pressure scale height ratio is constant throughout the atmosphere and equal to 1.0. When the convective flux exceeds three per cent of the total flux ($\sigma\, T_{\text{eff}}^4$) the purely radiative solution is abandoned and the equation

$$\nabla_{\text{act}} = (d\log T/d\log P)_{\text{act}} \tag{3}$$

is integrated inward by the Runge–Kutta method.

The second part of the atmosphere solution starts at the point where $T = T_{\text{eff}}$. In this section the mass and radius are allowed to vary while the luminosity is held constant. The photospheric boundary values are then

$$M_{\text{ph}} = M_{\text{total}}, \ R_{\text{ph}} = R_{\text{total}}, \ L_{\text{ph}} = L_{\text{total}}, \ P_{\text{ph}} = P_{T_{\text{eff}}}.$$

The structure equations

$$d \log T / d \log P = \nabla,$$

$$d \log R / d \log P = -rP/GM_r\rho, \qquad (4)$$

$$d \log M_r / d \log P = -4\pi r^4 P / GM_r^2$$

are integrated inward by the Runge–Kutta method to a point where the mass equals 97 per cent of the total mass. This is the point chosen as the fitting point for the interior solutions. There are two reasons for choosing this particular value. First, the convection is adiabatic below this point and it is not necessary to carry over the mixing length theory into the interior solution. Second, the variation of the quantities $\log P$ and $\log r$ with respect to $\log M_r$ (the independent interior variable) are small enough to insure a reasonable number of interior mass shells.

The radiative surface opacities (Rosseland mean) are those given by Auman (1967) and include the following sources:

Continuous opacity: H_2O, H, H^-, H_2^+, He, He^+, He^-

Line opacity: Balmer lines of II

Rayleigh scattering: H, H_2.

According to Auman the inclusion of water vapor as an opacity source increases the Rosseland mean opacity by a factor of between 10 and 30 in the neighborhood of 2500°. This has the effect of shifting an evolutionary track downward and to the right in the H–R diagram. Since the evolutionary tracks of Ezer and Cameron for 0.1 and 0.2 $M_\odot$ appear, in the H–R diagram, to be shifted approximately 400° to the left of the observational points given by Limber (1958) and shown in Table I, it is hoped that the inclusion of water vapor opacity will help to reduce some of this difference. At the present time we have not separated out the effects of different composition (Ezer and Cameron use $X = 0.739$, $Y = 0.240$, and $Z = 0.021$) and a different mixing length to pressure scale height ratio (Ezer and Cameron use $1/H_p = 2.0$). The latter parameter can have a large effect on the position of the evolutionary track, shifting it to lower temperatures for lower values of $1/H_p$.

The Auman opacity table has been coupled to the Los Alamos opacity table to provide a continuous run of opacity throughout the star.

TABLE I   Observations of low mass stars (according to Limber).

Star	$\log (M/M_\odot)$	$\log (L/L_\odot)$	$\log T_{\text{eff}}$ (°K)
Kr 60A	-0.572 ± 0.016	-2.01 ± 0.06	3.491 ± 0.028
Kr 60B	-0.788 ± 0.016	-2.40 ± 0.07	3.462 ± 0.026
Ross 614B	-1.116 ± 0.208	-3.37 ± 0.24	3.431 ± 0.040

The atmospheric integrations that have been carried out in the region of $T_{\text{eff}} = 2500°$ and $\log g$ in the range 1 to 5 indicate that the radiative gradient can become unstable as high in the atmosphere as $\tau = 0.1$ but the convection will not become efficient until values of τ between one and ten are reached.

III. EQUATION OF STATE

In order to compute the gas pressure and density to an accuracy of 1 per cent only the following species need to be considered:

$$H, H^+, H^-, H_2, H_2{}^+$$

$$He, He^+, He^{++}$$

$$M_i, M_i{}^+ \text{ (heavy elements)}.$$

Density effects on the ionization equilibrium have been treated in the manner suggested by Rouse (1964). The Saha equation is modified to include a factor which will force a shift, at a constant temperature, towards increased ionization as the density increases. The Saha equation then is written as

$$P_e \cdot N_{r+1}/N = K_r(T) \cdot \phi(r_0) \tag{5}$$

where N_{r+1} is the concentration of particles in the $r + 1^{\text{th}}$ ionization stage, N_r is the concentration of particles in the rth ionization stage, P_e is the ideal electron gas pressure, $K_r(T)$ is the standard zero density equilibrium constant and $\phi(r_0)$ is a function whose inverse is related to the probability that an ion in the rth ionization stage can exist in a mean atomic volume of radius r_0. The mean atomic radius is defined by the equation

$$r_0 \equiv (3/4\pi N_a)^{1/3} \tag{6}$$

where N_a is the concentration of all particles in the gas except electrons.

Rouse suggests a first order form for $\phi(r_0)$ to be:

$$\phi(r_0) = \exp[(\gamma a_n/r_0)^m] \tag{7}$$

where γ and m are constants to be determined and a_n is the classical Bohr orbit radius of the electron in question. We choose the constants in such a way as to give the exponent of Eq. (7) the following physical interpretation: the bracketed quantity will be the ratio of the most probable quantum mechanical volume of the atom in the rth ionization stage to the mean atomic volume of the gas. The value of m then is 3. The values of γa_n for the elements of importance are given in Table II. The value for H^- is given by Chandrasekhar (1947) and for all the heavy elements by Slater (1960).

TABLE II   Most probable radius.

Element	(ångströms)
H	0.795
H^-	2.12
H_2	1.18
H_2^+	1.30
He	0.475
He^+	0.400
Ne	0.32
N	0.53
O	0.45
C	0.66
S	0.82
Si	1.06
Fe	1.22
Mg	1.32
Ni	1.07
Ca	2.03
Al	1.21
Na	1.55
K	2.06

The molecular equilibrium equations for H_2 and H_2^+ are dealt with in the same manner as the atomic equations with γa_n replaced by one half the ground state internuclear distance plus the radius of a neutral hydrogen atom.

The independent variables for the ionization equilibrium calculation are N_e, the number of free electrons and T (Vardya 1965). Since the corrections to the Saha equations are functions of the ion pressure, which is the final result of the computation, an iterative solution is employed.

In Figure 1 the gas pressure is plotted as a function of electron pressure for log T values between 3.4 and 4.2. The abrupt decrease in slope of these curves is due to the pressure ionization effects.

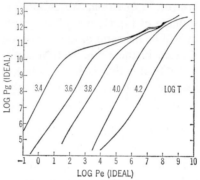

FIGURE 1   Log of the ideal gas pressure vs log of the ideal electron pressure. The lines are plotted for constant temperature.

Electron degeneracy both partial and complete in the equation of state calculation is taken into account in the standard way (Wrubel 1958) with only the non-relativistic case used for the low mass star problem.

IV. EVOLUTIONARY TRACKS

A preliminary evolutionary sequence of stellar models for a 0.1 $M_\odot$ star has been calculated by the Henyey method to test the atmosphere and equation of state formulation discussed in Sections II and III. The fitting of the atmospheres to the interior models is done by the method of triangles (Kippenhahn 1967). This sequence is for an object in which energy is produced by gravitational contraction only; nuclear reactions have not been included. The evolutionary track is started at 4.0 $R_\odot$ and the initial model is obtained by the fitting method (Sears 1965).

At the present time the calculations have been carried to the point where the radius is 0.53 $R_\odot$ representing an age of 3.5×10^6 years with respect to the initial model. The properties of some of the models are given in Table III.

In the case of the initial model, pressure ionization and electron degeneracy effects are not noticeable. The concentrations of ionized hydrogen and doubly ionized helium at the centre are 0.94 and 0.67 respectively. For the final model, the degeneracy parameter (α) is approximately -1 and pressure ionization effects are beginning to be noticeable. The concentrations of ionized hydrogen and doubly ionized helium are 0.98 and 0.70 respectively. Energy transport in the interior is by convection at all times.

TABLE III   Models.

Log T_{eff}	Log $L/L_\odot$	$R/R_\odot$	Log T_c	Log ρ_c
3.412	−0.202	4.00	5.405	−1.567
3.424	−0.361	3.12	5.499	−1.299
3.446	−0.669	1.98	5.679	−0.782
3.473	−1.137	0.98	5.942	−0.018
3.493	−1.626	0.53	6.175	0.723

The path of the evolutionary sequence, in the *HR* diagram, along with the track for 0.1 $M_\odot$ of Ezer and Cameron (1966) and the observational points of some *M* dwarfs (Table II) are shown in Figure 2. It is seen that the evolu-

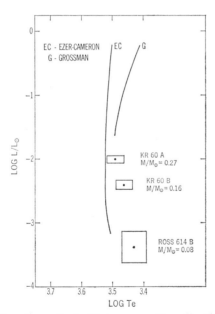

FIGURE 2   The theoretical Hertzsprung–Russell diagram for a 0.1 $M_\odot$ star during the gravitational contraction stage. Included are the *M* dwarf observations of Limber with the box sizes equal to the probable errors and the evolutionary track for the 0.1 $M_\odot$ star of Ezer and Cameron.

tionary track computed here is shifted about 300° in the direction of closer agreement with the observations. Also, for a given radius the models here have a higher central temperature than those of Ezer and Cameron (about 5×10^4 on the average) along with the lower effective temperature. Taking these two effects in consideration, there will be a greater possibility that stability on a hydrogen burning main sequence can be achieved with these models. Ezer and Cameron were not able to achieve stability at 0.1 $M_\odot$ for the composition $X = 0.739$, $Y = 0.240$, and $Z = 0.021$. This may be entirely a composition effect since the work of Hayashi indicates that the chance for stability increases as X decreases, although the increased surface opacities will certainly help the situation by lowering the luminosities of the models.

References

AUMAN, J., and BODENHEIMER, P. 1967, *Ap. J.* **149**, 641.
CHANDRASEKHAR, S. 1947, *Scientific Monthly* **64**, 313.
EZER, D., and CAMERON, A. G. W. 1966, *A.J.* **71**, 384(A).
HAYASHI, C., and NAKANO, T. 1963, *Progr. Theoret. Phys.* **30**, 460.
HENYEY, L. G., VARDYA, M. S., and BODENHEIMER, P. 1965, *Ap. J.* **142**, 841.
KIPPENHAHN, R., WEIGERT, A., and HOFMEISTER, E. 1967, in *Methods in Computational Physics: Volume 7 Astrophysics*, ed. B. ADLER (New York: Academic Press), Chap. III.
LIMBER, D. N. 1958, *Ap. J.* **127**, 363.
ROUSE, C. A. 1964, *Ap. J.* **139**, 339.
SEARS, R. L. and BROWNLEE, R. R. 1965, in *Stars and Stellar Systems*: Vol. 8, *Stellar Structure*, ed. L. H. ALLER and D. B. McLAUGHLIN (Chicago: University of Chicago Press), p. 575.
SLATER, J. 1960, *Quantum Theory of Atomic Structure, Vol. I* (New York: McGraw Hill).
VARDYA, M. S. 1965, *M.N.R.A.S.* **129**, 205.
VARDYA, M. S. 1966, *M.N.R.A.S.* **134**, 347.
WRUBEL, M. H. 1958, in *Handbuch der Physik: Volume 51*, ed. S. FLUGGE (Berlin: Springer), Chap. I.

The nature of low-mass 'dark' companions

SHIV S. KUMAR

Leander McCormick Observatory, University of Virginia

Abstract

It is pointed out that the lower limit to the mass of a main-sequence star is different from the final lower limit to the mass of a star. The value of the former is $M \simeq 0.1\ M_\odot$, while the latter has a value $M \simeq 0.001\ M_\odot$. Since the masses of the so-called dark companions lie in the range 0.002–0.05, it is concluded that these objects originated as stars, and that they are now approaching the black-dwarf stage after having missed the main-sequence stage of stellar evolution.

THE EVOLUTION of stars of very low mass ($M \simeq 0.1 M_\odot$) is fairly well understood now, and it can confidently be said that the lower limit to the mass of a main-sequence star lies in the range $0.07–0.1 M_\odot$, the exact value depending upon the chemical composition (Kumar 1963). A star with mass less than the limiting mass will not go through normal stellar evolution, and after shining for a brief period, it will become a completely degenerate object or a black dwarf. Each component of the visual binary L726–8 has a mass $\simeq 0.04 M_\odot$ and it is clear that these objects are going through the pre-black-dwarf contraction phase. Both components are expected to become extremely faint objects in the next few hundred million years. This is also the fate of all the intrinsically faint ($M_v > 17$) red objects such as the van Biesbroeck's star $+4°4048B$.

After a low-mass star has become a black dwarf, its existence can be established only if it is a member of a double (or multiple) system. Its presence will cause a perturbation in the motion of the visible component (or components) of the system and a study of the perturbation leads to a determination of the orbit of the 'dark' companion. Dark companions have already been detected around eight stars (61 Cygni, Ci 2354, BD $+20°2465$, Lalande 21185, BD $+5°1668$, η Cas, PGC 588, and Barnard's Star) within a distance of 5 parsecs from the sun. These companions have masses in the range 0.002–0.05. These objects are not completely dark as yet, for they are slowly approaching the black-dwarf stage and they still have a non-zero luminosity. For example, a low-mass star with $M = 0.01$ stops contracting at a radius $R \simeq 0.1 R_\odot$, and it may have an effective temperature T_{eff} as high as $2000°K$ when it starts approaching its cooling curve in the H–R diagram. With these values of R and T_{eff} we find that its luminosity is $L \simeq 2 \times 10^{-4} L_\odot$.

In the last few years, some workers have confused the lower limit of 0.1 with the final lower limit to the mass of a star. The observed stars such as the two components of L726–8 are known to have masses less than $0.1 M_\odot$ and the dark companions also have masses less than $0.1 M_\odot$. Therefore, the two limits should not be confused. The minimum mass that a star can have is dictated by the processes of star formation while the minimum mass on the main sequence is a result of stellar evolution. The theories of star formation have not enabled us to compute accurately the minimum mass of a star at the time of its formation. However, one can make an estimate of this mass by the following arguments. Hayashi and Nakano (1963) found that a star of one solar mass will be formed out of the interstellar medium only if its density is $\rho > 2 \times 10^{-18}$ gm/cm^3. If the star has a density less than this value, then it will expand back into the interstellar medium. For a star with a mass less than one solar mass, the value of the critical density is higher than 2×10^{-18}. For example, the critical density for a star of mass $0.1 M_\odot$ is $\simeq 10^{-15}$, while the mean density in the interstellar medium is 1.2×10^{-24} (Allen 1963). As we go down to even lower masses, the value of the critical density for star formation goes up and consequently, it would seem very difficult to form stars with masses $M \simeq 0.001$. We may say that the lower limit to the mass of a star is probably as low as 0.001.

While the exact value of the minimum mass is not known at all, it seems that the masses of the known black-dwarf companions lie above this limiting mass. One additional argument for the stellar nature of the unseen companions may be given here. The orbits of the companions are highly eccentric; for example, Lippincott (1967) gives an eccentricity of 0.9 for the orbit of the companion to Ci 2354. Further, the mass ratios of the primary (visible) to the secondary (dark) are in the range 10–50. These two properties of the orbits of the dark companions are characteristic of double (or multiple) stars and therefore they provide us with additional support for our proposal that the dark companions originated as stars.

Finally, it should be mentioned that the presence of at least eight black-dwarf companions and quite a few intrinsically faint red stars within only 5 parsecs from the sun implies that the space density of stars of very low mass is very high. It is very likely that the number of low-mass black dwarfs in the Galaxy is larger than the visible stars.

References

ALLEN, C. W. 1963, *Astrophysical Quantities.*
HAYASHI, C., and NAKANO, T. 1963, *Progress of Theoretical Physics* **30**, 460.
KUMAR, S. S. 1963, *Ap. J.* **137**, 1121.
LIPPINCOTT, S. L. 1967, *A.J.* **72**, 1349.

Convection in degenerate stars*

W. B. HUBBARD

California Institute of Technology, Pasadena

Abstract

The possibility of convective instability in the interiors of low-mass degenerate stars is explored. It is found that all hydrogen-rich white dwarfs with effective temperatures greater than about 1000°K are probably convective, and the less massive ones are probably convective at effective temperatures of a few hundred degrees.

I. INTRODUCTION

ALTHOUGH THE temperature gradient is always very small in a conventional white dwarf, due to the very high thermal conductivity of the dense matter in its interior, such is not necessarily the case for the low mass hydrogen-rich degenerate stars which are presumably formed directly from interstellar matter without undergoing nuclear evolution. In low mass degenerate stars the thermal conductivity can be low enough for a substantial temperature gradient to be maintained, and for convection to be maintained even at fairly low luminosities. It is the purpose of this discussion to estimate the location of the boundary between convective and conductive white dwarfs, and to comment on the implications.

II. DEFINITION

For the purposes of this discussion, we will consider only those stars which are in the state of electron degeneracy, such that the electron fermi energy is much greater than kT through the bulk of the star. The ions in the stellar matter may be either strongly or weakly coupled, depending on whether their average interaction energy is large or small compared to kT.

* Supported in part by the National Science Foundation GP–7976 formerly GP–5391 and by the Office of Naval Research Nonr–220 (47).

In the case of white dwarfs of mass comparable to a solar mass, one can verify that electron conduction is much more efficient for energy transport than photon conduction, in the degenerate core. Although the photon opacities are not well known in the case of lower mass stars, electron conduction may still dominate, since the temperature is much less than the electron plasma temperature,

$$T_p = \frac{\hbar \omega_{p,e}}{k}$$ (1)

where $\omega_{p,e}$ is the electron plasma frequency. The condition $T \ll T_p$ is particularly well satisfied in low mass white dwarfs, and thus the index of refraction for the photons around the peak of the Planck distribution would be imaginary, which should effectively impede energy transmission. For the purposes of this discussion, we therefore assume that the conductivity is determined by the electron conductivity.

Hereinafter, low mass white dwarfs ($M < .2\ M_\odot$) are discussed, such that the pressure equation of state can be approximated by $P \propto \rho^{5/3}$, where ρ is the mass density. Thus the structure of the white dwarfs is approximated by polytropes of index 3/2. In the case of very low mass white dwarfs ($M < 0.05\ M_\odot$), deviations from perfect Fermi gas behavior are so large that the polytrope of index 3/2 is no longer a good approximation, although qualitative results can still be obtained.

III. TEMPERATURE GRADIENTS IN WHITE DWARFS

The total luminosity passing through a sphere of radius r is given by

$$L = -K \frac{dT}{dr} 4\pi r^2$$ (2)

where K is the thermal conductivity and T is the absolute temperature. Let us consider a polytrope of total mass M, central density ρ_c, and average density $\bar{\rho}$. Let the point at which the density equals the average density be $\bar{r}$, and let the fraction of M enclosed within $\bar{r}$ be f. We then approximate Eq. (2) by

$$fL \simeq -K(\bar{\rho}) \frac{T(\bar{\rho}) - T(\rho_c)}{\bar{r}} 4\pi \bar{r}^2,$$ (3)

which we rewrite as

$$L \simeq K(\bar{\rho}) 4\pi \bar{r} \frac{[P(\rho_c) - P(\bar{\rho})]T(\bar{\rho})}{\frac{1}{2}[P(\rho_c) + P(\bar{\rho})]} \nabla,$$ (4)

where

$$\nabla = \frac{\frac{1}{2}[P(\bar{\rho}_c) + P(\bar{\rho})]}{T(\bar{\rho})} \frac{T(\rho_c) - T(\bar{\rho})}{P(\rho_c) - P(\bar{\rho})}$$

$$\simeq \frac{d \ln T}{d \ln P} \tag{5}$$

The adiabatic temperature variation, ∇_{ad}, is given by 0.4 in the case of weakly coupled matter, or 0.3 for strongly coupled matter (Hubbard 1968). It is sufficient here to set $\nabla_{ad} = 0.35$. We then have the critical luminosity at which the star is marginally unstable to convection by setting $\nabla = \nabla_{ad}$ in formula (4). However, it still remains to estimate the characteristic temperature $T(\bar{\rho})$ in the relation. Ordinarily, one obtains the internal temperature of a white dwarf in terms of the luminosity by applying an atmospheric boundary condition. Since the atmospheres of low mass white dwarfs are complex and not well studied, we content ourselves with establishing upper and lower limits for the critical luminosity. First, since we are by hypothesis considering a degenerate body, we must have

$$T(\bar{\rho}) < T_d(\bar{\rho}) = 1.2 \times 10^5 (\bar{\rho}/\mu_e)^{2/3} \tag{6}$$

where T_d is the degeneracy temperature, at which the ideal gas equation of state and the zero temperature equation of state give the same result for the pressure. On the other hand, the temperature must be greater than the Debye temperature, given by

$$T_D(\bar{\rho}) = 1.74 \times 10^3 (2/\mu_e) \bar{\rho}^{1/2} \tag{7}$$

(Van Horn 1968), or the star would be unobservable. Using the two limiting values (6) and (7), together with constants appropriate to a polytrope of index 3/2, we thus obtain for the critical luminosity

$$10^4 R K(\bar{\rho}) \left(\frac{2}{\mu_e}\right) \bar{\rho}^{1/2} < L < 7.3 \times 10^5 R \left(\frac{\bar{\rho}}{\mu_e}\right)^{2/3} K(\bar{\rho}), \tag{8}$$

where R is the radius of the star.

For strongly coupled matter, we have

$$K(\bar{\rho}) = 7.2 \times 10^7 (\bar{\rho}/\mu_e)^{4/3} \tag{9}$$

(Hubbard 1967; Hubbard and Lampe 1968); while for weakly coupled matter we obtain

$$K(\bar{\rho}) = 4.7 \times 10^8 \frac{\bar{\rho} T(\bar{\rho})}{A} G \tag{10}$$

(Hubbard 1966), where A is the atomic weight and G is a logarithmic factor which can be set equal to about 0.5. Here we will only consider weak coupling for hydrogen stars and thus set $A = 1$. Inserting the relations (9) and (10) in Eq. (8), the critical luminosity is bracketed by

$$5.4 \times 10^{11} R \rho_c^{11/6} \mu_e^{-7/3} < L < 1.5 \times 10^{12} R(\rho_c/\mu_e)^2, \tag{11}$$

for strong coupling, or

$$4.8 \times 10^9 R \rho_c^2 < L < 4 \times 10^{12} R \rho_c^{7/3} \tag{12}$$

for weak coupling.

Using the polytropic relations between central density and radius, the equations (11) and (12) can be converted into expressions involving only the radius and luminosity. However, for objects of mass much less than 0.05 $M_\odot$ which are not well approximated by $n = 3/2$ polytropes, equations (11) and (12) give better results since for such objects the radius is much less for a given central density than would be the case for a polytrope.

In terms of the radius, we thus obtain:

$$\frac{1.08 \times 10^{-17}}{\mu_e^{11.5}(R/R_\odot)^{10}} < \frac{L}{L_\odot} < \frac{8.7 \times 10^{-18}}{\mu_e^{12}(R/R_\odot)^{11}} \tag{13}$$

(strong coupling), and

$$\frac{2.8 \times 10^{-20}}{\mu_e^{10}(R/R_\odot)^{11}} < \frac{L}{L_\odot} < \frac{1.9 \times 10^{-18}}{\mu_e^{35/3}(R/R_\odot)^{13}} \tag{14}$$

(weak coupling), where the luminosity and radius are now expressed in solar units.

By using the standard relation between luminosity, radius, and effective temperature T_e, we then obtain

$$3.33 \log T_e - 1.92 \log \mu_e - 15.4 < \log \frac{L}{L_\odot} <$$
$$3.39 \log T_e - 1.85 \log \mu_e - 15.4 \tag{15}$$

(strong coupling);

$$3.39 \log T_e - 1.54 \log \mu_e - 15.8 < \log \frac{L}{L_\odot} <$$
$$3.47 \log T_e - 1.56 \log \mu_e - 15.4 \tag{16}$$

(weak coupling).

In Figure 1 are plotted the inequalities (15) and (16), together with cooling tracks for degenerate bodies of various masses. The two lines for white dwarfs marked 0.22 $M_\odot$ and 0.88 $M_\odot$ are for $\mu_e = 2$, and the corresponding dotted lines are from relation (15). It can be seen that for white dwarfs which have undergone nuclear evolution, convection probably never occurs unless such stars can be produced with masses less than about 0.10 $M_\odot$. The line marked 0.07 $M_\odot$ is for a hydrogen-rich star, and corresponds to the most massive hydrogen-rich white dwarf possible (Kumar 1963; Hayashi and Nakano 1964). The curve marked 0.1 $R_\odot$ corresponds to the largest possible radius for a degenerate star (DeMarcus 1958) and includes objects ranging in mass from about 0.01 $M_\odot$ down to 0.001 $M_\odot$. The point marked 'J' is a tentative position for Jupiter based on the infrared observations of Low (1966). The dotted lines in this portion of the graph correspond to relation (16). The spread between the two lines is larger for weak coupling than strong coupling because two powers of the unknown internal temperature enter rather than one.

IV. DISCUSSION

In Figure 1, any degenerate object found above the upper dotted line must certainly be convective, and any object observed below the lower dotted line must certainly be conductive. If we guess that the true relation lies about midway between the two dotted lines, then we conclude that any hydrogen-rich degenerate star will be convective if it has an effective temperature greater than about 1000 °K. For even lower masses than 0.01 $M_\odot$, the convection probably persists for effective temperatures of a few hundred degrees. The case for convection in such low mass objects is even stronger than appears from relations (15) or (16) since the radius of such an object is smaller than the polytropic model would indicate. Furthermore, such objects will be

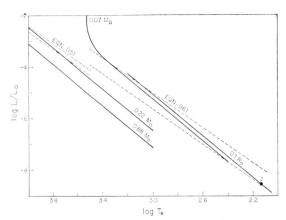

FIGURE 1 Temperature luminosity diagram for white dwarfs. Degenerate stars above the upper dotted lines are certainly convective, while those degenerate stars observed below the lower dotted lines should be conductive.

strongly coupled at their typical degeneracy temperature, and the adiabatic temperature gradient ∇_{ad} is even smaller than 0.3 due to the effects of partial pressure ionization (Hubbard 1968). Thus, for example, an object of the size of Jupiter (0.001 $M_\odot$) is probably completely convective at an effective temperature of about 150 °K.

The observable consequences of the convection in low mass hydrogen dwarfs may not be marked. Since low mass degenerate stars have small density gradients except at the surface the density rises to a rather large value only a short distance below the atmosphere, and such dense matter has such a large heat capacity that rather slow convective motions (of the order of a few meters per second) are probably sufficient to transport the required amount of energy. Thus emission lines and other manifestations of violent convective motions are probably not to be expected. Unlike standard white dwarfs ($\mu_e = 2$), however, the atmospheric abundances should be characteristic of the object as a whole. Moderate magnetic fields may be generated. The effect of the convection on the evolutionary time scales depends on the atmospheric structure, and remains to be explored.

References

DeMarcus, W. C. 1958, *A.J.* **63**, 2.
Hayashi, C., and Nakano, T. 1964, *Progress of Theoretical Physics*, 30, 460.
Hubbard, W. B. 1966, *Ap. J.* **146**, 858.
Hubbard, W. B. 1967 *Ph.D. Thesis*, University of California, Berkeley.
Hubbard, W. B. 1968, *Ap. J.* **152**. 745.
Hubbard, W. B., and Lampe, M. 1969, *Ap. J. Suppl.*, No. 163 (in press).
Kumar, S. S. 1963, *Ap. J.* **137**, 1121.
Low, F. J. 1966, *A.J.* **71**, 391.
Van Horn, H. M. 1968, *Ap. J.* **151**, 227.

Discussion

ALLER   You mentioned Jupiter being in convective equilibrium; planets are probably out of bounds for our discussion here, but they certainly represent a limiting case for low-mass objects. Now does this mean that the bulk of the body of the planet is actually in convective equilibrium? Or do you have a solid transition zone? Just what happens as we push our luck to the limits in calculating these models.

HUBBARD   I cannot really say whether the molecular hydrogen zone would be in convective equilibrium also, although there is indication from the model atmospheres of Trafton that the convective instability starts near the surface and proceeds into the interior. I would be willing to say, with reasonable assurance, that the metallic hydrogen zone is probably in convective equilibrium.

ALLER   In other words, we have a liquid star.

HUBBARD   A liquid core, with possibly a solid metallic hydrogen envelope around it.

Structure and stability of fully convective red dwarfs

M. GABRIEL

Berkeley Astronomical Department, University of California

Abstract

Models have been computed for Kr 60 A (0.27 $M_\odot$) and Kr 60 B (0.16 $M_\odot$) taking into account the electrostatic interactions between gas particles. Both are found to be fully convective except for a thin atmosphere in radiative equilibrium. The vibrational stability of the three first modes has been examined. Both stars are found to be unstable with e-folding times short compared to their lifetimes.

I. INTRODUCTION

IT WAS established by the works of Limber (1957), Kaminisi (1960), and Hayashi and Nakano (1963) that the main-sequence stars with mass less than about 0.27 $M_\odot$ are fully convective except for a thin atmosphere in radiative equilibrium and that they can be, in first approximation, represented by polytropic models of index $n = 3/2$. As a result of their low central condensation ($\rho_c/\bar{\rho} \simeq 6$), the growth of the amplitude of the first mode of adiabatic oscillation is low from the center to the surface. It follows that the destabilizing effects of the nuclear reactions can easily result in a vibrational instability. We have shown (Gabriel 1964) that these stars, represented by polytropic models, are unstable.

On the other hand, in a discussion of the effects on the vibrational stability of the energy exchange connected with convection, Boury, Gabriel and Ledoux (1964) concluded that the instability is appreciably reinforced when $\Gamma_3 = 1 + (\partial \ln T)/(\partial \ln \rho)_s$ is less than 5/3.

In order to consider this effect, we have constructed models taking into account the ionization of H and He and the dissociation of H_2. The characteristic time of the instability is then considerably reduced and becomes short compared to the life time of the stars.

II. THE MODELS

(a) *The thermodynamical properties of the gas*

In the stars under study, the interactions between particles affect significantly the thermodynamical properties of the gas, especially in the H and He ionization zones. In such a situation, we have to use a homogeneous theory in which the whole thermodynamics is deduced from the partition function of the system, in order to assure the compatibility of all the equations of the problem and its internal coherence. However, this kind of approach is actually possible only if we adopt a simple model to describe the interactions. We have used the results of works based on the concept of rigid spheres and described by Fowler (1955). Then the partition function Q is given by

$$\ln Q = V \left\{ n_e \left[\lambda + D(y) \right] + \sum_i n_i \left[1 + \ln B_i + \right. \right.$$

$$\left. \ln \frac{1}{n_i} \left(\frac{2\pi m_i kT}{h^2} \right)^{3/2} \right] - \sum_{j,k} \frac{n_j n_k}{\sigma_{jk}} v_{jk} + \tag{1}$$

$$\left. \frac{1}{3} \frac{e^2}{R_D kT} \sum_j n_j z_j^2 \, \Omega \left(\frac{a}{R_D} \right) + n_{H_2} \frac{ST - E}{T} \right\}$$

with

$$\Omega(x) = 3x^{-3}(\ln \, (1 + x) - x + \tfrac{1}{2}x^2) \tag{2}$$

$$R_D^{-2} = \frac{4\pi e^2}{kT} \left[\sum_i n_i z_i^2 - \frac{\partial n_e}{\partial \lambda} \right] \tag{3}$$

$$\frac{4\pi}{3} a^3 \sum_i n_i = 1 \tag{4}$$

$$v_{jk} = \frac{1}{2} (v_{jj} + v_{kk}), \quad v_{jj} = \frac{16\pi}{3} \, cd_j^3 \tag{5}$$

$$D(y) = \frac{V(\lambda, 3/2)}{V(\lambda, 1/2)} \tag{6}$$

$$y = V(\lambda, 1/2) = \frac{n_e}{2} \left(\frac{2\pi m_e kT}{h^2} \right)^{-3/2} \tag{7}$$

$$V(\lambda, n) = \frac{1}{\Gamma(n + 1)} \int_0^\infty \frac{z^n dz}{e^{\lambda + z} + 1} \tag{8}$$

where λ is a parameter which is a function of n_i and the temperature.

The notations have the following meaning:

V:	volume of the system
n_e:	number of free electrons per cm^3
n_i:	number of type i atoms or ions per cm^3
n_{H2}:	number of H_2 molecules per cm^3
B_i:	the partition function
m_j:	the mass of the particles of type j (electron or ion)
n_i, n_k:	the number of particles of type j or k per cm^3
z_j:	the charge of type j particles
v_{jk}:	the excluded volume for an interaction between type j and k particles
σ_{jk}:	a parameter equal to 1 if $j \neq k$ and equal to 2 if $j = k$
R_d:	the Debye radius
a:	the mean radius of the particles
d_j:	the radius of the outer electronic orbit
S:	the entropy per mole of H_2
E:	the energy per mole of H_2

On the right-hand side member of Eq. (1), the first bracket refers to the electronic gas, the third term gives the excluded volume correction, the fourth the electrostatic correction and the last term refers to the molecular hydrogen.

The mean radius 'a' can be interpreted as the radius of a sphere, centered on a particle, outside of which the charge distribution can be considered as continuous (Ecker and Kroll 1963). The parameter c is chosen so that the ionization of each type of atom is nearly complete when the fundamental level merges with the continuum. The expressions used for the entropy and for the energy of H_2 are those given by Fickett and Cowan (1955).

All the thermodynamical properties of the gas can now be deduced from (1) with simple thermodynamical formulae.

(b) Chemical composition and opacity

Models have been constructed for Kr 60 A and B with masses respectively equal to 0.27 $M_\odot$ and 0.16 $M_\odot$. These two stars have been chosen because they have the best known masses amongst red dwarfs.

Since Kr 60 B is a flare star which has emission lines in its quiet spectrum and because the system has a low velocity perpendicularly to the galactic plane (Delhaye 1953, Vyssotsky 1961), these stars probably belong to the young population I. Therefore, we have adopted the following chemical composition:

$$X = 0.6$$

$$Z = 0.044,$$

which allows us to use easily the Cox and Stewart's opacities (1965).

(c) *The nuclear reactions*

As the central temperatures of such small mass stars are rather low ($T_c \simeq 8 \cdot 10^6 K$ for Kr 60 A and $T_c \simeq 6.6 \times 10^6 K$ for Kr 60 B), the nuclear reactions proceed through the $p - p$ chain:

$$H^1 + H^1 \quad \rightarrow H^2 + \beta^+ + \nu$$

$$H^2 + H^1 \quad \rightarrow He^3 + \gamma$$

$$He^3 + He^3 \rightarrow He^4 + 2H^1,$$

and the third reaction is very slow. If the initial abundance of He^3 is zero, the He^3 reaches its equilibrium abundance at the center in a time span of 10^9 years for Kr 60 A and 10^{10} years for Kr 60 B. Thus, the third reaction of the $p - p$ chain does not contribute to the energy generation in Kr 60 B and probably very little in Kr 60 A. However, in the latter case, we consider two extreme situations:

(1) we suppose that He^3 does not burn at all (Kr 60 A)$_{CI}$,

(2) we suppose that He^3 has reached its equilibrium abundance (Kr 60 A)$_{CC}$.

(d) *Characteristics of the models*

In stars with outer convective layers, one can proceed in two different ways to build the model. We may choose a likely value l of the mean free path of the turbulent eddies or we may impose on the model the luminosity obtained from the observations keeping l as a parameter. Here we have adopted the second alternative. In the outer layers when $\Delta - \Delta_{ad} > 0.01$, the Böhm–Vitense theory has been used to represent the structure of the convective zone.

The models are found to be wholly convective and they differ only slightly from the $n = 3/2$ polytrope. The properties of the models are presented in Tables I and II.

In Table I, columns 1, 2, and 3 give respectively the central density ρ_c, temperature T_c, and pressure p_c; column 4 gives the maximum value of the radiative flux to the total flux at the level r; column 5 gives the ratio of the mixing length to the pressure scale height and column 6 gives the effective temperature of the models $T_{eff}^4 = L/(4\pi\sigma R^2)$.

In Table II, the first column gives the observational T_{eff} taken from Limber (1957); the second one the surface temperature of the models; the third and the fourth columns give respectively the temperature and the optical depth τ_{SNC} at the surface of the convective zone; the fifth and the sixth columns give respectively the temperature and the fraction of the total flux carried out by convection at the point $\tau = 2/3$.

TABLE I Properties of the models.

	$M/M_\odot$	ρ_c	T_c	p_c	$L_R(r)/L(r)_{max}$	l/Hp	T_{eff} model
(Kr60A)$_{CC}$	0.27	97.7	7.35×10^6	0.43×10^{16}	0.748	0.22	3346
Kr60A)$_{CI}$	0.27	120.2	7.92×10^6	1.25×10^{17}	0.974	0.26	3431
(Kr60B)	0.16	213.8	6.56×10^6	1.95×10^{17}	0.298	0.29	3305

TABLE II Properties of the outer layers of the models.

	$(T_{eff})_{obs}$	$T_{\tau=0}$	T_{SNC}	τ_{SNC}	$T_{\tau=2/3}$	$(L_t^*/L)_{\tau=2/3}$
(Kr60A)	3150 ± 200	2808	2989	0.188	3321	0.204
(Kr60A)	3150 ± 200	2879	3048	0.171	3367	0.257
(Kr60B)	2950 ± 225	2780	2970	0.202	3233	0.267

From column 4 of Table I, it can be seen that for $(Kr\,60\,A)_{CI}$, $[L_R(r)/L(r)]_{max}$ is nearly equal to 1; thus the maximum mass of wholly convective models is very near to 0.27 $M_\odot$, a result which is in good agreement with Hayashi's result (Hayashi and Nakano, 1963).

Table II shows that convection extends up to very low optical depths and that it still carries a large amount of energy at $\tau = 2/3$. Hence, convection is expected to affect significantly the energy distribution of the surface flux of these stars in such a way that they will look redder than giants with the same effective temperature. Therefore, we have to be very cautious in deriving the effective temperatures of red dwarfs by comparing their spectra with those of red giants.

K

III. THE VIBRATIONAL STABILITY

The coefficient of vibrational stability σ' is given by

$$2\sigma_a{}^2 J_a \sigma' = -\int_0^{Ma} \left(\frac{\delta T}{T}\right)_a \left[\delta\epsilon - \frac{d(\delta L_R + \delta L_t{}^*)}{dm}\right]_a dm \qquad (9)$$

$$-\int_0^{Mt} \left(\frac{\delta\rho}{\rho}\right)_a (\Gamma_3 - 5/3) \left[\delta\epsilon_2 + \delta\left(\frac{\bar{C}_r}{\rho}\frac{dp}{dr}\right)\right]_a dm$$

with

$$J_a = \int_0^M (\delta r)^2 \, dm \qquad (10)$$

where $J_a = 2\pi/P$, P being the period of the pulsation, ϵ is the energy generation rate, $L_R(r)$ the radiative luminosity, $L_t{}^*(r)$ the total convective flux times $4\pi r^2$, $\bar{C}_r$ the mean radial velocity of turbulence, δ the symbol for the Lagrangian perturbation, ϵ_2 is the rate of dissipation of kinetic energy of turbulence into heat per unit mass.

In the equilibrium configuration ϵ_2 is given by

$$\epsilon_2 = -\frac{\bar{C}_r}{\rho}\frac{dp}{dr} \qquad (11)$$

The subscript 'a' means that the quantities are calculated in terms of the adiabatic solution. To compute the eigen-functions of the adiabatic problem, one needs a boundary condition at the surface of the star. To derive it we have supposed that the model is connected at its outermost point ($\rho \simeq 10^{-9}$ gr/cm^3) to a parallel isothermal atmosphere and we have required the pulsation energy to keep a finite value (Simon, 1957). The periods, the relative displacements $\delta r/r$ and the relative perturbation of the pressure $\delta p/p$ are nearly the same as those for polytropic models.

The coefficient of vibrational stability has been computed for the fundamental mode and for the first two overtones; their period of pulsation and the values of $\sigma_a{}^2$ are given in Table III.

The perturbation of the energy generation rate is given by

$$\delta\epsilon = \delta\epsilon_{H,H} + \delta\epsilon_{H,D} + \delta\epsilon_{He^3,He^3} \qquad (12)$$

with

$$\frac{\delta\epsilon_{ij}}{\epsilon_{ij}} = \frac{\partial \ln \epsilon_{i,j}}{\partial \ln \rho}\frac{\delta p}{p} + \frac{\partial \ln \epsilon_{i,j}}{\partial \ln T}\frac{\delta T}{T} + \frac{\delta X_i}{X_i} + \frac{\delta X_j}{X_j} \qquad (13)$$

$$\frac{\delta X_H}{X_H} = \frac{\delta X_{He^3}}{X_{He^3}} = 0 \tag{14}$$

$$\frac{\delta X_D}{X_D} = \frac{(K_{H,D}\rho X_H)^2}{(K_{H,D}\rho X_H)^2 + \sigma^2} \left[\left(\frac{\partial \ln \epsilon_{H,H}}{\partial \ln T} - \frac{\partial \ln \epsilon_{H,D}}{\partial \ln T} \right) \frac{\delta T}{T} \right.$$

$$\left. + \left(\frac{\partial \ln \epsilon_{H,H}}{\partial \ln \rho} - \frac{\partial \ln \epsilon_{H,D}}{\partial \ln \rho} \right) \frac{\delta\rho}{\rho} \right] \tag{15}$$

TABLE III   Values of $\sigma_a{}^2$ and P.

	Fundamental mode		First overtone		Second overtone	
	$\sigma_a{}^2$	P_{min}	$\sigma_a{}^2$	P_{min}	$\sigma_a{}^2$	P_{min}
(Kr60A)cc	1.25281 10^{-5}	29.6	5.79579 10^{-5}	13.75	1.22066 10^{-4}	9.48
(Kr60A)ci	1.54093 10^{-5}	26.7	7.13855 10^{-5}	12.4	1.50292 10^{-4}	8.55
Kr60B	2.76885 10^{-5}	19.9	1.28341 10^{-4}	9.25	2.69481 10^{-4}	6.38

$K_{H,D}$ being related to the number $n_{H,D}$ of reactions $D(H^1, \gamma)He^3$ per gr and per sec by

$$n_{H,D} = \frac{K_{H,D}}{2m_H} \rho X_D X_H \tag{16}$$

The perturbation of the radiative luminosity is given by the usual expression

$$\frac{\delta L_R}{L_R} = 4\frac{\delta r}{r} + \left(4 - \frac{\partial \ln \mathcal{H}}{\partial \ln T} \right) \frac{\delta T}{T} - \frac{\partial \ln \mathcal{H}}{\partial \ln \rho} \frac{\delta\rho}{\rho} + \frac{\dfrac{d}{dr}\left(\dfrac{\delta T}{T} \right)}{\dfrac{1}{T}\dfrac{dT}{dr}} \tag{17}$$

where $\mathcal{H}$ is the opacity coefficient. The total convective luminosity can be written

$$L_t{}^* = 4\pi r^2 e_p \rho T \frac{\bar{T} - T}{\bar{T}} \bar{C} \tag{18}$$

with

$$C_p = \frac{1}{\rho T} \left\{ p + \rho \left[\frac{\partial E}{\partial \ln T \partial \ln p} \dfrac{\dfrac{\partial \ln p}{\partial \ln \rho}}{\dfrac{\partial \ln p}{\partial \ln T}} - \frac{\partial E}{\partial \ln \rho} \right] \dfrac{\dfrac{\partial \ln p}{\partial \ln T}}{\dfrac{\partial \ln p}{\partial \ln \rho}} \right\} \tag{19}$$

where C_p is the generalized specific heat at constant pressure per gram, $\bar{C}$ is the mean absolute radial velocity of turbulence and E is the energy per gram.

As to the perturbation of the terms related with convection, we used the results of Boury, Gabriel and Ledoux's paper (1964) and we took

$$\delta \frac{\overline{T} - T}{\overline{T}} = \delta \frac{\bar{\rho} - \rho}{\bar{\rho}} = 0$$

$$\delta \bar{C} = \delta t = 0 \quad \text{if} \quad \tau_t > P \tag{20}$$

$$\frac{\delta \bar{C}}{\bar{C}} = \frac{1}{3} \frac{\delta \rho}{\rho}; \quad \frac{\delta t}{t} = \frac{\delta r}{r} \quad \text{if} \quad \tau_t \leqslant P$$

$\tau_t = t/\bar{C}$ being the mean life-time of the turbulent eddies.
This yields for L_t^*

$$\frac{\delta L_t^*}{L_t^*} = 2 \frac{\delta r}{r} + \frac{\delta C_p}{C_p} + \frac{\delta \rho}{\rho} + \frac{\delta T}{T} + \frac{\delta V}{V} \tag{21}$$

Following Cowling (1938) we took for

$$\frac{\delta \epsilon_2}{\epsilon_2} = 3 \frac{\delta \bar{C}}{\bar{C}} - \frac{\delta t}{t} \tag{22}$$

Taking into account the relation

$$\bar{C}_r = \frac{\bar{\rho} - \rho}{\bar{\rho}} \bar{C} \tag{23}$$

the perturbation of $(\bar{C}_r/\rho) \, dp/dr$ is given by

$$\frac{\delta \left(\dfrac{\bar{C}_r}{\rho} \dfrac{dp}{dr} \right)}{\dfrac{\bar{C}_r}{\rho} \dfrac{dp}{dr}} = \frac{\delta \bar{C}}{\bar{C}} - \left[\sigma_a^2 \frac{r^3}{GM(r)} + 2 \right] \frac{\delta r}{r} \tag{24}$$

The results are shown in Table IV.

In Table IV, the first column gives $J_a = \int_0^M (\delta r)^2 \, dm$, the second one $\sigma_\epsilon' = \int_0^{Ma} (\delta T/T)_a \, \delta\epsilon \, dm$, the third one $\sigma_L' = \int_0^{Ma} (\delta T/T)_a \, d(\delta L)_a$, the fourth one $\sigma_t' = \int_0^{Mt} (\delta \rho/\rho)_a \, (\Gamma_3 - 5/3)(\delta\epsilon_2 + \delta(\bar{C}_r/\rho \, dp/dr))_a \, dm$, the fifth gives the coefficient of vibrational stability and the last one gives the characteristic time of the instability.

The energy dissipation by progressive waves in a corona and by turbulent viscosity has been evaluated. Their influence is found to be negligible.

IV. CONCLUSIONS

There are two possible interpretations of these results depending on the observational tests:

(1) The main-sequence stars of very small mass, which we consider as being fully convective, are stable. This implies that our idea concerning convection and its effects on the oscillations lead us to wrong results or that an unknown stabilizing effect plays an important role in these stars.

(2) These stars are indeed unstable and our work is correct, but the instability has not been observed yet.

TABLE IV Values of the coefficients of vibrational stability.

		J_a	σ_ϵ'	σ_L'
fundamental mode	(Kr60A)$_{CC}$	9.217×10^{52}	2.26×10^{33}	5.56×10^{32}
	(Kr60A)$_{CI}$	8.273×10^{52}	1.04×10^{33}	3.39×10^{32}
	Kr60B	2.470×10^{52}	4.54×10^{32}	1.24×10^{32}
1st harmonic	(Kr60A)$_{CC}$	3.786×10^{52}	1.56×10^{33}	2.40×10^{34}
	(Kr60A)$_{CI}$	3.391×10^{52}	7.16×10^{32}	1.69×10^{34}
	Kr60B	1.005×10^{52}	3.19×10^{32}	7.12×10^{33}
2nd harmonic	(Kr60A)$_{CC}$	1.597×10^{52}	1.05×10^{33}	1.65×10^{35}
	(Kr60A)$_{CI}$	1.432×10^{52}	4.80×10^{32}	1.04×10^{35}
	Kr60B	4.287×10^{51}	2.16×10^{32}	4.18×10^{34}

		σ_t'	$-\sigma'$	$\|\sigma'-\|^1$ years
fundamental mode	(Kr60A)$_{CC}$	1.10×10^{33}	1.2×10^{-15}	2.75×10^7
	(Kr60A)$_{CI}$	9.65×10^{32}	6.5×10^{-16}	5.11×10^7
	Kr60B	4.31×10^{32}	5.6×10^{-16}	5.99×10^7
1st harmonic	(Kr60A)$_{CC}$	5.49×10^{34}	7.4×10^{-15}	4.51×10^6
	(Kr60A)$_{CI}$	4.96×10^{34}	6.9×10^{-15}	4.83×10^6
	Kr60B	1.92×10^{34}	4.8×10^{-15}	6.91×10^6
2nd harmonic	(Kr60A)$_{CC}$	3.49×10^{35}	4.8×10^{-14}	6.97×10^5
	(Kr60A)$_{CI}$	3.14×10^{35}	4.9×10^{-14}	6.83×10^5
	Kr60B	1.22×10^{35}	3.1×10^{-14}	1.09×10^6

We think that the latter possibility is not to be excluded since the amplitude of the oscillations probably always remains small. Those stars have a relatively dense atmospheres in which the radiative relaxation time is of the same order as the period of the fundamental mode. It seems very likely that the pulsation will take a progressive character in the atmosphere as soon as the maximum pulsational velocity rises above the photospheric sound velocity by a factor of a few units. The shock waves which will then appear in the external layers will stabilize the amplitude to a finite value. Since the period is short, the sound velocity will be reached for very small surface amplitudes, $(\delta r/r)_s$ being then respectively equal to 10^{-2}, 6×10^{-3} and 3.5×10^{-3} for the first three modes. In that case the energy dissipated by shock waves will heat a corona. However, the amount of energy furnished by the pulsation to that corona is larger than the energy it radiates (computed by Unno's method, 1963) as far as the density at its bottom is lower than $\alpha^{-2}10^{-10}$, α being the ratio between the maximum pulsation velocity to the photospheric sound velocity. Since this value appears quite large, a large fraction of the energy furnished to the corona will contribute to the ejection of matter. If all that energy were devoted to mass loss, those stars will lose half of their mass in about $3\alpha^{-2} 10^{+10}$ years. If α has a value of a few units, mass loss can be a significant factor of the evolution of the stars of the lower main sequence. Let us finally note that the instability appears before those stars reach the main sequence since the destabilizing effects of convection are large enough to destabilize the models $(\sigma_t{}' > \sigma_L{}')$.

References

BOURY, A., GABRIEL, M., and LEDOUX, P. 1964, *Ann. d'Astrophys.* **27**, 92.
COWLING, T. G. 1938, *M.N.* **98**, 528.
COX, A. N., and STEWART, J. M. 1965, *Ap. J. Suppl.* **11**, 22.
DELHAYE, J. 1953, *Comptes rendus de l'Academie des Sciences de Paris* **237**, 294.
ECKER, G., and KROLL, W. 1963, *Physics of Fluids* **6**, 22.
FICKETT, W., and COWAN, R. 1955, *J. Chem. Phys.* **23**, 1349.
FOWLER, R. M. 1955, *Statistical Mechanics* (Cambridge Univ. Press).
HAYASHI, C., and NAKANO, T. 1963, *Prog. Theor. Phys.* **30**, 463.
KAMINISI, K. 1960, *Publ. A.S.J.* **12**, 336.
LIMBER, D. N. 1957, *Ap. J.* **127**, 336 and 387.
SIMON, R. 1957, *Bull. de l'Ac. Roy. de Belgique* **43**, 5e serie, 471.
UNNO, W. 1965, *Publ. A.S.J.* **17**, 205.

Discussion

COX   Can you say as to how sensitive your stability calculations are to your treatment of convection?

GABRIEL   Well, I have two cases. One is used in the interior where the life time of the turbulent eddies is long as compared to the period. Then we use the first set of terms in the interior. In the outer layer the life-times of the eddies become shorter than the period, and we use the other terms. But we cannot do anything when both the period and life-time are of the same order of magnitude.

KIPPENHAHN   I wonder if the convection in your models really gives a big contribution to the stability or not. It seemed quite convincing that all your models are unstable and have nice clean periods of about 30 minutes. Do you have any idea why we do not see these stars pulsate?

GABRIEL   There are two possibilities. The stars are not unstable and something is wrong with our picture of convection. There is also a possibility that the instability has not yet been observed. Since these stars have dense atmospheres, the relaxation time of the radiation is of the order of the period of the fundamental mode. And so we can expect to have a shock wave when the amplitude of the pulsation exceeds the sound velocity. Those amplitudes are still quite small.

AUMAN   The value of Z that you use is 0.044. Would it effect the stability any if you take the value of Z which is one half that.

GABRIEL   The answer is no, if the star remains fully convective.

SUGIMOTO   I wonder why the star becomes unstable. From the last table it seems just a nuclear instability. Then, if the star is unstable, the eigenfunction of pulsation has a larger amplitude near the central region. Does your eigenfunction have high value at the central region because of convection? Is convection essential or not?

GABRIEL   Yes, it is essential.

Section IV

Section IV

Faint, metal-poor, subluminous, and red degenerate stars*

JESSE L. GREENSTEIN

Mount Wilson and Palomar Observatories
Carnegie Institution of Washington, California Institute of Technology

---

Abstract

Discovery of white dwarfs among the blue proper-motion stars has been quite successful. Attempts to add to the small list of known, red degenerate stars have been less so. A large number of red, proper-motion stars have been found to have large ultraviolet color excess. Spectra have shown them not to be degenerate; their peculiarity consists in weakened lines after consideration of the blanketing by lines. Only extremely low-metal abundance can produce the $U - B$ excess. These stars must have high velocity and be quite subluminous. The red-degenerate stars become rare below absolute magnitude $M_v > +15.5$.

THROUGH THE many years of surveys of the white dwarfs searched for among the faint blue stars and the proper-motion stars of blue color, highly successful techniques have been developed by Eggen and myself (Eggen and Greenstein 1965a, 1965b, 1967, to be referred to as EG I, II, III). As an example of results, stars were selected merely by blue color, at moderate to high galactic latitudes, by Greenstein (1966). Sixteen per cent of the 80 stars observed spectroscopically were white dwarfs at apparent mags. brighter than 14.5. Of 30 stars fainter than 14.5, 40 per cent were white dwarfs of spectral types DO, DB, DA (the most common). In addition, Eggen and Greenstein concluded that essentially all Lowell proper motion stars of color class −1 were white dwarfs; clearly, it is hardly necessary to observe such stars! The percentage of success, however, fell rapidly at yellower color class (Lowell scale) and,

*This research was supported by the U.S. Air Force under Grant AFOSR 68-1401, monitored by the Air Force Office of Scientific Research of the Office of Aerospace Research.

especially at $+1$, large numbers of high-velocity, weak-lined G stars were found. The same high yield of degenerate stars was encountered in the Luyten proper-motion catalogs among the stars of Luyten color class b (and sometimes a) but the f, g, k stars proved to be largely subdwarfs.

The evolutionary theory of white dwarfs predicts large numbers of cool degenerate stars, since the cooling life-time, at constant radius, goes approximately as $L^{-5/7}$. The number, per unit absolute magnitude interval, per volume of space should rise nearly exponentially. Furthermore, among the bright, nearby white dwarfs of known trigonometric parallax (Table 2 in EG I), the distribution of colors shown in Table I was obtained.

TABLE I   Colors and luminosities of known red degenerates.

$$-1.0 < U - V < -0.5, n = 7, \langle M_v \rangle = +11.9$$
$$-0.5 < U - V <\ \ 0.0, n = 7, \langle M_v \rangle = +13.1$$
$$\ \ 0.0 < U - V < +1.0, n = 3, \langle M_v \rangle = +14.5$$
$$+1.0 < U - V < +2.0, n = 2, \langle M_v \rangle = +15.4$$

After extensive search (Table I in EG III), we achieved a modest success, increasing the number of red degenerate stars with $U - V > +1.0$ to 7 (including the 2 parallax stars). The number of disappointments had been enormous, but understanding the origin of these disappointments has now been a valuable experience. I will describe later, in detail, the problem of the 'Eggenites', but for the moment will only show its nature. As is well known, the UBV color system has been very productive and has proved useful for recognizing the metal-poor F and G subdwarfs. From the differential line blanketing with respect to metal-rich stars, it permitted determination of the M_v of equivalent main-sequence stars. In the UBV system values of $\delta(U - B)$ up to 0.35 mag. were found for F and G 'subdwarfs', but the cooler stars showed small $\delta(U - B)$, and a 'guillotine' near $B - V = +0.70$ mag. was supposed to exist. Any stars of larger $B - V$ with a large ultraviolet excess were assumed to be either red degenerate stars, or quasi-stellar objects. However, in EG II, III lists of weak-line stars are given with $\delta(U - B)$ of 0.30 mag; most of these stars had been observed in the hope that they would be red degenerates. The statistics of $\delta(U - B)$ among stars with $B - V > 0.50$ mag. are given in Table II. It is not certain that all 19 of the stars in the lower line are, in fact, genuinely very subluminous. I have had extreme difficulty in distinguishing spectroscopically between a weak-line faint sdM and a DM. In addition, a large $U - B$ excess may arise from a composite spectrum (e.g. a dK and a DA star, or a DM and a faint yellow DC), or randomly combined types in unresolved binaries. To get the above

TABLE II Frequency distribution of $\delta(U - B)$.

| | $\delta(U - B)$ | | | | | |
	$\leqslant 0.20$	0.21–0.29	0.30–0.39	0.40–0.49	$\geqslant 0.50$	Total
Spectra, sd or id or 'normal'	15	14	6	7	0	42
Red degenerates	3	3	3	4	6	19

table, I have given the 'degenerate star' category every possible benefit of doubt, assuming the star to be degenerate if the lines are very weak, or λ4226 of Ca I is very broad and shallow. Notice that many such degenerate stars would not be recognized by large $\delta(U - B)$ alone, since only one-half have $\delta(U - B)$ larger than 0.40 mag, and a few sdG do have measured values of $\delta(U - B)$ as large as 0.40 mag. Eggen (1968a) noted the existence of large numbers of stars with $\delta(U - B) \leqslant 0.30$ mag at $B - V > 0.7$ and at first suggested that most of them were red degenerate stars. The above statistics, however, suggest that a purely colorimetric method will fail unless $\delta(U - B) > 0.50$ mag. But so restrictive a search would also exclude two-thirds of the true yellow and red degenerates.

In consequence, special efforts have been made to observe proper motion stars whose *UBV* data suggest a luminosity which would permit us to deduce an excessively large tangential motion. Eggen has described, in this conference, his methods of selecting stars from the Lowell survey, and from the LTT list, for photoelectric observation. I observed as many of them as I could, spectroscopically. Nineteen such spectra have been obtained, with $\delta(U - B)$ up to 0.52 mag. The $\langle B - V \rangle = 0.77$, $\langle \delta(U - B) \rangle = 0.30$. Of all these stars, one seems to be a genuine DC (LTT 375 = L 651–57). It has $B - V = +0.65$, $U - B = -0.32$, $\delta(U - B) = 0.51$ (close to the blackbody line, and an extreme value). It is close to vMa 2, a classic DG star in $B - V$, but has a larger $\delta(U - B)$. But LTT 375 has no visible lines, i.e. it is a DC of fairly high effective temperature, while the continuum of vMa 2 is depressed in the ultraviolet by its strong, unresolved and broadened metallic lines. There is little doubt that LTT 375 is a yellow white dwarf, with $T_e \approx 6000°$K, but it is disappointing to find it the sole success out of 19 spectra taken. In addition, it is *not* a very low-luminosity red degenerate like such classic stars as W 489.

The unexpectedly large number of cool stars with ultraviolet excesses is related to observations by Woolf, Wallerstein, and Sandage (1965). They observed 15 relatively bright proper motion stars with $\langle B - V \rangle = 0.96$, $\langle \delta(U - B) \rangle = 0.25$, and found, at low dispersion, one peculiar object (a W UMa star), several stars with apparently weakened lines, and λ4226 weak in many. The mean transverse motion depended on the luminosity-calibration method. If $B - V$ is taken uncorrected as indicating M_v, they find $V_T = 290$ km/sec; if $\Delta(B - V) = \delta(U - B)$ (i.e. a line-blanketing correction at 45° slope), $V_T = 181$ km/sec. Thus, the bright stars they observed are a Population II group of as large a $\delta(U - B)$, at $B - V$ near $+1.0$, as the intermediate and very weak-lined subdwarfs are at $B - V$ near $+0.4$. The colorimetric approach, then, among proper motion stars, is probably unlikely to establish by itself unique criteria for all the red-degenerate stars. Two types of red degenerates could still be found: those with low metal surface content and no line blanketing should lie close to the black-body line (i.e. $\delta(U - B) > 0.40$); and those with very low luminosity, i.e. $R \leqslant 10^{-2} R_\odot$, and $T_e < 5000°$ ($M_v > +15$) should be found by the impossibly large space motion derived on the assumption that they are ordinary K or early M main-sequence stars.

The selection of stars we observed come from the size of the proper motions. The reduced proper motion

$$H \equiv m + 5 \log \mu \tag{1}$$

normally gives a regression, for Population II, of

$$M = H - 2.9, \tag{2}$$

and the relation between the luminosity and the tangential motion is

$$M = H + 8.39 - 5 \log V_T. \tag{3}$$

The total space motion $V = (3/2)^{1/2} \langle V_T^2 \rangle^{1/2}$, for uniform distribution of objects and space motions. Typical V_T for groups of extremely weak-line subdwarfs reach 250 km/sec. (The mean radial speed $\langle \rho \rangle = 2/\pi \langle V_T \rangle$, and $\langle V \rangle = 4/\pi \langle V_T \rangle$.) Let us consider the one successful discovery, LTT 375 ($=$ L 651–57). It has $V = 14.6$, $\mu = 0''.6$, or $H = 13.6$ and $M = 10.7$, from Eq. (2), far too bright for a yellow degenerate star. With color $U - V = +0.33$, assuming the white-dwarf nature, we expect $M_v \approx +14$ (from Table I) or $V_T \approx 40$ km/sec. On that assumption, the degenerate star has quite moderate space motion.

Let us examine in detail a group of stars for which Eggen (1968a) had UBV photometry, and for which I had spectra, on scales of 90 or 190 Å/mm. Figure 1 and Table III contain data on these 23 stars; among them, one (LTT 7983) is a rediscovered DA; LTT 3967 is a double, and its color is presumed to be unreliable; LTT 9491 lacks photometry, although it is a DC. For several others, the ultraviolet excess $\delta(U - B)$ is so small (or negative) that it would be unwise to include them in the statistics. This removes

LTT 2535, 7511, 14983, leaving 17 good cases, with $B - V > 0.50$, and a mean excess $\delta(U - B) = 0.29$. Relevant data are given in Table III, where the second column gives Eggen's decision as to whether the star was a high velocity (his $T(2)$ value) or white dwarf (WD). The column headed $\Delta(B - V)/\delta(U - B)$ was derived from Eqs. (1) and (3), as follows: I assumed $V_T = 400$ km/sec, and determined M_v. I then found the change in $B - V$ color, $\Delta(B - V)$, required to shift the star onto the Hyades main sequence, and took the ratio. The mean value of the slope is 0.67, omitting the uncertain values and the ratio for the DC, but including negative values. Henceforward, this slope, φ, of the blanketing line was taken as correct. From

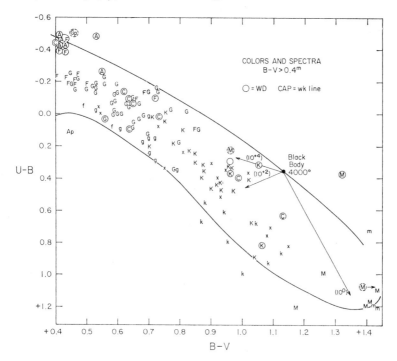

FIGURE 1 The observed colors and spectra of stars observed as possible yellow or red degenerates. In circles, type of white dwarf (broken circles, uncertain as to white dwarf nature). Capitals are subdwarfs, lower-case letters apparently normal, crosses are colors available, but no spectra. The black-body line is that of Matthews and Sandage, but the vectors emanating from the 4000°K point are those computed by Tsuji for line-blanketing effects for hydrogen/metal ratio = $(10°)$ $A_\odot$ and lower metal abundances (10^2) $A_\odot$, (10^4) $A_\odot$. Notice the large number of stars with $\delta(U - B)$ up to 0.3 mag. at all colors, of which only a small fraction are white dwarfs.

TABLE III Candidate stars for red degenerates.

Name	Eggen T(2)	$(B-V)$	$\delta(U-B)$	μ	$\phi=\dfrac{\Delta(B-V)}{\delta(U-B)}$	Spect.	$(B-V)_G$	$M_v(G)$	$\log V_T$
LTT125	600, WD	+0.56	+0.32	0″.20	0.8	sdG	+0.76	5.6	2.68
375	1800, WD	0.62	0.51	0.60	(1.6)	DC	0.96	(6.6)	(3.06)
424	540, WD	0.77	0.40	0.24	0.2	sdG	1.04	7.0	2.45
458	500, WD	0.53	0.23	0.26	1.8	sdG	0.68	5.2	2.87
1561	(610)	0.85	0.29	0.31	0.9	sdK	1.04	7.0	2.67
2535	730, WD	0.70	0.05	0.88	((14))	sdK	0.73*	5.5*	3.26*
3966		0.53	0.25	0.24	−0.3	sdG	0.70	5.3	2.30
3967		0.83*	0.23*	0.23	—	sdK } sdG	—	—	—
4210		0.56	0.26	0.31	1.1	sdG	0.73	6.6	2.41 }
4211	600, WD	1.00	0.30	0.31	0.8	—	1.20	7.5	2.66
4697	670, WD	0.97	0.30	0.30	1.2	sdK	1.17	7.4	2.74
4985		0.75	0.32	0.47	0.6	sdK	0.96	6.6	2.59
5074	WD	0.95	0.44	0.58	0.3	sdGp	1.24	7.7	2.49
7511		0.70	0.10*	0.32	—	sdG	0.77*	5.7	2.06 }
7512		0.96	0.42	0.32	−0.1	sdK	1.24	7.8	2.32
7983	WD	0.23	0.66	0.29	(1.2)	DA	—	—	—
9353		1.01	0.22	0.68	0.9	sdK	0.80	5.8	2.62
9372		1.02	0.47	0.40	0.1	sdK	1.34	8.2	2.36
9491				0.26	—	DC	—	—	—
9765	780, WD	0.87	0.19	0.75	2.6	sdK	1·00	6.8	2.90
14824		0.61	0.14	0.35	0.6	sdG	0.70	5.3	2.56
14983	800	0.87	−0.07*	0.49	—	sdK	—	—	—
15023	WD?	0.81	0.33	0.34	−0.1	sdK	1.03	6.9	2.36

*These stars have photometry uncertain, or are close doubles, or have peculiar (small) $\delta(U-B)$.

$\varphi \times \delta(U - B)$ a new color $(B - V)_G$ is derived, for a star supposedly equivalent to a Hyades star of normal metal-line strength. This $(B - V)_G$ then provides a new $M_v(G)$, and finally, $\log V_T$ is rederived from Eq. (3). The mean $\log V_T = 2.59$, or 390 km/sec, indicating that the process is self-consistent. The method is not completely based on circular reasoning, since its goal is to find the slope of the blanketing line for late-type stars, and to note exceptional cases. The one white dwarf, LTT 375, has the largest meaningful $\log V_T$ (over 1000 km/sec); LTT 2535 must be omitted because of its small $\delta(U - B)$. The slope, φ, is very flat, and the intersection of the blanketing line with the normal star locus is uncertain until the latter flattens out in the early M stars. With $\Delta(B - V) = \frac{2}{3}\,\delta(U - B)$, we find the intersection of this line and the $(U - V; B - V)$ locus to be at $(B - V)_G = 1.80\,(B - V) + 1.20\ \delta(U - B) - 0.98$. Since $dM_v = 4.7\ d(B - V)$, we have $dM_v = 5.6\ d(\delta(U - B))$, or $d \log V_T = -1.12\ d(\delta(U - B))$. Thus, errors of about 0.06 in $\log V_T$ will arise from the photometry and the shallow slope of the line.

An approximate set of relations can be derived. If $(U - B)$, $(B - V)$ are observed, the corrected values for red stars, after differential line-blanketing is removed are:

$$(B - V)_G = (B - V) + \tfrac{2}{3}\,\delta(U - B) = (B - V) + \phi\delta(U - B); \qquad (4)$$

$$(U - B)_G = 1.80\,(B - V) + 1.20\,\delta(U - B) - 0.98 \qquad (5)$$

These equations are not rigorously self-consistent since the real relations are not linear. Then, if the Hyades main-sequence color-magnitude relation is approximated by a linear formula, and the $(B - V)_G$ corrected for blanketing is used,

$$M_v = 4.7\,(B - V) + 2.1, \qquad (6)$$

and we obtain

$$M_v(G) = 4.7\,(B - V) + 4.7\,\phi\delta(U - B) + 2.1. \qquad (7)$$

Thus, if $\phi = 0$ (Eggen's assumption), as compared to our $\phi = \frac{2}{3}$, with a mean value of $\delta(U - B) = 0.29$, his absolute magnitudes are estimated 0.9 mag. too bright. Similarly, $\Delta \log V_T = -\frac{1}{5}\Delta M_v$, and the space motions derived are 0.19 in $\log_{10}$ too large (about 55 per cent). Since the slope of 0.67 seems about correct, from theoretical grounds also, it is likely that only those cases for which Eggen's V_T exceeds 1.55×500 km/sec, or about 800 km/sec, are certain to be degenerate stars. The errors in deduced $\log V_T$ are about 0.27 times the error in ϕ.

Is there any justification for this blanketing line, beyond the observational evidence that stars can be forced onto the main sequence, with $V_T = 390$ (or $V = 495$ km/sec)? The expected dispersion in radial velocity is 250 km/sec. I have just completed a preliminary set of measurements, at 90 Å/mm, of radial velocities of 22 of these stars. The accuracy is about ± 20 km/sec per plate, but will be higher after we develop a set of standard wavelengths for blended lines. According to the present results, the mean absolute value $\langle|\rho|\rangle = 112$ km/sec; the $\langle\rho\rangle = -20$ km/sec may depend on the limited sample, which does not cover the sky symmetrically. Note, however, that $\langle V_T\rangle = \pi/2\,\langle|\rho|\rangle$, so that the radial velocities suggest $\langle V_T\rangle$ of only 180 km/sec, rather than 390 km/sec if on the main sequence. This discrepancy would require that the stars lie, on the average, 1.6 mag. below the main sequence. The largest measured radial velocities are -237 km/sec for a sdG

283

(extremely weak-lined) and a poorly determined velocity of -220 km/sec for a sdMp star. These, while large, do not suggest any extraordinary large space motion for this group. The alternatives facing us are:

1. The stars have the blanketing corresponding to metal-poor stars even at a low temperature. The $\Delta(B - V) = \frac{2}{3}\delta(U - B)$. Eggen's luminosities are too bright by 1 mag. Most can be forced onto the main sequence with a space motion of about 500 km/sec. Some faster moving objects would not be permanent members of the Galaxy.

2. They lie below the main sequence by 2 or 3 mags and have nearly the same line blanketing as the metal-rich stars, i.e. $\Delta(B - V) \ll \frac{2}{3}\delta(U - B)$. Their space motions are then 160 km/sec, their surface gravity is either ten times normal (but their spectra look normal), or their mass is 0.1 that of normal stars of the same temperature.

3. Many are red, partially degenerate stars, in spite of apparently normal spectra. The broad lines expected at high surface gravity may be buried under a low-pressure shell. The frequency of such red degenerate stars is high, but their mean radius is 0.1 $R_\odot$ instead of 0.01 $R_\odot$, and their space motion is about 50 km/sec.

Astrophysical estimate of blanketing

The enormous atomic-line blanketing in K stars, and the complex band-blanketing by TiO in M stars have never been computed. Line-free models from $T_{\text{eff}} = 5700°$ to $4000°$ can be computed, but with low-metal abundances, become difficult as H^- decreases in importance. As the hydrogen-to-metal ratio increases from its solar value $A_\odot$, to $10^{+2} A_\odot$ (as in G subdwarfs) the P_e/P_g ratio varies about as $A^{-1/2}$. Scattering by H_2 and H increases, metal opacity decreases. The Rayleigh scattering makes integration of model atmospheres difficult, and also reduces the ultraviolet emergent flux, producing a quite complex behavior of the UBV unblanketed colors (see Figure 2).

Because of the strange properties of these weak-lined stars, a first approximation to the problem has been worked out by Tsuji. He computed three models, at $T_e = 4000°$, $\log g = +4$, and $A = A_\odot$, $10^2 A_\odot$, $10^4 A_\odot$. (The latter is essentially metal-free.) The models gave F_λ's which were integrated to predict UBV colors, in Figure 3. The depression of the $U - B$ color (see Figure 2) by Rayleigh scattering is quite marked in the metal-poor stars. Predicting the strength of the lines would be an enormously complex computation, so that only an approximate blanketing theory was used. The line strengths may be scaled in proportion to $A^{-1/2}$, when changing abundance, from a basic model for a star with $T_e = 4000°$ and $A_\odot$. To simulate this normal star's colors, Tsuji based himself on the wavelength distribution of line blanketing in the sun. After checking that the latter produced the correct

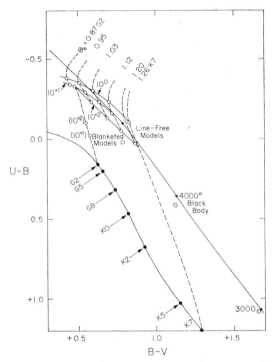

FIGURE 2 Tsuji's computation of *UBV* for the available line-free models, from $\theta_e = 0.87$ to 1.26, are shown with metal abundances as labelled. The dashed lines link models of the same θ_e and different A. Approximate spectral types on the normal color–color relation can be linked, for $A_\odot$, with the line-free models at the correct θ_e, by lines sloping slightly to the left and up. Tsuji's black-body curve lies below the Matthews and Sandage one, and is indicated by two open circles.

UBV colors of the sun, from a solar model, the line strengths were scaled up (because of the lower T_e and greater strength of neutral lines). This produced a blanketing vector, from the 4000° model, down to a point on the normal $U - B$, $B - V$ locus where stars of spectral type K6, $T_e = 4000°$ would be located. Thus, the distribution of mean line strength $\langle l_\lambda \rangle$ was determined empirically for $A_\odot$; for models with A, $A^{-1/2} \langle l_\lambda \rangle$ was added to the continuous absorption. Then, after computing the expected blanketing, F_λ, the *UBV* magnitudes and colors were predicted. The model with $10^4 A_\odot$ had negligible line blanketing; that with $10^2 A_\odot$ appreciably less than did $A_\odot$. Since the starting points differed, the net result is fairly complex. Figure 3 shows the *UBV* colors for the models, before and after blanketing correction. Note that the shift from a black-body at 4000°K is up and to the left, but the

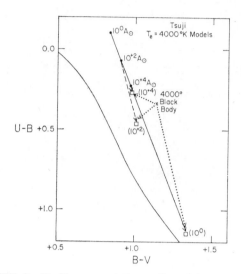

FIGURE 3 Tsuji's computations of model atmospheres at 4000°K (near type K7). As the metal abundance is decreased, the models lose ultraviolet flux by scattering. The line blanketing at $A_\odot$, and other values, are shown with open squares. The theoretical blanketing vector from a line-free model to the main sequence has a slope of about 0.88 at $A_\odot$, as compared to the empirical value of 0.67; at $10^2 A_\odot$ the slope is 0.71. Note, however, that the $B - V$ differs from that of a black body only about $- 0.15$ mag for $10^4 A_\odot$ and $10^2 A_\odot$, and by $+ 0.20$ mag for $A_\odot$. Values of $\delta (U - B)$ of 0.50 to 0.35 mag. will be encountered at 4000°K if the metal abundance is sufficiently low.

blanketing correction is far down, and slightly to the right. The final result is that the *UBV* colors can be represented by a vector with respect to the black-body, which is nearly straight downwards, and long, for $A_\odot$, and up to the left, and short for $10^4 A_\odot$. Thus, if we can generalize from computations at only one temperature, stars with large metal deficiencies will have *UBV* colors mostly between the black-body and main-sequence curves, and occupy the region of the 'Eggenites'. The locus of constant T_e is a crescent-shaped, nearly vertical curve, concave to the right. Objects with $10 A_\odot$ to $10^2 A_\odot$ would have $\delta(U - B)$ like those now commonly observed. Objects close to the black-body line may have even lower metal abundances.

The spectroscopic properties

Other effects of large A may be observable on higher-dispersion spectra. A strong line will vary as $W \propto \sqrt{(N \Gamma/k)}$; $N \propto A^{-1}$; $k \propto A^{-1/2}$; $\Gamma \propto P_g$, and $P_g \propto A^{+1/2}$. Therefore, W will be unchanged. For weak lines, $W \propto$

$N/k \propto A^{-1/2}$, so the background of weak lines will be reduced. In consequence, we would expect to see pressure-broadened lines relatively easily; the ionization level will also be higher, and differ at a given τ, once scattering becomes an important fraction of the opacity. Thus, persistence of Ca II resonance lines, as compared to Al I lines, to low temperatures and a predominance of a few strong lines like Ca I may be expected. Further, TiO bands would be weakened, as compared to MgH, a phenomenon already commented on in such cool stars as G 95-59, G 43-53, and G 100-54B (EG II, Table I).

The spectroscopic features, at low dispersion which lead to the classifications given in Table III and plotted in Figure 1 should be explained. Owing to the extremely weak metallic lines near $B - V = 0.4$, the low-dispersion spectra show mainly hydrogen lines, the K line of Ca II, and sometimes, weakly, the G band and a few of the strongest Fe I lines. If only the hydrogen and the K lines are seen, I call the star, for this investigation, sdF, although it is probably an extreme sdG like HD 19445, while if the G band is visible, and there are traces of other metallic lines, it is here called sdG. If the G band, Fe I lines, and Ca I become strong, it may be hard to distinguish a low-metal K star from a normal one. But if Ca I is by far the strongest metallic line except for Ca II, I call the star sdK. (The hydrogen lines Hγ and Hδ may still be visible.) The fact that most stars plotted in Figure 1 as subdwarfs have appreciable $\delta(U - B)$ suggests that the spectroscopic separation is successful. It is difficult to distinguish an idG (slightly weak-line) from a sdK star, at this scale. A normal K star has a background of a large number of barely-visible, low-excitation lines marginally visible, and can be separated from idG successfully. The M stars are noted by the disappearance of the G band, the strength and broadening of Ca I and the presence of Cr I; in the normal early M stars TiO is visible, but in the metal-weak, MgH is seen as a broad, shallow band. The persistence of hydrogen lines in stars of quite advanced type ($B - V \approx 0.8$) is notable; some stars were therefore first classified as sdF. We cannot distinguish horizontal-branch F stars either from extreme sdF or sdG types.

An attempt to add to the number of red degenerate stars

In EG III, a table and illustrations of the spectra are given for the well-established, red degenerate stars. In spite of extensive observations, complete data establishing the reality and spectral characteristics for a single, additional, red degenerate does not yet exist! The lists of the most promising candidates are derived from Eggen (1968a, b). In Eggen (1968b) his Table 3 give 58 southern stars with suspected $V_T > 500$ km/sec (a reasonable upper limit to stars bound gravitationally in our galaxy) among the 10 with

$\delta(U - B) \geqslant 0.40$. I have 5 spectra, and one is a DC star (the same LTT 375 mentioned earlier). I investigated, statistically, the sample of 58 of similar objects, for which I had spectra, including the same single DC. For all of them, using my blanketing rule, I derived V_T from

$$\log V_T = 0.2 \, V_E + \log \mu + 1.26 - 0.94 \, (B - V) - 0.63 \, \delta(U - B). \quad (8)$$

The range of $\log V_T$ was from 2.20 to 3.51; 17 were probable dM stars, for which this method does not work; 11 were probably very subluminous, or degenerate, with indicated values of $V_T > 1000$ km/sec; 13 had $V_T < 400$ km/sec, and were high-velocity stars; the last 17, therefore, had such high velocities (400–1000 km/sec) that (1) either the method or the photometry is erroneous, or (2) the stars are at least 1 mag below the main sequence.

The stars for which I already had spectra are discussed in my Table III. There are only 9 which do not overlap with Table III of Eggen (1968b), just discussed. There are weak-lined G or K stars; 6 stars had deduced $V_T < 400$ km/sec, and 3 had $400 < V_T < 1000$ km/sec. While the ratio of very high-velocity stars is lower, it is within the sampling error. A final statistical note can be obtained from the distribution of $\delta(U - B)$ for all stars published in EG (I–III), among which there were 19 red degenerates $(B - V \geqslant 0.50)$ and 41 sd or 'normal' stars. The overlap range of $\delta(U - B)$ is so large (see Table II) that two-thirds would not be recognized by $\delta(U - B)$ alone. Consequently, if we count all objects with $\delta(U - B) \geqslant 0.40$ (about the maximum for subdwarfs) and multiply by three, we might have an estimate which is good to within a factor of two. Eggen (1968a, Table 5) contains 107 yellow LTT stars (color classes f to k) with $\mu > 0''.20$. There are 80 stars of $B - V \geqslant 0.50$, and $\delta(U - B)$ occurs for 8 of them. The total possible yellow and red degenerates, after applying my correction factor, based on the distribution of $\delta(U - B)$, would be 22 per cent, without regard to the size of the space motion. The fraction of very large apparent motions is about one-sixth, so our final estimate is that the total number of yellow proper-motion stars that will be degenerates is 3 or 4 per cent.

The stars that are isolated as possible (southern) red degenerates are listed in Table IV. Probably most are, in fact, white dwarfs. They were selected from stars with color class g and k in the Luyten catalogs of stars of motion $>0''.2$, and should be a good finding ground for yellow or red degenerate stars. The $U - V$ colors are distributed as follows:

$U - V$	-0.5 to 0.0	0.0 to $+0.5$	$+0.5$ to $+1.0$	>1.0
n	4	3	2	2

Thus, although Eggen's selection has yielded 11/58 or 19 per cent degenerate stars, only 4/58 or 7 per cent were yellow or red degenerates! My method

TABLE IV New possible late-type degenerates.

LTT	V_E	$B-V$	$U-B$	$\delta(U-B)$	μ	log V_T	Spectr. or remarks
329	14.53	+0.45	−0.57	0.57	0″.32	2.89	Large $\delta(U-B)$
375	14.60	0.65	−0.32	0.51	0.60	3.05	DC; spectrum available
560	15.95	0.50	−0.42	0.45	0.27	3.13	V_E seems too faint
934	13.90	0.43	−0.44	0.44	1.05	3.38	Good case
1424	13.56	0.49	−0.20	0.22	0.50	3.07	
1534	14.71	0.16	−0.55	0.67	0.30	3.25	Probably DA
1679	14.92	0.74	+0.30	0.02	0.45	3.08	
2981	14.09	0.66	−0.17	0.37	2.05	3.51	Good case
4124	14.32	0.99	+0.68	0.14	1.26	3.20	
5622	15.20	0.80	+0.07	0.36	1.11	3.38	Good case
6885	13.98	0.76	+0.27	0.09	1.06	3.31	Small $\delta(U-B)$, high vel.

excluded stars with the *UBV* colors of ordinary dM stars, and therefore may have excluded DM stars, which normally have $\delta(U-B)$ near 0.40 mag. and $U-V$ near 1.5 mag. The reddest star in my Table IV has only small $\delta(U-B)$. Nevertheless, spectra and parallaxes of stars in Table IV are badly needed. The high percentage of 'Eggenites' (17/58 or 29 per cent) suggests their statistical importance. Further parallax measures are needed on such critical objects as LTT 4985, G 14-24 (where a good Cape parallax of 0″.074 disagrees with the normal spectrum, and the reasonable space motion, $V_T = 320$ km/sec). If the blanketing slope of 0.67 is used, $(B-V)_G$ becomes +1.05, $M_v(G) = +7.0$, and the photometric parallax is 0″.0068. (We discussed this controversial object in EG I.)

If the correction for line blanketing given here is accurate and line or band weakening is the source of the $\delta(U-B)$, the effective temperature of some of these stars may be quite low. In that case, far infrared photometry should be informative. However, for the present sample, $T_e > 4000°$K gives about the correct *UBV* colors, and it is only stars with $B-V > 1.00$ that will prove to be very cool. I hope to continue spectroscopic surveys of these objects, both with higher dispersion, and on an enlarged statistical basis, to obtain a better idea of the frequency of red degenerates and of the faintest absolute magnitudes at which they can be found, for comparison with the theory of a solid lattice formation, and the drop in specific heats. At present, rough theoretical predictions place the lower limit for a star with 0.01 $R_\odot$ near $M_v = +15.5$, in agreement with observation.

References

EGGEN, O. J. 1968a, *Ap. J. Suppl.*, No. 142.
EGGEN, O. J. 1968b, *Ap. J. Suppl.* (in press).
EGGEN, O. J., and GREENSTEIN, J. L. 1965a, *Ap. J.*, **141**, 83.
EGGEN, O. J., and GREENSTEIN, J. L. 1965b, *Ap. J.*, **142**, 925.
EGGEN, O. J., and GREENSTEIN, J. L. 1967, *Ap. J.*, **150**, 927.
GREENSTEIN, J. L. 1966, *Ap. J.*, **144**, 496.
LUYTEN, W. J. 1957, *Univ. of Minn. Obs.* (Southern).
LUYTEN, W. J. 1961, *Univ. of Minn. Obs.* (Northern).
WOOLF, S. C., WALLERSTEIN, G., and SANDAGE, A. R. 1965, *Pub. A. S. P.*, 77, 370.

Discussion

STULL   How early in spectral type do these stars between the main sequence and white dwarfs occur?

GREENSTEIN   Well, this whole technique has really been looking for what are called K stars, but some of them are classified by me as G. Now there is still an argument going on whether the classical G subdwarfs are or are not in fact on the main sequence. Some people claim they might be a magnitude below. Groombridge 1830, depending on what you do about the blanketing, would be an early example that was supposed to be a G8 star, but is about K0.

SALPETER   As you mentioned, these red stars cannot very well be stars below about $0.07 \, M_{\odot}$. But, could they not possibly be stars of mass just a little bit greater than this limiting mass, say 0.12, something a little bit greater than 0.07 which are already partially degenerate or fully degenerate. If you take a 0.12 star, which is not burning hydrogen, but has gone through its maximum temperature and is cooling down, its surface temperature would be higher than that of a main-sequence star of the same mass and it would then finally cool down to something red, in a line below the main sequence, but nowhere near as far below as ordinary white dwarfs and also the pressure broadening would not be great because the radii are a little bit down. I would like to ask two questions. One of them is: purely spectroscopically, would that be a possibility? And the second question is: if they were to be stars like that, then they clearly cannot be stars which started off on the main sequence with that mass and evolved by themselves; they would all have to be remnants of, say, globular cluster stars that have lost all their mass except the remaining 0.12. This is only quantitatively more drastic than ordinary white dwarfs. If this is so, then they will be very hydrogen poor because usually the innermost $0.12 \, M_{\odot}$ of a globular cluster remnant certainly should have no hydrogen left in it and may already be carbon rich, and metal poor, like the original star was.

GREENSTEIN If you can compute such objects, it would be very interesting, because, you may remember, while it is very controversial, the surface gravity of horizontal branch stars keeps wobbling around but sometimes you get a statistical impression that their masses are as low as 0.3 $M_\odot$. Now it is a very, very difficult problem to get spectroscopic surface gravities. Dr. Savedoff did it for RR Lyrae stars; I forget what he got, but most people have been getting low values, so that is very interesting. The lower the mass, the better the surface gravity, the better the broadening problem. Essentially, surface gravity and hydrogen to metal ratio enter in the same way, so if you cut the mass by a factor of 10 below the main sequence equivalent you gain about that much in the broadening, which is the same as cutting the metal deficiency by 10. Now, as to hydrogen deficiency—this is a very peculiar question. You always see hydrogen lines, until the temperature drops awfully low. The interesting point, however, is that it has been a known peculiarity for many years, even of the G subdwarfs and certainly of some of these, that the strongest feature in the spectrum is the G band of CH (hydrocarbon). It is always present at almost any temperature and in certain very metal-poor stars; one star analyzed by Baschek, in fact, has a step in the spectrum where the CH band occurs. So the possibility of carbon enrichment is not at all remote. We already, by the way, know a new group of stars, which may be spoken about by somebody else, which have main-sequence carbon enrichment.

OSTRIKER There's one question and one comment. The question is: At about what luminosity would you say the surprising absence of white dwarfs begins to show up?

GREENSTEIN Roughly $+13$ or $+14$. I showed you all the known stars at the beginning. There are roughly 200 known white dwarfs of spectral type DA. Of course, there are probably at most 8 or 9 red degenerate objects with radii less than 0.03 $R_\odot$. The deficiency really starts the minute you run out of the DA's which is around $+12$ or $+13$.

OSTRIKER The comment is that it seems possible that the red or white dwarfs cool faster than expected.

GREENSTEIN Well, that's what I hope that people will talk about, a very rapid cooling time.

Physical processes in white dwarfs*

H. M. VAN HORN

Department of Physics and Astronomy, C. E. Kenneth Mees Observatory
University of Rochester

Abstract

At temperatures T and densities ρ characteristic of the interiors of white dwarfs, the ratio

$$\Gamma = \frac{E_{coul}}{k_B T} = 2.28 Z^2 \left(\frac{\rho}{A \times 10^6 \text{ g . cm}^{-3}}\right)^{1/3} \left(\frac{10^{7\circ}\text{K}}{T}\right)$$

where A, Z are the atomic mass and charge, respectively, is large enough to cause crystallization of the plasma into a Coulomb lattice ($\Gamma \sim \Gamma_m \simeq 50 - 125$). For $\Gamma < \Gamma_m$, in the liquid phase, we have developed an interpolation formula for the equation of state of the interacting ion plasma in the (uniform) background of electrons:

$$\frac{(P - P_0) V}{N k_B T} = -0.113 \, \Gamma^{3/2} \left[\frac{1}{(1 + 0.142 \, \Gamma)^{1/2}} + \frac{1.54}{1 + 0.575 \, \Gamma^{3/2}}\right],$$

where P_0 is the pressure of the non-interacting ion gas at the given temperature and density. The transition to the solid phase is accompanied by the release of latent heat of crystallization, $T \delta S \sim N k_B T$ and this, together with the rapid decrease of the heat capacity of the solid phase at low temperatures, where the higher vibrational modes of the lattice can no longer be excited, leads to a large departure of the cooling law for white dwarfs at low temperatures from that expected for a perfect gas of ions. We also consider briefly the effects of crystallization on the transport properties of white dwarfs and show in particular that for a "typical" white dwarf with $\rho \sim 10^6$ g . cm^{-3}, $T \sim 10^{7\circ}$K one may expect contributions to the viscosity from the electrons and ions of the order $\eta^{(elect)} \sim 10^7$ poise, $\eta^{(liq)} \sim 10^3$ poise, $\eta^{(sol)} \sim 10^{10} - 10^{14}$ poise.

* This work was supported by the National Science Foundation under grants GP6174 and GP8120.

I. PHYSICAL CONDITIONS IN WHITE DWARF INTERIORS

AS HAS BEEN pointed out by Salpeter (1961, 1968; see also Hamada and Salpeter 1961), Coulomb interactions do not have a strong effect upon the equation of state in white dwarfs and therefore have relatively minor influence on the structure of these stars. The electrostatic terms do have significant effects upon the energy balance of white dwarfs, however, and these are the effects that I shall principally discuss.

Let us first consider briefly the physical conditions that exist in a 'typical' white dwarf—composed, say, of pure C^{12}—with central temperature $T_c \sim 10^7 °K$ and central density $\rho \sim 10^6$ g . cm^{-3}. Such a star would have a mass $\sim 0.4\,M_\odot$. The Fermi energy of the electrons, E_F, is then of the order of 1 Mev, which is large in comparison with the thermal energy $k_B T \equiv \beta^{-1} \sim$ 0.9 kev; with the K-shell ionization energy of C^{12}, $I_K \sim 0.5$ kev; and with typical Coulomb interaction energies

$$E_{coul} = (Ze)^2/a_s \sim 30 \text{ kev} \tag{1}$$

where $(\frac{4}{3}\pi a_s{}^3)^{-1} = \rho/AH$ is the number density of the ions. (Here k_B is Boltzmann's constant, $H = 1.66044 \times 10^{-24}$ g is the unit of atomic mass, and A is the atomic weight of an ion.) By virtue of the two inequalities $E_F \gg I_K$, $E_F \gg E_{coul}$ the plasma is therefore fully ionized, and the electron density is, to a high degree of approximation, uniform.

Under these conditions the partition function for the system of electrons plus ions can be written as a product of a partition function for the *non-interacting* electron gas multiplied by a partition function for the *interacting* ion gas in a uniform background of negative electronic charge. The thermodynamic functions for the system as a whole then become simply the sums of the thermodynamic functions for the electron and ion subsystems. Since we have $E_F \gg k_B T$, the main contribution to the pressure is provided by the degenerate electrons, and, as is well known, the equilibrium configuration of a white dwarf is therefore well approximated by the structure of a self-gravitating, zero-temperature, non-interacting electron gas (Chandrasekhar 1957). As the thermodynamic properties of a semi-degenerate, semi-relativistic non-interacting electron gas have been accurately calculated by Grasberger (1961) I shall not consider the electrons any further, and I shall discuss only the effects of the temperature-dependent Coulomb interactions, which involve the ions.

As has been pointed out by Salpeter (1961) and independently by Abrikosov (1961) and by Kirzhnits (1960), in the limit of zero temperature the ions in the plasma form a crystalline lattice, and the magnitude of the total Coulomb interaction energy per ion is of the order of that given by Eq. (1). At very high temperatures, however, such that

$$\Gamma = \frac{E_{\text{coul}}}{k_B T} = 2.28 \, Z^2 \left(\frac{\rho}{A \times 10^6 \, \text{g} \cdot \text{cm}^{-3}} \right)^{1/3} \left(\frac{10^7 \, {}^\circ \text{K}}{T} \right) \ll 1 \qquad (2)$$

the electrostatic term is quite different and much smaller, since the ions in this case are not localized near the points of a perfect lattice, but are instead able to move relatively freely and form a weakly interacting 'gas'. A transition between this Coulomb 'gas' or 'liquid' and the Coulomb lattice that exists when $\Gamma \gg 1$ evidently must occur at some intermediate Γ-value, Γ_m, that one would have expected *a priori* to be of order unity. The recent Monte Carlo calculations by Brush, Sahlin, and Teller (1966; hereafter referred to as BST) indicate, however, the much higher value $\Gamma_m \approx 125$, and additional semi-theoretical estimates by Mestel and Ruderman (1967) and by myself (Van Horn 1968) lead to $\Gamma_m \sim 50$–60. Since none of these results seems obviously better than the other, and since the difference in Γ_m leads only to an uncertainty of about a factor of two in the resulting temperature of crystallization, I shall make use of the latter value, which has perhaps somewhat more theoretical justification.

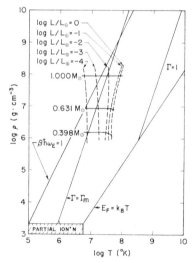

FIGURE 1   Temperature–density plane for pure C^{12} composition.

The applicability of these considerations to white dwarfs is indicated in Figure 1, which shows the log ρ–log T diagram for a composition of pure C^{12}. The line labeled '$E_F = k_B T$' marks the approximate boundary between the non-degenerate and fully degenerate regions for the electron gas. Throughout the entire plane except the small hatched region in the lower left-hand corner the plasma is fully ionized. I shall not discuss the region in which the ionization is incomplete. The line marked '$\Gamma = 1$' is the boundary of the region in which the Coulomb interactions first begin to become important, and the line labeled '$\Gamma = \Gamma_m$' indicates the locus of points where crystallization first sets in. I have also drawn in the evolutionary tracks of the central regions of three white dwarf models, with compositions of Fe^{56} and different masses, which were computed by Savedoff, Vila, and myself (Savedoff, Van Horn, and Vila, 1968), and on which the luminosities of the cooling star models are indicated. These curves show clearly that crystallization must take place in such objects in precisely the observed range of white dwarf luminocities ($-4 \lesssim \log L/L_\odot \lesssim -2$). Also shown on this diagram is a line marked '$\beta \hbar \omega_E = 1$'. This line separates the region in which the ion lattice is in its ground state ($\beta \hbar \omega_E \gg 1$) from the region in which it is more nearly a classical (non-quantum) solid ($\beta \hbar \omega_E \ll 1$). The vibrational energy $\hbar \omega_E$ is the level spacing for an Einstein solid, in which each ion moves independently of all the others in the harmonic oscillator potential produced by the electrons contained within its own Wigner–Seitz sphere of radius a_s, and is given by

$$\hbar \omega_E = \left(\frac{\hbar^2}{AHa_s^2} \cdot \frac{Z^2 e^2}{a_s} \right)^{1/2} = 0.194 \times \frac{2Z}{A} \left(\frac{\rho}{10^6 \text{ g} \cdot \text{cm}^{-3}} \right)^{1/2} \text{kev} \sim 0.2 \text{ kev.} \quad (3)$$

The last result holds for conditions in our 'typical' white dwarf. As we shall see, this line is of interest because it marks the boundary of the region in which rapid cooling can occur and is of importance for the more massive white dwarfs.

II. THERMODYNAMIC PROPERTIES OF CRYSTALLIZING WHITE DWARFS

Let us now consider the effects of the Coulomb interactions upon the thermodynamic functions of the ions. In the limit $\Gamma \ll 1$ the partition function can be evaluated analytically in the Debye–Hückel approximation, and as shown in BST the entropy S, for example, can be written in the form

$$\frac{S - S_0}{Nk_B} = -\frac{1}{2\sqrt{3}} \Gamma^{3/2}, \quad (4)$$

where S_0 denotes the entropy of a perfect gas of (non-interacting) ions at the given temperature and density. In the opposite extreme where $\Gamma \geqslant \Gamma_m$ the partition function can again be calculated analytically, if the crystalline phase is assumed to be an Einstein solid, and the total entropy becomes

$$\frac{S}{Nk_B} = \tfrac{3}{4} \beta\hbar\omega_E \coth\left(\tfrac{1}{2}\beta\hbar\omega_E\right) - 3\ln\left[2\sinh\left(\tfrac{1}{2}\beta\hbar\omega_E\right)\right], \qquad (5)$$

where $\beta^{-1} = k_B T$, and $\hbar\omega_E$ is given by Eq. (3). Within the limits of the Einstein approximation, Eq. (5) gives an excellent expression for the entropy throughout the range of Γ-values appropriate to the solid phase. However, it must be emphasized that at very low temperatures ($\beta\hbar\omega_E \gg 1$) the Einstein solid is not an adequate approximation, and for practical calculations one must instead use the Debye model for the solid (Kittel 1956; Mestel and Ruderman 1967; Van Horn 1968). In the present discussion I shall consider only the Einstein model, but *solely* for purposes of illustration.

From Eq. (5) the specific heat of the solid phase can easily be found and is given by

$$C_v = T\frac{\partial T}{\partial S}\bigg|_v = 3Nk_B\left[\frac{\tfrac{1}{2}\beta\hbar\omega_E}{\sinh\left(\tfrac{1}{2}\beta\hbar\omega_E\right)}\right]^2 \longrightarrow$$

$$\begin{cases} 3Nk_B, & \beta\hbar\omega_E \ll 1 \\ 3Nk_B \cdot (\beta\hbar\omega_E)^2 \exp\left(-\beta\hbar\omega_E\right), & \beta\hbar\omega_E \gg 1 \end{cases} \qquad (6)$$

Two important consequences follow from this result. First, the specific heat of the ion lattice at high levels of excitation ($\beta\hbar\omega_E \ll 1$) is *twice* that of a perfect gas, so that, in this regime, cooling times are twice as long as had been computed previously. This result is the same for both the Einstein model and the Debye model of the solid phase. Second, at very low temperatures, such that $\beta\hbar\omega_E \gg 1$, Eq. (6) shows that the specific heat drops very rapidly to zero. This result is qualitatively the same for both the Einstein and the Debye models, but the Debye specific heat falls off only as $(\beta\hbar\omega_E)^{-3}$ rather than exponentially with the inverse temperature, so that the details are somewhat different at low temperatures. The significant feature in either case, however, is that the ions gradually drop into the ground state as the temperature decreases, and can no longer contribute to the heat capacity of the lattice. Thus the total energy available to be radiated is proportionately much smaller at lower temperatures, and the white dwarfs in this regime will cool with increasing rapidity toward invisibility. This fact was previously pointed out by Mestel and Ruderman (1966) and has been discussed more recently by Ostriker (1967, 1968) and myself (Van Horn 1968).

The parameter $\beta\hbar\omega_E$ that completely characterizes the solid phase can be written in the form

$$\beta\hbar\omega_E = 0.198\,\frac{\Gamma}{A^{1/2}Z}\left(\frac{\rho}{A \times 10^6\,\text{g}\cdot\text{cm}^{-3}}\right) \qquad (7)$$

and near the melting point of the lattice is less than unity for ions with $Z \gtrsim 6$.

Under these conditions Eq. (5) for the entropy can be approximated by its high-temperature value. Since the entropy in the liquid phase just above the melting point should be only slightly different from that in the solid we then have, for $\Gamma_m \gtrsim \Gamma \gg 1$,

$$\frac{S - S_0}{Nk_B} = \tfrac{3}{2} - \tfrac{3}{2}\ln\Gamma - \ln\left(\frac{e}{3}\sqrt{\frac{2}{\pi}}\right) + \frac{\delta S}{Nk_B} \qquad (8)$$

where the first three terms are the leading terms in the expansion of Eq. (5) for $\beta\hbar\omega_E \ll 1$. The last term in Eq. (8) is the entropy difference between the liquid and solid phases at $\Gamma = \Gamma_m$. This term cannot presently be calculated theoretically but must be taken from the numerical work of BST, which gives

$$\delta S \approx + \tfrac{3}{4} Nk_B \qquad (9)$$

This value agrees quite well with the order of magnitude estimate, $\delta S \sim Nk_B$, obtained from plausibility arguments, and I shall assume the result given by Eq. (9) to be correct. The discontinuity in the entropy at the transition between the solid and liquid phases gives precisely the latent heat of crystallization, $T\delta S$, which represents a quantity of energy that must be emitted in the transition to the solid phase. This is not a negligible source of energy, since—as Eqs. (6) and (9) show—it amounts to about one-quarter of the entire thermal energy store of the star in these late evolutionary stages. The release of this latent heat of crystallization, which begins abruptly when the temperature in the isothermal core of a white dwarf first falls below the crystallization temperature at the center, thus provides a source of energy that acts to stabilize white dwarfs of given chemical composition on a 'crystallization sequence' parametrized by the stellar mass. Such a 'crystallization sequence' is directly analogous to the hydrogen-burning Main Sequence; in the latter case the evolution of a star is temporarily halted at a particular central temperature and density (for a given mass) by the release of nuclear energy, while in the former case this slowing-down is caused by emission of the latent heat.

In Figure 2, I have plotted several of these crystallization sequences (calculated with $\Gamma_m = 52$) for white dwarfs composed entirely of He^4, C^{12}, O^{16}, Si^{28}, or Fe^{52} (the last case chosen simply because $A = 2Z$ for it also). Also shown are lines of constant radius for three Chandrasekhar white dwarf models, and as can be easily seen, for a given stellar mass, elements of higher atomic weight crystallize at higher luminosities. The reason for this is simply that the crystallization temperature for a given central density is proportional to $Z^{5/3}$, as shown by Eq. (2).

Figure 2 can also be used to show the effect of the uncertainty in Γ_m upon the location of the crystallizing sequences. If Γ_m were as large the estimate of

125 given in BST, rather than our value of 50–60, the Ne²⁰ crystallization sequence (which would lie very slightly above the O¹⁶ sequence in Figure 2) would move down only to the position of the C¹² sequence shown in this figure.

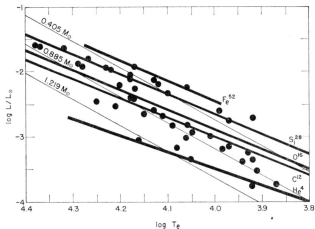

FIGURE 2 Crystallization sequences for white dwarfs composed entirely of He^4, C^{12}, O^{16}, Si^{28}, and Fe^{52} as calculated by Van Horn (1968).

Also shown in Figure 2 are the observations of Eggen and Greenstein (1965), showing the 'two sequences' of white dwarfs. When I originally began this study of crystallization it was with the hope of being able to place some restrictions upon the chemical composition of these objects by identifying the two observed sequences of white dwarfs with theoretical crystallization sequences corresponding to different values of Z. However, the recent work of Weidemann (1968) has thrown some doubt upon the reality of the two observed sequences, due to the large effects of blanketing in these stars. On the other hand, it is interesting to note that the predicted crystallization sequences for the astrophysically reasonable compositions of C^{12} and O^{16} agree quite well with the apparent position of the fainter group of white dwarfs in the H–R diagram. At present one can only conclude that it is not yet clear whether the compositions of the white dwarfs can be inferred from the existing observational data.

To conclude this section, let us finally examine briefly the equation of state. In the Debye–Hückel limit ($\Gamma \ll 1$) we have as in Eq. (4)

$$\frac{(P - P_0) V}{N k_B T} = -\frac{1}{2\sqrt{3}} \Gamma^{3/2} \tag{10}$$

where P is the true pressure of a system of N ions confined to a volume V,

L

and P_0 is the corresponding pressure for a perfect gas. For an Einstein solid ($\Gamma \geqslant \Gamma_m$) the pressure becomes

$$\frac{PV}{Nk_BT} = -\tfrac{3}{10}\Gamma + \tfrac{3}{4}\,\beta\hbar\omega_E \coth\left(\tfrac{1}{2}\,\beta\hbar\omega_E\right), \tag{11}$$

and for the liquid phase, with $\Gamma_m \gtrsim \Gamma \gg 1$,

$$\frac{(P - P_0)\,V}{Nk_BT} = -\tfrac{3}{10}\,\Gamma + \tfrac{3}{4}, \tag{12}$$

where the constant term includes the pressure difference between the solid and liquid phases at constant Γ as computed by BST. At present there exist no analytical formulae for the intermediate regime where Γ is neither large nor small compared to unity, and I have simply invented an interpolation formula which approaches Eq. (10) for $\Gamma \ll 1$, approaches Eq. (12) for $\Gamma \gg 1$, and has the correct numerical value, according to BST, at $\Gamma = 1$.

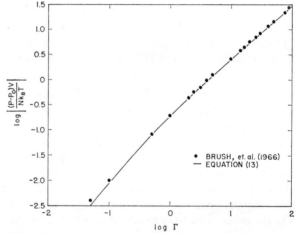

FIGURE 3   Interpolation formula for the pressure correction [equation (13)] as a function of Γ. Numerical data from BST are shown as solid dots.

This formula is plotted in Figure 3 and is given by

$$\frac{(P - P_0)\,V}{Nk_BT} = -0.113\,\Gamma^{3/2}\left[\frac{1}{(1 + 0.142\,\Gamma)^{1/2}} + \frac{1.54}{(1 + 0.575\,\Gamma)^{3/2}}\right]. \tag{13}$$

As can be seen from Figure 3, the fit to the numerical data of BST is quite good over the entire range of Γ-values except for a slight deviation for $\Gamma < 0.1$ where the electrostatic correction in any case is small.

III. TRANSPORT PROPERTIES

In addition to the effects on the energy balance, the Coulomb interactions in a white dwarf can also significantly influence the processes of energy and momentum transfer in the stellar interior. Two studies of the transport properties of the stellar plasma, including electron–ion correlations but not the ion–ion interactions, have recently been carried out by Abrikosov (1964) and by Hubbard (1966). As Hubbard has shown, the thermal conductivity K_e and viscosity η_e of the electronic component of the plasma depend upon the parameter Γ. This is due to the dependence of the transport coefficients upon the electron mean free path, which—under conditions typical of the interior of a white dwarf—is primarily determined by the distribution function of the ions. In the non-relativistic, strongly degenerate limit Hubbard gives

$$
\left.
\begin{aligned}
K_e &= 4.57 \times 10^{16} \left(\frac{\rho}{A \times 10^6 \text{ g . cm}^{-3}}\right) \left(\frac{T}{10^7 {}^\circ \text{K}}\right) G_\Gamma (\kappa_F) \text{ erg . cm}^{-2} \\
&\qquad\qquad\qquad\qquad\qquad\qquad\qquad\qquad \times \text{sec}^{-1} . ({}^\circ \text{K . cm}^{-1})^{-1} \\
\eta_e &= 4.15 \times 10^6 \left(\frac{\rho}{A \times 10^6 \text{ g . cm}^{-3}}\right)^{5/3} \kappa_F{}^2 H_\Gamma (\kappa_F) \text{ poise} \\
\kappa_F &= (9\pi Z/4)^{1/3}
\end{aligned}
\right\} \quad (14)
$$

where ρ is in g . cm^{-3} and T is in $^\circ$K, and the quantities G_Γ and H_Γ are complicated functions of Γ. In the two limits $\Gamma \ll 1$ or $\Gamma \geqslant \Gamma_m$ (for the case of an Einstein solid) these functions can be calculated analytically (Hubbard 1966, 1967) and are given by the following expressions:

$$
\left.
\begin{aligned}
\text{For } \Gamma \ll 1: G_\Gamma (\kappa_F) &= \left[\ln \left(1 + \frac{4\kappa_F{}^2}{3\Gamma}\right)\right]^{-1} \\
H_\Gamma (\kappa_F) &\approx \tfrac{1}{2} G_\Gamma (\kappa_F)
\end{aligned}
\right\} \quad (15)
$$

$$
\left.
\begin{aligned}
\text{For } \Gamma \geqslant \Gamma_m: G_\Gamma (\kappa_F) &= \frac{\Gamma}{4\kappa_F{}^2}\left(1 + \frac{\kappa_F{}^2}{\Gamma}\right) \\
H_\Gamma (\kappa_F) &\approx G_\Gamma (\kappa_F)
\end{aligned}
\right\} \quad (16)
$$

In general, however, for the intermediate Γ-values appropriate to the liquid phase, the functions G_Γ, H_Γ must be obtained by numerical integration over the ion pair distribution functions tabulated by BST. Hubbard (1967) has given a simple interpolation formula applicable in this regime, but his expression does not take account of the discontinuity in the transport coefficients during the phase change. This effect comes about because, in the transition from the liquid to the solid phase, the distribution function of the ions changes discontinuously to a function with a perceptibly smaller r.m.s. vibration amplitude, as is apparent from Figure 17 of BST. Since this results in about a fifty percent larger mean free path in the solid phase, the thermal conductivity in the solid at the melting point is larger by a factor $\sim\frac{3}{2}$ than in the liquid phase at the same temperature and density. This is consistent with both the experimental and theoretical results for normal metals. For this reason I have developed the following alternative interpolation formulae for G_Γ and H_Γ, which reduce to equations (15) for $\Gamma \ll 1$ and which give values at $\Gamma = \Gamma_m$ that are $\frac{2}{3}$ of those obtained from Eqs. (16) for the solid phase:

$$
G_\Gamma^{(\text{liq})}(\kappa_F) = \left\{ \ln \left[1 + \frac{4(1 + \Gamma)\kappa_F^2}{3\Gamma + \frac{2}{3}\Gamma^2} \right] \right\}^{-1}
$$

$$
H_\Gamma^{(\text{liq})}(\kappa_F) = G_\Gamma^{(\text{liq})}(\kappa_F) \cdot \frac{\kappa_F^2 + 3\Gamma}{2\kappa_F^2 + 3\Gamma}.
$$

(17)

Eqs. (16) and (17) together thus provide an estimate of the effects of the electron–ion interactions upon the transport properties of the electron gas.

The transport properties of the ions in the plasma may also be significant in white dwarfs, and I shall finally discuss briefly some of the processes which may be important here. The conduction of heat in a crystallizing white dwarf, for example, can occur by means of lattice vibrations, or 'phonons', as well as by electron transport, and this must in principle be taken into account. For our 'typical' white dwarf, however, the ratio of the 'phonon conductivity' to the thermal conductivity of the electrons is

$$
\frac{K_{ph}}{K_e} \sim \frac{C_{ph}. \, U_{ph}. \, l_{ph}}{Z \, C_e . \, U_e . \, l_e} \sim \beta\hbar\omega_E \cdot \frac{l_{ph}}{Z . \, l_e},
$$

(18)

where C, U, and l are respectively the heat capacity, particle velocity, and mean free path of the phonons or electrons. Except in the special and highly improbable case of an *extremely perfect* crystal, we have $l_{ph} \ll l_e$, because of the high degeneracy of the electrons, so that this ratio in general is much smaller than unity. Phonon conduction can therefore be neglected for all practical purposes.

A considerably different situation may obtain in the case of momentum transport, however, since in analogy with normal solids one would expect an increase in the viscosity by many orders of magnitude during crystallization. In the liquid phase, the viscosity due to ion–ion interactions can be calculated to within a factor or order unity by means of a simple model due to Andrade (1934). In this model a microscopic description of the process of momentum transfer between two adjacent layers of ions in the fluid is used to calculate the coefficient of viscosity just above the melting point, and the fact that the viscous interaction must be thermally activated is used to derive the result

$$\eta^{(\text{liq})} \approx \frac{1}{3\pi} \frac{AH\omega_E}{a_s} \exp\left(\frac{\Delta E}{k_B T} - \frac{\Delta E}{k_B T}\bigg|_{\Gamma_m}\right)$$

$$= 1.4 \times 10^3 \frac{Z}{A^{1/3}} \left(\frac{\rho}{10^6 \text{ g} . \text{ cm}^{-3}}\right)^{5/6} \exp[0.1\,(\Gamma - \Gamma_m)] \text{ poise} \quad (19)$$

In this equation ΔE is the activation energy for self-diffusion, which has been estimated by de Wette (1964) for a one-component Coulomb plasma to be

$$\frac{\Delta E}{k_B T} \approx 0.1\,\Gamma \qquad (20)$$

The analytic form of equation (19) has been tested experimentally for a large variety of normal liquids and has been found to be in excellent agreement with the experimental results, so that this formula should provide a good approximation right up to the point of crystallization.

In the solid phase, the nature of the flow becomes a much more complicated, time-dependent phenomenon, which depends upon the applied shearing stress as well as upon the temperature. At present there is no commonly accepted theory of the plastic flow, or 'creep', of solids, and probably the best that can be done is simply to make use of the experimental ratio of viscosities in the liquid and solid phases for normal metals:

$$\eta^{(\text{sol})} = \frac{\text{applied shearing stress}}{\text{time rate of strain}} \sim (10^7 - 10^{11}) \times \eta^{(\text{liq})}\,(\Gamma_m). \qquad (21)$$

This result is completely inadequate for all but the crudest of estimates of the magnitude of steady-state viscous effects in the solid phase, and it is inapplicable in cases where the strain rate is so high that flow takes place by local rupture of the solid. However, Eq. (21) does provide some indication of the importance of crystallization for the magnitude of the viscosity; for our 'typical' white dwarf equations (14), (19) and (21) lead to $\eta_e \sim 10^7$ poise, $\eta^{(\text{liq})} \sim 10^3$ poise, $\eta^{(\text{sol})} \sim 10^{10} - 10^{14}$ poise. In the solid phase, Hubbard's viscosity may therefore be a gross underestimate, and this would have some very interesting consequences, for example, for the differentially rotating white dwarf models computed by Ostriker and Bodenheimer (1968).

References

ABRIKOSOV, A. A. 1961, *Soviet Physics—JETP*, **12**, 1254.
ABRIKOSOV, A. A. 1964, *Soviet Physics—JETP*, **18**, 1399.
ANDRADE, E. N. da C. 1934, *Phil. Mag.* (*Ser.* 7), **17**, 497; *ibid.*, 698.

BRUSH, S. G., SAHLIN, H. L., and TELLER, E. 1966, *J. Chem. Phys.*, **45**, 2102.
CHANDRASSEKHAR, S. 1957, *Stellar Structure* (New York: Dover), p. 357ff.
DE WETTE, F. W. 1964, *Phys. Rev.*, **135**, A287.
EGGEN, O. J., and GREENSTEIN, J. L. 1965, *Ap. J.*, **141**, 83.
GRASBERGER, W. H. 1961, 'A partially degenerate, relativistic, ideal electron gas',
    UCRL-6196.
HAMANDA, T., and SALPETER, E. E. 1961, *Ap. J.*, **134**, 683.
HUBBARD, W. B. 1966, *Ap. J.*, **146**, 858.
HUBBARD, W. B. 1967, personal communication.
KIRZHNITS, D. A. 1960, *Soviet Physics—JETP*, **11**, 365.
KITTEL, C. 1956, *Introduction to Solid State Physics* (New York: Wiley), p. 125ff.
MESTEL, L., and RUDERMAN, M. A. 1967, *M.N.R.A.S.*, **136**, 27.
OSTRIKER, J. P. 1967, personal communication.
OSTRIKER, J. P. 1968, this volume.
OSTRIKER, J. P., and BODENHEIMER, P. 1968, *Ap. J.*, **151**, 1089.
SALPETER, E. E. 1961, *Ap. J.*, **134**, 669.
SALPETER, E. E. 1968, this volume.
SAVEDOFF, M. P., VAN HORN, H. M., and VILA, S. C. 1968, to be published.
VAN HORN, H. M. 1968, *Ap. J.*, **151**, 227.
WEIDEMANN, V. 1968, in press.

Discussion

COX   Since the thermal properties of the material are affected, one might expect perhaps a lowering of the adiabatic exponents. And this would have important stability implications. I wonder if you, or anyone else, has investigated these adiabatic exponents.

VAN HORN   No, not at all.

HUBBARD   I have. I'll talk about it when I present my paper.

KUMAR   How does the crystallization affect the evolution of low-mass degenerate stars?

VAN HORN   That is a problem that I have been very much interested in, but have not been able to do anything with as yet. It may be that it would be significant. If you ran the white dwarf sequence all the way down to where the main sequence ends, that would be roughly following out the crystallization lines, and it would intersect at about a place characteristic of something like perhaps a thousandth of the solar mass. Now, beyond that I have not really been able to put in any numbers. So, if you want to go as low as a thousandth of the solar mass, then it might be important.

KIPPENHAHN   How do the viscosity and conductivity behave when you put numbers into your equations?

VAN HORN There is an interesting point here. If you take only the contribution of the electrons, which is what Dr. Hubbard has calculated, then you get a jump of about a factor 1.5 when you go from the liquid phase to the solid phase. On the other hand, these calculations have not included what one might call the ion viscosities. One expects these to be small in the liquid phase and some numbers that I put in for typical cases for the white dwarfs look something like this: The electron viscosity that might be about 10^7 just at the crystallizing boundary under the typical conditions we talked about here. In the liquid phase the numbers might be something like 10^4, and in the solid phase I'm really waving my hands—I don't know. It's not quite clear what one means by viscosity for a solid. But if one takes, for example, the viscosities that one would get by looking at creep rates for normal metals and simply taking the ratio between the two cases, then this comes to be something like 10^{11} to 10^{15}. So it might be very important; on the other hand, it might be that one would be operating in such a regime that the viscosities are not that greatly increased over the liquid phase. You might be operating in the failure regime for the crystal, for example.

Luminosity function and space density
of white dwarfs

VOLKER WEIDEMANN

University of Kiel, Germany

LINES OF CONSTANT cooling time have been calculated for white dwarfs (consisting of 50% C and 50% O) according to the theory of Mestel and Ruderman (1967) and drawn in an improved H–R diagram. A constant white dwarf birth rate then predicts a theoretical luminosity function $\log \phi \ (M_{bol}) = $ const. $+ 2/7 \ M_{bol}$. Empirical luminosity functions derived in different ways show excellent agreement with this prediction in that range of luminosities for which the observational material is reliable ($11 < M_v < 13$). Figure 1a gives the relation obtained from Luyten's material (1958) comprising 130 white dwarfs with $m_{pg} < 16$. The luminosity function is obtained by multiplying the observed white dwarf fraction of all stars with the general luminosity function derived from the Bruce Proper Motion Survey. Figure 1b refers to more than 1000 white dwarfs with $\mu > 0''.04$/year, $m_{pg} < 21$ from Luyten's Palomar survey (1963–1966). Luyten's color classes b to k are converted to the M_{pg} scale by determining the average reduced motion $H = m + 5 + 5 \log \mu$ for each color class separately for those stars only which are brighter than a completeness limit $\bar{m}_{pg}$, and by setting $\Delta \langle H \rangle$ equal to $\Delta \langle M_{pg} \rangle$. The M_{pg} scale is then fixed by forcing the corresponding M_v's to cover the observed range from Eggen and Greenstein's (1965) color-magnitude diagram. Although the luminosity function (b) does not depart from the theoretical relation as far as (a) for the fainter and redder objects ($M_v > 13$), it cannot be decided if this is due to the incompleteness of the material or because of the fact that redder white dwarfs cool faster after reaching the Debye temperature in their interiors. Greenstein at this conference indicated that a large fraction of red proper motion stars detected by Luyten and Eggen and claimed to be white dwarfs, do not show spectroscopic evidence of degeneracy, but rather constitute a new class of subluminous stars between the main sequence and the white dwarf region. The results seem to point to a real deficiency of red degenerate stars. If on the other hand, one confines the survey to the well-observed white dwarfs of known distance only (Eggen–Greenstein material 1965) and takes the volume effects into account, then the deficiency of red degenerate stars disappears as may be seen from Figure 1c, or can be taken from the fact that within 10 parsecs there are more red than blue degenerate stars.

The space density of white dwarfs is estimated in different ways and turns out to be considerably higher than hitherto assumed. Table I shows the

TABLE I   Space density of white dwarfs and contribution to the mean density in the solar neighborhood.

Source	Number of w.d. within 10 pc		Number of w.d./pc^3		Mass density (g/cm^3)
	N_{10} (16.5)	N_{10} (15.5)	Z (16.5)	Z (15.5)	$M_{\mathrm{b}} = 16.5$
Luyten–Palomar	56	29	1.3×10^{-2}	0.7×10^{-2}	0.6×10^{-24}
Eggen–Greenstein	38	20	0.9×10^{-2}	0.5×10^{-2}	0.4×10^{-24}
$r < 10$ parsec	81	43	1.9×10^{-2}	1.0×10^{-2}	0.8×10^{-24}
Sandage–Luyten	200	103	4.8×10^{-2}	2.5×10^{-2}	2.0×10^{-24}

results which were obtained by determining the present white dwarf birth rates from best fits to the relations by Mestel–Ruderman as shown in Figure 1, and then by multiplying them with the cooling ages of 5×10^9 years (columns marked with 15.5) or 10^{10} years (columns marked by 16.5). (These figures characterize the limiting bolometric magnitudes to which white dwarfs with 0.6 solar masses would cool down if the Mestel-Ruderman theory applies all the way. It should be pointed out, however, that the derived space densities do not change if accelerated cooling renders these limiting magnitudes meaningless). The numbers given in the third line were derived from the fact that 8 white dwarfs within 6 parsecs seem to constitute a complete sample, whereas the last line was obtained from Sandage and Luyten's (1967) observation that between blue stars at high galactic latitude there is one white dwarf/ square degree down to $B = 18$ mag.

With most probably 100 to 200 white dwarfs within 10 parsecs, the mass contribution in the solar neighborhood is 1 to 2×10^{-24} g/cm^3. It may be higher and could completely account for the missing mass in the solar neighborhood. For a detailed account see *Zeitschrift f. Astrophys.* 1967, **67**, 286.

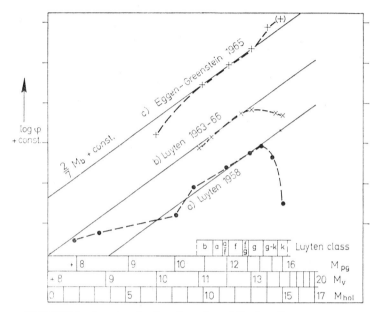

FIGURE 1 Luminosity functions for white dwarfs.

References

EGGEN, O. J., and GREENSTEIN, J. L. 1965, *Ap. J.*, **141**, 83.
LUYTEN, W. J. 1963–1966, *On the Frequency of White Dwarfs in Space*, *Publ. Astron. Obs. Univ. Minn.*, 3, No. 13–18.
MESTEL, L., and RUDERMAN, M. A. 1967, *M. N.*, **136**, 27.
SANDAGE, A., and LUYTEN, W. J. 1967, *Ap. J.*, **148**, 767.

Discussion

EGGEN I think from a very much larger body of data that I discussed in my paper, it is fairly well fixed that the space density of the DA white dwarfs cannot be much different than 10^{-3}/cubic parsec.

WEIDEMANN My results are not much different from this number.

EGGEN   My second comment is that in searching for the red white dwarfs, we are looking at the same volume of space, but since the luminosities drop off rather rapidly, we would certainly expect to find the less numerous discovery rate. And a third thing, you have to be a little careful of Luyten's white dwarfs, for he has a prejudice against velocity distribution. And we already know that for the blue white dwarfs, at least, that they can have very small space velocity. So, he is eliminating, I am sure, quite a few possible stars. That is, all these estimates are absolute lower limits.

WEIDEMANN   Yes, I agree.

OSTRIKER   What mass did you use in translating the numbers of white dwarfs to mass density?

WEIDEMANN $0.6 \, M_\odot$. I would now say that $0.65 M_\odot$ is a better value for the average mass.

A fine analysis of white dwarf spectra and mass–radius relation*

SATOSHI MATSUSHIMA and YOICHI TERASHITA

*Department of Astromony, The Pennsylvania State University
University Park, Pennsylvania*

Abstract

For the purpose of determining the masses and radii of a large number of hydrogen-rich white dwarfs, the effective temperatures and surface gravities of these objects are determined through detailed analyses of photometric and spectrographic observations of their spectra. The method is based on the calculation of various observable quantities for a grid of non-gray model atmospheres including the effect of Lyman and Balmer-line absorptions on the temperature stratification. In comparison with models without line blanketing, the values of T_e determined on the basis of the UBV colors of the blanketed models are about 10 per cent lower, but no appreciable change is found in log g. Instead, a careful treatment of Balmer-line blocking for the computation of the emergent flux, using more exact Stark profiles, leads to an increase of about 0.5 in log g.

The perturbation effect on electrons in the higher bound energy states due to the increase in ion and electron density is important in determining the continuum near the Balmer discontinuity. However, the increase in electron density produced by 'preionization' is found to be generally unimportant in the model calculations, being less than 2 per cent. A change in H to He ratio may produce an appreciable effect on observable features. For example, the reduction of the H content by a factor of 10 is found to be equivalent to a decrease of 0.5 in log g for the surface gravity obtainable from a comparison of observed color indices and those predicted from the models. On the other hand, if we assume a pure hydrogen atmosphere, the corresponding change in the surface gravity is found to be 0.2 in log g. The change in hydrogen content in either direction does not seem to affect the resulting T_e appreciably. The solar H to He ratio assumed in the present calculations appears to be a lower limit for normal type DA stars.

Using the above determined T_e and g, the radii and masses are determined for about 30 normal type DA stars for which the luminosities

*This research was supported by the National Science Foundation under grant NSF–GP 8058.

are known. Special attention is directed to the determination of these values for 40 Eri B, using observations of six-color photometry and photoelectric scanning in addition to UBV colors and hydrogen line profiles. The independent determinations of T_e and g from these four observations give sufficiently good agreement except in the case of the UBV colors, which predict values lower than the mean values by about 10 per cent in T_e. The mean values are $T_e = 16,000°K$ and $g = 5.9 \times 10^7$ cm/sec². From these values, we find the radius of 40 Eri B to be, $R = 0.0143 \, R_\odot$ and the mass $M = 0.44 \, M_\odot$ in excellent agreement with the observed value of $(0.43 \pm 0.04) \, M_\odot$. A comparison of these values with the theoretical mass–radius relations for zero-temperature degenerate stars predicts the core of this star as being composed of elements slightly heavier than helium on the average. Moreover, the above mass and radius predict the Einstein redshift of 20 km/sec, which is in good agreement with the observed value of 21 ± 1.9 km/sec.

As in the case of 40 Eri B, the value of T_e determined for other DA stars appears to be considerably higher than those previously estimated through equivalent black-body approximations. As a consequence, the resulting radii are systematically smaller than the previously assumed values, thereby eliminating some difficulties encountered in comparisons with theoretical mass–radius relations. The presently estimated masses and radii for 30 DA stars seem to generally satisfy the theoretical relations for heavy element models, although considerable deviations are noted for some objects. Possible causes and conditions for the stars showing large discrepancies from the theoretical mass–radius relations are considered. While the possibility remains that the atmospheric chemical composition might vary within the DA stars, especially between the upper and lower luminosity sequence groups, there is an indication that a few objects lie far above the theoretical relations. There is also an indication that the presently adopted *UBV* system might involve some systematic errors.

1. INTRODUCTION

IN AN EARLIER paper (Terashita and Matsushima 1966, hereafter called Paper I), we have attempted to determine the effective temperature and surface gravity for the DA-type stars. The method is based on comparisons of observed *UBV* colors and hydrogen line profiles with those predicted from a grid of flux-constant model atmospheres. The resulting temperatures are found to be systematically larger and the surface gravities smaller than those previously estimated through more approximate methods. In particular, the effective temperatures are, on the average, several thousand degrees higher than those estimated by Greenstein and Trimble (1967) through linear interpolations between the observed $(U - B, B - V)$ relation for the main sequence stars and that computed for black bodies. The discrepancy appears to be even more serious when the average value of surface gravities obtained in Paper I is compared to that determined by Greenstein and Trimble using their observations of gravitational redshift of hydrogen lines. Thus, Greenstein and Trimble have questioned the validity of a more elaborate procedure based on the construction of a large number of non-gray model atmospheres.

The most obvious approximation in the previous models is the neglect of the line blanketing effect. While we included the line absorption in computing colors for a given model, the back warming effect on the temperature stratification was not taken into account. Less obvious, but apparently large, errors were introduced for the computation of emergent flux and colors by overestimating the Balmer-line absorption through the use of an approximate wing formula by Griem (1960) for the Stark broadening.

Thus, we have reconstructed a grid of basic models including the blanketing effect of both the Lyman and Balmer series lines and using more exact wing formulae given by Griem (1967). As a result, it is found that the errors due to the use of the approximate formula for computing colors are unexpectedly large and that the previously noted discrepancies in masses and radii appear to be removed (Terashita and Matsushima 1967). The analysis has since been extended to include all of the normal DA stars and the previous values of T_e and log g are revised. The resulting mass–radius relation seems to be consistent with that theoretically predicted by Hamada and Salpeter (1961). Moreover, since we have the most reliable knowledge of the mass and parallax for 40 Eri B, more detailed calculations have been made for this star, using additional data of six color photometry by Kron, Feinstein and Gordon (1966) and photoelectric scanning spectra obtained by Oke (1963).

The details of the above investigations will be published elsewhere. In the present report, we attempt to examine the source of errors in the previous models and critically review the inherent problems and uncertainties which may be characteristic of the low luminosity, high pressure atmospheric conditions in general. Specifically, aside from the question of line blanketing, we discuss the uncertainties involved in the chemical composition of white dwarfs and a possibly large change in electron density due to the perturbation on bound electrons caused by the fluctuations of the surrounding Coulomb field. We then consider possible implications of the mass–radius relation resulting from our investigation. It is hoped that the following calculation would serve to give more insight into the present limitations of the interpretation of white dwarf spectra on the basis of the model atmosphere method.

II. LINE BLANKETING EFFECT

The basic models for the present investigation consist of 15 DA models; 12 given by combinations of log $g = 7$ and 8 and $T_e = 7000, 8000, 10,700,$

15,000, 20,000, and 25,000°K, and 3 additional models for $T_e = 15,000$°K
with log $g = 6$ and 9 and $T_e = 12,000$°K with log $g = 8$. In addition to
these, 7 more models for $T_e = 14,000, 15,000, 16,000$°K and log $g = 7, 7.5$,
and 8 are computed specially for more detailed analyses of the spectra of
40 Eri B. The details of the method of model calculation and the results in
comparison with earlier models without line blanketing will be published
elsewhere.

The mode of variation in the emergent flux distribution, F_ν, with decreasing
T_e for models with log $g = 8$ is shown in Figure 1. The total flux for each

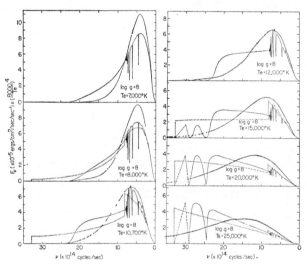

FIGURE 1 Comparisons of the emergent flux distributions for
various T_e with log $g = 8$, resulting from the models including line
blanketing (solid curves), those without blanketing (dotted line curves)
and black body radiators (smoothed curves). Each curve is normalized
such that the area under the curve is equal to $(\sigma/\pi) \times (8000°K)^4$.

model in the figure is normalized to that for $T_e = 8000$°K, so that $\int_0^\infty F_\nu \, d\nu$
$= (\sigma/\pi) \times (8000°K)^4$. The smooth curves are the Planckian energy distributions for corresponding T_e normalized in the same manner, and the
dotted-line curves represent the same quantities resulting from models
without line blanketing. The broken lines near the Lyman limit for high-
temperature models indicate the present approximation for treating the
upper Lyman lines. Due to a strong concentration of flux in the visual
region of the spectrum, Balmer lines are the main source of blanketing in the
low-temperature models, whereas this role is played by Lyman lines in high-

temperature models, in which the main part of flux is shifted to the ultraviolet region.

The effects of line blanketing on *UBV* colors and the use of the more accurate wing formula may be best examined by comparing the locations o the improved and the earlier models in the $(U - B, B - V)$ diagram shown in Figure 2. The new positions are connected by iso-T_e and iso-log g lines and

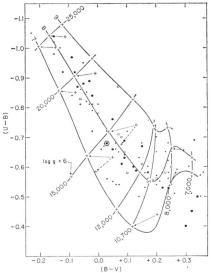

FIGURE 2   Comparison of $(U - B, B - V)$ relations predicted by the present models with those obtained from models without line blanketing. The equi-log g curves are given for log g = 9, 8 and log g = 7 as indicated at the top of each curve. The observed colors for those DA stars with known luminosities are shown by larger dots than the others. The high and low luminosity sequences are distinguished by the solid and open dots, respectively. For further details, see Section II.

the resultant displacement of the position of each model is indicated by an arrow in the figure. The change is seen to be such that when the diagram is used to determine the values of T_e and log g of a star from its observed colors, the improved calculations predict systematically smaller values for T_e and larger values for log g except for the lower temperature regions where the situation is somewhat complicated. In order to separate the effect of line blanketing and the change due to the improvement in the computation of line-absorptions, we have computed the *UBV* colors for two of the earlier models without blanketing ($T_e = 15,000°$K and log g = 7 and 8) using the new wing formula. The resulting changes in the positions of these models are shown by broken arrows in Figure 2. Thus, we see that the effect of the improvement of the wing formula is such that when the revised models are used the surface gravity is increased by nearly a factor of 3, whereas the inclusion of line blanketing lowered the effective temperature by about 1000°K. The direction of these changes may be understood at least qualitatively from the following consideration. Since the improved formula predicts considerably lower values of absorption than the previously used formula, the flux in the line-influenced region of the spectrum tends to become larger. As is discussed later, the high pressure lowers the last visible line of the Balmer series while the 'pre-ionization' effect flattens the discontinuity. As a result, the U-magnitude is much less affected by the Balmer lines in the whie dwarf atmospheres as compared to the main-sequence. Hence, if we consider that only B-magnitude is increased appreciably by the change in the wing formula, the $(U - B)$ index should be less negative, while the $(B - V)$ becomes more negative. The result is that the point for each model in Figure 2 is moved toward the lower-left direction, or along the iso-T_e curve. On the other hand, the 'back-warming' effect due to the inclusion of line blanketing yields higher temperature for a given optical depth, so that the continuum would represent a higher temperature compared to a model without blanketing. Consequently, a model with line blanketing predicts the color indices corresponding to a higher temperature than those given by a line-free model of the same effective temperature. As a result, the position in the $(U - B, B - V)$ diagram is shifted upward nearly along the iso-gravity curves.

Summarizing the above result, it should be emphasized that the effect of the overestimate of the wing absorption, leading to an error of about 0.5 in log g but no appreciable error in T_e, is far greater than the error due to the neglect of the line blanketing effect which amounted to a change of only 10 per cent in T_e. The overestimate due to the use of the earlier wing formula by Griem (1960) may be seen in Figure 3, which is a reproduction of Figure 3 in Paper I, where the theoretical emergent fluxes for the earlier unblanketed models with $T_e = 15,000°K$ are plotted against wavelength. The upper curves showing sharp discontinuities are computed without the Balmer lines, and the broken lines indicate the depression of the continuum due to the pre-ionization effect without including the lines (see Section V). The lower curves are obtained with including both pre-ionization and the lines. The over-estimate of the wing absorption is most clearly seen in the 'continuum' between H_β and H_γ lines as compared to the continuum without the lines. We note that the two lines are separated by about 500 Å, which should be sufficiently large to show a line free region between these two lines even though the pressure broadening is quite large for the models under

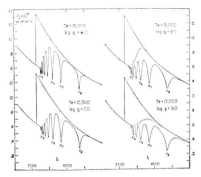

FIGURE 3   Theoretical flux distribution near the Balmer limit for models with $T_e = 15,000°K$ and various values of log g.

consideration. That is, we should see at least a portion of the flux curve between H_β and H_γ lines coinciding with the continuum without the lines.

Thus, although we were aware of the error introduced by the use of the approximate wing formula, its effect on the surface gravity determined from the three color indices was unexpectedly large. This, in turn, leads to correspondingly smaller values of mass obtainable from the result of Paper I, whereas the line blanketing effect appears to produce little change in masses, as we shall discuss later. In fact, an inspection of Figure 2 indicates that the increase of 0.5 in log g results from an increase of only 0.035 in B-magnitudes or only 3.5 per cent increase in the flux in B region. This implies that a great precision is required in computing the line profiles for determining colors in the line-influenced regions. At the same time, apparent in the above result is the limitation in the determination of the surface gravity, and hence, the values of mass, from observed color indices.

III. CONTINUUM OF 40 ERI B

Among many DA stars, we have for 40 Eri B the most reliable observational data, including the trigonometric parallax and the mass determined from its binary motion. In addition to the usual *UBV* colors and the profiles of hydrogen lines obtained by Greenstein (1960), photoelectrically scanned spectra taken by Oke (1963) and six color photometric observations by Kron, Feinstein, and Gordon (1966) are available. These data provide useful information for examining the accuracy of our calculations, and more detailed comparisons are made particularly for this star. In this report, only a brief summary of the important results will be given.

1. *UBV colors*

The $(U - B)$ and $(B - V)$ indices are shown by a circled solid dot near the center of Figure 2. In order to facilitate a finer interpolation, the models for $\log g = 7.5$ are computed for $T_e = 14,000$, 15,000 and 16,000°K, also for $\log g = 7.0$ and 8.0 with $T_e = 14,000$ and 16,000°K. Interpolations in the diagram give the values of $T_e = 14,700$°K and $\log g = 7.75$. With these values, the mass may be computed from the relation

$$\log M/M_\odot = 12.50 - 4\log T_e + \log g - 0.4\, M_b \qquad (1)$$

where M_b denotes the bolometric magnitude. Using the bolometric correction of -1.24 (see Figure 8) for the absolute visual magnitude, $M_v = 11.0$ (Greenstein 1960), we find $M/M_\odot = 0.49$ as compared to the observed value of 0.43 ± 0.04 (Popper 1954). As will be shown later, we believe that the error in T_e should be much less than that in $\log g$, so that the 15 per cent discrepancy is most likely due to the uncertainty in the determination of $\log g$ from *UBV* colors. In fact, the following calculations indicate that the above value of T_e is a lower limit, and hence, any correction in T_e should lower the mass. If we interpret the error to be entirely due to the error in g-value, then the necessary correction is 0.07 in $\log g$ and perhaps we must allow this much uncertainty in this particular method, as mentioned in the previous section. We shall return to this point in Section VI(2).

2. *Six-color observations*

In order to compare with the observations of Kron *et al.*, U, V, B, G, R, and I magnitudes are computed for the 9 basic models mentioned above, and the corresponding values are determined for the 3 additional models with $T_e = 17,000$°K through extrapolations. For comparisons with theoretical colors, it is convenient to convert the original data, which are based on the 'standard' systems of Stebbins and Kron (1956), to the 'absolute' system also established by Stebbins and Kron (1964). The details of the method of these calculations will be published elsewhere.

The solid dots in Figure 4 indicate the resulting values on the magnitude scale plotted against the effective inverse wavelength, and the open dots represent the observed values. Following the definition of Stebbins and Kron (1964), theoretical values are also scaled such that the mean of B, G, and R magnitudes become equal to zero. The I magnitudes should be weighted less since the Paschen lines are not included in the computations.

The comparisons in Figure 4 indicate that the computed colors depend little on $\log g$, and that an interpolation by sight in Figure 4 gives a well-defined effective temperature. Thus, we find the best fit can be obtained for $T_e = 16,700$°K.

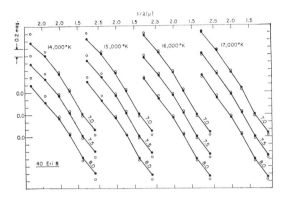

FIGURE 4 Comparison of six color observations by Kron *et al.* (1966) of 40 Eri B (open dots) with the theoretical values (solid dots) computed for various T_e and log g. The zero point of the ordinate axis is chosen such that the mean of the B, G, and R magnitudes becomes equal to zero.

3. *Monochromatic flux*

The photoelectric scanning observations by Oke (1963) with the Palomar 200-inch telescope provide useful data for determining the monochromatic flux variation of the continuum with wavelength as well as detailed profiiles of the Balmer lines for 40 Eri B. (We are grateful to Dr. Oke for placing at our disposal a copy of his original tracings.) In Figure 5, the theoretical flux computed in magnitude units is plotted against the inverse wavelength for each of the 9 basic models together with the extrapolated values for the 3 models with $T_e = 17,000°$K as before. The zero point in the ordinate is chosen so as to give the best fit to the visual side of H_γ line.

The theoretical flux gradient in the visual side of H_γ line shows only a small difference among the models included here, and in fact, any one of the 12 models seems to give almost equally good agreement within an accuracy of fitting by sight. As is expected, however, the flux at the short wavelength side of the H_γ line ($\lambda^{-1} = 2.37 \mu^{-1}$) shows a large dependence on the surface gravity due to differences in depth of the continuum formed by the overlapping of the wings of $H\gamma$ and H_δ lines. At this wavelength, the models with larger g seem to show better agreement for a given T_e. The observed gradient of the ultraviolet flux appears to be much larger than the theoretical predictions unless an effective temperature higher than 20,000°K is assumed, in which case the Balmer jump becomes too small compared to the observations.

The number listed near each broken line gives the computed Balmer jump (defined here by the difference between the flux at $\lambda^{-1} = 2.385$ and $2.70 \mu^{-1}$)

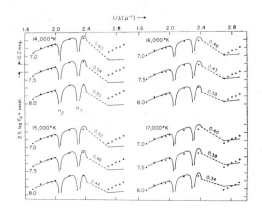

FIGURE 5 Comparison of the photoelectric scanning observations by Oke (1963) for 40 Eri B with the monochromatic flux distributions computed for various models.

in magnitude units. The corresponding value taken from Oke's measurement is 0.34 mag. However, recently a number of investigators have pointed out the necessity of a systematic correction to the value of the Balmer jump measured on the basis of the old calibration of Vega. For example, on the basis of new measurements of Vega by Bahner (1963), Mihalas (1966) suggests that the ultraviolet flux beyond the Balmer jump should be lowered so as to increase the discontinuity by 0.06 mag. A similar conclusion is reached in the extensive study by Heintze (1968). Even larger corrections are suggested by Whiteoak (1966) and Hayes (1967). Thus, we adopt the observed Balmer jump to be 0.40 mag.

Thus, considering the comparisons of both the Balmer jump and the flux at $\lambda^{-1} = 2.37\ \mu^{-1}$, we find the best fitting value of T_e to be about 16,000°K and $\log g = 7.8$. This result should be weighted less because of some uncertainty remaining in the ultraviolet.

IV. HYDROGEN LINE PROFILES

1. *Observational data*

From Oke's spectrum scanning observations of 40 Eri B used in the previous section, we can obtain excellent profiles at least for H_γ and H_δ lines. These data together with the detailed profiles published by Greenstein (1960) provide another independent means to determine the values of T_e and $\log g$. Greenstein's measurements were made on the photographic plates taken with the 200-inch coudé spectrograph. Thus, we have determined theoretical profiles of these two lines for the 9 models and again extrapolated profiles for the 3 high-temperature models.

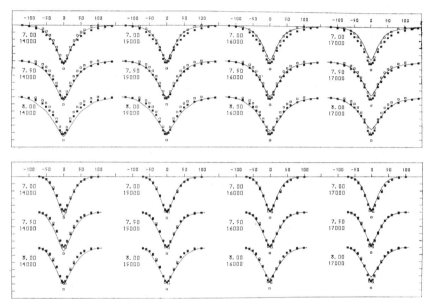

FIGURE 6   Hydrogen line profiles of 40 Eri B obtained by Oke (1963) (filled square) and by Greenstein (1960) (open squares), compared to the theoretical predictions for various models. The upper diagram is for H_γ, and the lower one is for H_δ.

The theoretical profiles are compared with observations in Figure 6, where Oke's profiles are plotted by filled squares and Greenstein's by open squares. While both measurements show satisfactory agreement for H_δ, Greenstein's profile for the H_γ line appears to be considerably shallower than Oke's except at the line center. The much deeper core obtained by Greenstein compared to that given by Oke may be the result of the lower resolution (10 Å band width) used in the spectrum scanner. We shall first consider Oke's data and return to discuss the discrepancy between the two observations.

2. *Determination of T_e and g*

The Stark broadened profiles of hydrogen lines depend strongly on both temperature and pressure, and hence, it is a function of the effective temperature and the surface gravity. Since the functional dependencies of T and P_e are different at each wavelength within a line, the observed profile of one line should, in principle, define a unique set of T_e and g specifying the best fitting model. Thus, in Paper I, we have attempted a quantitative comparison of the residual intensity at each point within a line to determine a self-consistent solution of T_e and g. However, this method has not succeeded, apparently due to the errors in both theory and observations (Figure 13 of Paper I). The best one can do, therefore, is to make a comparison by sight over the entire profile and to find a statistical mean for that line. As a consequence, it is difficult to determine both T_e and $\log g$ simultaneously, since, for example, for $T_e \gtrsim 12,000°$K the decrease in line intensity due to the increase in T_e is compensated by an increase due to a larger $\log g$, and the observed profile appears to yield multi-values for the best fitting set of T_e and $\log g$. This point is well demonstrated in Figure 6, in which we see the theoretical profile increases as either T_e or $\log g$ increases. As a result, it appears that the models falling approximately along a diagonal line from the upper left to the lower right in the figure predict equally good agreement. Hence, it seems necessary to know either T_e or $\log g$ in order to determine the other quantity.

In the case of 40 Eri B, the accurate knowledge of its mass and luminosity may be used to solve this problem in the following manner. Through interpolations of the 12 theoretical profiles of H_γ and H_δ lines shown in Figure 6, we determine a grid of profiles with finer divisions of $\log g$ and T_e. Specifically we take an interval of 0.2 in $\log g$ and $500°$K in T_e and compute the profiles for the resulting 42 models ranging in $\log g$ between 7 and 8 and T_e from 14,000 to 17,000°K. Using a Cal-Comp plotter, the theoretical profiles are plotted with the observed profile. Then, the best fitting value of $\log g$ is determined for each temperature. Due to the considerable asymmetry of both the observed and theoretical profiles, the determination of $\log g$ for a given T_e is made for the short and long wavelength sides of the line center independently. The asymmetry occurs mainly because of the difference in overlapping of the wing portion with neighbor lines (see Figure 13), which defines a 'continuum'. As a result, we find without exception for both H_γ and H_δ lines that for a given T_e, the fit to the short wavelength side always predicts a lower value of $\log g$ than that to the long wavelength side. The mean of the two $\log g$ values thus obtained for each T_e are plotted in Figure 7, with the lower and upper values given by vertical bars on each point. Hence, the bar gives some measure of the accuracy although it does not directly correspond to the usual definition of errors.

We then find the mean values can be well represented by a straight line in

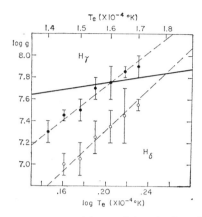

FIGURE 7 Values of T_e and $\log g$ giving the best fit to the observed line profiles of H_γ (upper broken line) and H_δ (lower broken line) for 40 Eri B. The heavy solid line represents the combination of T_e and $\log g$ which gives the observed mass of the same star.

Figure 7. Thus, we find the following expression for the broken line representing the H_γ profile;

$$\log g = 7.82 \log T_e = 25.125. \tag{2}$$

Similarly, from the H_δ profile we obtain,

$$\log g = 9.48 \log T_e = 31.765. \tag{3}$$

Using the observed mass and luminosity of 40 Eri B, we may obtain from Eq. (1) a similar but independent relation between T_e and $\log g$. To do this, one must find the bolometric corrections. The bolometric correction, ΔM_b, is computed for the grid of our revised models and the results for the series of $\log g = 8$ models are shown by filled dots in Figure 8, in comparison with the values computed by Strom and Avrett (1965) for main-sequence stars (open dots, $\log g = 4$). The differences between the values for $\log g = 8$ and 7 are negligibly small for T_e higher than 10,000°K. If we plot ΔM_b against $\log T_e$, we find that for $T_e > 11,500°$K, the following expression gives ΔM_b within an accuracy of 2 per cent, which is sufficient for the present investigation:

$$-\Delta M_b = 6.078 \log T_e = 24.060. \tag{4}$$

Substituting the above equation into Eq. (1), we have

$$\log \frac{M}{M_\odot} = -0.4 \, M_v - 1.57 \log T_e + \log g + 2.87, \tag{5}$$

or with the observed values of $M/M_\odot = 0.43$ and $M_v = 11.0$ mag., we have a relation

$$\log g = 1.57 \log T_e + 1.16. \tag{6}$$

Equation (6) is represented by the solid line in Figure 7. The intersections of this line with the two broken lines give the best values of $\log g$ and T_e determined from H_γ and H_δ profiles. Or by solving Eqs. (2) and (6) simultaneously, we find for H_γ, $T_e = 16,030°K$ and $\log g = 7.758$, and for H_δ line, $T_e = 17,990°K$ and $\log g = 7.836$. The rather large difference in T_e

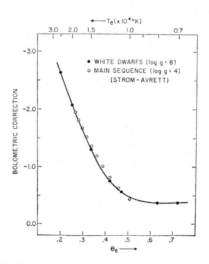

FIGURE 8   The bolometric correction as a function of the reciprocal effective temperature obtained from the models with $\log g = 8$. For comparison, the bolometric corrections for the main-sequence models without line blanketing are shown by open dots.

found between the two determinations is reflected by the nearly parallel nature of the two broken lines in Figure 7, whereas the small variation of $\log g$ is due to the slow increase of the solid line with respect to $\log T_e$. Ideally the three lines should intersect at one point, defining one set of T_e and $\log g$. In addition to observational errors, the discrepancy may be partly due to the inaccuracy in theoretical values. For the computation of the H_δ profile, it is often necessary to extrapolate Griem's (1964) table of the Stark broadening function for higher ion density, whereas for the H_γ line the table covers sufficiently large values of the ion density. Thus, the theoretical profile for H_δ line may involve larger errors than that for H_γ. Considering as well larger uncertainties implied by larger bars in Figure 7, we place one-half weight on the result obtained from the H_δ profile and take a weighted mean of $T_e = 16,600°K$ and $\log g = 7.78$ as the final values determined from line profiles for 40 Eri B.

3. Discussion

Let us now return to consider the discrepancy between the H_γ profiles of Oke and Greenstein. If we apply the above method to Greenstein's profiles, we obtain an effective temperature as high as $20,000°K$ or a value much lower than $\sim 12,000°K$, at which temperature hydrogen lines under white dwarf conditions reach a maximum intensity. Either value seems to be too far different from the values determined from other observations.

In order to examine whether Greenstein's H_γ profiles are systematically lower than theoretical predictions, we compare the equivalent width of H_γ line, $W (H_\gamma)$, for about 50 normal DA stars published by Eggen and Greenstein (1965a, b) with theoretical values in Figure 9. Using the values of T_e

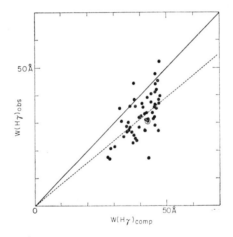

FIGURE 9 Comparison of the observed H_γ equivalent width for about 50 DA stars with the values computed for the best fitting models determined from UBV colors.

and $\log g$ obtained from *UBV* colors for each star (Section II, Figure 2), theoretical equivalent widths are obtained through interpolation of the values computed for the grid of basic models. The broken line in Figure 9 is drawn so as to include an equal number of points in the upper and lower parts of the line. The scatter of the points then indicates that the observed $W (H_\gamma)$ is, on the average, about 25 per cent smaller than the theoretical values. This is in accordance with the report of Eggen and Greenstein

(1965a) that for 40 Eri B (marked by a double circle in Figure 9) they find $W(H_\gamma) = 31.3$ Å as compared to Oke's value of 42 Å obtained from his scanning spectrum.

Greenstein's (1960) detailed profiles of hydrogen lines include 16 norma type DA stars. Again using T_e and $\log g$ determined from *UBV* colors, theoretical profiles are obtained for each of the 16 stars through interpolation of the profiles computed from the grid of models. The results are compared in Figure 10. We see that the observed profiles of the H_γ line are smaller

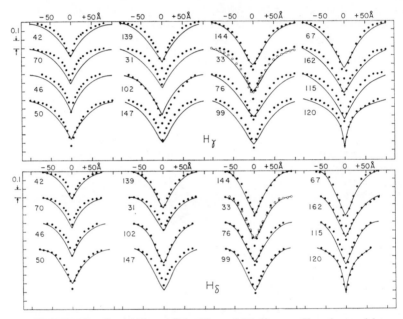

FIGURE 10   Comparison of the H_γ and H_δ line profiles observed by Greenstein (1960) (open dot) with the theoretical profiles for the best fitting models as in Fig. 8. For the star, 33 (40 Eri B), Oke's measurements are indicated by the solid dots.

than theoretical predictions, while the discrepancies appear to be greatly reduced for the H_δ.

We can consider various causes for this discrepancy. First, if the values of $\log g$ obtained from *UBV* colors are systematically in error, we then have to assume a mean $\log g$ unacceptably low. As a matter of fact, the apparently good agreement with computed profiles of H_γ found in Paper I is due to the fact that $\log g$ values estimated there are, on the average, lower by about 0.5

as we discussed in Section II.* On the other hand, if our values of T_e involve systematic errors, the necessary correction should tend to lower the values. We note that the value T_e obtained from *UBV* colors for 40 Eri B is the lowest value of the four independent determinations. It should also be pointed out that the stars in Figure 10 are arranged in the order of decreasing T_e from upper left (24,700°K) to lower right (9500°K). Hence, even if the T_e values are indeed too high, we should then observe good agreement at least for a certain portion of the diagram, but there does not seem to be such a region seen in the figure.

As a possible error in the basic theory, one might doubt the validity of our choice of chemical composition, for which the solar abundance ratios are taken. This point is discussed in detail in Section VI, where we note two important results relevant to the present argument. First, the increase of hydrogen content up to 100 per cent may decrease the line broadening, but at the same time, it increases the surface gravity determined from the observed colors. Hence, the two effects seem to compensate, resulting with little change in the predicted profiles. Second, the effect of decreasing the hydrogen content is such that it lowers the values of log g obtainable from *UBV* colors and increases the line broadening due to higher pressure for the same optical depth produced by lower opacity. Hence, again the compensation of the two effects makes it unlikely that a good agreement is obtained. Some uncertainty may remain with the metal abundances. The effect of metal abundances on the continuous absorption may be small for the ranges of temperature under consideration. From the present result on the hydrogen line blanketing effect, the metal line blanketing may not change T_e by more than ~ 1000°K. However, a possibly large error might enter in the theoretical determination of the continuum if a large number of metal lines overlap the hydrogen line wings. This point requires further investigation.

Finally, there might be a systematic error in the empirical determination of the continuum on the intensity tracings of the photographic plates. While Oke's tracings seem to show a well-defined and smooth continuum on both sides of the H_γ line, an apparently strong blend of the H_δ wing with the H_γ might produce a difficulty in locating a high point between H_γ and H_δ (see Figure 13), thus introducing a systematic error such that the continuum for H_γ is taken relatively too low. In fact, we find that if we lower the

*It should be noted that while the use of approximate wing formula resulted in overestimates of line absorption in the computations of UBV colors in Paper I, we used the more exact formula by Griem (1964) for determining the detailed profiles of H_γ, which is not so much different from the newer values (Griem 1967) used in the present investigation.

continuum of Oke's tracing by about 5 per cent, the resulting profile of H_γ shows excellent agreement with that of Greenstein.

V. PRE-IONIZATION EFFECT

Before we summarize the results of foregoing determinations of T_e and log g, we shall consider the effect of depression of the continuum due to high pressure (pre-ionization). Possible changes in observable features resulting from an erroneous choice of chemical composition will be discussed in the following section.

The influence of perturbing electric fields from surrounding ions on a bound electron results in the escape of the electron from its bound state, which causes very high, normally bound states to become unbound. While the perturbation affects the many levels near the series limits with varying probabilities, a simple treatment of the problem is to consider that a finite value of m exists, which designates the principal quantum number for the highest energy level for which the electron is still bound. It is assumed that m is uniquely determined for a given electron or ion density, N (Schatzman 1958). The result is essentially the same as a lowering of the ionization potential by a small amount ΔE, which for hydrogen can be written as:

$$\Delta E = \frac{E}{m^2}. \tag{7}$$

Then, following many investigators (see Drawin and Felenbok (1965) for a complete reference), we may write the relation between m and N as

$$6 \log m = C - \log N. \tag{8}$$

Somewhat different values are given for the additive constant C by different investigators, ranging from 21.48 (Kolesov 1964) to 22.2 (Pannekoek 1938).

The lowering of the ionization potential of hydrogen atoms affects a model calculation in two ways. The first is the smooth depression of the Balmer discontinuity due to the shift of the series limit toward the longer wavelength region by an amount which depends on the electron density at each layer. The result is similar to the depression of the continuum due to the overlapping of line wings of higher Balmer lines, in which case the value of m is given by the Inglis-Teller formula

$$7.5 \log m = 23.26 - \log 2N. \tag{9}$$

For $C = 21.5$, we find that m given by Eq. (8) is smaller than that given by (9) for log N less than 15.66. Thus, Eq. (9) becomes more important than (8) near the surface region of relatively cool DA stars.

The first effect discussed above is taken into account in our calculation of the emergent flux and is found to be important in determining the U and B magnitudes. The situation is well demonstrated by the depression of the Balmer discontinuity shown by the broken lines in Figure 3.

The problem of our present interest is therefore the second effect which has not been considered in our model calculation mainly because of considerable increase in the amount of computations. Namely, that effect is the increase in electron density due to the lowering of ionization potential. The increase in electron density produces a change in line absorption as well as the continuous opacity, especially in the absorption due to H^- since it is directly proportional to the electron pressure. Thus, using Eq. (8) with $C = 21.5$, the $T - P_e - P_g$ relations are computed by Travis (1967). Figure 11 shows the resulting $P_e - P_g$ relation (broken lines) in comparison with that for the

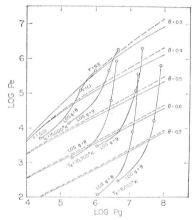

FIGURE 11   Log P_e as a function of log P_g for various reciprocal temperatures as indicated in the figure. The increase due to the lowering of the hydrogen ionization potential by the perturbation from the surrounding electric field is shown by the broken line as compared to the case neglecting pre-ionization (solid lines). The connected open dots represent approximate $T - P_g - P_e$ conditions prevailing in various approximate model atmospheres of white dwarfs.

case without pre-ionization. In the same figure, the actual conditions prevailing in the white dwarf atmospheres are indicated by the curves connecting open circles for six models taken from Paper I. The optical depth at $\lambda 5000$ is given for each circle. Thus, we see that the increase in the electron pressure due to pre-ionization is at most a few per cent. Since we cannot expect the accuracy of opacity calculation better than a few per cent, we conclude that we can neglect pre-ionization in computing the $P_e - P_g$ relation although it is important in the determination of the emergent flux.

VI. CHEMICAL COMPOSITION

Perhaps the largest uncertainty involved in the analysis of white dwarf spectra at the present stage is in our meager knowledge of the chemical composition of these stars. The fact that DA spectra show only hydrogen lines while no hydrogen line is visible in the DB-type is generally interpreted as due to the fundamental difference in H to He ratio between the two classes. Then, what about the abundances of other elements? In addition to the observational limitation with low dispersion spectra, the strong pressure broadening may greatly reduce the detectability of absorption lines, and the possible existence of highly broadened molecular bands might also wash out many weak lines. Thus, the application of the ordinary method of abundance determination does not seem to be feasible.

1. *Effect of reduced hydrogen content*

In the present analysis of DA spectra, we have simply assumed the solar values for the H to He and H to metal ratios. However, the existence of the DA_{wk} group, in which hydrogen lines appear extremely weak, might imply relatively wide ranges of hydrogen content even among the normal DA stars. Thus, it is important to carefully examine the effects of changing the abundance ratios on various observable features. Such an investigation is now in progress, and in the present report we discuss only the preliminary results.

First, we consider the effect of reducing the hydrogen content from 87.4 per cent in number to 10 per cent. Then, assuming that the fraction of hydrogen removed is replaced by helium, the H to He ratio by number is taken to be 1/9. As is expected (Matsushima and Terashita 1964), the result of iterative computations for flux constant models with $T_e = 15,000°K$ and $\log g = 7$ and 8 has shown that the change in T for the same τ is less than 0.2 per cent for layers above $\tau = 2$, as compared to the models with the solar abundances. Hence, avoiding the iterative procedure, only one model is computed for each set of T_e and $\log g$ with the decreased abundance of hydrogen. Thus, the emergent monochromatic flux and *UBV* magnitudes are computed for each of the 9 basic models used for the finer analysis of the spectra of 40 Eri B.

The resulting change in the model for $T_e = 15,000°K$ and $\log g = 8$ is shown in Figure 12. Despite an order of magnitude increase in P_g for a given τ, the corresponding change in P_e is relatively small, being less than a factor of 2 for the layers above $\tau = 1$, as a result of the much higher He to H ratio. The effect of the increase in P_e is well demonstrated in the intensified Stark broadening as seen in Figure 13, where the emergent monochromatic flux is plotted against wavelength for both the model with the solar abundance

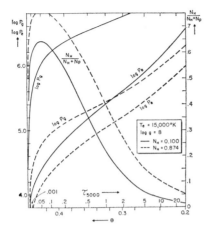

FIGURE 12   Comparison of models for $T_e = 15,000\,°K$ and $\log g = 8$ for differing hydrogen contents as indicated in the figure.

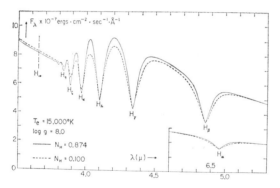

FIGURE 13   The change in the emergent flux distribution with wavelength due to the reduction of hydrogen content (broken-line curve) as compared to the model with solar abundance (solid curves).

(solid curve) and that with the reduced hydrogen content (broken line). As a result of strengthening the wing absorption, the continuum between the lines in the blue region is greatly lowered, while the ultraviolet continuum is increased and no change is seen in the continuum between $\lambda\lambda 5200$ and 6200.

While the increase in ultraviolet flux may at first appear surprising, it can be understood from the following consideration. It should be remembered that in computing the model for reduced hydrogen content, we ignored a possible change in the temperature stratification and used the same $T - \tau$ (at $\lambda 5000$)

M

relation as that obtained for the solar abundance ratio. Hence, the flux near λ5000 may remain practically unchanged. Due to the increase in P_e, however, the relative contribution of H⁻ to the total opacity becomes larger. Since H⁻ absorption (however small in the temperature range under consideration) relative to the others is greater in the visual than in the ultraviolet region, the effective change in opacity is such that at each τ_{5000}, $\kappa_{UV}/\kappa_{5000}$ decreases. As a result, the ultraviolet opacity at a given depth (or temperature) in the hydrogen reduced model becomes smaller than that in the basic model.

The resulting effect on UBV colors is such that U increases, B decreases, and there is no change in V. Thus, the location of each model in the ($U - B$, $B - V$) diagram moves toward the upper right direction as shown in Figure 14. A circled cross in the figure represents the observed color indices of

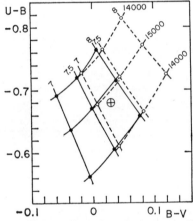

FIGURE 14 The change in the ($U - B$, $B - V$) diagram resulting from the reduction of the hydrogen content (broken lines) compared to the models with solar abundance as in Fig. 13.

40 Eri B. We see practically no change in T_e, but log g is changed by more than 0.5 for the same position in the color diagram. Thus, as one might expect qualitatively, a reduction of hydrogen content is, in effect, equivalent to a decrease in the surface gravity. The result in Figure 14 indicates that the effect of the assumed underestimate of hydrogen content is such that the prediction of the surface gravity for a star will be about a factor of 4 too small, while the error would not affect the temperature determination. On the other hand, we argued in Section II that the value of log g obtained from UVB colors cannot be much lower than the adopted value, hence the above result indicates that the hydrogen abundance cannot be much lower than the assumed value.

The above point is further confirmed by a comparison of the H_γ line profiles between the two groups of models. In Figure 15, H_γ profiles obtained from the models with low hydrogen content are shown by broken lines and Oke's observed profile is added for a comparison. Due to the increase in P_e at the same optical depth as a result of decreasing opacity, the hydrogen line is considerably more broadened than before. Hence, as in the case of *UBV* colors, a fit to the broken line profile tends to predict smaller surface gravity. The discussion in Section III indicating that the theoretical profile of H_γ may not be much larger than that predicted from our models would then imply that the adopted hydrogen abundance is a lower limit.

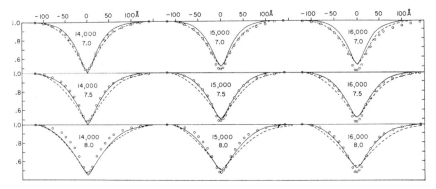

FIGURE 15   The same comparison as in Figures 12/13 for the H_γ line profiles. The open dots represent the observed profiles by Oke (1963).

2. *Pure hydrogen atmosphere*

Similar calculations are being carried out for models with increased hydrogen content. The preliminary results for pure (100 per cent) hydrogen atmospheres reveal that a comparison between the predicted and observed color indices yields values systematically larger by about 0.2 for log g and no appreciable change for T_e, as compared to the values obtained for the models with the solar abundances. The change in log g is equivalent to an increase of 0.2 in the logarithmic mass, and hence, it would provide a means to estimate the hydrogen content of a star if its mass is accurately known. An interesting example is the case of 40 Eri B. In Section III, we obtained log $g = 7.75$ from the color indices, whereas the comparison of the monochromatic flux distribution and the hydrogen line profiles predicted log $g = 8.0$ and 7.78, respectively. Hence, if we assume a pure hydrogen atmosphere for this star, the color indices should predict log $g = 7.95$. On the other hand, the value of log g given by a comparison of line profiles should not be much different, as discussed below; thus, the hydrogen content cannot be much larger than the present value. It should be pointed out, however, that unfortunately the above change in log g would result from increases (or less negative) of less than 0.02 mag. in $(U - B)$ and less than 0.015 mag. for $(B - V)$. Thus, the differences are comparable to the observational accuracy. However, since the change is systematic for the grid of models under consideration, the above result should have significant implications for a statistical analysis.

Although the computations have not been completed, one would expect that a comparison of hydrogen line profiles would probably produce no significant change for models with increasing hydrogen content. The theoretical line profiles will be slightly narrower than those obtained for the models with the solar abundances for the same T_e and log g. However, the value of log g obtained from the colors for a particular star will be systematically larger. As a result, the predicted profiles for each individual star would be practically unchanged, and it will be difficult to remove the discrepancies shown in Figure 10, as pointed out before.

Summarizing the above result, we tentatively conclude that the solar hydrogen content assumed in the present calculations is probably a reasonable value for the normal type DA stars.

3. *Effect of increasing helium content*

If the atmosphere is not entirely composed of hydrogen, it is natural to assume that the remaining element would be mostly helium. Then it is interesting to examine how much helium some of the hottest DA stars could contain before helium lines would become detectable in their spectra; thus, setting an upper limit for the helium abundance in these stars. Such a check may be made by computing the central depth of one of the strongest helium lines for a series of models obtainable by reducing the He and H ratio from that assumed in the present investigation. The calculation is not completed, and we present only the preliminary result in this paper.

As the strongest He I line observed in DB stars, we choose $\lambda 4472$ $(2\,^3P - 4\,^3D)$ neglecting a possible blend with the forbidden line, $\lambda 4470$ $(2\,^3P - 4\,^3F)$ as in the case of hot main-sequence stars. The Stark profile $S(\alpha)$ for the core of this line is tabulated by Griem (1964) and the Holtzmark distribution, $S(\alpha) \propto \alpha^{-5/2}$, is assumed for the wing. Thus, the profiles are computed for various models by changing the surface conditions as well as the He to H ratio. The resulting profiles for the models with normal He abundance (12.7 per cent) as well as one-fifth of that value are shown in Figure 16. While the line predicted from $T_e = 15{,}000°$K seems to be too shallow to be observable, the line appears to be detectable even with a relatively low helium abundance for $T_e = 20{,}000$ and $25{,}000°$K. In fact, only for two of the five DB stars for which the He line profiles are given by Greenstein (1960) are the observed profiles larger than the theoretical profile for $T_e = 25,000°$K, $\log g = 7$ and $N_{He} = 12.7$ per cent. This might imply that the helium content in DB stars is not necessarily much larger than that in main-sequence stars. This point requires further investigation, since in that case, one would expect the coexistence of hydrogen lines as well as helium lines, as in the case of DO type.

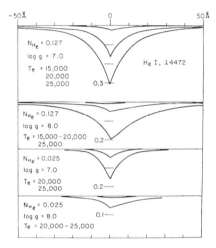

FIGURE 16   The He I, λ 4472 profiles computed for various models with He to H ratios of 0.15 and 0.026.

The depth at the line center and at 8 Å off the center are plotted against helium abundance for various models in Figure 17. The helium abundance is expressed in units of 12.7 per cent, and the remaining composition is assumed to be entirely hydrogen in accounting for the continuous opacity. According to Greenstein (1958), we set the limit of detectability of a line to be about 5 per cent absorption at the center of the line. This limit is indicated by the broken line in the figure. We note that the effective temperature of a few normal type DA stars are undoubtedly higher than 20,000°K, so that the helium content for these stars could not be more than that in main-sequence stars.

The above result is consistent with the previous one in that the hydrogen abundance assumed in the present calculation is a lower limit. The possible overestimate of helium content does not seriously affect the overall result, since the contribution of helium to opacity is still small in the ranges of

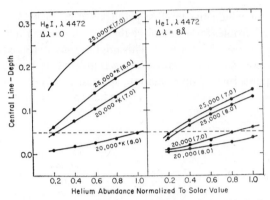

FIGURE 17 Variation of the central depth of the He I, λ 4472 line due to differing He to H ratios. The He content is indicated on the abscicca in units of the solar value.

temperature under investigation. However, we wish to emphasize that the above calculations do not take into account the possible effect of the hydrogen to helium content on the temperature stratification under the condition of flux constancy. Moreover, we have no quantitative basis to extend the above result to cooler DA stars. These are interesting problems for our future investigation, and further computation might require some modifications of the present results.

VII. MASS–RADIUS RELATION

1. *Mass and radius of* 40 *Eri B.*

The results of the four different determinations of T_e and $\log g$ for 40 Eri B reported in Sections III and IV are summarized in Table I, together with the

TABLE I Mass and radius of 40 ERI B.

Basic Data	T_e (°K)	$\log g$	ΔM_b	$R/R_{\odot}$	$M/M_{\odot}$
UBV	14,700	7.75	−1.25	0.0155	0.49
Six-colors	16,700	—	−1.61	0.0141	—
Scanner	16,000	7.8	−1.47	0.0145	0.48
Line profile	16,600	7.78	−1.58	0.0141	(0.43)
Mean	16,000	7.77	−1.47*	0.0143*	0.44*

*These values are obtained from the mean values of T_e and $\log g$, but they are not significantly different from the means of the 4 values in each column.

mean values in the last row. In taking the mean, the values obtained from the line profiles could be given twice the weight of the other three, since both H_γ and H_δ line profiles are used for the determination. In this case, we find $T_e = 16,120°K$ and $\log g = 7.77$, which turns out to be the same as the straight mean value. On the other hand, one might wish to place zero weight on the line profile values on the grounds that the observed mass is used in the computations. Then, we have $T_e = 15,800°K$ but again the value of $\log g$ is unaltered.

For a given set of T_e and $\log g$, the radius may be computed from the equation;

$$\log \frac{R}{R_\odot} = 8.47 - 2 \log T_e - 0.2 (M_v + \Delta M_b), \tag{10}$$

and the mass is given by,

$$\log \frac{M}{M_\odot} = -4.44 + \log g + 2 \log \frac{R}{R_\odot} \tag{11}$$

or directly from Eq. (1). The resulting R and M together with the adopted bolometric correction are listed in Table I. The mean mass and radius are computed from the mean T_e and $\log g$ rather than averaging the four values in the same column. However, the difference between the two procedures is insignificantly small.

We wish to emphasize the importance of the fact that different methods of averaging give essentially the same value for the mean T_e and $\log g$, and hence the mean radius and mass. Thus, the precise agreement of the mean mass with the observed values of 0.43 ± 0.04 (Popper 1954) is not a consequence of the use of the observed mass in the determination of T_e and $\log g$ from line profiles.

It is interesting to compare the above result to theoretically predicted relations between mass and radius for zero-temperature degenerate stars. The solid curves in Figure 18 denote the mass-radius relations computed by Hamada and Salpeter (1961) for pure compositions of He, Mg and Fe without hydrogen envelopes. The broken curves represent the same relation for Fe with H envelopes. The position of 40 Eri B given by the presently adopted values of mass and radius is indicated by a filled circle with error bars. The error bar for mass is due to Popper's estimate and that for radius is based on the accuracy of parallax determination. The change in radius due to the use of lower temperatures is indicated by a circled cross ($T_e = 14,700°K$ as determined from the *UBV* colors) and a circle ($T_e = 12,000°K$, the semi-black-body temperature according to the method of Eggen and Greenstein (1965a)).

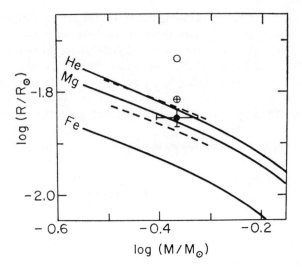

FIGURE 18 The mass–radius relation of 40 Eri B (solid dot with error bars) compared to those predicted by theoretical computations for zero-temperature degenerate models with various compositions. The solid curves are due to Hamada and Salpeter (1963) and the broken-line curves represent the same relation for Fe and H envelopes. For comparison, we show the positions computed for the cases; $T_e = 14,700°$K, $\log g = 7.77$ (circled cross), and $T_e = 12,000°$K, $\log g = 7.77$ (open circle).

Thus, we see that the mean values of T_e and $\log g$ based on the four different methods seem to satisfy both the observed mass (and also the observed gravitational red shift as will be shown later), and the theoretical mass-radius relation for a core composition heavier than He without a significant hydrogen envelope; thus, eliminating some difficulties involved in the previous interpretations of these data (Mestel 1965). At the same time, the present result suggests that the temperature scale previously estimated for 40 Eri B should be raised by 2500 to 4000°K.

Figures 2–6 and Eqs. (2) and (3) show that all of the four observations are much more sensitively dependent on the temperature than on the gravity, and hence, the T_e determination is correspondingly more accurate than $\log g$. In fact, the close agreement among the last three values of T_e in Table I implies that the deviation of about 8 per cent from the mean in the case of *UBV* colors may be considered as about the maximum uncertainty in the temperature determination. Moreover, we note that the radius does not depend strongly on the temperature. This is because an error in T_e is largely compensated by a corresponding error in ΔM_b in the opposite sense for T_e larger than $10,000°\mathrm{K}$. From Eqs. (4) and (10), we can write

$$\frac{\Delta R}{R} = -0.78 \frac{\Delta T_e}{T_e} \tag{12}$$

Thus, the above estimated uncertainty in T_e produces an error of only 6 per cent in the radius and from Eq. (11), 12 per cent in the mass. On the other hand, the apparently large deviation of the mass obtainable from *UBV* colors from the mean value is largely due to the difference in the value of $\log g$. We see in Figure 2 that for $T_e = 15,000°\mathrm{K}$, an error of 0.01 mag. in either $(U - B)$ or $(B - V)$ will result in an error of about 0.1 in $\log g$ or 25 per cent in g, whereas the corresponding error in T_e is less than 2 per cent. Also from the result in Section VI we see that a small uncertainty in the chemical composition could, in general, produce an error of 0.1 in $\log g$.

2. High and low luminosity sequences

Eggen and Greenstein (1965a) have shown that white dwarfs tend to form two separate sequences in a $(M_v, U - V)$ diagram. In Figure 19, we show

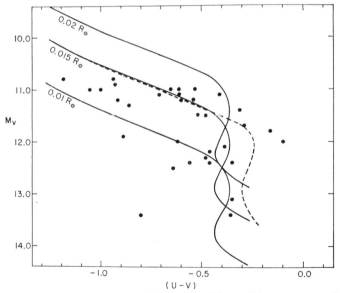

FIGURE 19 M_v vs. $(U - V)$ relation. The solid curves are the theoretical relations for $R = 0.01, 0.015$ and $0.02\ R_\odot$ on the basis of the models with $\log g = 8$. The broken-line curve for $R = 0.015\ R_\odot$ is based on the models with $\log g = 7$. Dots represent the observed values for DA stars.

theoretical (M_v, $U - V$) diagram for various R and $\log g$ together with the observed positions of DA stars except for the DA_{wk} group. The noticeable feature is the sharp drop of the visual magnitude, which occurs at slightly different ($U - V$) between $\log g = 8$ and 7. Since for white dwarfs, the effect of Balmer lines is not appreciable for either the U or V-magnitude as discussed before, this S-shaped turn of the theoretical curves appears to be due to the effect of the Balmer discontinuity on the temperature (see Figure 13). This may also explain the sudden break in the upper sequence at ($U - V$) $= -0.1$, in view of the fact that all of the members of the upper sequence are of DA type except for one unclasisfied object (see Figure 5 of Eggen and Greenstein).

Greenstein and Trimble (1967) have recently made a statistical analysis of their measurements of the Einstein redshifts of 53 white dwarfs. In this analysis, they assume that all of the DA stars for which the luminosity is not available belong to the lower sequence, thereby estimating the luminosities on the basis of the empirical M_v vs. ($U - V$) relation. This assumption is rather unfounded, since we see in Figure 19 or Figure 5 of Eggen and Greenstein (1965a) that the concentration of DA stars in the upper sequence is definitely pronounced. One of the reasons for introducing this assumption appears to be that otherwise the median radius estimated on their temperature scale seems to yield a value too high to be compatible with the theoretical mass-radius relations for heavy-element configurations. As mentioned earlier in this paper, the temperatures adopted by Greenstein and Trimble are determined through a linear interpolation between the main-sequence and black-body curves in the ($U - B$, $B - V$) diagram. However, we note that the difficulty can be greatly reduced if one applies the present temperature scale to those DA stars for which the luminosities are known.

The positions of those DA stars for which the luminosities are available are indicated by larger filled dots (upper sequence) and open dots (lower sequence) in Figure 2. Those include 12 objects used in the analysis of Greenstein and Trimble. Using the values of T_e determined from Figure 2, we show in Figure 20 a modification of Figure 2 of Greenstein and Trimble. In Figure 20, the previous positions are indicated by open circles and the revised ones by solid dots, the difference being shown by broken arrows. When the present method of temperature determination is not applicable (mostly DA_{wk} stars), the values given by Greenstein and Trimble are used and indicated by dots. Thus, we see that the median radius is reduced by about 25 per cent. If we further assume that the error found in the temperature determination from colors for 40 Eri B indicate a mean systematic error, the above reduction will increase to about 35 per cent.

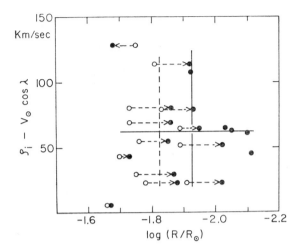

FIGURE 20 A redshift vs. radius diagram. The observed redshifts corrected for the solar motion are due to Greenstein and Trimble (1967). The solid dots indicate the radii based on the present temperature scale, while the open dots are the radii obtained by Greenstein and Trimble. When the present temperature scale is not applicable, the radii given by these authors are used and are shown by solid dots.

3. *The Einstein redshift*

Although the observations seem to involve a large error, it is interesting to compare with observations of the gravitational redshift predicted from the masses and radii determined in the present investigation. The Einstein redshift, k, expressed as a velocity is given by

$$k = 0.635 \frac{M/M_\odot}{R/R_\odot} \text{ km/sec.} \tag{13}$$

In the case of 40 Eri B, a reliable value has been obtained by Popper (1954) from the measurements of 42 hydrogen lines on 27 spectrograms. The result is, $k = 21 \pm 1.9$ km/sec. If we substitute the mean values of $M/M_\odot$ and $R/R_\odot$ listed in Table I into Eq. (13), we find $k = 20.0$ km/sec, showing an excellent agreement with the observed value. Earlier predictions of 16 km/sec (Greenstein 1958) and 17 km/sec (Popper 1954) appear to be the result of overestimates of the radius due to considerable underestimates of the effective temperature based on approximate calculations, as compared to the present results.

Greenstein and Trimble's measurements include 7 DA stars, for which the radial velocities are accurately known since they are members of the Hyades

TABLE II Comparison of predicted redshift with observations.

No.	Star	$T_e(\times 10^{-4})$	$\log g$	$\log \dfrac{R}{R_\odot}$	$\log \dfrac{M}{M_\odot}$	$k_{\text{pred.}}$	$k_{\text{obs.}}$	$\dfrac{k_{\text{pred.}}}{k_{\text{obs.}}}$
26	HZ 4	1.30	8.45	-1.86	0.29	89	69	1.44
28*	LB 1240	1.27	7.75	-2.02	-0.73	12	52	0.23
30	HZ 10	1.14	8.15	-1.73	0.26	62	43	1.44
31	HZ 2	1.90	8.15	-1.87	-0.03	44	27	1.63
39	HZ 7	1.88	8.40	-1.93	0.10	68	79	0.86
42	HZ 14	2.46	8.20	-1.95	-0.14	41	65	0.63
139	W 1346	1.91	7.90	-1.85	-0.24	26	38	0.68

*Low-Luminosity sequence star.

cluster. For those, they give directly determined values of the gravitational redshift. In Table II, we compare the predicted values of k with the observed values for these stars. For each star listed by the catalogue number of Eggen and Greenstein (1965a, 1967), the values of T_e and $\log g$ determined through interpolations in Figure 2 are listed in the third and fourth columns. The next two columns give the radius and mass obtained from Eqs. (10) and (11). Using these values of R and M, we find the values of $k_{\text{pred.}}$ from Eq. (13), as listed in the seventh column. The observed values, $k_{\text{obs.}}$, are listed in the eighth column. The ratio of $k_{\text{pred.}}$ to $k_{\text{obs.}}$ given in the last column gives an indication of the deviation of these values. The comparison of $k_{\text{obs.}}$ and

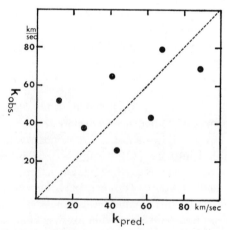

FIGURE 21 Comparisons of the Einstein redshift observed by Greenstein and Trimble (1967) with the values predicted in the present investigation.

$k_{\text{pred.}}$ is also made in Figure 21. We see a generally good agreement between the two values except for the star, EG28. This is the only object of this group belonging to the low luminosity sequence, for which a large deviation is also seen from the theoretical mass-radius relation (see Figure 22). Another object showing a relatively large discrepancy, EG42, is the hottest star of those for which we have determined the mass and radius, with $T_e = 24{,}600°K$. In this region of the color diagram the error in the determination of $\log g$ can be very large since the equi-$\log g$ curves tend to become very close to each other (Figure 2). A possible systematic error in the calibration of the *UBV* system would also be very large at large negative values of $(U - B)$. In fact, at $(U - B) = -1.0$, the transformation relations obtained by Karyagina and Kharitonov (1965) predict values of $(U - B)$ about 0.1 mag. less negative than those presently used (Matthews and Sandage 1963).

The mean of the seven predicted values of k is 49 km/sec as compared to the mean of $k_{\text{obs.}} = 53$ km/sec. If we exclude the star EG28 and take a mean of the remaining values, we find $k_{\text{pred.}} = 55$ km/sec which is in good agreement with the corresponding mean of $k_{\text{obs.}} = 54$ km/sec. It is interesting to note that the two objects, EG26 and EG30, show good agreement despite the fact that they deviate far from the theoretical mass–radius relations for heavy elements, as will be discussed below.

4. *Mass–radius relation for DA Stars*

Using the values of T_e and $\log g$ determined from Figure 2, the masses and radii are determined for the DA stars with known luminosities. These stars are represented by larger dots in Figure 2. The resulting mass-radius relations are compared to the theoretical relations in Figure 22. As in Figure 18, the curves for He, Mg, and Fe are due to Hamada and Salpeter (1961), and the curve for H is taken from Chandrasekhar's (1939) relation for $\mu_e = 1$. The broken line represents the same relation for neutron stars as given by Hamada and Salpeter. When a star apparently belongs to the lower sequence, it is indicated by an open circle or an open triangle, which will be discussed below. The high-temperature objects with $(U - V) < -0.8$ and $M_v < 11.5$ are indicated by filled dots, though it does not necessarily imply that they belong to the upper sequence (see Figure 19). Also, those objects with $T_e < 10{,}000°K$ are parenthesized indicating the values are less reliable. Although the dispersion of points is somewhat large, we can see an apparent concentration at $\log(M/M_\odot) \sim -0.20$ and $\log(R/R_\odot) \sim -1.85$. This value for the mass is smaller by about 40 per cent than the median mass derived by Greenstein and Trimble (1967) on the basis of the statistical value of gravitational redshifts. The difference is about 20 per cent, if the correction of 0.03 suggested by them is made on θ_e.

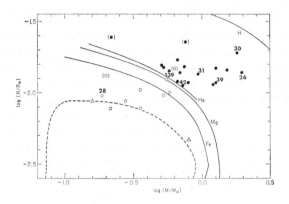

FIGURE 22 Mass–radius relations for 20 DA stars compared to the theoretical predictions as in Figure 18. The upper sequence and the lower sequence stars are distinguished by solid and open dots, respectively. Those objects with uncertain values are parenthesized, and the triangles denote the stars for which the spectral classification is uncertain. The stars for which the Einstein redshifts are determined by Greenstein and Trimble are indicated by the EG catalogue number. The broken-line curve represents the mass–radius relation for neutron stars as given by Hamada and Salpeter.

The present result does not only fail to confirm the usually assumed relation that the lower-sequence stars have masses greater than the upper-sequence stars, but the opposite is found to be the case. This point may also be seen qualitatively in the color diagram in Figure 2. In this diagram we cannot see a clear separation between the upper (dots) and the lower (open circles) sequence stars. Then, the average surface gravity should not differ significantly between the two groups, and consequently, through Eqs. (10) and (11), the upper-sequence stars are expected to be more massive than the lower-sequence. This trend, which contradicts the theoretical mass–radius relations, does not seem to be altered by possible modifications that could be made on the computed color relations within the framework of the present investigation.

A possible explanation may be that although these objects are classified as DA stars, they might be different from the ordinary DA stars in the atmospheric chemical composition, and consequently the theoretical color relations are no longer valid. As we discussed before, a decrease of H to He ratio by as much as a factor of 10 does not seem to alter the value of T_e resulting from observed color indices. Hence, different choices of the hydrogen content within the above range may not change the radii appreciably but only the masses. However, a decrease in the hydrogen content might be accompanied by a considerable increase in some of the heavier elements as well as helium, as is often seen in a hydrogen deficient star. This effect, which might be important for the calculation of continuous opacity, is not taken into account in the present investigation. In fact, for the two objects indicated by triangles, EG101 and EG165, the spectral classification is somewhat uncertain, according to Eggen and Greenstein. The H_γ equivalent widths of these stars appear to be too small to be a normal DA star, and they might be more closely related to the DA_{wk} group.

The seven stars included in the comparison of gravitational redshifts in Table II are denoted by their EG numbers in Figure 22. Despite the unusual positions of EG26 and EG30 in the mass–radius diagram, the good agreement between the predicted and observed redshifts for these objects requires special consideration. Even if the temperature determination involves some errors, its effect on the radius will be small as discussed before. Hence the masses of these two objects appear to be too large to be compatible with the theoretical relations for heavy element configurations. For the lower sequence star, EG28, the predicted redshift definitely disagrees with the observed value. If we combine the present radius and the observed redshift, we obtain $\log (M/M_\odot) = -0.10$, in which case EG28 falls precisely on the Mg line.

Finally, it may be of interest to note that these stars fall on the mass–radius relation for neutron stars as shown by the broken line in Figure 22. These neutron star models differ from more typical ones in that the central density is not high enough ($\rho_c \sim 3 \times 10^{14}$ g/cm^3), and a small neutron core is surrounded by an extensive electron-ion shell whose radius is almost as large as that of ordinary white dwarfs. The possible existence of such a configuration has been suggested by Tsuruta and Cameron (1966). However, since a stability consideration (Harrison, Thorne, Wakano, and Wheeler 1965) may prohibit such neutron stars from existing in nature, the possibility that the above objects can be identified with them is rather doubtful.

We wish to thank Dr. J. L. Greenstein for his encouragement and advice and Drs. G. E. Kron and J. B. Oke for providing their data before publication.

Note added in proof (January 1969): On the basis of futher investigation, taking into account the detailed variation of hyrodgen content, the values listed in Table 1 have been somewhat revised (Matsushima and Terashita 1969). The mean values of the surface parameters of 40 Eri B are found to be $T_e = 15,000°K$, $\log g = 7.72$, $R/R_\odot = 0.015$, $M/M_\odot = 0.43$, and the hydrogen content in number, $N_H = 0.90$.

References

BAHNER, K. 1963, *Ap. J.*, **138**, 1314.

CHANDRASEKHAR, S. 1939, *An Introduction to the Study of Stellar Structure* (Chicago: University of Chicago Press).

DRAWIN, H., and FELENBOOK, P. 1965, *Data for Plasmas in Local Thermodynamic Equilibrium*, Gauthier-Villars, Paris.

EGGEN, O. J., and GREENSTEIN, J. L. 1965a, *Ap. J.*, **141**, 83.

EGGEN, O. J., and GREENSTEIN, J. L. 1965b, *ibid.*, **142**, 925.

EGGEN, O. J., and GREENSTEIN, J. L. 1967, *ibid.*, **150**, 927.

GREENSTEIN, J. L., 1958, *Hb. d. Phys.*, Vol. 50, p. 160, Springer-Verlag, Berlin.

GREENSTEIN, J. L. 1960, *Stellar Atmospheres*, ed. J. L. Greenstein (Chicago: University of Chicago Press), chap. XIX.

GREENSTEIN, J. L., and TRIMBLE, V. L. 1967, *Ap. J.* **149**, 283.

GRIEM, H. R. 1960, *Ap. J.* **132**, 883.

GRIEM, H. R. 1964, *Plasma Spectroscopy* (New York: McGraw-Hill Book Co.).

GRIEM, H. R. 1967, *Ap. J.* **147**, 1092.

HAMADA, T., and SALPETER, E. E. 1961, *Ap. J.*, **134**, 683.

HARRISON, B. K., THORNE, K. S., WAKANO, M., and WHEELER, J. A. 1965, *Gravitation Theory and Gravitational Collapse* (Chicago: University of Chicago Press).

HAYES, D. 1967, *Thesis*, U.C.L.A.

HEINTZE, J. R. W. 1968, Preprint.

KOLESOV, A. K. 1964, *Soviet Astron.*, **8**, 185.

KRON, G. E., FENSTEIN, A., and GORDON, K. C. 1966, Preprint.

MATSUSHIMA, S., and TERASHITA, Y. 1964, *Ap. J.*, **140**, 285.

MATSUSHIMA, S., TERASHITA, Y., 1959, *Ap. J.*, **156**, in press.

MATTHEWS, T. A., and SANDAGE, A. R. 1963, *Ap. J.* **138**, 30.

MESTAL, L. 1965, *Stellar Structure*, ed. L. H. ALLER (Chicago: University of Chicago Press), chap. V.

MIHALAS, D. 1966, *Ap. J. Suppl.* **13**, 1.

OKE, J. B. 1963, Paper presented at the Cleveland meeting of the AAS and private communication.

PANNEKOEK, A. 1938, *M. N.*, **90**, 694.

POPPER, D. M. 1954, *Ap. J.* **120**, 316.

SCHATZMAN, E. 1958, *White Dwarfs* (Amsterdam: North-Holland Publishing Co.).

STEBBINGS, J., and KRON, G. E. 1956, *Ap. J.*, **123**, 440.

STEBBINGS, J., and KRON, G. E. 1964, *ibid.*, **139**, 424.

STROM, S. E., and AVRETT, E. H. 1965, *Ap. J.*, **140**, 1381.

TERASHITA, Y., and MATSUSHIMA, S. 1966. *Ap. J. Suppl.*, **8**, 461.

TERASHITA, Y., and MATSUSHIMA, S. 1967, *A. J.*, **72**, 832.

TRAVIS, L. D. 1967, *M.S. Thesis*, Univ. of Iowa.

TSURUTA, S., and CAMERON, A. G. W. 1966, *Canadian J. Phys.*, **43**, 1616.

WHITEOAK, J. B. 1966, *Ap. J.* **144**, 305.

Evolution of white dwarf stars

SAMUEL C. VILA

Institute for Space Studies
NASA Goddard Space Flight Center

Abstract

The thermodynamic properties of a plasma of heavy ions immersed in a neutralizing uniform background of electrons has been recently calculated by Brush, Sahlin, and Teller. These properties are used to obtain estimates for the cooling times of white dwarf stars whose ion component is undergoing condensation.

It is found that ion condensation increases the cooling times by a factor between one and two at luminosities higher than $10^{-4} L_{\odot}$. At lower luminosities there is a sudden decrease of luminosity, related to the decrease of the specific heat of a lattice at temperatures below the Debye temperature; but when this occurs the stars are too faint to be observable.

IT IS KNOWN that in a white dwarf star, at sufficiently low temperatures, the ions form a rigid lattice structure. At higher temperatures they adopt a state intermediate between the arrangement in a rigid lattice and a perfect gas. The thermodynamic properties of this intermediate state have been recently calculated by Brush, Sahlin, and Teller (1968) by a Monte Carlo study of a plasma of heavy ions immersed in a neutralizing uniform background. Their results and sequences of models for white dwarfs, calculated previously by the author, are used to obtain estimates for the changes in the cooling rates of white dwarfs caused by ion condensation.

The evolutionary sequences of models used correspond to 0.4 and 1.0 solar masses and compositions of pure oxygen and iron. These were chosen to cover the possible range in masses and compositions for realistic models of white dwarfs. These sequences of models have been obtained by use of a computer program for studying stellar evolution. We assume complete ionization, the matter being composed of a non-interacting Fermi gas of electrons and a perfect gas of ions. The opacities used were obtained by use of the Los Alamos opacity program.

The pressure-density relation and the entropy of the electron component were taken from the tables by Grasberger (1961). The cooling times are presented in Figure 1. The range of luminosities studied is given by

$$0 < \log \frac{L}{L_\odot} < -4$$

and the zero of time is chosen at $L = L_\odot$.

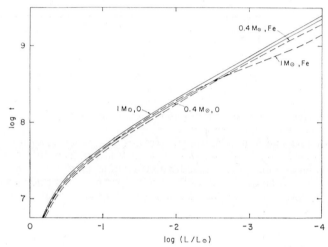

FIGURE 1 Logarithm of the evolution time in seconds for 0.4 and 1.0 solar masses models of oxygen and iron compositions.

It appears from the results obtained that the cooling times down to a given luminosity do not depend strongly on composition, except at low luminosities, where the major part of the heat released comes from the ion component which is inversely proportional to the molecular weight of the ions. The cooling times show only a small dependence on the mass.

This is explained by the fact that, for a given luminosity, the electron component of a smaller star is less degenerate and the electron specific heat per gram is higher than in the case of a more massive one. The higher specific heat of a lesser degenerate electron gas compensates for the mass factor in the cooling time formula:

$$dt = -\frac{Mc_v \, dT_c}{L}$$

These sequences of models and corresponding cooling times represent an improvement on the 'old' estimates (see Schwarzschild 1958) in the following respects: (a) The electron specific heat has been included; this is particularly important for the initial stages of cooling where it is still comparable to the ionic specific heat. (b) The degenerate–nondegenerate region of the star that controls the rate of energy release has been calculated using detailed equations of state and opacities.

Two important factors have not been included in these models. They are: (c) Energy loss by neutrinos. It has been shown by the author (1967) that for $L < 10^{-2} L_\odot$ the neutrino luminosity is $L_\nu < 0.1 L$. That makes the later cooling times practically independent of neutrino losses, since L_ν decreases faster than L with a decrease in temperature. (d) Ion condensation effects. Detailed evolutionary calculations, taking these into account, are being made by the author and will appear elsewhere. Their effect is estimated below.

The main change due to ion condensation is to be expected in the cooling times. The structure changes for zero temperature white dwarfs have already been studied by Hamada and Salpeter (1961) and were found not to be greater than a few per cent. An estimate of the effect on the cooling times has been made on the assumption that the structure of the models is not changed by the condensation of the ions, but only the specific heats are affected. This procedure is supported by work presented in a recent paper by Mestel and Ruderman (1967), where it is shown that the gravitational energy release is absorbed by the increase of the electron exclusion energy due to the density increase and it is only the thermal energy that is available for radiation.

The departures from perfect gas behavior for the ion component are measured by the parameter.

$$\Gamma = \frac{(Ze)^2}{kT} \left(\tfrac{4}{3} \pi \eta \right)^{1/3}$$

where η is the number of ions per cm³. To estimate the specific heat of the ions, we have made the following assumptions:
(a) $\Gamma < 1.0$; we used the perfect gas formula;
(b) $1.0 < \Gamma < 100$; we used the results of Brush, Sahlin, and Teller;
(c) $\Gamma > 100$; we assumed a rigid lattice and used the Debye approximation for the specific heat. The Debye temperature was taken from Carr's work and is given by:

$$\Theta = 1.74 \times 10^3 \frac{2Z}{A} \rho^{1/2}$$

In Figure 3, we present a comparison of the partial specific heats of the electron and ion components for oxygen at $\rho = 10^6$ gm/cm³.

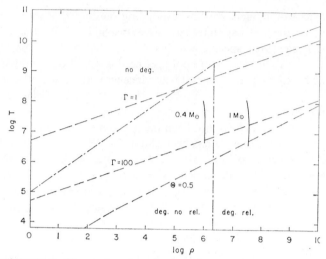

FIGURE 2 Illustrating the equation of state for oxygen. Γ and Θ are defined in the text. The evolutionary tracks correspond to the center conditions of oxygen stars in the time interval corresponding to $0 < \log{(L/L_\odot)} < -4$.

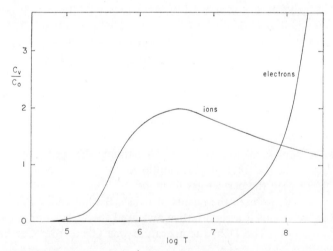

FIGURE 3 Specific heats of electron and ion components for oxygen at $\rho = 10^6$ gm/cm³ as function of temperature. The perfect gas specific heat, $C_0 = \frac{3}{2}\,k/\mu_I H$ is taken as unit.

An estimate of the change in the cooling rates of the stellar models has been made in the following way; for each of the models with no condensation, we have obtained the density and temperature at the spherical shell that encloses one half of the star's mass. For these, we have calculated the ratio

$$r = \frac{C_v \text{ (ions as cond.)} + C_v \text{ (electrons)}}{C_v \text{ (ions no cond.)} + C_v \text{ (electrons)}} - 1$$

This is presented in Figure 4 for the sequences of stellar models considered.

The main conclusions are:

(a) There are only slight changes in the first stages of cooling. This is due to the still important electron specific heat and to the gradual onset of condensation in the Brush, Shalin, and Teller mode .

(b) The evolutionary times are increased by a factor no larger than two in all cases studied. The ratio r increases gradually to a maximum value and decreases more sharply afterwards. In the range $0 < \log (L/L_\odot) < -4$ the maximum was attained only by the more massive stars.

(c) Although r decreases after the maximum and the cooling rate will eventually fall below the noncondensation values due to the drop in the specific heats in the Debye approximation, this will happen at luminosities too low to be observable in real stars.

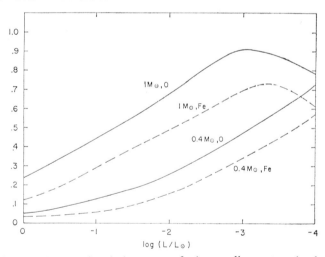

FIGURE 4 Fractional increase of the cooling rates (under the assumption of the ion of condensation) as a function of the star's luminosity.

The work reported in this paper was supported by the National Aeronautics and Space Administration under Grant No. NGR-33-015-036. The author would also like to thank Dr. Robert Jastrow for his hospitality at the Goddard Institute for Space Studies.

References

BRUSH, S. G., SAHLIN, H. L., and TELLER, E. 1968, *J. of Chem. Phys.*, **45**, 2102.

CARR, W. J. 1961, *Phys. Rev.*, **122**, 1437.

DEBYE, 1912, *Ann. Physik*, **39**, 789.

GRASBERGER, W. H. 1961, *A Partially Degenerate, Relativistic, Ideal Gas* (U.C.R.L. 6196).

HAMADA, T., and SALPETER, E. E. 1961, *Ap. J.*, **134**, 683.

MESTEL, L., and RUDERMAN, M. A. 1967, *M.N.*, **136**, 27.

SCHWARZSCHILD, M. 1958, *Structure and Evolution of the Stars* (Princeton Univ. Press).

VILA, S. C. 1967, *Ap. J.*, **149**, 613.

On the cooling of white dwarfs

JEREMIAH P. OSTRIKER and LEON AXEL

Princeton University Observatory

Abstract

We have recalculated the cooling times of white dwarfs using recent results on the specific heat of degenerate matter. The typical ($\approx 1 \, M_\odot$) white dwarf found by Greenstein and Trimble (1967) will fade to $\approx 10^{-3.5} \, L_\odot$ ($M_v \approx 13$) in less than 10^{10} years at which point the star's interior temperature falls below the Debye temperature, and cooling proceeds very rapidly. Thus, relatively few white dwarfs will be found fainter than $M_v \approx 13$–14 and those formed early in the history of the galaxy will now be invisible.

I. INTRODUCTION

FOLLOWING SALPETER'S (1961) argument for crystallization in the interior of white dwarfs, Mestel and Ruderman (1967) have pointed out that the specific heat under these circumstances is due to lattice vibrations rather than particle motions in a gas. When the interior temperature falls below the Debye temperature ($\Theta_D \approx 10^{7} \, ^\circ\mathrm{K}$) the lattice specific heat falls rapidly and it might be expected that the stars will cool more quickly than has previously been computed. We have recomputed the cooling times both for classical white dwarfs and rapidly rotating models using specific heats calculated from the classical Debye approximation.

II. CALCULATIONS

We assume that the only important energy source is the thermal energy of the ions. The crystallization that occurs on cooling will also release heat; Van Horn (1968) has investigated the effects this will have on the cooling process. Our calculations are concerned with cooling *subsequent* to short range crystallization. An additional contribution to the specific heat from the thermal component of the electron gas (cf. Mestel 1965) is unimportant

357

until very low temperatures are reached and will be neglected. Gradual contraction will release gravitational energy, but this is absorbed by the internal energy (cf. Mestel and Ruderman 1967). We shall also neglect nuclear energy sources, which, it has been shown (Schatzman 1958), will lead to instability. And finally, neutrino losses are unimportant in the temperature range considered.

To calculate the cooling, we need to know the energy generation per unit mass. Consistent with our assumptions, this is

$$\epsilon = -\frac{dU}{dt} = -\frac{dU}{dT}\frac{dT}{dt},\qquad(1)$$

or

$$\epsilon = -\frac{C_v}{A}\frac{dT}{dt}\qquad(2)$$

where T is the temperature, U the internal energy, C_v the specific heat per particle which in general depends on density and temperature, A is the mean atomic mass, and t the time. To find the total luminosity, we integrate equation (2) over the star:

$$L = \int \epsilon\, dM = -\frac{M}{A}\int_0^1 C_v\frac{dT}{dt}\, dm \approx -\frac{M}{A}\frac{dT_c}{dt}\int_0^1 C_v\,(T_c, \rho)\, dm.\qquad(3)$$

M and m are the total and fractional stellar mass. In the last step we have replaced the local interior temperature T by T_c, the central temperature, since low luminosity white dwarfs are approximately isothermal over most of their mass, due to the high thermal conductivity of a degenerate electron gas. To find the time to reach any given temperature, we integrate equation (3) over time, giving

$$t\,(T_c, M) = \frac{M}{A}\int_{T_c}^{T_{cry}} \frac{dT_c}{L\,(T_c, M)}\int_0^1 C_v\,(T_c, \rho)\, dm.\qquad(4)$$

In earlier calculations, C_v was taken to be $3/2\ k$ per particle, the value for a perfect gas. However, Mestel and Ruderman (1967) have shown that the ions in a typical white dwarf form a lattice, and that therefore the specific heat for a lattice applies over all resonable temperature ranges. For a lattice the specific heat is $3\ k$ per particle at high temperatures, but at temperatures below a characteristic Debye temperature, the specific heat falls to zero as T^3. If the electron contribution had been included, the behavior close to zero would go as T rather than as T^3. The gas and lattice heat capacities are shown in Figure 1. For the latter we have applied the Debye approximation using

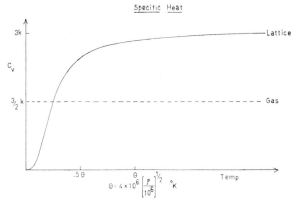

FIGURE 1 The broken line shows the specific heat per particle as a function of temperature for a monatomic gas, the solid line for a lattice in the Debye approximation.

numerical values given by Jahnke, Emde and Losch (1960), Table 53a. According to Mestel and Ruderman (1967), the Debye temperature is given by

$$\Theta_D = 4 \times 10^3 \sqrt{\rho} \; ^\circ K, \tag{5}$$

where μ_e is assumed to be two, giving $\Theta_D \approx 10^7$ for typical conditions. In subsequent calculations we shall use the Debye specific heat with Θ_D taken from Eq. (5) evaluated at the half-mass point; that is, we crudely perform the inner integration in Eq. (4) with a one point 'quadrature' formula. Subsequent, more accurate integrations show that the results are very insensitive to this numerical approximation.

The luminosity of a white dwarf is determined by radiative transfer through the outer nondegenerate layers. It has been shown by Mestel (1952) and Schwarzschild (1958) that, for a Kramers-type opacity, the luminosity is given by a relationship of the form

$$L = KMT_c{}^\alpha, \tag{6}$$

where K and α depend on the opacity. In our calculations the luminosity-temperature-mass relationship is taken from calculated white dwarf models, particularly those of Vila (1966, 1967), and L'Ecuyer (1966), who kindly furnished some additional unpublished results. We also used the published results of Lee (1950), Rose (1966), and Hayashi *et al.* (1962), summarizing the data in Figure 2. For T_c greater than $2.75 \times 10^7 {}^\circ K$, α was taken as 2.5.

Finally, A was taken to be 14 amu, corresponding to a mixture of C^{12} and O^{16}. Using Eq. (6) we are now able to integrate Eq. (4) to obtain $t(T_c)$ for white dwarfs of various masses. Inverting this relation and substituting back into Eq. (6) gives $L(t)$, the desired cooling curve.

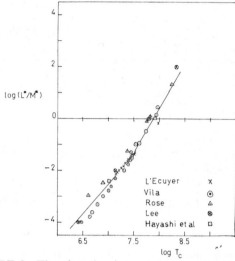

FIGURE 2 The plotted points represent indicated theoretical calculations; the line shows the approximate relation adopted in this paper. L^* and M^* are the luminosity and mass, respectively, in solar units, and T_c the central temperature.

III. RESULTS

The results for classical white dwarfs are plotted in Figure 3. At first, luminosity falls in the manner described by Mestel and Ruderman (1967). When the internal temperature reaches the Debye temperature, cooling becomes

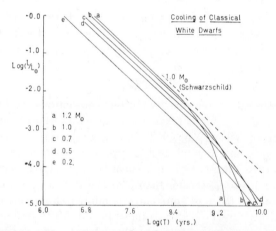

FIGURE 3 The cooling of classical white dwarfs of a range of masses: luminosity in solar units versus time in years.

rapid and all models will reach very low luminosities in moderate times although the abruptness of the transition is slightly exaggerated in our calculations since we neglect the variation of Θ_D throughout the interior. For comparison, we also show the cooling curve of a 1 $M_\odot$ white dwarf with Kramers' opacity taken from Schwarzschild (1958).

The 'typical' mass, however, is ≈ 1 $M_\odot$, as deduced from Greenstein and Trimble's (1967) analysis of Einstein gravitational shifts observed in various samples of white dwarfs. Examination of Figure 3 shows that the cooling curves for such 'typical's tars follow the power law dependence ($L \propto t^{-1.4}$) derived by Mestel (1952) and Schwarzschild (1958), until the luminosity reaches $\approx 10^{-3.5}$ $L_\odot$ ($M_v \approx 13$). After this point cooling proceeds rapidly, the total cooling time being less than 10^{10} years. It follows that white dwarfs fainter than $M_v \approx 13$–14 should be relatively scarce* and also that many of the white dwarfs formed early in the evolution of the galaxy may now be invisible.

In Figure 4, similar curves are plotted for some differentially rotating white dwarf models of Ostriker and Bodenheimer (1968). Approximate correction was made for the variation of brightness over the surface corresponding to the variation in gravity. For these stars the cooling times, which depend primarily on the mean density rather than the mass, are characteristically quite short, occasionally as brief as $\approx 10^8$ years.

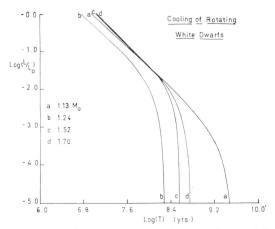

FIGURE 4   Like Figure 3, but for rapidly rotating models.

In Figure 5, the cooling is shown in the H–R diagram for classical white dwarfs. Evolution proceeds along curves of constant radius, some of which are shown, labelled by the corresponding mass in solar masses. Some curves

* This suggestion has been confirmed by recent observational work (Greenstein 1969).

described by stars of the same age are also indicated and labelled by the common logarithm of the time. The rotating white dwarfs will cool approximately along the 1 $M_\odot$ line, but at about the same rate or faster than the 1.3 $M_\odot$ classical white dwarf. Table I presents the cooling times for classical and rapidly rotating white dwarfs.

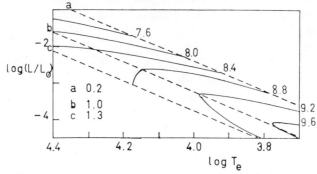

FIGURE 5 Lines of constant cooling time for classical white dwarfs of varying masses are indicated by the solid lines. The numbers at their right give the logarithm of the time in years. The dotted lines give the cooling tracks for white dwarfs of the indicated solar masses.

TABLE I Cooling times of white dwarfs.

$M/M_\odot$	$J/10^{50}$ (c.g.s.)	log t_c (years)
1.3	0	9.0
1.2	0	9.3
1.0	0	9.7
0.7	0	10.0
0.5	0	>10.0
1.13	6.16×10^{-2}	9.5
1.24	1.32×10^{-1}	8.3
1.52	6.93×10^{-1}	8.6
1.70	1.44	8.8

These results can perhaps explain the observation that there are faint white dwarfs in clusters that, by the earlier calculations would have to be much older than the clusters. For example, Eggen and Greenstein (1965) report a white dwarf in the Hyades with luminosity $\approx 10^{-4} L_\odot$. By the old estimates, this corresponds to an age of about 5×10^9 years, but the Hyades cluster is considered to be only about 4×10^8 years old. For the classical white dwarfs, the new specific heats give a lower age, but even for a 1.2 $M_\odot$ it is still about

1.6×10^9 years. However, the rotating models can easily cool down within the necessary times. The very sudden loss of luminosity will probably be slowed at very low luminosities by such things as the influx of interstellar matter onto the surface.

Also, the presence of substantial numbers of white dwarfs at luminosities too low to be observed would help to explain the discrepancy between the estimate of the mass present in the galactic plane as given by the Oort limit and the mass observed in stars and interstellar matter.

Weidemann (1967) has shown that simply taking the increased atomic mass will greatly increase the number of faint white dwarfs expected; the rapid cooling below the Debye temperature and the possibility of the rapidly rotating white dwarfs further increase the expected numbers.

A more detailed presentation of this material will be submitted for publication elsewhere.

Support for this work was provided in part by grants from the Air Force Office of Scientific Research under contract AF 49(638)-1555 and National Science Foundation grant NSF-GP279.

References

EGGEN, O. J., and GREENSTEIN, J. L. 1965, *Ap. J.*, **141**, 83.
GREENSTEIN, J. L. 1969, *Comments on Astrophy. Space Phys.*, **1**, No. 2.
GREENSTEIN, J. L., and TRIMBLE, V. L. 1967, *Ap. J.*, **149**, 283.
HAYASHI, C., HOSHI, R., and SUGIMOTO, D. 1962, *Prog. Theoret. Phys.*, Suppl. No. 22.
JAHNKE, E., EMDE, F., LOSCH, R. 1960, *Table of Higher Functions* (New York: McGraw-Hill Book Co., Inc.).
L'ECUYER, J. 1966, *Ap. J.*, **146**, 845.
LEE, T. D. 1950, *Ap. J.*, **111**, 625.
MESTEL, L. 1952, *M.N.R.A.S.*, **112**, 583.
MESTEL, L. 1965, *Stars and Stellar Systems*, ed. by L. H. ALLER and D. B. McLAUGHLIN (Chicago: U. of Chicago Press) Chap. 5, Section 2.
MESTEL, L., and RUDERMAN, M. A. 1967, *M.N.R.A.S.*, **136**, 27.
OSTRIKER, J., and BODENHEIMER, P. 1968, *Ap. J.*, **151**, 1089.
ROSE, W. K. 1966, *Ap. J.*, **146**, 838.
SALPETER, E. E. 1961, *Ap. J.*, **134**, 669.
SCHATZMAN, E. 1958, *White Dwarfs* (Amsterdam: North Holland Publ. Co.).
SCHWARZSCHILD, M. 1958, *Structure and Evolution of the Stars* (Princeton, N.J.: Princeton U. Press).
VAN HORN, H. M. 1968, *Ap. J.*, **151**, 227.
VILA, S. C. 1966, *Ap. J.*, **146**, 437.
VILA, S. C. 1967, *Ap. J.*, **149**, 613.
WEIDEMANN, V. 1967, *Zs. f. Ap.*, **67**, 286.

High-luminosity white dwarfs

H.-Y. CHIU

Institute for Space Studies
Goddard Space Flight Center, NASA, New York

Abstract

In this paper we studied the evolution of high luminosity white dwarfs. We assume that the prewhite dwarf phase is characterized by a temperature greater than 10^8 °K and a density of the order of 10^5 g/cm^3 for a mass of 0.8 $M_\odot$. This corresponds roughly to the physical condition of the central star of a planetary nebula. It is the current belief that the central star of a planetary nebula evolves to become a white dwarf. We have found that the plasma neutrino process will dissipate energy of such a star quickly so that it cools down rapidly to become a white dwarf. If neutrino processes were not present, then the time of evolution for stars of mass greater than 0.5 $M_\odot$ through the high luminosity ($\sim 10 \to 10^3 L_\odot$) and high surface temperature phase ($\sim 10^5$ °K) is around 10^{5-6} years. On the other hand, when neutrino processes are taken into account the time of evolution is around 10^{3-4} years. For stars of mass less than 0.5 $M_\odot$ the evolution is not affected. The lack of population of stars in this stage (which has been called 'ultraviolet dwarf' by Stothers) indicates the existence of neutrino processes.

THE WHITE DWARFS are characterized by their exceedingly high mean density ($> 10^5$ g/cm^3) and low luminosity ($< 1 L_\odot$). At such high densities hydrogen is unstable (Lee 1950). It is therefore believed that the white dwarf population represents highly evolved stars whose main composition is helium or the elements made in more advanced stellar evolution stages.

Recent observations on planetary nebulae further cast new light on the origin of white dwarfs. The nuclei of planetary nebulae (n.p.n.) are also dense stars with very high luminosities ($\sim 10^4 L_\odot$) and high surface temperatures ($\sim 10^5$ °K) (Seaton 1966). These two figures give a radius of around 0.07 $R_\odot$

which is roughly seven times the radius of a white dwarf. They are too dense to be ordinary stars and yet not dense enough to be white dwarfs, with central temperatures of the order of $3 \times 10^8 °K$ which represent an advanced state of stellar evolution. This value of central temperature is obtained by using stellar models of a mass of 1 $M_\odot$ with a surface temperature of $10^5 °K$. The structure is dominated by a semidegenerate gas pressure and an electron scattering or electron conduction opacity, and the resulting model is almost independent of stellar composition. From these considerations, it may be concluded that the n.p.n. are the most probable candidates for the pre-white dwarf population.

Knowing the evolution properties of proto-white dwarf stars, one may study further the evolution properties of white dwarfs. We have found that when the temperature of a proto-white dwarf of a mass less than 0.8 $M_\odot$ drops to or below $10^8 °K$, the structure of the proto-white dwarf can be very well approximated by Chandrasekhar's classical model (1960) or the improved model of Salpeter (1961); in both Chandrasekhar's and Salpeter's models the temperature is assumed to be zero in obtaining the mechanical structure. Only a small modification needs to be taken into account near the surface where degeneracy is removed by a finite temperature. The evolutionary track of a 0.736 $M_\odot$ proto-white dwarf is shown in Figure 1 with some observation points.

One therefore expects that if white dwarfs are produced at a constant rate one should see a whole distribution of white dwarfs along the dotted line with luminosity from $10^2 L_\odot$ downwards. For convenience let us refer to the white dwarfs with temperature greater than $10^5 °K$ as the 'ultraviolet dwarfs'. The population of stars of the same type (e.g. a specific temperature range) in a steady stream of stellar evolution tracks is inversely proportional to the evolution time of stars of this type. This is shown in Table I indicated by entries marked 'no ν'. R. Stothers (1966) has shown that there is a definite lack of population of stars in the ultraviolet dwarf stage.

What causes this discrepancy? After excluding other impossible premises like massive white dwarfs may be born dead (e.g. the stars discussed by S. Kumar, which have masses less than 0.1 $M_\odot$, etc.) we concluded that the lack of ultraviolet dwarf population is caused by an acceleration of the evolution in the ultraviolet dwarf stage due to the emission of plasma neutrinos (Chin, Chiu, and Stothers, 1966).

The nature of plasma neutrinos has been discussed extensively previously (Adams, Ruderman, and Woo 1963). Briefly speaking, in a stellar plasma the relation between photon angular frequency ω and its wave number vector k is as follows:

$$(\hbar\omega)^2 = (\hbar\omega_0)^2 + (\hbar k c)^2 \tag{1}$$

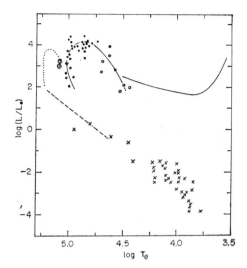

FIGURE 1 H–R diagram, adapted from Harman and Seaton. Open and filled circles represent some observed central stars of planetary nebulae. Crosses represent some observed white dwarfs. Circles with crosses represent the two evolutionary stages of our model for a central star of 0.736 $M_\odot$. The dotted line represents, schematically, the theoretical evolutionary track in the central star region. The dashed curve represents the evolutionary track of the 0.736 $M_\odot$ star through the Gap. The solid curve on the right refers to the horizontal branch and the red-giant branch of a typical globular cluster.

TABLE I Logarithms of evolutionary times from $L = 100\,L_\odot$ to luminosities indicated. Y represents helium content in the envelope. The rest is Russell's mixture.

$M/M_\odot$		$1\,L_\odot$		$3\,L_\odot$	$5\,L_\odot$	$10\,L_\odot$
		$Y = 0.9$	$Y = 0$	$Y = 0.9$	$Y = 0.9$	$Y = 0.9$
1.33	No ν	6.4	6.4	6.1	5.9	5.7
	ν	6.1	5.3	5.4	5.0	4.5
1.08	No ν	6.4	6.3	6.0	5.8	5.6
	ν	5.7	4.6	4.7	4.3	3.8
0.736	No ν	6.2	6.2	5.9	5.7	5.5
	ν	5.2	4.4	4.5	4.2	—
0.6	No ν	6.2	6.6	5.8	5.7	—
0.4	No ν	6.1	6.5	—	—	—
	ν	6.1	6.5	—	—	—

N

where ω_0 is the plasma frequency characteristic to a medium. A photon thus behaves as if it has a rest mass $\hbar\omega_0/c$. This makes it possible for a photon to decay into two neutrinos via the direct electron neutrino interaction:

$$\gamma \to \nu + \bar{\nu} \tag{2}$$

It may be remarked that the direct electron neutrino interaction is implied in the currently accepted weak interaction theory (Feymann and Gell Mann theory), but has not been observed yet.

We have computed the evolution of the proto-white dwarfs with and without the neutrino processes. Figures 2, 3, and 4 show a comparison of the neutrino and optical luminosities of stars of masses $0.405\ M_\odot$, $0.736\ M_\odot$, and $1.08\ M_\odot$. One sees that the neutrino luminosity can be as much as 1000 times greater than the optical luminosity in case of 0.7 and $1.08\ M_\odot$ stars but the effect of neutrino processes is nearly absent in the case of $0.405\ M_\odot$ stars. This is due to the different density temperature conditions. The distribution of neutrino energy production rates in the two models $0.736\ M_\odot$ and $0.405\ M_\odot$ are shown in Figure 5. The resulting calculation of the lifetime is shown in Table I, in the entries marked with 'ν'. The cooling time of a star of $0.736\ M_\odot$ is shown in Figure 6.

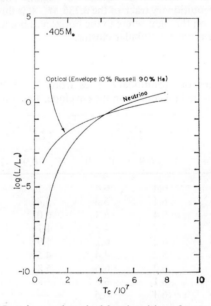

FIGURE 2   Neutrino and optical luminosities of a star of $0.405\ M_\odot$.

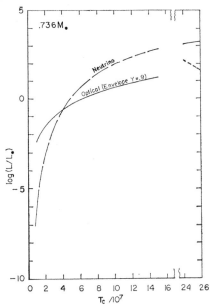

FIGURE 3   Neutrino and optical luminosities of a star of 0.736 $M_\odot$. *Y* represents helium content in the envelope.

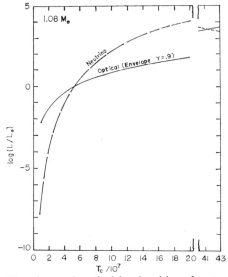

FIGURE 4   Neutrino and optical luminosities of a star of 1.08 $M_\odot$. *Y* represents helium content in the envelope.

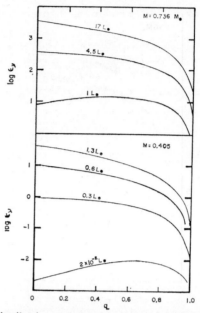

FIGURE 5 Distributions of neutrino energy loss rates in erg/g-sec as a function of mass fraction for stars at various stages through the Gap.

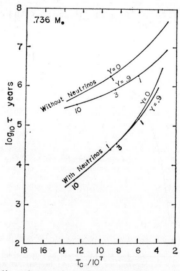

FIGURE 6 Cooling times of a star of $0.736\,M_\odot$. Y represents helium content in the envelope. Numbers along the curves indicate the optical luminosities in units of $L_\odot$ at corresponding points.

Table II summarizes the comparison with observations. The ultra-violet dwarf population is much reduced in high luminosity ends, but this will be most difficult to prove because the theoretical population of ultraviolet dwarfs (without ν) is already small. One has to introduce a cutoff in the surface temperature for the ultraviolet dwarfs. Choosing the criteria $U - V < -1.4$, we have arrived at the figure for the lifetime in Table II. We therefore conclude that the absence of high luminosity white dwarfs in the ultraviolet dwarf domain is due to plasma neutrino emission process.

TABLE II Comparison with observation.

Object	$L/L_\odot$	$M/M_\odot$	Lifetime (yrs)		
			Observa-tional	Theoretical (ν)	Theoretical (no ν)
Planetary central star	10^2–10^4	$\leqslant 1$	3–5×10^4	$\sim 10^4$	1×10^6
Ultraviolet dwarf	1–10^2	$\gtrsim 0.7$	$<5 \times 10^5$	2×10^5	2×10^6
Nova	1–10^2	~ 0.1–0.5	several 10^6	4×10^6	4×10^6
U Gem star	~ 1	~ 0.7	several 10^5	—	—
White dwarf	<1	~ 0.7	$>10^6$	$>1 \times 10^6$	$>2 \times 10^6$
		~ 0.3	$>10^6$	$>4 \times 10^6$	$>4 \times 10^6$

Discussion

KIPPENHAHN I have always had an uneasy feeling about stars which are not too massive, but their neutrino luminosity is much higher than the optical luminosity. In our calculations we find that whenever the neutrinos became important, they cooled the region in such a way that the contribution of neutrino luminosity was always very small. I think that the neutrino losses keep the luminosity small and that the evolution never brings the white dwarf to such high temperatures.

SALPETER At Cornell, we have carried out a comparison of evolutionary calculations for a few different masses with and without neutrinos, and I would say that the ratio of neutrino luminosity to optical luminosity is certainly depressed over what you get in a simple calculation. It is still true that, if you go to a massive enough star, this ratio of $L_{neutrino}$ to $L_{optical}$ gets above unity. But it is still only about 5 or so at a mass of about 0.9 $M_\odot$. Although qualitatively Dr. Chiu is right, quantitatively I would disagree with him. I am pessimistic about being able to tell, from star counts, whether there are neutrino losses or not, because we are not likely to have white dwarf masses as large as 0.9 or higher. And I want to raise a second point, which unfortunately makes that even harder. If you do not know absolute distances, but merely have to rely on a number count against apparent magnitude without knowing the distances, then unfortunately the decrease in time is counterbalanced by the fact that the neutrinos indirectly also raise the optical luminosity. And if you raise the optical luminosity by a certain factor, even though you decrease the time scale somewhat, it raises the number again just because you see to farther distances. If you do a star count just purely by accident, even for a star of about 0.9 solar mass, you will get just about the same count with and without neutrinos. But the distance for a given apparent magnitude will be different.

CHIU   Dr. Hayashi several years ago presented a very interesting calculation. He found that the moderate amount of neutrino emission changed the central conditions and his results are similar to those of Kippenhahn. However, I am talking about a white dwarf, which is already in a state of degeneracy, and Kippenhahn's comments do not apply to it.

In reply to Dr. Salpeter's question, I agree that one should really use star clusters and not use individual stars, and that is what Stothers did.

References

ADAMS, J. B., RUDERMAN, M. A., and WOO, C. H. 1963, *Phys. Rev.*, **129**, 1383.
CHANDRASEKHAR, S. 1960, *An Introduction to the Study of Stellar Structure* (Dover Press) chap. XI.
CHIN, C. W., CHIU, H. Y., and STOTHERS, R. 1966, *Ann. Phys.*, **39**, 280.
CHIU, H. Y. 1966, *Ann. Rev. Nucl. Sc.*, **16**, 591.
LEE, T. D. 1950, *Ap. J.*, **111**, 625.
SALPETER, E. E. 1961, *Ap. J.*, **134**, 669.
SEATON, M. J. 1966, *M.N.R.A.S.*, **132**, 113.
STOTHERS, R. 1966.

Formation and evolution of white dwarfs in close binary systems

R. KIPPENHAHN and A. WEIGERT

Göttingen University Observatory

Abstract

White dwarfs can be formed in certain close binary systems and it is possible to follow up the evolution of the original primary from the main sequence to the white dwarf stage by model calculations. From observation it seems that the process is fairly frequent; in some cases we are even witnessing the mass exchange leading to white dwarf formation.

I. MODEL CALCULATIONS FOR THE PROCESS

WE WILL describe and discuss here how white dwarfs can be formed in close binary systems by mass exchange between the two components. An example for this process is shown in Figures 1, 2, and 3 which are based on computations carried out by our group (Kippenhahn, Kohl, Weigert 1967). At the beginning of the computed evolution (A), two main sequence stars of $2\,M_\odot$ and $1\,M_\odot$, separated by $6.6\,R_\odot$, form a close binary system. The primary leaves the main sequence after exhaustion of its central hydrogen and increases in volume until filling its critical Roche lobe. At this moment (D), the mass exchange starts during which the more evolved star transfers most of its mass to the other component. The mass exchange in this system stops (K), when only $0.26\,M_\odot$ are left from the $2\,M_\odot$ which formed the star at the beginning. Since we have assumed conservation of total mass of the system and conservation of orbital angular momentum, the separation of the system after mass exchange is much larger than at the beginning (Figure 1), corresponding to the extreme value of the final mass ratio. When the mass loss stops, the secondary component of $0.26\,M_\odot$ is in the red giant region of the H–R diagram. The internal structure of this red giant star is rather unusual (see Figure 3). At the onset of mass exchange, the star consisted of a helium

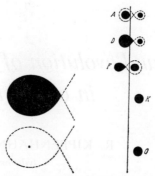

FIGURE 1   The binary system of initially $2\,M_\odot + 1\,M_\odot$ is shown here in different phases of the computed evolution from the main-sequence phase (A) through the phase of mass exchange (from D to K) to the final white dwarf stage of the original primary (O). The dashed curve gives the Roche lobe, the black areas show the volume occupied by the stars. The vertical line indicates the axis of rotation through the center of mass. (Kippenhahn, Kohl, Weigert 1967).

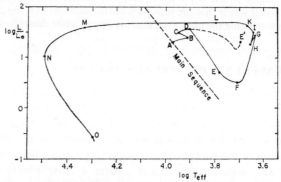

FIGURE 2   The evolutionary track in the H–R diagram is shown for the original primary in a close binary system of initially $2\,M_\odot + 1\,M_\odot$ as computed by Kippenhahn, Kohl, Weigert (1967). The letters A to O correspond to the different phases shown in Figure 1. The star starts with $2\,M_\odot$ on the main sequence (dashed line), loses $1.74\,M_\odot$ during mass exchange (points D to K) and ends up as a white dwarf of $0.26\,M_\odot$ (point O). Without mass loss, it would have evolved to point E'.

core and a hydrogen rich envelope. Almost all of this envelope has then been stripped off during mass exchange, after which the helium core is left, now surrounded only by a very small amount of hydrogen rich material. Although this remaining envelope contains only about 4 per cent of the total mass of the star, it occupies 99.7 per cent of the radius. The small helium

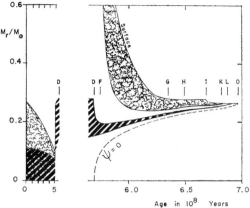

FIGURE 3   The variations are shown which occur in the interior of the original primary (initial system $2\ M_\odot + 1\ M_\odot$, see Figures 1 and 2) during its evolution. The mass M_r is plotted (from the center, $M_r = 0$, up to $M_r = 0.6\ M_\odot$) over the age in units of 10^8 years. At the beginning, the surface is far above this graph; it moves into the picture only at about $t = 6 \cdot 10^8$ years (upper heavy line), when the star has already lost an appreciable amount of mass. The hatched areas indicate regions of hydrogen burning ($\epsilon_H > 10^2$ erg g^{-1} sec^{-1}). In 'cloudy' regions, the energy is transported by convection. A line of constant degeneracy parameter ψ is dashed. The letters A to O correspond to the points on the evolutionary track in Figure 2 (Kippenhahn, Kohl, Weigert 1967).

core has already all the internal properties of a white dwarf; but due to the enormously extended and diluted envelope, the star still has the surface properties of a red giant. In the following phase of evolution, the extended envelope contracts and, in the H–R diagram, the star leaves the red giant region and moves towards the white dwarf region (from K to O in Figures 1 and 2). In Figure 3 the star's internal structure is indicated. Even in the white dwarf stage, there is still a hydrogen burning shell source between the helium core and the envelope; this hydrogen rich envelope occupies about half of the white dwarf's radius.

This process of white dwarf production will be discussed in the following. The mass exchange starts, when the primary increases its volume and fills its critical Roche lobe. For the evolution of stars which on the main sequence have convective cores, we know three phases of increasing volume during which the mass exchange can begin. These three phases are indicated in Figure 4. If the primary star starts with mass loss on part (a) of the evolutionary track, then it has not yet finished central hydrogen burning. This burning is continued after mass exchange and the resulting star is not too different from normal main-sequence stars. We, therefore, skip this case here and confine ourselves to the two cases in which white dwarfs can be formed.

The evolution described in the beginning (Figures 1–3) corresponds to case (b), where the mass loss starts after exhaustion of central hydrogen, but before the onset of helium burning in the original primary. Meanwhile, two other examples for this case have been calculated by Refsdal and Weigert (1969). Here, the initial systems were chosen as 1.8 $M_\odot$ + 0.7 $M_\odot$ separated by 5.1 $R_\odot$, and 1.4 $M_\odot$ + 1.1 $M_\odot$ separated by 3.4 $R_\odot$. In both of these systems, the evolution was found to resemble very closely that described above for the 2 $M_\odot$ + 1 $M_\odot$ system. The original primaries ended up as white dwarfs of 0.265 $M_\odot$ and 0.268 $M_\odot$ respectively.

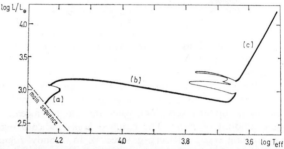

FIGURE 4   The evolutionary track of a 5 $M_\odot$ star, as computed by Kippenhahn, Thomas and Weigert (1965). The three phases in which the star's radius increases are indicated by a heavier line. Phase (a) is that of central hydrogen burning. In phase (b), the star has already finished central hydrogen burning and moves from the main sequence into the red giant region where central helium burning starts. During that burning, the evolutionary track consists of several loops, depending on the star's chemical composition and mass. That burning is finished when the star moves upwards along the Hayashi-line on part (c) of the track.

Finally, in case (c), the mass loss sets in after the completion of central helium burning but before the onset of carbon burning. This case has been investigated by D. Lauterborn at the Göttingen Observatory. He started with a system of 5 $M_\odot$ + 2 $M_\odot$, separated by 300 $R_\odot$. This distance of the components has been chosen in such a way that the primary of 5 $M_\odot$ will fill its critical Roche lobe after exhausting the helium in the center. After the mass exchange, only about 1 $M_\odot$ was left from the original primary which then consisted of a central C–O core and a He envelope, while the original hydrogen envelope was almost completely stripped off. In the most recent calculations, the star starts already to move from the red giant region to the left in the H–R diagram and will also finally become a white dwarf. The separation was increased to about 700 $R_\odot$.

II. COMPARISON WITH OBSERVATION

After we have seen that several cases of white dwarf formation in close binary systems can be calculated, the question arises whether this process occurs only in computers or whether it also occurs in nature.

The answer would be easy if we could directly observe systems which look like our computed systems after mass exchange, i.e. the primary being a main-sequence star and the secondary a white dwarf at a distance of about 50–700 $R_\odot$. Unfortunately, such systems will normally not be detected; they are too close to be separated visually and too wide to be detected as spectroscopic binaries. For the observed binaries containing white dwarfs such as the Sirius system, we do not yet dare say whether the white dwarf has been formed by our process, since these pairs are wider than those we get. We could theoretically obtain systems of larger final separation by relaxing the conditions for the change of orbits with mass exchange. Different results could be expected, for instance, in the case of highly eccentric orbits where mass transfer occurs essentially near periastron only, or for certain (arbitrarily assumed) anisotropies of mass transfer and mass loss of the whole system.

To obtain indirect conclusions, one has first to see whether there are now unevolved systems which, according to their parameters, can be expected to undergo the process in the future. In Table I, a list of detached systems is taken from Kopal (1959). With our present knowledge about the evolution of single stars and of close binaries, we can try to predict the future of these systems or—in order to use more our professional terms—we can cast their horoscopes.

In all of them, the future expansion of the faster evolving primary must finally start mass exchange. We have first to ask in which of the systems the primary will reach its Roche lobe after exhaustion of central hydrogen. To determine this for any system in Table I, we have compared the critical radius R_{cr} of the Roche lobe with the maximum radius R_{max} which can be achieved by the primary during hydrogen burning. R_{cr} is given by Kopal. For R_{max}, the product $R_1 . f$ was taken where R_1 is the observed radius of the primary. The factor f gives the maximum expansion of a single star of the same mass during hydrogen burning, relative to its zero age main-sequence radius R_0; f depends on the mass and is taken from earlier calculations (c.f. Figure 7 of Kippenhahn, Kohl, Weigert 1967). For $R_{max} < R_{cr}$, the system was assigned to case (b), otherwise to case (a) (Table I, third column). Only two systems of Kopal's Table VII-I do not appear in our Table I; they have extreme values of M_1 and are not of interest here. (We use $R_{max} = f . R_1$ instead of $R_{max} = f . R_0$; this is to allow for a possible scaling factor between the theoretical and the real initial radii. Thus, we overestimate R_{max} when the observed primary is already somewhat evolved; a system assigned here to case (a) might then really follow case (b), but not vice versa since we cannot expect $R_1 < R_0$.)

TABLE I Detached binary systems from a list of Kopal (1959; p. 484). M_1 is the mass of the primary. The last column indicates, whether mass exchange will start before (case a) or after (case b) exhaustion of central hydrogen.

System	$M_1/M_\odot$	Case
V 805 Aqu	1.85	b
σ Aqu	6.8	a
TT Aur	6.7	a
WW Aur	1.92	b
Ar Aur	2.55	b
β Aur	2.33	b
AH Cep	16.5	a
Y Cyg	17.4	a
V 477 Cyg	2.4	b
RX Her	2.1	b
TX Her	2.1	b
UV Leo	1.36	b
U Oph	5.3	a
V 451 Oph	2.3	b
AG Per	5.1	b

Secondly, we have to make sure that after mass exchange the star will immediately go into the white dwarf region. As has been shown by Kippen-hahn, Kohl, and Weigert (1967), original primaries will never achieve helium burning but rather become helium white dwarfs if they experience mass loss during part (b) of the evolutionary track (Figure 4) and if they have initial masses less than 2.8 $M_\odot$. We can see from Table I that there are enough systems observed which fulfil these two conditions.

A more direct indication that the process really takes place comes from certain semidetached systems. The semidetached binaries with total masses larger than about 3.5 $M_\odot$ can apparently be understood as being in the final stages of mass exchange which started during central hydrogen burning (Kippenhahn, Weigert 1967; Giannone, Kohl, Weigert 1968); but in semidetached systems with total masses of around 2.5 $M_\odot$, the secondaries are by far too overluminous and have radii too large for stars with central hydrogen burning after mass exchange (Refsdal and Weigert 1968). However, these low-mass systems can be explained by assuming that they are in the stage of a mass exchange which started after the exhaustion of central hydrogen in the original primary and which has already reversed the mass ratio. Indeed, such a mass transfer is rather violent only at the beginning and slows down toward the end, thus providing an appreciable chance for observing this phase. The computations carried out recently by Refsdal and Weigert at the University of Nebraska show that the secondaries in systems of $M_1 + M_2 = 2.5\ M_\odot$ have about the right overluminosities and radii (see Figure 5) during the

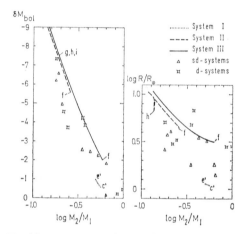

FIGURE 5 For binary systems with total masses around 2.5 $M_\odot$, the overluminosity δM_{bol} (compared with a main sequence star of the same mass) and the radius R of the now secondary star are plotted over the mass ratio. The triangles and stars correspond to observed semi-detached systems and to observed detached systems with giant second-aries respectively. The solid and dashed lines show the calculated variations of δM_{bol} and R for the now secondary components during the phase of rather slow mass exchange in two systems; later on these two stars become white dwarfs (Refsdal and Weigert 1969).

slow phase of a case (b) mass transfer. Before the calculated mass exchange definitely ceases, it is always interrupted and the system is detached for a while. During this time, the theoretical systems seem to resemble the observed detached systems with overluminous secondaries (KO Aql Systems, see Table VII-VI of Kopal). Although one has still to prove that the number of observed objects is in agreement with the time scales of the mass exchange, we feel rather sure, that at least in the semi-detached systems with total masses of about 2.5 $M_\odot$ we are now witnessing the formation of white dwarfs.

III. THE WHITE DWARF MODELS

To calculate the process of white dwarf formation in close binary systems can also be important for studying the white dwarf phase of single stars. It is a cheap way to get white dwarf models having undergone some kind of reasonable evolution which might have some similarities with the real formation of white dwarfs out of single stars. The mass exchange can be considered as a kind of artificial aging of stars. Being bound to a companion for their life seems to bring them sooner to the stellar graveyard than single ones. Also for single stars, we would expect a small hydrogen envelope to be left on the surface and in some cases hydrogen burning might still take place in the outer layers. We, therefore, have investigated some further properties of the white dwarf which we had obtained by mass exchange. We will discuss here the secular and the vibrational stability of our white dwarf model.

The white dwarf at state (O) in Figures 1, 2, and 3 has a hydrogen burning shell which is rather thin. One therefore would expect that thermal pulses can occur as encountered previously in other cases by Schwarzschild and Härm (1965) and Weigert (1966). The evolution of the white dwarf of 0.26 $M_\odot$ after state (O) has been followed by Kippenhahn, Thomas, and Weigert (1968). Indeed, two thermal pulses occur before the hydrogen shell source completely stops burning. The two pulses follow each other within 600 years. During this phase of secular instabilities, the star moves through the H–R diagram as indicated in Figure 6 by the solid line. This is a direct continuation of the evolutionary track which is given in Figure 2. After these pulses, the hydrogen burning shell source finally dies out and the evolutionary track approaches a line of constant radius for the cooling white dwarf.

D. Lauterborn investigated the vibrational stability of the white dwarf in order to find out whether the hydrogen burning shell can cause pulsations. The fundamental period of the white dwarf model (log $L/L_\odot = -0.30$, log $T_{\mathrm{eff}} = 4.33$) was found to be 65.2 sec. The calculated stability coefficient turned out to be negative, i.e. the model is vibrationally stable. This can be understood from the structure of such a model. Since we cannot expect hydrogen burning to occur in the very atmosphere, some layers of hydrogen rich matter must still exist above the shell source. These layers contained 1.3 per cent of the mass of the treated model. However, they occupy about half of the star's radius. The amplitude of pulsation will then be large only in the envelope and we have a large damping, while the amplitude is rather small at the shell source which drives the pulsation. Thus, the driving mechanism

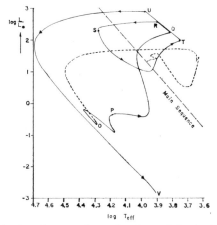

FIGURE 6 During the two thermal pulses of the hydrogen burning shell source, the 0.26 $M_\odot$ star moves in the H–R diagram along the complicated solid line starting at point O. From T to U and, in the second pulse, from Q to R, the star's radius reaches the radius of the critical Roche lobe. The evolutionary track ends in the white dwarf region (V) after the shell source has died out. The dashed line indicates the foregoing evolutionary track of this same star (compare Figure 2), from the main sequence (dotted line) to point O. (Kippenhahn, Thomas, Weigert 1968).

cannot overcome the damping since the shell source, although at a high mass value, is geometrically too deep in the interior. There is also not much hope to get similar models for a white dwarf pulsating due to an excitation from the shell source. Different models were constructed in which the shell source was brought geometrically further outwards by diminishing the mass of hydrogen in the envelope. But then the contribution of the shell source to the total luminosity becomes smaller and the driving mechanism is again reduced.

There is another point of interest. When the white dwarf later on cools down along a line of constant radius, it crosses the extrapolated cepheid strip. Corresponding models were also tested by Lauterborn for vibrational instability. Even in these white dwarfs, the driving mechanism of cepheids (kappa mechanism) is working although it does not seem that it can overcome the damping. According to Lauterborn (1968), a model of log $L/L_\odot = -1.56$ and log $T_{eff} = 4.15$ has a fundamental period of 39 sec. The growth rate of the amplitude turned out to be in the order of 10^{13} periods, i.e. a remarkable amplitude of pulsation will not be built up in reasonable time scales.

A part of the research reported here was supported by Deutsche Forschungsgemeinschaft, Schwerpunktprogramm Stellarastronomie.

References

GIANNONE, P., KOHL, K., and WEIGERT, A. 1967, *Z. f. Astrophys.*, **67**, 41.

KIPPENHAHN, R., THOMAS, H.-C., and WEIGERT, A. 1965, *Z. f. Astrophys.*, **61**, 241.

KIPPENHAHN, R., and WEIGRET, A. 1967, *Z. f. Astrophys.*, **65**, 251.

KIPPENHAHN, R., KOHL, K., and WEIGERT, A. 1967, *Z. f. Astrophys.*, **66**, 58.

KIPPENHAHN, R., THOMAS, H.-C., and WEIGRET, A. 1968, *Z. f. Astrophys.*, **69**, 265.

KOPAL, Z., *Close Binary Systems.* (London, Chapman and Hall, 1959).

LAUTERBORN, D. 1968, to be published *in Astronomy and Astrophys.*

REFSDAL, S., and WEIGRET, A. 1969, to be published.

SCHWARZSCHILD, M., and HÄRM, R. 1965, *Ap. J.*, **142**, 855.

WEIGERT, A. 1966, *Z. f. Astrophys.*, **64**, 395.

Discussion

SALPETER   I have a couple of questions, one of them on the observational side. In the globular cluster diagram, I presume if you start off with two stars which are not very different in mass, and both reasonably close to the main-sequence turn-off point, what you would then predict from your mechanism is to see some things which are somewhere near the main sequence but above where the turn-off should be. I remember there are some such observed, but I am curious whether you have tried to make estimates of numbers that you expect and how they compare with observations.

KIPPENHAHN   No, we have not. There is already a classical paper about this by McCrea, who suggested these processes. The difficulty is that one has to assume quite a number of close binaries and, as far as I know, the globular clusters are deficient in eclipsing binaries.

SALPETER   My second question is: Does this process go on a second time, because the star which is now more massive will eventually also become a red giant. Is it too far separated to put mass back?

KIPPENHAHN   In this case the system is widely separated and we do not know all the evolution of the other star. But mass exchange may take place more than once!

GINGERICH   Your model may work for Sirius for two reasons, one of which is that Strom and I have independently shown that Sirius has a much higher abundance of metals compared to Vega. Although this work has been criticized by Warner, I think our rebuttal to him still leaves Sirius over-populated with metals. The other thing is that your runaway burning process is the first time I have seen an evolutionary process that is fast enough to explain Ptolemy's observations of Sirius as being a red star along with Antares and Aldebaran.

KRAFT   I just want to say that your very calculations show why you do not find any eclipsing binaries in the globular clusters, for they would have disappeared already.

COX   I do not understand exactly what your reasons were for saying that your models turned out to be pulsationally stable. Was it that the amplitude was too small in the hydrogen burning regions?

KIPPENHAHN   Yes.

On a possible explanation of novae and U Geminorum systems

ANNIE BAGLIN and EVRY SCHATZMAN

Institut d'Astrophysique, Paris, France

NOVAE AND U Geminorum stars seem to have a tendency to appear in short-period binaries, and, in some cases, pulsations with very short periods of the blue component have been seen. The red component seems to have a very large radius and even fills the lobe of the inner Lagrangian surface. Schatzman's theory of novae attributes the explosion to a resonance between the orbital motion and an intrinsic mode of pulsation of one of the components. We would like to show here that, in terms of gravity modes, both the resonance and the short-period pulsation can be found at least in some of these systems.

We shall study here the well known Nova DQ Herculis system which has been extensively studied by Walker (1956, 1958, 1961) and by Kraft (1959, 1964) and for which many parameters are known from observations; they are listed in Table I. Our aims are to find the 71 second oscillation in the white dwarf and its instability and, at the same time, a longer period of the 4^h and 34^m responsible for the resonance.

I. THE SHORT PERIOD IS THE PERIOD OF A RADIAL MODE

It has been suggested by Kraft that the 71^s period corresponds to the fundamental mode of a dense object. This led him to the conclusion that the mass of the white dwarf was approximately 0.12 $M_\odot$. However, there exists another possibility for such a period. For more massive white dwarfs the general relativity correction to the classical hydrostatic equilibrium has to be taken into account. This shortens the frequency which has a maximum around 1 $M_\odot$, so that a 1.1 $M_\odot$ star composed of carbon almost reaching the instability limit (Baglin 1966b) has a 71^s period fundamental radial pulsation. This large mass seems very unlikely as it would lead to a 1.8 $M_\odot$ red companion which would have a luminosity large enough to be seen. And it is not

TABLE I   The nova DQ Herculis system.

Mass ratio	0.6
Period of the Blue Star	71^s
Period of the orbital motion	$4^h 34^m$
Radius of the Lagrangian surface	2.4×10^{10} cm
Magnitude of the red companion	$M_v > 9$
Magnitude of the blue companion	$M_v = 8.5$

observed. The relation between temperature, mass and luminosity for a white dwarf can be written in the following form:

$$\frac{T}{T_0} = \left(\frac{L}{M}\right)^\alpha \tag{1}$$

where L and M are in solar units. The values of T_0 and α depend on the chemical composition of the atmosphere and on the central temperature. For $T_c < 10^7 °K$, Schatzman (1952) gives log $T_0 = 7.67$, $\alpha = 0.349$ for pure hydrogen, and for 10 per cent Russell mixture log $T_0 = 7.6$, $\alpha = 0.32$. For $T_c > 10^7 °K$, diffusion is the main opacity process and log $T_0 = 7.6$, $\alpha = 0.5$.

The observed luminosity is:

$$\log \frac{L}{L_\odot} = -1.6 \tag{2}$$

and a value of 0.12 $M_\odot$ for the mass gives a central temperature around 1.6×10^7 degrees. This object will very likely have a hydrogen burning shell in the envelope or at the limit between the degenerate and non-degenerate region. But, as soon as a burning shell is present, the outer layers are very extended due to the high temperature of the hydrogen burning shell, and the radiative damping cancels the destabilizing effect of the nuclear energy production. This is due to the very rapid increase of the amplitude of the fundamental radial mode in the very tenuous and extended outer layers. So that the oscillation of 71^s will certainly not show up. On the other hand it seems to be completely impossible to find a resonance with the orbital motion in the system.

One can think of having long-period non-radial gravity modes in the white dwarf. As Cowling (1941) pointed out, the frequencies of the g modes are proportional to the difference between the actual and the adiabatic gradient when this difference is small. This is the case in isothermal white dwarfs and, the $T = 0$ degenerate matter being isentropic, a cold white dwarf would have g modes of zero frequency. The frequency of the g_1 mode for $l = 2$ have been computed for white dwarfs of various central temperatures and densities. At $T_c = 1.5 \times 10^7 °K$ the period is of the order of 100^s for 0.1 $M_\odot$.

Instead of assuming an isothermal degenerate core one can think of a highly luminous object which has evolved more or less adiabatically if the evolutionary time scale is short ($\tau < 10^7$ years) as pointed out by Baglin (1966). These models with an adiabatic core have also been checked. The g_1 mode is slower than in the isothermal case but the period is still of the order of 20^m. If the red companion has a mass of 0.2 $M_\odot$ and is fully convective, then the period of its fundamental radial oscillation is of the order of 15^m. As most of the star is adiabatic except the atmospheric layers the g modes are very slow. The frequency is given by the radiative envelope only. Using the models computed by Ezer and Cameron (1967) for a 0.2 $M_\odot$ star on the main sequence, one gets an estimate of $Pg_1 = 4 \times 10^7$ seconds. High order harmonics could have an even longer period.

II. THE SHORT PERIOD IS A PERIOD OF A g_1 MODE

The order of magnitude of the g_1 mode of an isothermal white dwarf of a central temperature around 10^7 degrees suggests that a 71^s period oscillation could be searched for in the non-radial modes. The frequency of the g modes depends strongly on the total mass and on the temperature of the object as shown in Table II. Under this hypothesis, we have two conditions for determining the model. The first one is

$$P_{g_1} = 71 \text{ seconds} \tag{3}$$

and the second is relation (1) in which the observed luminosity given by (2) is introduced. Both curves are plotted in a log T–log M diagram (Figure 1). They intersect at

$$\log T_c = 7.06$$

$$\log \frac{M}{M_\odot} = -0.64$$

The eigenfunctions of these dense models have not yet been computed, but referring to similar results obtained for polytropes, it seems very likely that the ratio of the amplitude at the surface to the amplitude at a reference point in the interior is much smaller for the g_1 mode than for the radial fundamental mode. This makes the vibrational instability of the g_1 modes most probable.

This hypothesis leads to a new system for Nova DQ Herculis: a 0.2 $M_\odot$ white dwarf given by the condition $P_{g_1} = 71^s$, and a 0.38 $M_\odot$ red companion given by the observed mass ratio. We have to find in this system some period around 4, 5 hours to account for the resonance. The white dwarf

TABLE II   Values of σ_g for different values of T_c and M.

Log $M/M_\odot$	T_c				
	10^6	5×10^6	10^7	1.5×10^7	2×10^7
− 0.02	0.0255	0.0758	0.116	0.162	0.200
− 0.10	0.0181	0.0621	0.102	0.138	0.182
− 0.35	0.0180	0.0547	0.0945	0.123	0.149
− 0.52	0.0176	0.0514	0.0894	0.121	0.146
− 0.84	0.0174	0.0501	0.0830	0.114	0.126

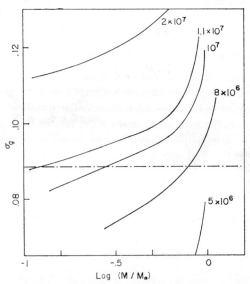

FIGURE 1   Values of σ_{g_1} as a function of M for different central temperatures.

cannot be pulsating with such a period. Neither the fundamental nor the gravity mode changes so rapidly from 0.12 $M_\odot$ to 0.23 $M_\odot$ (see Table II). The red companion, if it lies on the main sequence, has already developed a radiative region containing approximately 25% to 30% of its total mass as given by the models computed by Ezer and Cameron (1967). The fundamental radial period is around 3^m. The g_1 mode frequency of such a model is given essentially by its radiative zone.

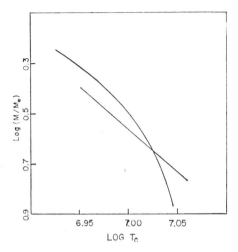

FIGURE 2 Determination of T_c and M for the white dwarf. Curved line is given by $\sigma_{g_1} = 2\pi/71$ and the straight line by equation (1).

In a rough approximation, using a polytropic density distribution in the radiative zone, and an exactly adiabatic convective zone, the frequency is given by

$$\sigma_{g_1} = \frac{GM}{R^3}\frac{3}{4\pi}\left(\frac{\rho_c}{\rho}\right)^3\left(\frac{1}{\gamma^*}-\frac{1}{\Gamma}\right)\left(\frac{r}{R}\right)_c^7\frac{1}{7K\alpha},\tag{4}$$

where

M is the total mass,

R the radius,

ρ_c the central density,

$\bar{\rho}$ the mean density,

$\gamma^* = \dfrac{d\log P}{d\log\rho}$ in the star,

$\Gamma = \dfrac{d\log P}{d\log\rho}$ adiabatic,

r is the distance to the center.

The index c means that the value is taken at the boundary of the convective region. K is the constant defined by $I = KMR^2$, where I is the moment of inertia. α is a constant given by $T_c = (\mu/R)(GM/R)$; here $\alpha = 0.4$.

Using the values corresponding to the models of Cameron and Ezer (1967) in (4), one gets

$$P_{g^*} = \frac{2\pi}{\sigma_{g_r}} = 14{,}400 \text{ sec} = 4\text{h}$$

These preliminary estimates suggest strongly that g_1 modes are responsible for both a short-period oscillation and the resonance in Nova DQ Herculis system. It is tempting to generalize this model to at least a part of the novae and U Germinorum systems.

This model has three interesting properties. (1) g_1 modes are very likely to be excited in close binaries, and novae and U Gem seem to always occur in such systems. (2) The explosion would occur in the red companion. Krzeminski (1965) showed that it does in U Gem. Some recent unpublished observations of SS Cygni by Walker seem to show the reverse situation. And this might suggest that all the systems do not behave the same way; this preliminary work shows that an important factor is the mass of the stars. This type of resonance between the g_1 mode exists only if the mass is smaller than 0.6 $M_\odot$; and we know at least one nova system TCrB in which the masses are much higher. (3) The explosion of the nova or the flashes in U Gem stars have a very pronounced non-spherical shape. Being due to non-radial oscillations, it seems very likely that the accumulation of nuclear fuel necessary to trigger the explosion will be non-spherically symmetric similar to the observed explosion.

References

BAGLIN, A., 1966, *Ann. d'Ap.*, **30**, 185.
BAGLIN, A., 1966b, *Ann. d'Ap.*, **29**, 103.
COWLING, T. G., 1941, *M. N.*, **101**, 367.
EZER, D., and CAMERON, A. G. W., 1967, NASA preprint.
KRAFT, R. P., 1959, *Ap. J.*, **130**, 110.
KRAFT, R. P., 1964, *Ap. J.*, **139**, 457.
KRZEMINSKI, W., 1965, *Ap. J.*, **142**, 1051.
SCHATZMAN, E., 1952, *Ann. d'Ap.*, **15**, 361.
WALKER, M. F., 1956, *Ap. J.*, **123**, 68.
WALKER, M. F., 1958, *Ap. J.*, **127**, 319.
WALKER, M. F., 1961, *Ap. J.*, **134**, 171

Discussion

OSTRIKER   I just want to say a word in defense of Kraft's model for DQ Herculis. The 71 second period, if you try to derive it from a zero temperature white dwarf, would imply a very low-mass white dwarf. The natural periods of it are not those of a zero temperature white dwarf. And correspondingly I would imagine that, for a moderate mass, you can have a degenerate interior with a non-degenerate envelope, and have a period of 71 seconds. I do not think you need an instability to excite this oscillation because the system underwent an explosion in this century and the radiative damping times are fairly long. So whatever set the system in motion in the past can be just dying out slowly at present.

The outer convection zone of low-temperature white dwarfs

K. H. BÖHM

Department of Astronomy, University of Washington

Abstract

The structure of the convection zone in the outer (nondegenerate) layers of two different models of van Maanen 2 has been studied using the modified form of the mixing length theory as suggested by Böhm and Stückl (1967). It is found that van Maanen 2 has an outer convection zone having a thickness of 8–22 km (essentially depending on the adopted effective temperature). The temperature at the lower boundary of the convection zone is of the order of $10^6 {}^\circ$K. The cooling time of a cool white dwarf of a given effective temperature is decreased considerably by the presence of an outer convection zone.

THOUGH IT HAS been known for a long time that the cooler white dwarfs can have outer convection zones (Schatzman 1958), there have been only a few studies of the structure of these convection zones. I should mention here the work by Myerscough (1966) who found that van Maanen 2 has a convection zone reaching the deeper layers of the atmosphere of the star. On the other hand, Miss Myerscough was mainly interested in the atmospheric parameters of van Maanen 2 and she did not study the structure of the whole convection zone, its thickness, and its influence on the thermal properties of the white dwarf.

A study of the complete outer convection zone of the cooler white dwarfs is of interest for the following reason: for a white dwarf of a given luminosity the typical cooling time depends essentially on the energy contents of the nondegenerate ions in its degenerate core. This energy content is proportional to the temperature of the core. Since the degenerate core can be considered as approximately isothermal (Mestel 1965) the core temperature is approximately determined by the temperature of the transition point from nondegenerate to degenerate matter (Schwarzschild 1958), which can be found by an inward integration starting in the atmosphere. Usually this integration is done assuming radiative equilibrium (Schwarzschild 1958, Mestel 1965). One gets a lower transition temperature if a sizeable fraction of the thin nondegenerate shell is in convective equilibrium. The lowering of the transition temperature leads to a proportionate lowering of the free energy content of the white dwarf and therefore finally to a decrease in the remaining cooling time.

We have calculated models of the outer hydrogen convection zones for two possible sets of atmospheric parameters of van Maanen 2 as given by V. Weidemann (1960).

The parameters are:

(1) $T_{eff} = 5790°K$, $g = 10^8$ cm sec^{-2} (Weidemann's 'standard' model)

(2) $T_{eff} = 5040°K$, $g = 3.16 \times 10^7$ cm sec^{-2}.

We have mainly used Weidemann's data for the chemical composition of the atmosphere, assuming a hydrogen abundance (by numbers of atoms) of about 2%, a helium abundance of 97.8% and normal C, N, O and Ne abundances. Following Weidemann (1960), we have used metal abundances which are lower by a factor 8×10^3 than in the sun. For this composition, we have carried out a new computation of the opacity as a function of T and ρ for the upper layers of the convection zone. For the deeper layers ($T > 1.5 \times 10^4°K$) the tables by A. N. Cox and J. N. Stewart (1965) have been used.

The stratification of the convection zone has been calculated using a slight modification (see Böhm and Stückl 1967) of the standard mixing length theory (Böhm–Vitense 1958). In this modification the mixing length is assumed to be equal to the pressure-scale height in general, but we propose that it should never be larger than the distance from the nearest boundary of the convection zone. The numerical procedure used was essentially the same as described by Böhm and Stückl (1967).

The computation gives a thickness of about 8 km (see Figures 1 and 2) for the model with $T_{eff} = 5790°K$, $g = 10^8$ cm. sec^{-2} and a thickness of about 22 km for the model of $T_{eff} = 5040°K$, $g = 3.16 \times 10^7$ cm. sec^{-2}. For both models the transition layer between the fully radiative atmosphere (with $F = F_{rad}$) and the fully convective layers (with $F \approx F_{conv}$) is of the order of 10 meters. The temperature at the lower boundary of the convection zone is of $1 \times 10^6°K$ for both models and is close to the transition temperature.

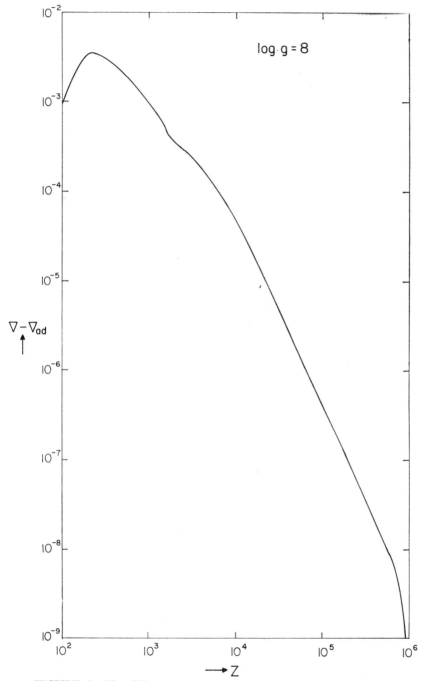

FIGURE 1 The difference between the actual gradient $\nabla = (P/T)$ (dT/dP) and the adiabatic gradient ∇_{ad} as a function of the geometrical depth z (in cm) for the convection zone in the model with $T_{eff} = 5790°K$; $g = 10^8$ cm sec^{-2}.

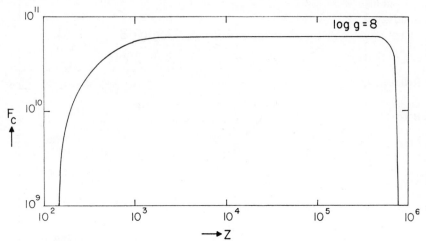

FIGURE 2   The convective energy flux F_c (in cgs units) as a function of z for the same model as in Figure 1.

The reduction of the core temperature (and consequently of the cooling time) due to the presence of the outer convection zone is about a factor 4.

The maximum values of the convective velocities are 83 m/sec for the higher (see Figure 3) and 62 m/sec for the lower temperature model.

I should like to thank Miss E. Stückl for considerable assistance in doing the computation. A detailed description of these results will be published in *Astrophysics and Space Science*.

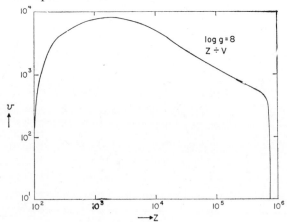

FIGURE 3   The convective velocity (in cm sec^{-1}) as a function of z for the same model as in Figure 1.

I am also grateful to Miss Myerscough for informing me about her results in advance of publication.

This study has been supported in part by NSF Grant No. GP-4984.

References

BÖHM, K. H., and SÜTCKL, E. 1967, *Zs. f. Ap.*, **66**, 487.
BÖHM-VITENSE, E. 1958, *Zs. f. Ap.*, **46**, 108.
COX, A. N., and STEWART, J. N. 1965, *Ap. J. Suppl.* **11**, 22.
MESTAL, L. 1965, The Theory of White Dwarfs, *Stars and Stellar Systems*, ed. L. H. ALLER and D. B. McLAUGHLIN (Chicago: University of Chicago Press,) **8**, 297.
MYERSCOUGH, V. 1966, unpublished
SCHATZMAN, E. 1958, *White Dwarfs* (Amsterdam, North Holland Publishing Co.)
SCHWARZCHILD, M. 1958, *Structure and Evolution of the Stars* (Princeton University Press, Princeton, N.J.).
WEIDEMANN, V. 1960, *Ap. J.*, **131**, 638.

Section V

o

Rapid contraction and flare-up of
protostars

C. HAYASHI, T. NAKANO, S. NARITA

Department of Physics, Kyoto University

N. OHYAMA

Department of Physics, Hiroshima University

Abstract

The rapid contraction phase of evolution, which precedes the pre-main-sequence phase of quasi-hydrostatic equilibrium, is investigated for an opaque protostar of one solar mass by solving the hydrodynamic equations of motion. The protostar is assumed to retain spherical symmetry. The results of computations show that the star flares up from a luminosity as low as $10^{-3} L_\odot$ to a peak luminosity $2 \times 10^3 L_\odot$ when the shock wave, which starts from the center, reaches sufficiently near the surface. Finally, comparison is made with infrared objects which have recently been discovered.

I. INTRODUCTION

IN THE PRE-MAIN-SEQUENCE STAGE of evolution, stars cannot be in quasi-hydrostatic equilibrium when the surface temperature is below about 3000°K. Thus, in the H–R diagram the stars in equilibrium are considered to evolve downwards nearly vertically. The initial highest luminosity depends on the stellar mass and it is estimated to be as high as 300 solar luminosities for a star of one solar mass. A review of recent works on this phase of evolution was given by Hayashi (1966).

However, it can be said that the stellar structure at the initial stage mentioned above has not yet been fully clarified and, consequently, there still remain some questions on the subsequent evolution. The reason for this is that the initial stage of the quasi-hydrostatic equilibrium may depend on the previous history of evolution, which has proceeded in the so-called forbidden region of quasi-hydrostatic equilibrium. Thus, we have to study the dynamic contraction stage of protostars and this has been of our great concern since the discovery of the forbidden region in 1961 (Hayashi, 1961).

In a previous work (Hayashi and Nakano, 1965), we have examined the dynamical and the thermal behavior of a protostar of one solar mass, which have once been formed from the interstellar medium. We have assumed for simplicity that the protostar retains both spherical symmetry and mass and that the structure is simply represented by the mean temperature $\bar{T}$ and the mean density $\bar{\rho}$ (or the mean radius). We have examined the time variations of $\bar{\rho}$ and $\bar{T}$ of a transparent protostar by comparing the time-scales of contraction, expansion, cooling and heating. As the cooling processes, we have taken into account the emission of radiation by H_2 molecules, heavy ions and grains of interstellar origin. As the number density of the grains we adopted $n_g = 10^{-13} n_H$ where n_H is the number density of hydrogen atoms.

The result of the above study is that, (1) the protostar is destined to contract continually if $\bar{\rho} > 10^{-18}$ g/cm^3 (or $R < 0.1$ light year) and (2) the protostar begins to be opaque (the optical thickness being unity) when

$$\bar{\rho} \simeq 10^{-13} \text{ g/cm}^3 \ (R \simeq 3 \times 10^4 \, R_\odot), \bar{T} \simeq 15^\circ \text{K}$$

$$\text{and } L \simeq 0.2 \, L_\odot \text{ for } M = M_\odot. \quad (1)$$

This initial state of the opaque protostar has been determined mainly by the cooling effect of grains and it is almost independent of its previous history. The pressure corresponding to the above density and temperature is too low by a factor of more than 20 to maintain gravitational equilibrium and, consequently, the star is in a state of free fall. After 100 years of free fall from this state, the star becomes sufficiently opaque and we have

$$\bar{\rho} = 10^{-11} \text{ g/cm}^3, \bar{T} = 100^\circ \text{K}, R = 5000 \, R_\odot \text{ and } L = 10^{-2} \sim 10^{-3} L_\odot. \quad (2)$$

We have adopted this stage as the beginning of the detailed numerical computations.

The qualitative feature of the evolution after the stage (described by Eq. (2)) has already been studied by Hayashi and Nakano (1965). The result is that, if the star has a density distribution such that the density decreases outwards, the inner mass shells contract faster and the pressure gradient near the center increases more and more. Then this pressure force becomes greater than the gravity force, the central region begins to bounce and this gives rise to a shock wave which propagates outwards. Thus, the star is expected to flare up suddenly when the shock wave approaches sufficiently near the surface. After this flare-up the star will finally settle down to the quasi-hydrostatic equilibrium. The purpose of the detailed computations is to find the quantitative features of the flare-up process and the subsequent evolution to the initial stage of the quasi-hydrostatic equilibrium, and to compare the results with a flared-up star, FU Ori, discovered by Herbig (1966) and also with a number of infrared objects recently discovered by many people.

In order to start the numerical computations, we have to specify the initial distributions of density, temperature and velocity as functions of the mass fraction $q \equiv m/M$. As the velocity distribution we have adopted the velocity of free fall from infinity. For the density distribution we have considered the two cases; one is the Emden solution of polytropic index $N = 1.5$ (relatively uniform distribution) and the other is that of an index 4.0 (centrally condensed distribution). For the temperature distribution, we have considered an isothermal case, an isentropic case and also an intermediate case.

Since the time of dynamical evolution is short, the energy flow by radiation is negligible in the bulk of the star, except for the surface region containing about one-tenth of the total mass. Then, we have first computed the evolution under the complete neglect of the radiation flow in order to see the difference in evolution due to the difference in the initial distributions of the density and temperature (Nakano, Ohyama and Hayashi 1968).

The neglect of radiation flow does not mean an adiabatic change since the dissipation of kinetic energy into thermal energy has been included in the hydrodynamic equation in terms of von Neumann and Richtmyer's artificial viscosity. Further, the effects of the dissociation of hydrogen molecules and the ionization of hydrogen atoms have been fully taken into account. The results of the computations show that the final equilibrium structure is quite insensitive to the initial distributions of the density and temperature.

Next, we have computed the evolution including the effect of radiation flow (Narita, Nakano and Hayashi 1968). We have found that the evolution is very similar to the case without the radiation flow, except for the surface region which determines the magnitude of the luminosity. The flare-up occurs from $L = 10^{-2} \sim 10^{-3} L_\odot$ to a peak luminosity $2 \times 10^3 L_\odot$ in a time as short as $1 \sim 10$ days and afterwards the luminosity decreases gradually. At present the computation has not been completed in a sense that the surface region has not yet attained the state of quasi-hydrostatic equilibrium. However, as far as the present results indicate, the luminosity and the effective temperature after the flare-up are in agreement with the observations of infrared objects.

II. EVOLUTION WITHOUT RADIATION

The fundamental equations to be solved numerically are as follows. First, in the Langrangian scheme where t (time) and m (mass contained in a sphere of radius r) are independent variables, the equations of motion are given by

$$\frac{\partial u}{\partial t} = -4\pi r^2 \frac{\partial}{\partial m}(P+Q) - \frac{Gm}{r^2}, \frac{\partial r}{\partial t} = u, \qquad (3)$$

$$\frac{1}{\rho} = \frac{\partial}{\partial m}\left(\frac{4\pi}{3}r^3\right), \qquad (4)$$

where Q is von Neumann and Richtmyer's artificial viscosity which is expressed as

$$Q = \begin{cases} \rho l^2 (\partial u/\partial r)^2 & \text{if} \quad \partial u/\partial r < 0 \quad \text{and} \quad \partial \rho/\partial t > 0, \\ 0 & \text{otherwise}, \end{cases} \qquad (5)$$

where l is a constant which is taken as nearly equal to the thickness of each shell when the star is divided into a large number of shells. Second, the equation of energy conservation is given by

$$\frac{\partial \epsilon}{\partial t} + (P+Q)\frac{\partial}{\partial t}\left(\frac{1}{\rho}\right) = -\frac{\partial}{\partial m}(4\pi r^2 F), \qquad (6)$$

where ϵ is the internal energy per unit mass and F is the outward energy flux which has been neglected in the present case. Taking into account the molecular dissociation and the atomic ionization of hydrogen, the expressions of P and ϵ as functions of ρ and T have been obtained.

The above Eqs. (3) to (6), which determine completely the time variations of r, u, ρ and T, have been transformed into difference equations which describe the evolution of each shell of the star. The computations have been made for four different cases of the initial distributions and the chemical composition. The adopted initial distributions and the results of computations for the onset stage of gravitational equilibrium are summarized in Tables I and II.

TABLE I   Initial distributions of the density and temperature.

Case	Composition	Density distribution	Temperature distribution	Central density (g/cm^3)	Central temp. (°K)
1	H only	$N = 1.5$	$T \propto \rho^{2/5}$	5×10^{-11}	200
2	H only	$N = 1.5$	$T = \text{constant}$	5×10^{-11}	100
3	H only	$N = 4.0$	$T \propto \rho^{1/5}$	5×10^{-9}	100
4	H (0.7) He (0.3)	$N = 4.0$	$T \propto \rho^{1/5}$	5×10^{-9}	100

The adopted distributions of the initial density are illustrated in Figure 1 for the cases $N = 1.5$ and 4. The time variations of the radii of typical shells from the initial stage to the final onset stage of gravitation equilibrium are shown in Figure 2 for case 1. The path of the shock front is denoted by the dashed curve. The velocity of this front is about a few km/sec. The distribution of the entropy (per one atomic mass) at five different stages are shown in Figure 3, which indicates the way of entropy increase by the passage of the

TABLE II Structure at the onset of gravitational equilibrium.

Case	Mean polytropic index	Stellar radius ($R_\odot$)	Central density (g/cm³)	Central temp. (°K)	Ejected mass ($M_\odot$)
1	4.1	170	5.1×10^{-2}	1.0×10^5	$<10^{-3}$
2	4.1	170	6.3×10^{-2}	1.0×10^5	$<10^{-3}$
3	4.3	200	4.8×10^{-3}	9.9×10^4	10^{-4}
4	4.6	230	3.2×10^{-3}	1.2×10^5	10^{-5}

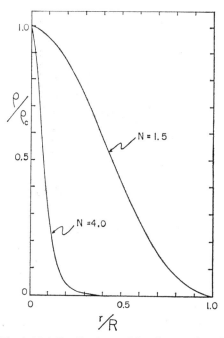

FIGURE 1 The initial distributions of density as a function of radius, which are represented by the Emden functions of index $N - 1.5$ and 4.0.

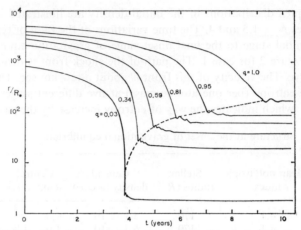

FIGURE 2 The time variation of the radii of typical shells from the initial stage to the onset stage of gravitational equilibrium (for case 1 without radiation flow). The fractional masses m/M of the shells are denoted by q.

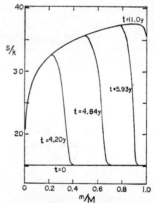

FIGURE 3 The distributions of the entropy S (per atomic mass) at four stages of case 1 without radiation flow.

shock wave. The center bounces at the time, $t = 3.7$ years (cases 1 and 2) or $t = 0.45$ years (cases 3 and 4), and finally the surface bounces at the time, $t = 10$ years, for all the cases when the shock wave reaches the surface.

The distributions of temperature, density and entropy at the onset stage of quasi-hydrostatic equilibrium are shown in Figures 4, 5 and 6. It is to be noticed that among the cases of pure hydrogen (cases 1, 2 and 3) the differences are very small in spite of considerably great differences in the initial

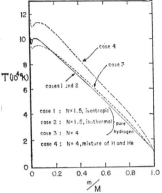

FIGURE 4   The temperature distributions at the onset stage of gravitational equilibrium for the cases without radiation flow.

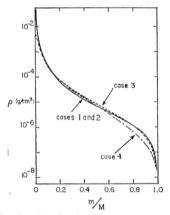

FIGURE 5   The density distributions at the same stage as in Figure 4.

distributions. This will be due mainly to the fact that the contraction factor of each shell (i.e. the ratio of the initial radius to the final radius) is so large that the kinetic energy or the gravitational energy of each shell, which has been finally dissipated into the thermal energy, is much greater than any of the thermal energy, the kinetic energy and the gravitational energy in the initial stage. It is to be noticed that in the final stage the entropy is low in the central region and thus the star has a nearly isothermal and dense core and a relatively extensive envelope (the stellar radius being about 200 $R_\odot$). The stellar radius will become smaller if the effect of energy flow in the envelope will be taken into account. The difference in the final structure between the

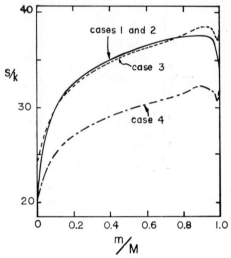

FIGURE 6   The entropy distribution at the same stage as in Figure 4.

case of pure hydrogen and the case of the mixture of hydrogen and helium is due to the fact that more energy is consumed in the ionization of helium in the latter case.

III. EFFECT OF RADIATION FLOW

In order to know the changes in the luminosity and the effective temperature (or more precisely, the spectrum of radiation) during the rapid contraction phase considered above, we have to take into account the energy flow term in Eq. (6). Since the envelope of the star is, in general, very tenuous and extensive, the change of the radius is appreciable in a shell whose optical thickness is unity. Thus, for the tenuous envelope we have to consider the spherical effect in the theory of radiative transfer. Namely, we have to calculate the directional radiation intensity I, which propagates in the direction s, using the equation.

$$\frac{dI}{ds} = \kappa \rho (B - I), \quad B = \frac{acT^4}{4\pi},$$
(7)

where κ is the mass absorption coefficient and $\kappa \rho B$ is the emissivity of the gas. The net outward flux F is given by the integral over all the solid angle

$$F = \int I(r, \theta) \cos \theta \, d\omega,$$
(8)

where $I(r, \theta)$ is the angular distribution of the intensity on a spherical surface of radius r.

In the inner region where the optical depth is sufficiently large, Eqs. (7) and (8) reduce to the formula

$$F = -\frac{4\pi}{3\kappa\rho}\frac{\mathrm{d}B}{\mathrm{d}r}.\tag{9}$$

We have calculated the flux F in the two different ways mentioned above. The interface of the outer and inner regions has been chosen in such a way that the difference in the calculated flux at this radius is within a certain accuracy limit. The value of F found in this way is to be inserted into Eq. (6).

The opacity κ has been expressed as a function of T and ρ. It is very sensitive to the temperature. Below 150°K, the effect of ice grains predominates and κ is about 1 cm²/g (Gaustad 1963; Hayashi and Nakano 1965). Between 150°K and 1200°K, mineral grains are important and κ is about 0.1 cm²/g. Above 1200°K the grains have been evaporated and κ takes a minimum value as small as 10^{-4} cm²/g at about 2000°K (Vardya 1964; Auman and Bodenheimer 1967) and it becomes large again at higher temperatures (Cox 1965). We have used the opacity values given by the authors referred to above. In some cases where the opacity values are different according to the authors, we have chosen some appropriate values. The opacity values used in our computations are illustrated in Figure 7.

The computations have been made with the same initial distributions of the density and temperature as case 4 (in the case without radiation flow) except for a minor change of the chemical composition (in the present case we have adopted $X = 0.70$, $Y = 0.28$ and $Z = 0.02$).

The time variations of the radius and the temperature of various shells are shown in Figures 8 and 9, and the variation of the luminosity is shown in Figure 10.* The time variations of the radius and the temperature are not different from the case without radiation flow, except for the surface region which contains about 5 per cent of the total mass. The essential difference is that at a time of about 2 years the propagation of the shock wave proceeds to such an extent that the optical depth of the front becomes suddenly as small as 10 and, consequently, the diffusion time of photons through the outer envelope becomes very small. Thus, the temperature of the outer envelope (containing about 4 per cent of mass) rises suddenly and this is accompanied by a sudden flare-up of the star to a luminosity as high as $2 \times 10^3 \, L_\odot$. The luminosity before the flare-up is as low as $10^{-3} \, L_\odot$ (cf. the luminosity at the initial stage of the opaque protostar is $0.2 \, L_\odot$) and this low luminosity is due to the low emissivity of the surface region. The rapid change of the luminosity

*The periodical changes in these diagrams, especially the periodic light variation, do not represent the real behavior. They are due to the mathematical procedure that the number of divisions is finite.

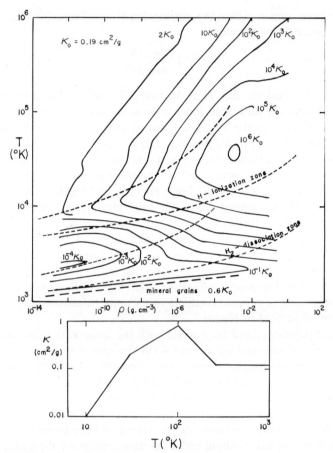

FIGURE 7 The mass absorption coefficient as a function of the density and temperature for the composition of population I stars.

by a factor as large as 10^6 occurs in a time of the order of 10 days. The computations have been continued for a period of 2.5 years after the flare-up. The result shows that the luminosity decreases very gradually.

Since the surface region where the optical depth is less than unity is very extensive, it is difficult to define accurately the effective temperature of the star. Considering this uncertainty, the evolutionary path in the HE diagram is shown by the shaded region in Figure 11. The rightward boundary of the shaded region corresponds to the temperature of a shell having the optical depth 2/3, while the leftward boundary corresponds to a region of somewhat higher temperature.

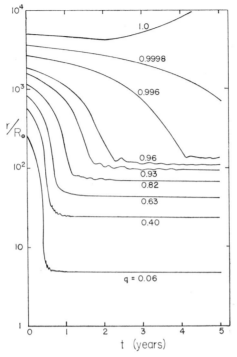

FIGURE 8   The time variation of the radii of typical shells in the case with radiation flow.

The computations have also been made for another case where the initial density distribution is represented by a polytropic index $N = 1.5$ instead of $N = 4.0$ and the other conditions are unchanged. The results are very similar to the case of $N = 4$. The flare-up occurs at $t = 6.7$ years (instead of 2.0 years in the case of $N = 4$) from a luminosity $10^{-2} L_\odot$ to a peak luminosity $2 \times 10^3 L_\odot$. Thus, the value of the peak luminosity is unchanged. The evolutionary path in the HR diagram lies also in the shaded region in Figure 11.

IV. DISCUSSIONS

After the flare-up of the protostar, the luminosity decreases gradually as shown in Figure 10. In order to follow the change in the luminosity up to the onset stage of gravitational equilibrium, the inclusion of the energy transport by convection will be essentially necessary. The time required for the stellar surface to settle down to gravitational equilibrium is of the order of 10 years,

412 Low-luminosity stars

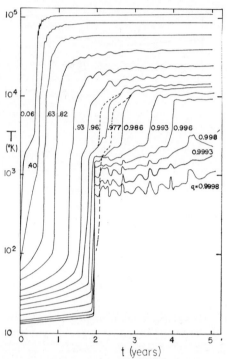

FIGURE 9 The time variation of the temperatures of typical shells in the case with radiation flow. The dashed curves show the results for two shells ($q = 0.977$ and 0.986) in case 4 without radiation flow.

which is the free-fall time of the surface layer from $R = 5000\ R_\odot$. If the luminosity $10^3\ L_\odot$ continues for a period of 10 years after the flare-up, this amount of radiation energy can be supplied by the loss of only $1/20$ of the total thermal energy of the star (see Figure 4 for the distribution of the thermal energy) which corresponds to the thermal energy of an outer envelope containing about 10 per cent of total mass. The entropy distribution has a peak at $q \simeq 0.9$ and we have found that this situation is not altered when the effect of the radiation is included. It is knows that the region of negative entropy gradient should be highly convective and the energy flow by convection proceeds in such a way that the region of negative entropy gradient disappears. At present, the computations including the convective energy flow have not been completed, but it is expected that the luminosity will not change appreciably until the surface region settles down to gravitational equilibrium. Further, it is also expected from the entropy distribution shown in Figure 6 that the effective temperature at the onset stage of gravitational

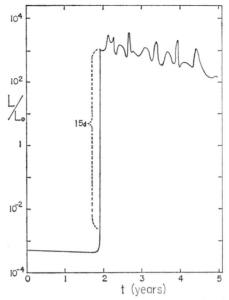

FIGURE 10   The time variation of the stellar luminosity in the case with radiation flow.

equilibrium is higher to a certain extent than that corresponding to the wholly convective structure.

The comparison with the observations of the infrared objects, for which both the luminosity and the effective temperature are known, is made in the H–R diagram, as shown in Figure 11. As the observational data, we have FU Ori (Herbig 1966), R. Mon (Low and Smith 1966) and a point source in the Orion nebula (Kleinmann and Low 1967). The last two infrared objects are interpreted as flared-up protostars which still have extended envelopes of relatively low temperatures. It is to be particularly noticed that in our computations we have assumed the initial density distributions which vanish exactly at the surface. In reality, however, initial protostars will have generally a diffuse envelope of small mass which is extended, for example, to a radius of $10^5\,R_\odot$. The free-fall time from $r = 10^5\,R_\odot$ is about 10^3 years. Then, it may be possible that the lifetime of a protostar as an infrared object is determined by the motion of the diffuse nebulosity surrounding the main body rather than by the evolution of the main body as considered by us.

The flare-up of FU Ori in 1936 by 5 ~ 6 magnitudes may be interpreted as follows. The main body of FU Ori had already flared-up long before 1936 along the path in the shaded region in Figure 11 and had been an infrared star until 1936. The surrounding nebulosity was undergoing free fall, and

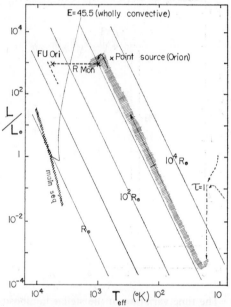

FIGURE 11   The evolutionary track (shaded region) of a protostar of one solar mass in the H–R diagram. The track in the pre-main-sequence phase of gravitational equilibrium which starts with a wholly-convective structure is shown by the solid curve. The positions of the three objects are denoted by the crosses and the dotted lines. The right cross for R Mon corresponds to the temperature of its main infrared radiation, while the left cross represents the position of the central star as interpreted by Low and Smith (1966).

when it reached the radius of about 1 A.U. the grains were evaporated by the radiation from the main body of luminosity $10^3 L_\odot$. The free-fall time at the radius of 1 A.U. is of the order of 0.1 year. Thus, the flare-up of FU Ori in 1936 may be due to the clear-up process of the observing nebulosity, which fell finally onto the main body or formed a rotating disk which will be regarded as a primordial solar system.

References

AUMAN, J. R., and BODENHEIMER, P. 1967, *Ap. J.*, **149**, 641.
COX, A. N., and STEWART, J. N. 1965, *Ap. J. Suppl.* **11**, 22.
GAUSTAD, J. E. 1963, *Ap. J.*, **138**, 1050.
HAYASHI, C. 1961, *Publ. Astron. Soc. Japan*, **13**, 450.
HAYASHI, C., and NAKANO, T. 1965, *Prog. Theor. Phys.*, **34**, 754.
HAYASHI, C. 1966, *Ann. Rev. Astron. Astrophys.*, **4**, 171.

HERBIG, G. H. 1966, *Vistas in Astronomy* (Pergamon Press), **8**, 109.
KLEINMANN, D. E., and Low, F. J. 1967, *Ap. J.* **149**, L1.
Low, F. J., and SMITH, B. J. 1966, *Nature*, **212**, 675.
NAKANO, T., OHYAMA, N., and HAYASHI, C. 1968, *Prog. Theor. Phys.* **39**, 1448.
NARITA, S., NAKANO, T., and HAYASHI, C. 1968 (to be published).
VARDYA, M. S. 1964, *Ap. J. Suppl.* **8**, 277.

Discussion

BODENHEIMER   Is there a large decrease in the central density when the bounce occurs?

HAYASHI   No, the central density does not decrease.

Theoretical models of a low-mass protostar

PETER BODENHEIMER

Lick Observatory, University of California, Santa Cruz

Abstract

Detailed calculations of the evolution of a protostar of 0.1 solar mass in dynamical collapse are presented. Two different opacity laws are investigated; also, the solution for 0.1 solar mass is compared with that for 1 solar mass. The calculations begin with a mean density of 10^{-12} g/cm³ and a central temperature of about 15°K, and they are terminated upon the development of a strong shock wave in the central region, which occurs at a central density of about 5×10^{-6} g/cm³ and a central temperature of 2700°K for the case of 0.1 solar mass.

ALTHOUGH THE theoretical calculations which have been made recently concerning the very early evolution of stars are quite uncertain, the results are of some interest in view of the recent infrared observations which suggest the existence of protostars in the Orion Nebula (Kleinmann and Low 1967; Becklin and Neugebauer 1967). This note summarizes some calculations, made in collaboration with John Gaustad, of a portion of the dynamic evolution of a protostar of 0.1 solar mass, beginning approximately at the point where the object has become opaque to radiation.

The problem was simplified by the assumption of spherical symmetry; rotation, magnetic fields and fragmentation were not considered. The complete equations of stellar structure, including hydrodynamic effects, were solved by a detailed numerical technique which is very similar to Henyey's method of solving equilibrium problems in stellar evolution. (Henyey, Forbes and Gould 1964; Bodenheimer 1968). Above a temperature of 1400°K Rosseland mean opacities (or slight extrapolations thereof) computed by Auman, including the contribution of water vapor, were used (cf. Auman and Bodenheimer 1967). Below this temperature two considerably different opacity laws were used in parallel calculations. In the first case a constant value of the opacity was assumed, based upon extrapolation of

Auman's results. (This law is referred to as the 'low' opacity in what follows.) In the second case, recent calculations by Gaustad and Kellman (1968) of the Rosseland mean opacity due to ice grains and mineral grains in the temperature range 10–1400°K were used. These opacities (referred to as the 'high' opacity) are temperature- and density-dependent and are roughly a factor of 100 to 1000 higher than the constant value used in the first case.

The calculation was started with a mean density of 10^{-12} g/cm³, with a slight degree of central condensation, with zero collapse velocity, with a nearly isothermal temperature distribution, and with a radius of about 5000 solar radii. The equation of state of an ideal gas was used, including radiation pressure; the hydrogen was assumed to be in molecular form at the start, and a normal solar composition was taken. The radiative diffusion equation was used to treat the energy transport even in the optically thin outer regions; hence, the position in the Hertzsprung–Russell diagram of these models is somewhat uncertain. The evolution starts off practically in free fall, the degree of central concentration of the mass increases continuously with time, and as a result the evolutionary time scale becomes much shorter at the center than in the outer regions.

Figure 1 gives a rough idea of the positions of the models in the theoretical

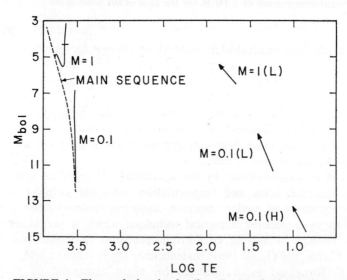

FIGURE 1   The evolution in the (log T_e, M_{bol}) plane of pre-main-sequence equilibrium configurations of 1.0 and 0.1 $M_\odot$ (solid curves) and of collapsing configurations of the corresponding masses (arrows). The symbols (L) and (H) indicate that low and high opacities, respectively, were used in the calculation.

H–R diagram. The difference in opacity between the high and low cases results in a difference of about 4 magnitudes in M_{bol}. For purposes of comparison a collapsing sequence for 1 solar mass with the low opacity is also illustrated, as are the pre-main-sequence equilibrium solutions for 1 and 0.1 solar mass. The effective temperatures fall in the range 10–50°K, and the direction of evolution of the model protostars is upward and to the left in this diagram, although the slopes of the arrows are not well determined. Note that the evolution time along these curves is approximately 20 years in all cases; this time depends only very slightly on mass and on the opacity law used. The short evolutionary time scale and the low luminosity makes observations of such objects an unlikely possibility.

The relatively small surface changes are in contrast to the large changes which occur during the same time period at the center of the collapsing configuration, as Figure 2 illustrates. First, we may note that the solutions for the low-opacity case and the high-opacity case lie very close together

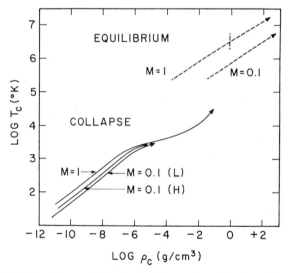

FIGURE 2 The evolution in the (log ρ_c, log T_c) plane (T_c and ρ_c are central temperature and central density, respectively) of pre-main-sequence equilibrium configurations (dashed lines) and collapsing configurations (solid lines) of 1.0 and 0.1 $M_\odot$. The upper ends of the dashed curves give approximate main-sequence positions. In the case of 0.1 $M_\odot$ the dashed curve represents fully convective evolution; in the case of 1.0 $M_\odot$ the vertical dashed line indicates the point where a radiative core develops. The symbols (L) and (H) have the same meaning as in Figure 1.

which indicates that the opacity has very little effect on the hydrodynamics. In both cases only a very small fraction of the released gravitational energy is radiated away. The higher opacity results in increased internal temperature at a given density, and the collapse becomes more and more nearly adiabatic at the center as the density increases. Second, the pressure gradient increases quite rapidly relative to gravity as the central density increases until at a central density of about 10^{-8} g/cm^3 it exceeds gravity, resulting in the formation of a region of deceleration at the center which grows in mass. When the central density has increased to 10^{-6} g/cm^3 a shock wave forms at the outer edge of the decelerating core. However, while the shock is in the process of formation, the central temperature rises above 1500°K, causing the dissociation of hydrogen molecules to begin, which results again in instability to gravitational collapse at the center. The calculation was terminated when the shock became well-developed. The change in opacity had no effect on the shock development. Third, in the parallel calculation (from the same initial density) for 1 solar mass the shock wave does not develop at this value of ρ_c; the small region of deceleration at the center in this case is eliminated by the occurrence of dissociation before the shock has a chance to form. A shock does occur in the case of 1.0 $M_\odot$ only when dissociation is complete at the center and T_c reaches 25,000°K. (Upper end of the collapse curve in Figure 2.) Work is now in progress on an efficient technique for calculating the behavior of these shocks.

We may make certain projections and somewhat speculative predictions on the basis of these results. (1) The properties of the central region of the 0.1 $M_\odot$ solution at $\rho_c = 10^{-5}$ g/cm^3 are very similar to those of the 1.0 $M_\odot$ solution at the corresponding point, and we may expect the further evolution in the case of 0.1 $M_\odot$ to follow very closely that of 1.0 $M_\odot$ in the (log ρ_c, log T_c) plane. (2) No 'bounces' (expansions) of the central regions of collapsing protostars have developed during any of the calculations made so far (cf. Hayashi and Nakano 1965). The suggestion that 'bounces' of the core will *not* significantly affect the evolution is confirmed by the more extensive calculations of Larson (1968) and by those of Hayashi (this volume), whose results show that if such bounces occur at all, they amount to only minor perturbations on the continued contraction. (3) If the previous statement is indeed true, then we may predict that when the protostar of 0.1 $M_\odot$ comes into hydrostatic equilibrium as a fully convective star, its position on the Hayashi track will be at a lower luminosity than previous estimates would indicate (Hayashi 1966). However, this initial equilibrium position is estimated to fall at a large enough radius so that the contraction time to the main sequence of about $1·5 \times 10^9$ years will be unaffected.

References

AUMAN, J. R., and BODENHEIMER, P. 1967, *Ap. J.*, **149**, 641.
BECKLIN, E. E., and NEUGEBAUER, G. 1967, *Ap. J.*, **147**, 799.
BODENHEIMER, P. 1968, *Ap. J.*, **153**, 483.
GAUSTAD, J., and KELLMAN, S. 1968, *unpublished*.
HAYASHI, C. 1966, *Annual Reviews of Astronomy and Astrophysics*, 4, (Palo Alto, Calif.: Annual Reviews, Inc.), 171.
HAYASHI, C., and NAKANO, T. 1965, *Prog. Theor. Phys.*, **34**, 754.
HENYEY, L. G., FORBES, J. E., and GOULD, N. L. 1964, *Ap. J.*, **139**, 306.
KLEINMANN, D. E., and LOW, F. J. 1967, *Ap. J.*, **149**, L1.
LARSON, R. B. 1968, *Thesis*, California Institute of Technology.

Discussion

HAYASHI   We made a computation of the dynamic contraction of a star of mass 0.05 $M_\odot$, and we found that the flare-up of the star occurs from a luminosity of $10^{-5} L_\odot$ to the peak luminosity of about $20 L_\odot$. Have you carried your calculations to the stage when the shock wave comes near the surface?

BODENHEIMER   No, the calculation was stopped after the shock wave had progressed a short distance outward. It has not reached the surface as yet. So I have not computed the final flare-up as Dr. Hayashi has.

The pre-Hayashi phase of stellar evolution

A. G. W. CAMERON

Belfer Graduate School of Science
Yeshiva University, New York, N.Y.
and
Institute for Space Studies
Goddard Space Flight Center, NASA, New York, N.Y.

Abstract

The structure and evolution of protostellar disks ('stellisks') are being studied. These are flattened fragments of a collapsed interstellar cloud. Their angular momentum properties are those appropriate to uniformly rotating spheres of variable density distribution. Corresponding to each angular momentum distribution there are two separate density distributions in which there is centrifugal equilibrium in the disk. One distribution is very flat, rotates approximately uniformly, and probably deforms into a close binary pair. The other is axially condensed and is the probable precursor of single stars and planetary systems. In stellisks of the latter kind with lower densities near the rim, one or a few separate rings form near the outer edge. Since stellisks are formed as a result of a dynamical collapse process, they have an initial stage of short-lived turbulence. This transports angular momentum outwards and increases the tendency for formation of outer rings. The structure perpendicular to the plane of the disk has also been studied. The disk cools and contracts at constant central pressure. For central temperatures $\leqslant 10^3 °K$, the disk has thermally-driven convection in the regions with surface densities $\geqslant 10^5$ gm/cm². The resulting turbulent viscosity causes mass inflow to form a central star on the lower part of the Hayashi track. Such a star will be in the T Tauri phase and a strong stellar wind will blow away the low density gas in the remainder of the disk. It is probable that an axially condensed stellisk with low surface densities cannot form a star. It is suggested that the unseen mass in the solar neighborhood consists largely of permanent stellisks which have negligible surface temperature but whose presence may be inferred from stellar occulations.

SOME TIME AGO I presented an analysis of the conditions in which an ordinary interstellar cloud could become unstable against gravitational collapse

(Cameron 1962). For a cloud of 10^3 $M_\odot$, an initial density of $\sim 10^3$ particles/cm^3 is required. High initial densities of this kind may be produced if the outer regions of an HI cloud are ionized by an O or B star, thus creating the necessary high surface pressure for the compression to high density. The gas must be predominantly compressed along the magnetic lines of force in order that the enhanced magnetic pressure should not halt the compression. In the subsequent collapse the magnetic field should not play an important dynamical role (Cameron 1962; Pneuman and Mitchell 1965).

It is to be expected that the resulting cloud will be turbulent and that the collapse will be roughly isothermal. The resulting density fluctuations should initiate fragmentation in the cloud. In a recent paper I concluded that the turbulent component of the internal angular momentum of a fragment should be comparable to that associated with an initial corotation of the interstellar cloud with its orbital motion about the center of the galaxy (Cameron 1968).

The spherical collapse of a protostar would form an object on the Hayashi track in the H–R diagram. However, considerations of the conservation of angular momentum in the collapsing fragment indicate that it should form a rotating flat disk with a radius of several tens of astronomical units, much larger than a star would possess on the Hayashi track (Cameron 1962, 1963). It has been the objective of the present work to discover how such a disk might dissipate to form a more compact star on the Hayashi track.

In order to relate the problem to the interstellar cloud conditions, the cloud fragments were assumed to be uniformly-rotating spheres. The equators of these spheres were assumed to have conserved their local angular momentum from the initial cloud corotation condition with its angular velocity of 10^{-15} radians/second. The density was assumed to vary linearly with radius from a central value to a surface value. The specific models used in the present calculations have two solar masses. In one sphere the density was uniform ('uniform sphere') and in the other sphere the density fell linearly from the central value to zero at the surface ('linear sphere').

Each sphere was divided into 50 cylindrical zones concentric about the axis of rotation, and the mass and angular momentum of each zone was calculated. The mass was then considered to have collapsed into a thin flat disk, and it was required that the mass in the disk be everywhere in centrifugal equilibrium with respect to the gravitational forces at that point in the disk. Radial pressure gradients were neglected since I was interested in the case in which thermal energies in the gas would be much less than the bulk rotational kinetic energies.

Gravitational potentials were calculated by the technique of a superposition of concentric spheroidal shells of varying density in the limit of zero eccentricity (Burbidge, Burbidge, and Prendergast 1960 *a*, *b*; Brandt 1960; Brandt and Belton 1962; Mestel 1963). With some trial mass distribution in the disk, the surface density in each zone was varied and the changes in angular momentum per unit mass required for circular motion for all the other zones were determined. A matrix was then inverted to determine what perturbations to introduce into the surface densities of all the zones so that the angular momentum required for circular motion in the model would approach the assigned values. The problem is highly non-linear and convergence to the centrifugal equilibrium condition is very slow. It would be helpful to develop improved mathematical techniques which would lead to faster convergence.

After the above technique had been developed, it was discovered that each of the two original spheres had two different surface density distributions for which it was in centrifugal equilibrium. The four solutions are shown in Figure 1. The flat solution for the uniform sphere is known classically (McMillan 1958); it is in uniform rotation. In addition there is an axially condensed solution. Mestel (1963) has discussed such axially condensed solutions which he shows can be produced by slight perturbations in the mass distribution of a uniform sphere. However, it has been shown by the present numerical techniques that the same mass and angular momentum distribution is consistent with each solution. In the axially condensed solution the angular velocity varies roughly inversely as the radial distance, so that there is a large amount of shear in the solution.

The flattened linear sphere exhibits a similar behavior, as shown in Figure 1.There is a nearly flat solution, which is nearly in uniform rotation, and which has a very sharp edge. The axially-condensed solution is more strikingly so. The irregularity of this solution near the axis is an artifact of the finite zoning and has no physical significance. This solution also has a condensed ring near the outer edge, which will be discussed in more detail below.

I believe that the existence of these two solutions has great cosmogonic significance. Hunter (1963) has examined the stability of the classical uniformly-rotating flat disk. He has found it to be unstable against both radial and non-radial perturbations. (Professor Gold has informed me that he and Bondi have found a very limited range of parameters in which this appears not to be true.) K. H. Prendergast (private communication) has shown numerically that a flat galactic disk with superposed random velocities of the individual mass points deforms into a corotating bar. It appears likely that the corotating disk will deform in a similar manner and will subsequently form a close pair of binary stars. No further calculations have been carried out with these flat solutions. Whether a collapsing interstellar could fragment forms a flat disk or an axially-condensed disk must depend upon subtle features of the dynamics of the collapse.

I refer to the axially-condensed protostellar disks as 'stellisks'. I assume that they are stable against perturbations, but no analysis of this point has been made, and one is badly needed.

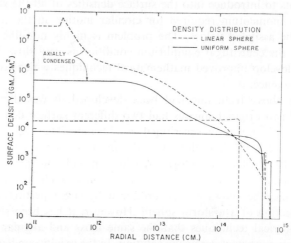

FIGURE 1 Surface density distributions for centrifugal equilibrium of disks formed from collapse of uniform and linear spheres.

When a stellisk is formed as a result of the dynamical collapse process, the gas will overshoot the position of centrifugal equilibrium, thus inducing a state of initial turbulence in the disk. The energy input into the turbulence arises from the released gravitational potential energy. Since there is no continuing source of energy input into the turbulence, it is short-lived. The largest eddy motions are broken into smaller motions after the gas has moved through a mixing length, which takes a small fraction of an orbital period.

I have estimated the angular momentum transfer between adjacent zones due to turbulent viscosity, using data appropriate to the stellisk models of Figure 1. It turned out that the angular momentum transfer would be sufficient to make adjacent zones corotate if they did not change their radial positions. In fact, the angular momentum transfer would spread the zones apart and increase the shear between them. The situation was treated very crudely by taking the inner five zones for a stellisk model and redistributing the angular momentum so that they would corotate at their given positions. This procedure was then applied to zones 2 to 6, 3 to 7, and so on in the model until the outer edge was reached. The models were then again relaxed to centrifugal equilibrium.

The justification for this procedure lies in the fact that the center of a cloud fragment will collapse faster than the surface layers, so that the turbulent redistribution of angular momentum will be completed near the center before the outer parts have finished collapsing.

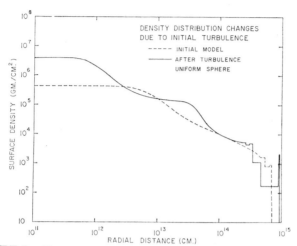

FIGURE 2 Changes in surface density distributions in the uniform sphere stellisk after dissipation of initial turbulence.

The model of the uniform sphere stellisk after dissipation of initial turbulence is shown in Figure 2; it is compared in the figure with the initial model of Figure 1. It may be seen that a slight further net central condensation of mass has occurred, except near the outer edge. A prominent outer condensed ring has formed, and a slight second ring-like perturbation is also present.

A similar behavior for the linear sphere is shown in Figure 3. In this case two prominent condensed outer rings are present.

It may be deduced from these figures that condensed rings will have an increasing tendency to form as the relative mass fraction in the outer layers is reduced and as the gradient of the angular momentum per unit mass in the outer layers is increased. The reality of the rings was tested by compressing and smoothing the outer zones of the linear sphere stellisk and relaxing again to centrifugal equilibrium. Again the two condensed rings appeared, but the outer one centered on a different zone. Thus the reality of the ring structure was demonstrated, but the uniqueness of the structure remains an open question. Calculations with a finer zoning mesh are needed for a further investigation of this question.

It seems likely to me that these rings will be unstable against non-radial perturbations, and that they will deform to form separate disk-like condensa-

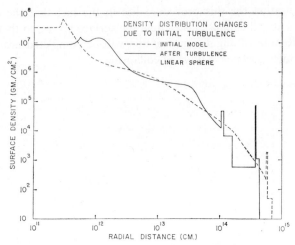

FIGURE 3   Changes in surface density distributions in the linear
sphere stellisk after dissipation of initial turbulence.

tions orbiting about the central disk. Such sub-disks seem likely precursors of
the giant planets of the solar system.

The collapse of the interstellar cloud fragment will, in the late stages,
lead to adiabatic heating of the inner parts of the stellisk to 10^4°K or higher.
There will be an initial period of rapid radiative cooling. Further dissipation
of the disk will depend on the operation of turbulent viscosity to transport
angular momentum outwards. The only apparent energy input source for
such turbulence would be thermally-driven convection. This would require
superadiabatic temperature gradients to exist perpendicular to the plane of
the disk. Hence the structure of the disk perpendicular to the plane was
investigated with the simplifying assumption that any column of the disk
could be considered part of an infinite plane of matter having the same local
conditions.

The structure of an isothermal column of this kind is well known (see for
example Mestel 1963). The central pressure of such a column is proportional
to the square of the surface density and is independent of the temperature.
The height of such a column is proportional to the temperature and inversely proportional to the surface density. A more realistic column will not
be isothermal owing to the presence of internal opacity, but the above
relations are always at least approximately true.

At temperatures in the general vicinity of 3000°K the opacity of stellar
material is very low and no convection should be present. It is worth noting
in passing that at 3000°K about 10^{-3} of the mass will be in the form of OH

under conditions of statistical equilibrium. Because of the low opacity there is no assurance of good interaction between radiation and matter, and it is an interesting hypothesis that population inversions in the OH molecule may be possible. P. M. Solomon and I are currently considering the question of whether this may be a suitable model for the maser-amplified OH emission from the compact galactic OH sources.

Below $2000°K$ the opacity becomes larger due to the presence of condensed solids and H_2O, NH_3, and CH_4 molecules. The solids are mainly particles of metallic iron and magnesium silicate. Electron microscope measurements of the structure of very primitive meteorites indicate metal and silicate particle sizes in the submicron range (E. Anders, private communication). These sizes are small compared to thermal wavelengths and allow Rosseland mean opacities to be calculated with negligible dependence on particle size. With opacities based on the particles alone, I have found that the surface density threshold for vertical convection lies between 10^5 and 10^6 gm/cm^2.

I am currently putting in the additional effects of molecular opacities. Such opacity contributions depend on both pressure and temperature. My rough guess is that the threshold for vertical convection will be lowered to between 10^4 and 10^5 gm/cm^2.

It may be seen in Figures 2 and 3 that surface densities in the stellisks lie above $10^{4.5}$ gm/cm^2 out to distances of several astronomical units (approaching 10^{14} cm). Hence these inner parts of the stellisks will be subject to dissipation via turbulent viscosity once the disks have cooled to $\sim 10^3°K$. This will lead to an inward flow of mass and an outward flow of angular momentum. Hence a good fraction (~ 1 $M_\odot$ in the present models) of mass will flow in to form a star on the Hayashi track.

This mass flow will carry with it all finely divided solid material. In fact, all material of less than planetary size will be carried inward into the star. The inner planets represent only $\sim 10^{-2}$ of the condensed solids that would be present. Hence the rapid dissipation of the disk limits the growth of the inner planets.

After the central star has formed, it will be in the T Tauri phase, emitting an extremely strong stellar wind, presumably due to a hot corona excited by the turbulent motions in the fully-convective Hayashi phase. This will sweep away the primitive atmospheres which were captured by the inner planets from the stellisk gases, and also it will sweep away the thinner gases remaining in the outer nonconvective part of the stellisk. During the early part of this T Tauri phase a great deal of dust will remain in the environment, which will thermalize much of the radiation of the central star, producing a secondary peak of infrared radiation in the spectrum of the star.

The general features of the architecture of the solar system appear to emerge from the above analysis. Detailed agreement is not to be expected owing to the arbitrary initial density distribution assumed.

It was indicated at the beginning that there should be a considerable spread in the total angular momentum per unit mass of these stellisks. For a given mass, the surface density varies inversely as the fourth power of the total angular momentum. Thus we should expect that if the angular momentum is increased by only a factor of 4 or 5, the surface density will become too small for thermally-driven convection to exist. Such stellisks probably cannot form stars except via the impossibly long process of dissipation by molecular viscosity. They are probably permanent. I wish to suggest that a good portion of the unseen one-third of the mass density in the solar neighborhood may consist of such stellisks.

It should also be noted that the first generation of stars to form in our galaxy presumably contained a negligible content of elements heavier than helium. Hence they would always have low opacity, and they could not form stars unless they were massive and compact. This may indicate that there is a lot of mass spherically distributed in the galaxy in the form of permanent stellisks.

Fleischer and Conti (1955) examined Palomar Sky Survey prints and concluded that there is a large number of small dark gaps in the star distribution in crowded regions of the Milky Way. They attribute these to dark 'globules'. If these have the average properties of Bok's globules, then they have a space density of 0.03 per cubic parsec. Although their conclusions were highly tentative, I regard them as permissive with respect to the above hypothesis. It is clear that further studies of this sort are greatly needed.

I wish to thank J. M. Greenberg, K. H. Prendergast, P. M. Solomon, P. Thaddeus, E. Anders, and J. A. Wood for helpful discussions on some aspects of this work. This research has been supported in part by the National Science Foundation and the National Aeronautics and Space Administration.

References

BRANDT, J. C. 1959, *Ap. J.*, **131**, 293.
BRANDT, J. C., and BELTON, M. J. S. 1962, *Ap. J.*, **136**, 352.
BURBIDGE, E. M., BURBIDGE, G. R., and PRENDERGAST, K. H. 1960a, *Ap. J.*, **132**, 282.
BURBIDGE, E. M., BURBIDGE, G. R., and PRENDERGAST, K. H. 1960b, *Ap. J.*, **132**, 640.
CAMERON, A. G. W. 1962, *Icarus*, **1**, 13.
CAMERON. 1963, *ibid.*, 339.
CAMERON. 1968, *Infrared Radiation Associated with Protostars*, to be published in Proceedings of London Conference on Infrared Astronomy (May 1967).
FLEISCHER, R., and CONTI, P. S. 1955, *Rensselear Obs. Mem.* (November, 1955).
McMILLAN, W. D. 1958, *The Theory of the Potential* (Dover Publishers, New York).
MESTEL, L. 1963, *M.N.R.A.S*, **126**, 553.
PNEUMAN, G. W., and MITCHELL, T. P. 1965, *Icarus*, **4**, 494.

Discussion

OSTRIKER I would like to make two comments, one of them is that I believe the rings are very real. If you have enough angular momentum it pays energetically to put the material in the form of rings. If you allowed for the finite thickness, then you might find more rings, which might be nice to form planets. The second point is that the centrally condensed disks, which formed out of roughly uniform spheres, may be stable with respect to the non-axisymmetric perturbations. Lynden-Bell carried out an investigation of the collapse of the uniform sphere, which would be the stage preceding this equilibrium disk, and found that it becomes unstable to the non-axi-symmetric mode during the collapse, before it would reach the flat stage. Now I think you can probably get around this just by starting with an initially more centrally condensed model. But the particular ones you pick will be likely to be unstable in the collapse phase.

CAMERON May I remark that this whole question of stability is one that demands a great deal of attention. I have been informed recently by Professor Gold that he and Bondi have been looking closely at the stability of the flat, uniformly rotating disk, and that they have found a narrow range of parameters in which they think it will be stable against perturbations. I simply report that as something which he authorized me to say.

VON HOERNER I would like to mention that about 15 years ago, we did a lot of work under von Weizsäcker on this problem, but with very crude means. But I think Weizsäcker was the first to show the tendency to build rings, although we could never confirm it numerically. But we have confirmed numerically the tendency to build a strong cusp in the center and finally a central mass.

SALPETER If these things are really as common as you say, and if they are 10–20 °K in temperature, then, if you looked in the plane of the galaxy they would subtend a large enough fraction of space so that you might have a chance of detecting them in the high frequency tail of the black-body radiation, in particular the higher rotational CN frequencies.

CAMERON I think it is rather unlikely, because you might expect something in the range of 10^{-6} to 10^{-8} of the total solid angle of the sky to be subtended by these. And that taken together with the small additional temperature over black body, I suspect, is not very favorable.

P

Section VI

Model atmospheres for cool dwarf stars

DUANE CARBON, OWEN J. GINGERICH, and
DAVID W. LATHAM

Smithsonian Astrophysical Observatory and Harvard College Observatory
Cambridge, Massachusetts

Abstract

We have extended our previous work with cool star atmospheres to include the infrared opacity due to water vapor and the pressure-induced dipole of molecular hydrogen. Eight dwarf models with effective temperatures ranging from 3500° to 1500°K are presented in detail. These models are idealized in that they do not include convection or absorption from many molecular species (e.g., CO, TiO, CH). Experiments with a phenomenological theory of convection have convinced us that the adiabatic temperature gradient is a good approximation in the convective zone of dwarf stars with effective temperatures less than 2500°K.

I. INTRODUCTION

WE APPROACHED the calculation of realistic cool star atmospheres by breaking the problem into component parts that could be tackled one at a time. Some of these problems have now been solved quite satisfactorily; other aspects we have handled by only rough approximations; and we have not even faced still further difficulties. Nevertheless, we believe that two of the chief goals of the model atmosphere approach are in sight: first, we are on the verge of being able to furnish a reasonably good pressure–temperature relation to serve as the boundary envelope for stellar interior calculations, and second, we believe that reasonable models are nearly in hand for abundance work and particularly for the C/O ratio.

The models presented here have been computed with the program discussed at the Trieste Colloquium on Late-Type Stars (Gingerich, Latham, Linsky, and Kumar 1966). The present calculations include the water-vapor opacity, which had not been available for the Trieste paper. Although the models

435

we shall describe are more realistic than any we have hitherto published, they are still highly idealized and are computed without convection and without consideration of several of the important opacity sources.

II. EQUATION OF STATE

Our method of integrating the equation of hydrostatic equilibrium has been described in the Trieste paper. In order to get the number densities of the five species that affect the opacity (electrons, H, H_2, He, H_2O), we consider 30 neutral atoms and their ions as well as H_2, H_2O, OH, CO, CO_2, SiH, MgH, SiO, MgO, ZrO, and TiO. The molecular constants required for the evaluation of the partition functions were obtained from Dolan (1965).

We assume that the atoms, ions, and molecules behave as a perfect gas in local thermodynamic equilibrium, and we use an iterative procedure to calculate the various ionization and dissociation equilibria. Although for simplicity we have limited the following discussion to diatomic molecular species, the method can easily be generalized to the triatomic case.

Letting the gas be composed of the neutral atoms, ions, and diatomic molecules of N elements, we can write for the ionization and dissociation equilibria

$$n_i^+ n_e / n_i = Z_i(T) \qquad i = 1, \dots N \qquad (1)$$

and

$$n_i n_j / n_{ij} = Z_{ij}(T) \qquad \begin{matrix} i = 1, \dots N \\ j = 1, \dots N, \end{matrix} \qquad (2)$$

where the $Z(T)$ are defined in terms of the appropriate partition functions. The total electron and particle densities can be expressed as

$$n_e = \sum_{i=1}^{N} n^+_i \qquad (3)$$

$$\frac{P}{kT} = n_e + \sum_{i=1}^{N} \left(n_i + n_i^+ + \sum_{j=1}^{N} \epsilon_{ij} n_{ij} \right), \qquad (4)$$

respectively, where $\epsilon_{ij} = 1$ if the molecule is to be included and $\epsilon_{ij} = 0$ otherwise. The fractional abundance of each element with respect to hydrogen is

$$\eta_i = \frac{n_i + n_i^+ + \sum_{j=1}^{N} \epsilon_{ij} n_{ij} + \epsilon_{ii} n_{ii}}{n_1 + n_1^+ + \sum_{j=2}^{N} \epsilon_{1j} n_{1j} + 2n_{11}} \qquad i = 1, \dots N, \qquad (5)$$

where the subscript 1 refers to hydrogen. Rewriting Eq. (5) as

$$n_i + n_i^+ + \sum_{j=1}^{N} \epsilon_{ij} n_{ij} = \eta_i \left(n_1 + n_1^+ + \sum_{j=2}^{N} \epsilon_{1j} n_{1j} + 2n_{11} \right) - \epsilon_{ii} n_{ii}$$

$$i = 1, \ldots N,$$

substituting it into Eq. (4), and rearranging it, we have

$$(n_1 + n_1^+) \sum_{i=1}^{N} \eta_i + n_{11} \left(2 \sum_{=1}^{N} \eta_i - 1 \right) = \frac{P}{kT} - n_e$$

$$+ \sum_{i=2}^{N} \epsilon_{ii} n_{ii} - \left(\sum_{j=2}^{N} \epsilon_{1j} n_{1j} \right) \left(\sum_{i=1}^{N} \eta_i \right)$$

Using Eqs. (1) and (2) for hydrogen, we find

$$n_1^2 \left[\frac{2 \sum_{i=1}^{N} \eta_i - 1}{Z_{11}(T)} \right] + n_1 \left\{ \left[1 + \frac{Z_1(T)}{n_e} \right] \left[\sum_{i=1}^{N} \eta_i \right] \right\}$$

$$= \frac{P}{kT} - n_e + \sum_{i=2}^{N} \epsilon_{ii} n_{ii} - \left(\sum_{j=2}^{N} \epsilon_{1j} n_{1j} \right) \left(\sum_{i=1}^{N} \eta_i \right). \quad (6)$$

The problem to be solved can be stated as follows: Given P and T, find self-consistent values for the various n_i, n_i^+, and n_{ij}. As the first step in a model we begin by assuming initial values for the n_e and n_i while ignoring all molecules except H_2. Equations (6), (1), and (2) can then be solved simultaneously for n_1, n_1^+, and n_{11}. If n_1^+ and n_{ij} are expressed in terms of n_i using Eqs. (1) and (2), Eq. (5) can be rewritten as

$$n_i = \frac{\eta_i \left(n_1 + n_1^+ + \sum_{j=2}^{N} \epsilon_{1j} n_{1j} + 2n_{11} \right)}{1 + \frac{Z_i(T)}{n_e} + \sum_{j=1}^{N} \epsilon_{ij} \frac{n_j}{Z_{ij}(T)} + \frac{\epsilon_{ii} n_i}{Z_{ii}(T)}} \qquad i = 2, \ldots N. \quad (7)$$

With the calculated values for n_1 and n_1^+, we can use Eq. (7) to find new values of the n_i. These will then provide a new estimate of n_e from Eqs. (1) and (3). With the n_i determined, we can also find the n_{ij} from Eq. (2).

The process can now be repeated including in Eqs. (5), (6), and (7) all the terms involving molecular number densities. This iteration is continued until convergence is obtained in both electron and molecular number densities. The convergence criterion for n_e is that the fractional change shall be no more than 10^{-4} from one iteration to the next. This is relaxed to 10^{-3} in the case of the molecules.

This iterative method converges in all the cases that we have investigated over a wide range of effective temperatures. In a typical model approximately 15 sec on the CDC 6400 is required for the calculation of the number densities and the integration of the hydrostatic equation at 120 optical depth points. If 15 or fewer species are used, this time can be reduced by more than a factor of 2.

Figure 1 shows the depth dependence of some interesting molecular species as calculated by our program for log g = 5, 3500° and 2500° models. Our results are similar to those found previously by Tsuji (1964).

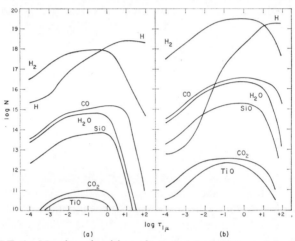

(a)   (b)

FIGURE 1 Number densities of neutral hydrogen and important molecular species vs monochromatic optical depth at 1 μ; (a) for T_{eff} = 3500°, log g = 5, (b) for T_{eff} = 2500°, log g = 5.

It should be noted that since n_e is found by summing over the ions, it is dependent upon how many atoms of the various species are bound up in molecules. Our calculation of n_e is an upper limit because we have not considered all the possible molecules such as KH that can tie up the important electron contributors, nor have we considered such negative ions as F^- and Cl^- that also remove electrons (e.g. Vardya and Kandel 1967). Nevertheless, we do not expect that the errors in n_e have seriously influenced the models we present here. The continuous opacity is dominated by scattering or H_2O in regions where errors in n_e (and hence in H^- or H_2^-) are likely to occur (cf. Kandel 1967).

An important advantage gained by the use of this iterative procedure is the relative ease of including additional molecules in the equation of state. To add another species it is necessary only to add terms in accordance with well-defined rules and to calculate new partition functions. No algebraic manipulations or major reprogramming are required.

III. OPACITY

Representative opacities at optical depth unity (at 1 μ) for 3500°, 2500°, and 1500°K models are shown in Figures 2–4. Because the opacity at 1 μ is smaller than the mean, the actual temperatures for these graphs are somewhat higher than are the effective temperatures of the models. Scattering is negligible at these depths and wavelengths, but nearer the surface Rayleigh scattering from H_2 predominates at wavelengths shortward of 7000 Å for the two cooler models. This may not be true in real stars, which would have other molecular opacities that we have ignored here, but in our models the scattering poses a severe computational problem at short wavelengths. As described in the Trieste paper, we have been able to achieve reasonable results with the integral equation solution to the scattering discussed by Avrett and Hummer (1965), and by Avrett and Loeser (1966).

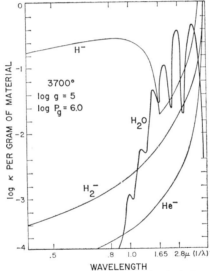

FIGURE 2   Opacities at optical-depth unity (1μ) for a model with $T_{\text{eff}} = 3500°$.

Details of many of the absorptions have been given in the Trieste paper and hence are merely listed here:

(a) Negative Hydrogen Ion.

(b) Neutral Hydrogen.

(c) Negative Helium Ion. The calculations by McDowell, Williamson, and Myerscough (1966) and more recently by John (1968) are 20 per cent lower at low temperatures and long wavelengths than are the values we have used formerly from Somerville (1965). The following

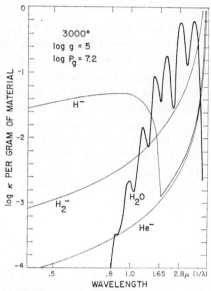

FIGURE 3   Opacities at optical-depth unity (1 μ) for a model with $T_{\text{eff}} = 2500°$.

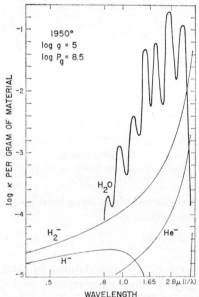

FIGURE 4   Opacities at optical-depth unity (1 μ) for a model with $T_{\text{eff}} = 1500°$.

polynomial approximation represents the values given by John to within 6 per cent (and to within 3 per cent over the most important range) for $0.3 \simeq \theta \simeq 3.0$ and $\lambda > 5000$ Å:

$$a_{\lambda,\theta} = 2.46 \times 10^{-4} + (-1.26 \times 10^{-7} + 5.67 \times 10^{-10} \lambda) \cdot \lambda$$
$$+ [-5.92 \times 10^{-4} + (5.12 \times 10^{-7} + 1.30 \times 10^{-9} \lambda) \cdot \lambda] \cdot \theta$$
$$+ [1.45 \times 10^{-2} - (8.30 \times 10^{-7} + 1.80 \times 10^{-10} \lambda) \cdot \lambda] \cdot \theta^2$$

where λ is in angstroms and a is per neutral helium atom and per unit electron pressure.

(d) Negative Molecular Hydrogen Ion. From the formulas of Dalgarno and Lane (1966) we have computed absorption coefficients for H_2^-; a comparison with those of Somerville (1964) shows that his coefficients should be raised by a factor of about 1.9 in the relevant temperature-wavelength region. We have used Somerville's polynomial approximation multiplied by 1.9.

(e) Metals.

(f) Rayleigh Scattering from H and H_2.

(g) Water Vapor. Our molecular H_2O opacity is based on Auman's (1967) extensive calculations. He tabulates in 100 cm^{-1} intervals the change in the harmonic mean opacity produced by adding water vapor, as a function of the ratio of continuous to water-vapor absorption. Linsky has fit polynomials to appropriate values in these tables for 25 wavelengths. These polynomials enabled us to include the seven H_2O band in our calculations. Linsky chose a microturbulent velocity of 8 km/sec, on the grounds that the effective increase in opacity provided by this mechanism might roughly approximate the pressure broadening, which was not included in Auman's calculation. However, as Auman has pointed out to us, the pressure broadening has a different wavelength dependence within the band than does the Doppler broadening, so our procedure is not strictly correct.

The H_2O opacity would be easier to use if it were provided in the form of a straight mean rather than as a harmonic mean. We propose to carry out numerical tests to determine which form is most accurate. possibly the division of each wavelength interval into regions of 'high' and 'low' absorption, with a double tabulation, would be the most satisfactory.

(h) Pressure-Induced Dipole of Molecular Hydrogen. Linsky has prepared an opacity subroutine for the pressure-induced dipole transitions of H_2 based on the discussion of Olfe (1961). Since the fundamental and first overtone vibration-rotation bands (at 2.2 and 1.16 μ, respectively) are essentially continuous opacities, they have been represented by a rectangular box approximation in which the effective bandwidth is proportional to $T^{1/2}$. In accordance with the results of Hare and Welsh (1958) the integrated opacity of the first overtone band has been set equal to 0.025 times that of the fundamental. The opacity from the pure rotational transitions has been calculated by using a polynomial approximation to the more exact expression given by Olfe.

A more extensive discussion of this opacity source is being prepared by Linsky.

IV. DISCUSSION OF MODELS

It should be stressed once again that the models that we present here are highly idealized. The opacity code we used did not include either the blanketing from atomic lines or the opacities from such species as CO, TiO, OH, CN, CH, VO, or ZrO. These opacities will play an important role in the determination of the emergent fluxes in the visible and near infrared. When the radiation leakage in this part of the spectrum is blocked by these additional opacity sources, the temperatures will rise in the deeper layers and the pressures will drop. We hesitate to predict whether this will have an appreciable effect on the H_2O bands without actually computing the models. In addition, we have ignored the role of convection. All the models described here have been driven to radiative equilibrium with a flux constancy of better than 1 per cent. Unless otherwise noted we have assumed a He/H ratio of 0.1 by number and solar abundances for the metals (Goldberg, Müller, and Aller 1960).

Figure 5 shows the emergent flux in magnitudes per unit wavelength vs reciprocal wavelength for 3500°, $\log g = 5$ models with and without water vapor. For the model without vapor the peak in the flux near 1.6 μ occurs because of the deep minimum in the H^- opacity; adding water does not block this radiation leak because the water vapor also has a window at this wavelength.

Three 2500°, $\log g = 5$ models are shown in Figure 6. For these cooler models the minimum in the H^- opacity has been filled in by H_2^- absorption. Thus the model without water no longer shows a peak in the flux at 1.6 μ. For the models with water our calculations probably overestimate the fluxes between the water-vapor bands since additional molecular opacities, particularly CO, may be very significant in the water-vapor windows.

Some interesting features of the model with 0.01 normal metal abundance should be mentioned. Oxygen is much less abundant, and as a result the H_2O bands are noticeably weaker. Since decreasing the metal abundances also decreases the number of free electrons, the opacity from the negative ions is reduced. Rayleigh scattering becomes very strong, and in the violet regions the source function does not saturate to the Planck function until mono-

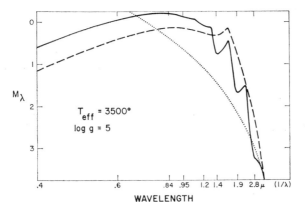

FIGURE 5 Emergent fluxes (ergs/cm³ sec sr) in magnitudes for $T_{eff} = 3500°$, $\log g = 5$: dashed line—model without H_2O; solid line—model with H_2O; dotted line—constant energy curve proportional to $2 \cdot 5 \log \lambda^2$.

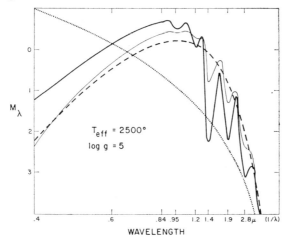

FIGURE 6 Emergent fluxes (ergs/cm³ sec sr) in magnitudes for $T_{eff} = 2500°$, $\log g = 5$: dashed line—model without H_2O; heavy solid line—model with H_2O and normal abundances; light solid line—model with H_2O and $0 \cdot 01$ normal metal abundancies; dotted line—constant energy curve proportional to $2 \cdot 5 \log \lambda^2$.

chromatic optical depth 300. The surface temperature of the model increases by over 150°, and the average pressure increases by an order of magnitude. The model without H_2O may not represent the limiting case of very low metals because the metals still contribute enough electrons that the negative ions are significant absorbers in the visible and ultraviolet.

Inclusion of the H_2 pressure-induced dipole opacity does not appreciably affect the 2500° model with normal metal abundances. However, this absorption considerably alters the emergent spectrum of the 2500° low metals model, as may be seen in Figure 7. The H_2 dipole transitions become more significant in going from normal to low metals for two reasons: first, the higher pressures of the low metals model increase the likelihood of the induced H_2 transitions and, second, the lowered oxygen abundance weakens the H_2O opacity. Since in this model the opacity from the two vibration-rotation bands is very much greater than the H_2O opacity at the same wavelengths, the features characteristic of the water-vapor spectrum have been obliterated. In addition, there has been a marked change in the slope of the visible and near-infrared spectrum. The discontinuous character of the infrared spectrum is due to the box model approximation used for the H_2 dipole opacity; in a more detailed treatment the edges of the bands would be less sharply defined.

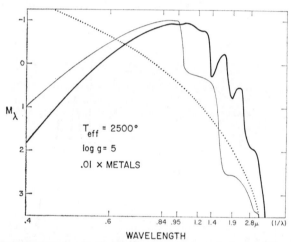

FIGURE 7   Emergent fluxes (ergs/cm³ sec sr) in magnitudes for $T_{eff} = 2500°$, $\log g = 5$, 0·01 normal metals: heavy solid line—model with H_2O; light solid line—model with H_2O and H_2 pressure-induced dipole; dotted line—constant energy curve proportional to $2·5 \log \lambda^2$.

Figure 8 shows the emergent fluxes, both with and without the H_2 dipole opacity, for a very cool dwarf model with an effective temperature of 1500°. The very high pressures make the H_2 dipole absorption stronger relative to the H_2O than was the case for the 2500° normal metals model. However, since the H_2 effective band widths are narrower for the 1500° model, the H_2 dipole bands do not overlap all the H_2O absorption features. The previous remarks about missing sources of molecular opacity in the near infrared and visible are even more germane here. Moreover, at these temperatures one

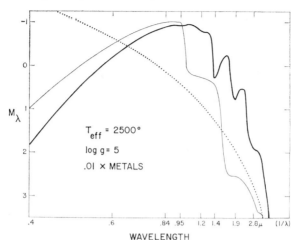

WAVELENGTH

FIGURE 8 Emergent fluxes (ergs/cm^3 sec sr) in magnitudes for $T_{\text{eff}} = 1500°$, log $g = 5$: heavy solid line—model with H_2O; light solid line—model with H_2O and H_2 pressure-induced dipole; dotted line—constant energy curve proportional to $2{\cdot}5$ log λ^2.

is entering the domain in which it may be necessary to consider the opacity produced by the formation of mineral grains.

Since the strength of the pressure-induced transitions of H_2 vary as $p_{H_2}^2$, this opacity source might produce significant luminosity effects. These will be most apparent in stars of very low effective temperature or in hotter stars with low oxygen abundance (or with a large C/O ratio).

We have made some exploratory calculations including convection for this set of dwarf models. In the $1500°$ model the convective instability begins around $\tau = 0.1$, and in the warmer models the instability is even closer to the surface (see Tables 1–8). The instability arises initially from the excitation and dissociation of H_2, with H ionization occurring much deeper. We consider all the species in our determination of the adiabatic gradient and specific heat. The necessary thermodynamic quantities are obtained by numerical differentiation of the number densities calculated by our equation-of-state program. We have adopted a phenomenological theory of convection described earlier by Latham (1964). At the present time, however, it is not coupled tightly enough to the Avrett–Krook temperature-correction procedure to allow us to iterate the models to consistency.

Nevertheless, we can state two important conclusions from our preliminary calculations. First, the convection in these dwarf models is very efficient because of the very high densities. For the cooler models our simple convective theory predicts a temperature gradient very nearly equal to the adiabatic gradient even near the top of the convective zone. Smaller mixing-length to scale-height ratios predict less efficient convection, but the 1500° model is so dense that even a mixing-length of 0.01 H in our theory predicts a gradient that is within a few per cent of the adiabatic gradient. We expect that for dwarf models cooler than 2500° the effect on the temperature-pressure relation will be independent of the convection theory, and we can employ the adiabatic gradient in our calculations.

Second, in these dwarf models the temperature difference between the rising and descending elements is at worst only a few tens of degrees. The velocities of the convective elements are only a few tens or hundreds of meters per second. This indicates that the use of a homogeneous model in which an average temperature characterizes each layer is not a bad assumption.

In conclusion, we feel that we are finally approaching the point where we can predict the physical properties of a wide variety of very cool atmospheres. In this context it is worthwhile to note the similarities between the Stratoscope observations reported by Woolf, Schwarzschild, and Rose (1964) and the emergent fluxes of the models we report here. An interesting example is the Stratoscope spectrum of R Leonis whose water-vapor bands, although affected by CO absorption, resemble those of our 2500° normal metals model. It will be necessary to consider additional molecular opacities before all the prominent features of the continuous spectrum can be explained. Nevertheless, differential abundance analyses should be possible even with an incomplete set of opacities, since the temperature–pressure relations are becoming reasonably good. It is imperative to include convection in cool dwarf models, but a bad theory will not necessarily mean disaster. Finally, we wish to emphasize that because of the extreme nongrayness of these stars any attempt to deduce their total energy from the visual spectrum alone is dangerous.

We wish to thank Dr. Jeffrey Linsky for numerous helpful criticisms of this paper, and for supplying us with subroutines for calculating the water-vapor and hydrogen pressure-induced dipole opacities, and Miss Barbara Welther, who prepared the polynomial approximation for He$^-$ for us. We would also like to thank the staff of the SAO Computing Center for their cheerful cooperation.

References

AUMAN, J. R. 1967, *Ap. J. Suppl.*, **14**, 171.
AVRETT, E. H., and HUMMER, D. 1965, *M.N.*, **130**, 295.
AVRETT, E. H., and LOESER, R. 1966, *Smithsonian Ap. Obs. Spec. Rep. No.* 201.
DALGARNO, A., and LANE, N. F. 1966, *Ap. J.*, **145**, 623.
DOLAN, J. F. 1965, *Ap. J.*, **142**, 1621.
GINGERICH, O., LATHAM., D. W., LINSHY, J., and KUMAR, S. S. 1966, *Colloquium on Late-Type Stars* (Trieste: Observatorio Astronomico di Trieste), p. 291.

GOLDBERG, L., MÜLLER, E. A., and ALLER, L. H. 1960, *Ap. J. Suppl.*, **5**, 1.
HARE, W. F. J., and WELSH, H. L. 1958, *Can. J. Phys.*, **36**, 88.
JOHN, T. L. 1968, *M.N.*, **138**, 137.
KANDEL, R. 1967, *Ann. d'Ap.*, **30**, 439.
LATHAM, D. W. 1964, *Smithsonian Ap. Obs. Spec. Rep. No.* 167, 199.
McDOWELL, M. R. C., WILLIAMSON, J. H., and MYERSCOUGH, V. P. 1966, *Ap. J.*, **144**, 827.
OLFE, D. B. 1961, *J.Q.S. R.T.*, **1**, 104.
SOMERVILLE, W. B. 1964, *Ap. J.*, **139**, 192.
SOMERVILLE, W. B. 1965, *ibid.*, **141**, 811.
TSUJI, T. 1964, *Ann. Tokyo Astr. Obs. Ser.* 2, **9**, 1.
VARDYA, M. S., and KANDEL, R. 1967, *Ann. d'Ap.*, **30**, 111.
WOOLF, N. J., SCHWARZSCHILD, N., and ROSE, W. K. 1964, *Ap. J.*, **140**, 833.

Note to the Tables of Models

Tables 1–8 give the data for the physical structure of the following nongray dwarf models.

T_{eff}	Description	Table
3500°	no water vapor	1
3500°	with water vapor	2
2500°	no water vapor	3
2500°	with water vapor	4
2500°	with water vapor and 0.01 normal metals	5
2500°	with water vapor, H_2 dipole and 0.01 normal metals	6
1500°	with water vapor	7
1500°	with water vapor and H_2 dipole	8

The various quantities given in these tables are described below.

1st column: optical depth at 10,000 Å

2nd column: temperature in °K.

3rd column: log (gas pressure in dynes/cm²).

4th column: log (electron pressure in dynes/cm²).

5th column: primary sources of electrons; the numbers listed against each element give the percentage of electrons contributed by that element.

6th column: log (opacity at 10,000 Å in cm²/g of stellar material).

7th column: primary sources of opacity at 10,000 Å; S2 stands for Rayleigh scattering due to molecular hydrogen, H^- for the negative hydrogen ion, H_2 for H_2^-, HE for He^-, ME for metals, H for hydrogen, SH for Rayleigh scattering due to atomic hydrogen, W for water, and PD for H_2 pressure-induced dipole.

8th column: log (density in g/cm³).

9th column: log (number of hydrogen atoms/cm³).

10th column: log (number of ionized hydrogen atoms/cm³).

11th column: log (number of hydrogen molecules/cm³).

12th column: log (number of water molecules/cm³).

13th column: log (adiabatic temperature gradient, $dT/d\tau$).

14th column: C in this column means that the layer is convectively unstable.

15th column: log (physical depth in cm).

TABLE 1

3500 DEGREE LOG G=5
EFFECTIVE TEMPERATURE = 3500   LOG(SURFACE GRAVITY) = 5.000   OPACITY AT 10000 ANGSTROMS   TEMP=   3 064

TAU	TEMP	LOG P GAS	LOG P ELE	PRIMARY SOURCES OF FREE ELECTRONS	LOG KAPPA	PRIMARY SOURCES OF 10000 OPACITY	LOG RHO	LOG H	LOG H+	LOG H2	LOG H2O	LOG ADIAB GRADIENT	LOG DEPTH
.0010	2836	4.234	-1.408	NA=81 CA=13 K = 3	-2.057	H=99 H2=0 S2= 0	-6.878	16.31	1.72	16.24	13.10	5.130	6.725
.0013	2836	4.301	-1.353	NA=82 CA=12 K = 3	-2.031	H=99 H2=0 S2= 0	-6.803	16.36	1.72	16.33	13.20	5.040	6.708
.0016	2836	4.369	-1.299	NA=83 CA=12 K = 3	-2.007	H=99 H2=0 S2= 0	-6.730	16.40	1.71	16.42	13.29	4.952	6.691
.0020	2836	4.439	-1.246	NA=83 CA=11 K = 3	-1.984	H=99 H2=0 S2= 0	-6.654	16.45	1.69	16.51	13.38	4.864	6.673
.0025	2836	4.509	-1.192	NA=84 CA=10 K = 3	-1.962	H=99 H2=0 S2= 0	-6.577	16.49	1.68	16.60	13.48	4.776	6.654
.0032	2835	4.581	-1.138	NA=85 CA=10 K = 4	-1.942	H=99 H2=0 S2= 0	-6.499	16.54	1.67	16.69	13.57	4.689	6.635
.0040	2835	4.654	-1.085	NA=85 CA= 9 K = 4	-1.922	H=99 H2=0 S2= 0	-6.420	16.58	1.66	16.78	13.66	4.604	6.614
.0050	2835	4.729	-1.031	NA=85 CA= 9 K = 4	-1.903	H=99 H2=0 S2= 0	-6.340	16.63	1.65	16.87	13.75	4.516	6.592
.0063	2836	4.804	-.977	NA=86 CA= 8 K = 4	-1.883	H=99 H2=0 S2= 0	-6.260	16.67	1.65	16.96	13.84	4.430	6.569
.0079	2837	4.880	-.923	NA=86 CA= 8 K = 4	-1.864	H=99 H2=0 S2= 0	-6.179	16.72	1.66	17.05	13.93	4.342	6.545
.0100	2840	4.957	-.868	NA=86 CA= 8 K = 5	-1.844	H=99 H2=1 S2= 0	-6.098	16.77	1.67	17.14	14.02	4.255	6.519
.0126	2844	5.034	-.814	NA=86 CA= 7 K = 5	-1.827	H=99 H2=1 S2= 0	-6.016	16.81	1.67	17.23	14.10	4.169	6.491
.0158	2849	5.113	-.758	NA=86 CA= 7 K = 5	-1.806	H=99 H2=1 HE= 0	-5.935	16.86	1.69	17.32	14.19	4.081	6.462
.0200	2855	5.191	-.701	NA=86 CA= 7 K = 6	-1.785	H=99 H2=1 HE= 0	-5.854	16.91	1.73	17.40	14.28	3.991	6.431
.0251	2863	5.269	-.643	NA=86 CA= 7 K = 6	-1.760	H=99 H2=1 HE= 0	-5.774	16.96	1.77	17.49	14.36	3.899	6.398
.0316	2870	5.347	-.583	NA=86 CA= 7 K = 6	-1.733	H=98 H2=1 HE= 0	-5.695	17.01	1.83	17.57	14.44	3.804	6.362
.0398	2882	5.423	-.525	NA=86 CA= 7 K = 6	-1.710	H=98 H2=1 HE= 0	-5.617	17.06	1.89	17.65	14.52	3.713	6.324
.0501	2892	5.499	-.462	NA=85 CA= 7 K = 7	-1.675	H=98 H2=2 HE= 0	-5.541	17.12	1.98	17.73	14.60	3.614	6.282
.0631	2896	5.573	-.397	NA=85 CA= 7 K = 7	-1.635	H=98 H2=2 HE= 0	-5.468	17.18	2.09	17.81	14.67	3.512	6.238
.0794	2913	5.645	-.330	NA=85 CA= 7 K = 7	-1.595	H=98 H2=2 HE= 0	-5.397	17.24	2.23	17.88	14.75	3.407	6.189
.1000	2933	5.715	-.259	NA=85 CA= 7 K = 7	-1.547	H=98 H2=2 HE= 2	-5.331	17.30	2.39	17.95	14.81	3.298	6.136
.1259	2958	5.781	-.185	NA=85 CA= 7 K = 6	-1.491	H=98 H2=2 HE= 2	-5.268	17.36	2.58	18.01	14.87	3.182	6.079
.1585	2989	5.844	-.107	NA=85 CA= 7 K = 6	-1.426	H=98 H2=2 HE= 2	-5.211	17.43	2.82	18.06	14.93	3.061	6.016
.1995	3023	5.902	-.026	NA=84 CA= 8 K = 6	-1.356	H=98 H2=2 HE= 2	-5.159	17.50	3.08	18.11	14.98	2.938	5.946
.2512	3061	5.957	.054	NA=84 CA= 8 K = 6	-1.284	H=98 H2=2 HE= 2	-5.111	17.57	3.36	18.16	15.02	2.815	5.867
.3162	3101	6.007	.134	NA=83 CA= 9 K = 6	-1.210	H=98 H2=2 HE= 2	-5.067	17.67	3.60	18.23	15.06	2.684	5.774
.3981	3109	6.057	.210	NA=83 CA= 9 K = 5	-1.128	H=98 H2=2 HE= 2	-5.017	17.71	4.01	18.26	15.12	2.564	5.664
.5012	3209	6.102	.310	NA=82 CA=10 K = 5	-1.028	H=98 H2=2 HE= 2	-4.999	17.79	4.41	18.27	15.12	2.427	5.524
.6310	3279	6.142	.417	NA=81 CA=11 K = 5	-.915	H=98 H2=2 HE= 2	-4.973	17.87	4.88	18.28	15.13	2.287	5.335
.7943	3300	6.178	.503	NA=77 CA=15 K = 3	-.815	H=99 H2=2 HE= 2	-4.956	17.95	5.39	18.27	15.13	2.148	5.022
1.000	3451	6.211	.597	NA=74 CA=18 AL= 3	-.704	H=98 H2=2 HE= 1	-4.946	18.04	5.95	18.26	15.11	2.013	0.000
1.259	3551	6.241	.690	NA=69 CA=21 AL= 4	-.597	H=99 H2=2 HE= 1	-4.942	18.12	6.53	18.24	15.07	1.888	5.006
1.585	3668	6.274	.783	NA=63 CA=23 MG= 5	-.489	H=99 H2=1 HE= 1	-4.946	18.20	7.17	18.20	15.00	1.769	5.303
1.995	3804	6.294	.877	NA=55 CA=25 MG= 9	-.382	H=99 H2=1 HE= 1	-4.957	18.28	7.86	18.14	14.88	1.660	5.478
2.512	3954	6.317	.976	NA=47 CA=25 MG=16	-.278	H=99 H2=1 HE= 1	-4.974	18.35	8.56	18.05	14.71	1.566	5.604
3.162	4122	6.339	1.087	NA=37 MG=25 CA=22	-.172	H=99 H2=0 ME= 1	-4.994	18.41	9.26	17.93	14.47	1.481	5.705
3.981	4311	6.360	1.222	MG=36 NA=28 CA=18	-.058	H=99 H2=1 ME= 1	-5.016	18.45	9.94	17.78	14.15	1.406	5.789
5.012	4522	6.379	1.378	MG=47 NA=20 CA=13	.058	H=99 H2=0 ME= 2	-5.039	18.48	10.61	17.60	13.76	1.348	5.860
6.310	4759	6.397	1.548	MG=55 NA=13 CA= 9	.172	H=99 ME=0 H2= 1	-5.059	18.50	11.27	17.39	13.32	1.318	5.922
7.943	5026	6.413	1.720	MG=58 NA= 8 SI=13	.271	H=99 H2=0 H2= 1	-5.077	18.51	11.93	17.15	12.86	1.319	5.977
10.00	5317	6.430	1.884	MG=55 SI=19 NA= 6	.351	H=99 ME=5	-5.092	18.51	12.58	16.91	12.44	1.340	6.030
12.59	5628	6.447	2.045	MG=47 SI=24 H =11	.424	H=98 ME=1	-5.104	18.51	13.19	16.67	12.04	1.350	6.080
15.85	5999	6.464	2.229	MG=35 SI=26 H =24	.556	H=99 ME=2	-5.118	18.50	13.74	16.44	11.64	1.316	6.129
19.95	6307	6.479	2.455	H =48 MG=22 SI=18	.656	H=98 ME=1	-5.124	18.49	14.18	16.22	11.23	1.220	6.172
25.12	6671	6.492	2.716	H =67 MG=13 SI=12	.832	H=98 ME=2	-5.136	18.47	14.58	16.02	10.81	1.075	6.207
31.62	7065	6.503	3.000	H =81 MG= 7 SI= 4	1.031	H=97 ME=2 H = 1	-5.151	18.45	14.92	15.79	10.39	.901	6.235
39.81	7485	6.511	3.287	H =89 SI= 4 MG= 2	1.238	H=96 NA=4 ME= 2	-5.168	18.43	15.22	15.58	9.98	.705	6.257
50.12	7931	6.517	3.568	H =93 SI= 2 MG= 2	1.445	H=94 H = 7 ME= 2	-5.188	18.41	15.50	15.38	9.58	.501	6.273
63.10	8404	6.522	3.839	H =96 C = 1 SI= 1	1.649	H=91 H =11 MF= 2	-5.208	18.39	15.76	15.18	9.20	.292	6.286
79.43	8905	6.526	4.098	H =97 C = 1 SI= 1	1.854	H=86 H =11 MF= 2	-5.231	18.39	16.00	14.99	8.83	.075	6.297

TABLE 2

3500 DEGREE LOG G=5 WITH H2O
EFFECTIVE TEMPERATURE = 3500   LOG(SURFACE GRAVITY) = 3500   OPACITY AT 10000 ANGSTROMS   TEMP= 3 083

TAU	LOG TEMP	LOG P GAS	LOG P ELE	PRIMARY SOURCES OF FREE ELECTRONS	LOG KAPPA	PRIMARY SOURCES OF 10000A OPACITY	LOG RHO	LOG H	LOG H+	LOG H2	LOG H2O	LOG ADIAB GRADIENT	LOG
.0010	2239	4.824	-2.059	NA=60 K =3H CA=1	-2.764	H =79 H--20 S2=1	-6.075	15.74	.00	17.24	14.10	5.474	6.479
.0013	2266	4.910	-1.949	NA=62 K =37 CA=1	-2.735	H =76 H--23 S2=1	-5.994	15.84	.00	17.32	14.18	5.361	6.456
.0016	2297	4.992	-1.829	NA=66 K =34 CA=2	-2.686	H =71 H--27 S2=1	-5.919	15.95	.00	17.40	14.25	5.240	6.432
.0020	2336	5.070	-1.699	NA=68 K =32 CA=2	-2.646	H =66 H--33 H2=1	-5.849	16.06	.00	17.46	14.32	5.108	6.408
.0025	2380	5.142	-1.566	NA=69 K =28 CA=2	-2.598	H =59 H--39 H2=1	-5.786	16.18	.00	17.53	14.39	4.966	6.383
.0032	2424	5.209	-1.430	NA=72 K =25 CA=2	-2.513	H =52 H--46 H2=2	-5.728	16.30	.00	17.58	14.44	4.823	6.359
.0040	2462	5.270	-1.317	NA=74 K =23 CA=2	-2.448	H =52 H--46 H2=2	-5.674	16.40	.00	17.64	14.50	4.692	6.335
.0050	2507	5.328	-1.194	NA=76 K =20 CA=3	-2.366	H =59 W =39 H2=2	-5.626	16.50	.01	17.68	14.54	4.546	6.310
.0063	2560	5.379	-1.061	NA=79 K =17 CA=3	-2.264	H =66 W =32 H2=1	-5.586	16.60	.04	17.72	14.58	4.382	6.287
.0079	2618	5.425	-.929	NA=81 K =15 CA=3	-2.150	H =73 W =25 H2=2	-5.553	16.73	.13	17.75	14.61	4.211	6.264
.0100	2668	5.465	-.816	NA=83 K =13 CA=4	-2.048	H =77 W =21 H2=2	-5.523	16.83	.33	17.78	14.64	4.061	6.243
.0126	2734	5.502	-.726	NA=82 K =11 CA=4	-1.965	H =81 W =14 H2=2	-5.495	16.90	.57	17.80	14.67	3.937	6.221
.0158	2754	5.530	-.628	NA=85 K =11 CA=5	-1.871	H =84 W =11 H2=2	-5.469	16.98	.91	17.82	14.67	3.802	6.199
.0200	2807	5.570	-.528	NA=85 K =10 CA=5	-1.766	H =87 W =8 H2=2	-5.448	17.07	1.34	17.84	14.70	3.658	6.178
.0251	2855	5.600	-.435	NA=85 V =11 CA=7	-1.669	H =89 W =6 H2=2	-5.430	17.15	1.77	17.85	14.72	3.526	6.156
.0316	2892	5.630	-.367	NA=85 CA=1 K =7	-1.601	H =90 W =5 H2=2	-5.408	17.20	2.05	17.87	14.74	3.428	6.133
.0398	2921	5.631	-.309	NA=85 CA=1 V =6	-1.545	H =91 W =5 H2=2	-5.384	17.25	2.29	17.89	14.76	3.343	6.108
.0501	2957	5.691	-.242	NA=85 CA=2 A =1	-1.478	H =92 W =4 H2=2	-5.362	17.31	2.58	17.91	14.77	3.246	6.080
.0631	3002	5.721	-.167	NA=84 CA=2 A =1	-1.399	H =93 W =4 H2=1	-5.342	17.37	2.93	17.92	14.78	3.136	6.050
.0794	3051	5.751	-.090	NA=84 CA=2 MG=1	-1.314	H =94 W =4 H2=1	-5.324	17.44	3.31	17.93	14.80	3.024	6.018
.1000	3100	5.779	-.016	NA=83 CA=3 K =1	-1.235	H =95 H--4 H2=1	-5.307	17.50	3.67	17.94	14.81	2.918	5.984
.1259	3141	5.807	.045	NA=82 CA=11 K =1	-1.171	H =96 H--3 H2=1	-5.254	17.54	3.97	17.95	14.82	2.828	5.945
.1585	3189	5.837	.114	NA=81 CA=14 K =1	-1.103	H =96 H--3 H2=1	-5.270	17.61	4.28	17.96	14.83	2.735	5.901
.1995	3230	5.866	.174	NA=78 CA=15 A =1	-1.035	H =97 H--2 H2=1	-5.253	17.67	4.61	17.97	14.83	2.641	5.851
.2512	3268	5.896	.239	NA=75 CA=19 A =1	-.966	H =97 H--2 H2=1	-5.237	17.72	4.95	17.98	14.82	2.548	5.791
.3162	3303	5.925	.306	NA=73 CA=19 A =1	-.895	H =98 H--1 H2=1	-5.222	17.76	5.32	17.98	14.81	2.454	5.720
.3981	3403	5.955	.372	NA=73 CA=21 A =1	-.824	H =98 H2=1 I =1	-5.208	17.85	5.69	17.98	14.81	2.362	5.631
.5012	3473	5.985	.441	NA=65 CA=22 A =1	-.749	H =98 H2=1 I =1	-5.197	17.91	6.11	17.97	14.79	2.268	5.515
.6310	3542	6.013	.510	NA=65 CA=22 AL=1	-.676	H =98 H2=1 I =1	-5.188	17.97	6.54	17.95	14.77	2.179	5.349
.7943	3628	6.044	.579	NA=61 CA=24 MG=1	-.603	H =98 H2=1 I =1	-5.181	18.03	6.99	17.93	14.70	2.092	5.059
1.000	3716	6.071	.649	NA=56 CA=26 MG=1	-.531	H =99 H2=1 I =1	-5.176	18.09	7.45	17.90	14.63	2.009	0.000
1.259	3819	6.100	.727	NA=50 CA=25 MG=18	-.455	H =99 H2=1 I =1	-5.175	18.15	7.95	17.85	14.52	1.928	5.082
1.585	3909	6.120	.814	NA=43 CA=25 MG=18	-.374	H =99 H2=1 TE =0	-5.183	18.20	8.50	17.78	14.36	1.850	5.394
1.995	4008	6.150	.912	MG=36 NA=24 CA=19	-.289	H =99 H2=1 TE =0	-5.179	18.25	9.04	17.69	14.17	1.776	5.581
2.512	4202	6.208	1.143	MG=43 NA=24 CA=15	-.111	H =99 ME =1 TE =0	-5.190	18.29	9.55	17.59	13.95	1.709	5.716
3.162	4355	6.233	1.284	MG=50 NA=17 CA=19	.013	H =99 ME =1 TE =0	-5.196	18.33	10.07	17.47	13.68	1.648	5.822
3.981	4531	6.256	1.436	MG=56 NA=12 CA=19	.083	H =99 ME =1 H2=0	-5.202	18.35	10.60	17.33	13.35	1.600	5.909
5.012	4734	6.279	1.595	NA=7 MG=58 SI=11	.172	H =99 ME =1 HE =0	-5.209	18.38	11.17	17.15	12.99	1.572	5.983
6.310	4968	6.301	1.747	NA=1 SI=16 MG=55	.247	H =99 ME =1 HE =0	-5.214	18.39	11.73	16.95	12.22	1.573	6.048
7.943	5266	6.324	1.830	SI=23 H =8	.312	H =92 ME =1	-5.215	18.40	12.90	16.55	11.90	1.571	6.107
10.00	5489	6.360	2.048	SI=25 H =7	.389	H =93 ME =1	-5.216	18.40	13.41	16.35	11.20	1.541	6.163
12.59	6004	6.390	2.241	MG=28 SI=22	.499	H =96 ME =1	-5.218	18.40	13.80	16.16	10.83	1.464	6.217
15.85	6264	6.419	2.473	MG=37 SI=16	.669	H =98 ME =1	-5.224	18.40	14.28	15.97	10.45	1.337	6.265
19.95	6762	6.438	2.613	MG=57 SI=10	.829	H =97 ME =1	-5.240	18.39	14.94	15.78	10.05	1.178	6.306
25.12	7144	6.437	3.091	H =84 SI=6	1.230	H =95 H =1	-5.255	18.37	15.58	15.58	9.66	.996	6.339
31.62	7556	6.437	3.568	H =94 C =2	1.435	H =93 H =1	-5.272	18.35	15.23	15.38	9.28	.794	6.364
39.81	7999	6.443	3.837	H =96 C =1	1.641	H =88 H =4	-5.292	18.33	15.38	15.19	8.90	.588	6.384
50.12	8474	6.447	4.096	H =97 C =1	1.849	H =84 H =13 ME=3	-5.313	18.31	15.99	14.80	8.54	.374	6.399
63.10	8983											.148	6.412
79.43													6.422

TABLE 3

2500 DEGREE LOG G=5

EFFECTIVE TEMPERATURE = 2500   LOG(SURFACE GRAVITY) = 5.000   OPACITY AT 10000 ANGSTROMS   TEMP= 3 080

TAU	TEMP	LOG P GAS	LOG P ELE	PRIMARY SOURCES OF FREE ELECTRONS	LOG KAPPA	PRIMARY SOURCES OF 10000A OPACITY	LOG RHO	LOG H	LOG H+	LOG H2	LOG H2O	LOG ADIAB GRADIENT	LOG DEPTH
.0010	1426	6.800	-4.046	K =92 NA= 6 RB= 2	-4.758	S2=99 H2= 1 HE= 0	-3.898	13.93	0.00	19.42	16.28	5.563	6.307
.0013	1429	6.891	-4.025	K =92 NA= 6 RB= 2	-4.757	S2=99 H2= 1 HE= 0	-3.807	13.99	0.00	19.52	16.37	5.471	6.284
.0016	1432	6.984	-3.961	K =92 NA= 6 RB= 2	-4.757	S2=99 H2= 1 HE= 0	-3.715	14.05	0.00	19.61	16.46	5.378	6.259
.0020	1434	7.079	-3.899	K =92 NA= 6 RB= 2	-4.756	S2=99 H2= 1 HE= 0	-3.621	14.11	0.00	19.70	16.55	5.284	6.233
.0025	1436	7.174	-3.843	K =92 NA= 6 RB= 2	-4.756	S2=99 H2= 1 HE= 0	-3.526	14.16	0.00	19.80	16.65	5.189	6.204
.0032	1439	7.270	-3.790	K =92 NA= 6 RB= 2	-4.755	S2=99 H2= 1 HE= 0	-3.430	14.22	0.00	19.89	16.74	5.092	6.172
.0040	1442	7.367	-3.696	K =92 NA= 6 RB= 2	-4.755	S2=98 H2= 2 HE= 0	-3.336	14.31	0.00	19.99	16.84	4.996	6.138
.0050	1445	7.464	-3.599	K =92 NA= 6 RB= 2	-4.752	S2=98 H2= 2 HE= 0	-3.241	14.40	0.00	20.08	16.93	4.900	6.101
.0063	1449	7.562	-3.515	K =92 NA= 6 RB= 2	-4.750	S2=98 H2= 2 HE= 0	-3.146	14.48	0.00	20.18	17.03	4.802	6.060
.0079	1453	7.659	-3.457	K =92 NA= 6 RB= 2	-4.749	S2=97 H2= 2 HE= 0	-3.049	14.54	0.00	20.27	17.13	4.704	6.014
.0100	1456	7.757	-3.438	K =92 NA= 6 RB= 2	-4.749	S2=97 H2= 3 HE= 0	-2.949	14.56	0.00	20.37	17.22	4.604	5.963
.0126	1489	7.855	-3.207	K =91 NA= 7 RB= 2	-4.740	S2=95 H2= 4 HE= 0	-2.862	14.78	0.00	20.46	17.31	4.507	5.905
.0158	1528	7.950	-2.959	K =90 NA= 8 RB= 2	-4.726	S2=92 H2= 7 HE= 1	-2.777	15.02	0.00	20.55	17.40	4.406	5.837
.0200	1568	8.043	-2.708	K =89 NA=10 RB= 2	-4.700	S2=87 H2=11 HE= 1	-2.696	15.25	0.00	20.63	17.48	4.297	5.757
.0251	1608	8.131	-2.478	K =87 NA=11 RB= 2	-4.662	S2=80 H2=17 HE= 3	-2.618	15.47	0.00	20.70	17.56	4.180	5.661
.0316	1634	8.214	-2.315	K =86 NA=12 RB= 2	-4.624	S2=73 H2=23 HE= 3	-2.542	15.62	0.00	20.78	17.63	4.065	5.544
.0398	1708	8.288	-1.959	K =84 NA=15 RB= 1	-4.495	S2=54 H2=38 HE= 7	-2.488	15.95	0.00	20.83	17.69	3.878	5.401
.0501	1790	8.342	-1.604	K =80 NA=18 RB= 1	-4.291	H2=53 S2=34 H+= 7	-2.454	16.27	0.00	20.87	17.72	3.637	5.245
.0631	1867	8.377	-1.232	K =76 NA=22 RB= 1	-4.001	H2=63 S2=17 H+=12	-2.442	16.61	.00	20.88	17.73	3.331	5.092
.0794	1934	8.397	-.889	K =72 NA=27 RB= 1	-3.682	H2=66 H-=18 S2= 8	-2.444	16.91	.00	20.88	17.73	3.010	4.965
.1000	2005	8.410	-.642	K =68 NA=31 RB= 1	-3.432	H2=65 S2=23 HE= 7	-2.448	17.13	.00	20.87	17.73	2.761	4.856
.1259	2093	8.419	-.470	K =65 NA=33 RB= 1	-3.250	H2=63 S2=27 HE= 6	-2.451	17.28	.00	20.87	17.72	2.579	4.746
.1585	2191	8.427	-.278	K =62 NA=36 RB= 1	-3.043	H2=61 S2=36 HE= 6	-2.457	17.44	.00	20.87	17.71	2.375	4.631
.1995	2267	8.433	-.080	K =58 NA=40 RB= 1	-2.822	H2=57 S2=36 HE= 5	-2.466	17.61	.00	20.86	17.70	2.159	4.513
.2512	2367	8.437	.120	K =54 NA=43 CA= 1	-2.590	H2=53 H-=41 HE= 5	-2.477	17.78	.00	20.85	17.69	1.938	4.394
.3162	2439	8.440	.320	K =52 NA=45 CA= 2	-2.374	H2=49 H-=45 HE= 5	-2.489	17.93	.00	20.83	17.68	1.726	4.273
.3981	2508	8.442	.504	K =49 NA=48 CA= 2	-2.198	H-=53 H2=42 HE= 4	-2.498	18.06	.01	20.82	17.67	1.553	4.141
.5012	2561	8.445	.654	K =45 NA=52 CA= 2	-2.020	H-=59 H2=40 HE= 4	-2.508	18.18	.07	20.82	17.66	1.335	4.041
.6310	2610	8.446	.745	K =42 NA=55 CA= 2	-1.857	H-=59 H2=37 HE= 4	-2.518	18.29	.17	20.80	17.65	1.220	3.770
.7943	2716	8.446	.893	K =39 NA=58 CA= 3	-1.678	H-=61 H2=35 HE= 4	-2.529	18.41	.50	20.79	17.63	1.045	3.432
1.000	2806	8.449	1.047	NA=61 K =35 CA= 3	-1.489	H-=63 H2=33 HE= 3	-2.551	18.54	1.23	20.78	17.62	.859	0.000
1.259	2895	8.450	1.201	NA=64 K =31 CA= 4	-1.300	H-=66 H2=31 HE= 3	-2.554	18.66	1.99	20.77	17.61	.672	3.355
1.585	2995	8.451	1.325	NA=67 K =28 CA= 4	-1.110	H-=69 H2=28 HE= 2	-2.557	18.78	2.76	20.75	17.60	.483	3.619
1.995	3091	8.451	1.497	NA=69 K =24 CA= 5	-.934	H-=74 H2=24 HE= 2	-2.579	18.89	3.47	20.74	17.59	.307	3.761
2.512	3200	8.452	1.647	NA=72 K =21 CA= 6	-.745	H-=77 H2=21 HE= 1	-2.592	19.01	4.23	20.73	17.56	.115	3.854
3.162	3324	8.452	1.807	NA=74 K =17 CA= 7	-.545	H-=80 H2=18 HE= 1	-2.617	19.13	5.04	20.70	17.53	-.085	3.920
3.981	3481	8.453	1.969	NA=75 K =14 CA= 9	-.341	H-=82 H2=17 HE= 1	-2.636	19.25	5.85	20.68	17.50	-.293	3.967
5.012	3606	8.453	2.132	NA=76 K =11 CA=10	-.138	H-=84 H2=15 HE= 1	-2.656	19.37	6.65	20.65	17.48	-.501	4.003
6.310	3804	8.453	2.292	NA=76 K = 9 CA=11	.062	H-=85 H2=13 HE= 1	-2.678	19.49	7.46	20.63	17.46	-.705	4.030
7.943	3924	8.453	2.440	NA=75 K = 7 CA=14	.247	H-=86 H2=13 HE= 1	-2.700	19.60	8.21	20.60	17.44	-.893	4.052
10.00	4100	8.454	2.585	NA=72 CA=17 K = 5	.428	H-=87 H2=12 HE= 1	-2.725	19.71	8.97	20.57	17.44	-1.076	4.071
12.59	4300	8.454	2.733	NA=68 CA=20 MG= 8	.612	H-=88 H2=11 HE= 1	-2.754	19.82	9.82	20.53	17.39	-1.256	4.086
15.85	4522	8.454	2.871	NA=62 CA=22 MG= 8	.784	H-=90 H2= 9 HE= 1	-2.787	19.92	10.56	20.48	17.33	-1.421	4.100
19.95	4748	8.454	2.996	NA=46 CA=24 MG19	.936	H-=91 H2= 8 HE= 1	-2.821	20.01	11.26	20.42	17.26	-1.560	4.112
25.12	4987	8.454	3.114	NA=37 MG28 CA=21	1.077	H-=92 H2= 7 HE= 1	-2.858	20.10	12.01	20.36	17.17	-1.683	4.123
31.62	5241	8.454	3.233	NA=28 MG37 CA=17	1.215	H-=93 H2= 6 HE= 1	-2.898	20.17	12.68	20.28	17.06	-1.796	4.135
39.81	5513	8.454	3.360	NA=20 MG49 CA=13	1.354	H-=94 H2= 5 HE= 1	-2.942	20.23	13.32	20.19	16.91	-1.903	4.146
50.12	5805	8.454	3.501	NA=14 SI=10 CA13	1.493	H-=95 H2= 4 HE= 1	-2.987	20.28	13.91	20.09	16.74	-2.003	4.157
63.10	6118	8.454	3.642	MG=49 SI=14 H = 4	1.631	H-=95 H2= 3 HE= 1	-3.033	20.31	14.46	19.96	16.55	-2.094	4.168
79.43	6451	8.454	3.795	MG=48 H = ? SI=12	1.766	H-=96 H2= 3 ME= 1	-3.079	20.34	14.97	19.82	16.33	-2.175	4.178

Model atmospheres for cool dwarf stars* *451*

TABLE 4

2500 DEGREE LOG G=5.0 WITH H2O
EFFECTIVE TEMPERATURE = 2500 LOG(SURFACE GRAVITY) = 5.000 OPACITY AT 5.000 OPACITY AT 10000 ANGSTROMS TEMP= 3 082

TAU	TEMP	LOG P GAS	LOG P ELE	PRIMARY SOURCES OF FREE ELECTRONS	LOG KAPPA	PRIMARY SOURCES OF 10000A OPACITY	LOG RHO	LOG H	LOG H+	LOG H2	LOG H2O	LOG ADIAB GRADIENT	LOG DEPTH
.0010	1393	5.568	-4.887	K =93 NA= 5 RB= 2	-3.547	H2= 6 H2- 0	-5.099	13.14	0.00	18.22	15.08	5.554	6.365
.0013	1400	5.679	-4.801	K =93 NA= 5 RB= 2	-3.537	S2= 6 H2- 0	-5.010	13.22	0.00	18.31	15.16	5.455	6.345
.0016	1414	5.769	-4.660	K =93 NA= 5 RB= 2	-3.515	S2= 5 H2= 0	-4.925	13.35	0.00	18.40	15.23	5.348	6.325
.0020	1434	5.856	-4.511	K =92 NA= 6 RB= 2	-3.487	S2= 5 H2= 0	-4.844	13.50	0.00	18.48	15.33	5.238	6.304
.0025	1456	5.940	-4.337	K =92 NA= 6 RB= 2	-3.455	S2= 5 H2= 0	-4.694	13.66	0.00	18.56	15.41	5.128	6.282
.0032	1483	6.021	-4.155	K =91 NA= 7 RB= 2	-3.424	S2= 5 H2= 0	-4.694	13.84	0.00	18.63	15.48	4.915	6.260
.0040	1510	6.099	-3.976	K =90 NA= 8 RB= 2	-3.389	S2= 4 H2= 0	-4.623	14.01	0.00	18.70	15.55	4.912	6.237
.0050	1537	6.176	-3.794	K =90 NA= 9 RB= 2	-3.356	S2= 4 H2= 0	-4.554	14.17	0.00	18.77	15.62	4.812	6.212
.0063	1567	6.251	-3.610	K =89 NA=10 RB= 2	-3.326	S2= 4 H2= 0	-4.487	14.35	0.00	18.83	15.69	4.710	6.187
.0079	1605	6.324	-3.395	K =87 NA=11 RB= 2	-3.281	S2= 3 H2= 0	-4.424	14.55	0.00	18.90	15.75	4.605	6.159
.0100	1649	6.395	-3.157	K =86 NA=12 RB= 1	-3.237	S2= 3 H2= 0	-4.366	14.77	0.00	18.99	15.81	4.499	6.131
.0126	1697	6.463	-2.917	K =84 NA=14 RB= 1	-3.192	S2= 3 H2= 0	-4.310	15.00	.00	19.01	15.86	4.395	6.100
.0158	1743	6.530	-2.646	K =82 NA=16 RB= 1	-3.152	S2= 2 H2= 0	-4.254	15.20	.00	19.07	15.92	4.297	6.067
.0200	1794	6.596	-2.463	K =80 NA=18 RB= 1	-3.107	S2= 2 H1= 1	-4.201	15.41	.00	19.12	15.97	4.199	6.031
.0251	1866	6.659	-2.166	K =77 NA=22 RB= 1	-3.052	S2= 2 H1= 1	-4.155	15.68	.00	19.17	16.02	4.091	5.991
.0316	1923	6.720	-1.872	K =73 NA=25 RB= 1	-2.993	S2= 2 H1= 1	-4.112	15.95	.00	19.21	16.06	3.983	5.948
.0398	1963	6.778	-1.598	K =69 NA=29 RB= 1	-2.933	S1= 1 H2= 2	-4.070	16.19	.11	19.25	16.10	3.875	5.900
.0501	1999	6.834	-1.346	K =66 NA=33 RB= 1	-2.874	H1= 8 H2= 3	-4.029	16.40	.33	19.29	16.15	3.768	5.860
.0631	2126	6.887	-1.094	K =61 NA=37 RB= 1	-2.786	H1=25 H2= 5	-3.994	16.64	.72	19.33	16.18	3.635	5.786
.0794	2264	6.934	-.848	K =56 NA=42 CA= 1	-2.675	H1=25 H2= 7	-3.964	16.86	1.07	19.36	16.21	3.482	5.722
.1000	2351	6.974	-.612	NA=47 CA= 1 K = 51	-2.547	H2=36 H2= 9	-3.939	17.05	1.71	19.38	16.24	3.318	5.655
.1259	2450	7.011	-.449	NA=51 CA= 1 K = 43	-2.431	H2=33 H2=10	-3.916	17.20	2.21	19.41	16.26	3.164	5.585
.1585	2504	7.041	-.281	NA=58 V = 39 CA= 2	-2.295	H1=25 H2=11	-3.897	17.34	.00	19.44	16.28	2.996	5.511
.1995	2552	7.069	-.164	NA=58 K = 36 CA= 2	-2.170	H1=12 H2=11	-3.886	17.46	.00	19.45	16.29	2.842	5.432
.2512	2613	7.091	-.012	NA=61 CA= 3 K = 32	-2.043	H1=12 H2=10	-3.859	17.57	.00	19.47	16.31	2.687	5.344
.3162	2657	7.112	.119	NA=67 K = 29 CA= 3	-1.906	H1= 6 H2=10	-3.845	17.68	.33	19.47	16.33	2.523	5.246
.3981	2757	7.131	.236	NA=70 V = 26 CA= 4	-1.778	H1= 6 M2= 9	-3.842	17.78	.72	19.47	16.33	2.371	5.134
.5012	2855	7.149	.340	NA=72 V = 23 CA= 5	-1.647	H1= 6 M2= 9	-3.842	17.87	1.71	19.36	16.33	2.222	4.994
.6310	2958	7.163	.451	NA=76 V = 19 CA= 5	-1.527	H1= 4 M2= 5	-3.842	17.96	1.71	19.32	16.33	2.078	4.803
.7943	2970	7.177	.557	NA=76 V = 15 CA= 5	-1.402	H1=25 M2= 7	-3.838	18.04	2.21	19.48	16.33	1.932	4.466
1.000	2988	7.190	.666	NA=76 CA=17 V = 3	-1.270	H1=87 H2= 8	-3.837	18.13	2.75	19.48	16.33	1.779	0.000
1.259	3096	7.201	.787	NA=78 V = 14 CA= 6	-1.121	H1=89 H2= 7	-3.839	18.23	3.98	19.47	16.33	1.609	4.746
1.585	3230	7.211	.913	NA=80 CA=10 V = 6	-.963	H1=90 H2= 6	-3.850	18.33	4.56	19.46	16.32	1.431	4.723
1.995	3325	7.220	1.025	NA=80 CA=11 K = 5	-.824	H1=92 H2= 6	-3.859	18.44	5.12	19.45	16.31	1.272	4.762
2.512	3436	7.227	1.154	NA=78 CA=13 K = 5	-.683	H1=93 H2= 5	-3.871	18.51	5.77	19.42	16.28	1.113	4.841
3.162	3555	7.233	1.293	NA=75 CA=16 K = 4	-.523	H1=94 H2= 4	-3.887	18.60	6.25	19.40	16.25	.943	4.611
3.981	3691	7.239	1.504	NA=71 CA=18 MG= 2	-.365	H1=95 H2= 3	-3.908	18.70	6.25	19.36	16.23	.774	4.121
5.012	3852	7.244	1.615	NA=69 MG=12 CA= 9	-.193	H1=96 H2= 3	-3.932	18.80	7.17	19.32	16.17	.600	5.211
6.310	3979	7.248	1.713	MG=54 SI=19 NA= 7	-1.059	H1=96 H2= 2	-3.958	18.89	8.54	19.27	16.10	.336	5.247
7.943	—	7.252	—	MG=34 H =33 SI=18	—	—	—	18.97	—	—	—	—	—
10.00	4135	7.256	1.809	NA=55 CA=24 MG=10	.193	H1=97 H2= 1	-3.989	19.05	9.23	19.20	16.01	.223	5.281
12.59	4415	7.259	1.908	CA=25 MG=16 NA= 9	.315	H1=97 H2= 1	-4.006	19.12	9.40	19.12	15.88	.120	5.313
15.85	4488	7.262	2.008	MG=39 CA=22 NA= 5	.430	H1=98 H2= 0	-4.061	19.18	10.61	19.02	15.71	.031	5.344
19.95	4688	7.266	2.111	MG=31 CA=19 SI=15	.537	H1=98 H2= 0	-4.101	19.23	11.40	18.91	15.51	-.045	5.375
25.12	4903	7.269	2.236	NA=32 CA=15 K = 6	.655	H1=98 H2= 0	-4.142	19.26	11.80	18.77	15.24	-.118	5.405
31.62	5164	7.272	2.385	MG=52 NA=15 CA=11	.780	H1=98 ME= 1	-4.186	19.29	13.04	18.59	14.89	-.178	5.435
39.81	5441	7.274	2.535	NA=56 SI=12 NA=11	.888	H1=98 ME= 1	-4.226	19.30	13.39	18.39	14.52	-.207	5.465
50.12	5735	7.277	2.682	MG=56 NA= 7	.980	H1=98 ME= 1	-4.260	19.31	14.12	18.18	14.16	-.208	5.495
63.10	6055	7.280	2.837	MG=46 SI=19 H =16	1.068	H1=98 ME= 1	-4.290	19.29	13.96	17.96	13.80	-.201	5.526
79.43	6401	7.283	3.022	MG=34 H =33 SI=18	1.178	H1=98 HE= 0	-4.318	19.28	14.59	17.74	13.43	-.219	5.557

TABLE 5

2500 DEGREE LOG G=5 WITH H2O, .01 METALS

EFFECTIVE TEMPERATURE = 2500   LOG(SURFACE GRAVITY) = 5.000   OPACITY AT 10000 ANGSTROMS   TEMP= 3 154

TAU	TEMP	LOG P GAS	LOG P ELE	PRIMARY SOURCES OF FREE ELECTRONS	LOG KAPPA	PRIMARY SOURCES OF 10000A OPACITY	LOG RHO	LOG H	LOG H+	LOG H2	LOG H2O	LOG ADIAB GRADIENT	LOG DEPTH
.0010	1565	6.697	-4.397	K =89 NA=10 RB=2	-4.650	S=79 K =21 H2=0	-4.047	15.57	0.00	19.28	14.14	5.591	6.418
.0013	1597	6.787	-4.201	K =88 NA=11 RB=2	-4.642	S=77 K =22 H2=0	-3.966	14.75	0.00	19.36	14.22	5.501	6.399
.0016	1600	6.878	-4.139	K =88 NA=11 RB=2	-4.641	S=77 K =22 H2=0	-3.876	14.81	0.00	19.45	14.31	5.409	6.378
.0020	1603	6.970	-4.079	K =87 NA=11 RB=2	-4.640	S=77 K =22 H2=0	-3.784	14.87	0.00	19.55	14.49	5.316	6.335
.0025	1607	7.064	-4.016	K =87 NA=11 RB=2	-4.638	S=77 K =23 H2=0	-3.691	14.93	0.00	19.64	14.49	5.222	6.305
.0032	1609	7.159	-3.960	K =87 NA=11 RB=2	-4.638	S=76 K =23 H2=1	-3.597	14.99	0.00	19.73	14.58	5.127	6.277
.0040	1632	7.254	-3.820	K =87 NA=11 RB=2	-4.632	S=76 K =23 H2=1	-3.507	15.12	0.00	19.82	14.68	5.031	6.247
.0050	1649	7.349	-3.659	K =86 NA=13 RB=1	-4.627	S=74 K =24 H2=1	-3.419	15.27	0.00	19.91	14.76	4.933	6.214
.0063	1660	7.443	-3.591	K =85 NA=14 RB=1	-4.615	S=73 K =25 H2=2	-3.332	15.42	0.00	20.01	14.85	4.836	6.214
.0079	1705	7.536	-3.349	K =84 NA=15 RB=1	-4.605	S=71 K =26 H2=2	-3.245	15.56	0.00	20.09	14.94	4.739	6.178
.0100	1718	7.629	-3.249	K =83 NA=15 RB=1	-4.591	S=70 K =27 H2=3	-3.155	15.66	.00	20.18	15.03	4.642	6.139
.0126	1747	7.722	-2.956	K =80 NA=19 RB=1	-4.583	S=68 K =28 H2=3	-3.071	15.83	.00	20.26	15.11	4.541	6.096
.0158	1792	7.812	-2.630	K =80 NA=19 RB=1	-4.561	S=66 K =28 H2=6	-2.991	16.02	.00	20.34	15.19	4.437	6.048
.0200	1843	7.899	-2.430	K =80 NA=19 RB=1	-4.524	S=64 K =28 H2=9	-2.915	16.23	.00	20.41	15.27	4.326	5.995
.0251	1900	7.981	-2.399	K =75 NA=24 RB=1	-4.477	S=53 K =28 H2=13	-2.847	16.45	.00	20.48	15.34	4.205	5.936
.0316	1945	8.057	-2.203	K =73 NA=27 RB=1	-4.425	S=47 K =28 H2=18	-2.781	16.62	.00	20.55	15.41	4.085	5.871
.0398	2002	8.127	-1.912	K =66 NA=32 RB=1	-4.367	S=39 K =24 H2=24	-2.723	16.81	.00	20.61	15.47	3.946	5.800
.0501	2082	8.189	-1.774	K =62 NA=37 RB=1	-4.237	H2=31 K =27 S=22	-2.675	17.01	.00	20.66	15.52	3.788	5.724
.0631	2137	8.239	-1.537	K =62 NA=37 RB=1	-4.088	H2=38 S=22 K =21	-2.640	17.22	.00	20.69	15.56	3.598	5.647
.0794	2215	8.278	-1.313	K =57 NA=43 RB=1	-3.913	H2=44 S=24 K =16	-2.516	17.42	.00	20.71	15.58	3.395	5.573
.1000	2264	8.309	-1.127	K =52 NA=47	-3.746	H2=44 S=23 K =10	-2.599	17.58	.00	20.73	15.60	3.210	5.503
.1259	2332	8.335	-1.002	K =49 CA=49	-3.630	H2=44 S=37	-2.582	17.69	.00	20.75	15.62	3.072	5.430
.1585	2385	8.358	-.870	NA=53 K =45	-3.360	H2=44 S=43	-2.568	17.81	.00	20.76	15.65	2.923	5.350
.1995	2442	8.379	-.738	NA=57 K =41 CA=2	-3.212	H2=43 HE=3	-2.558	17.93	.02	20.77	15.65	2.771	5.264
.2512	2511	8.396	-.601	K =61 NA=36 CA=3	-3.056	H2=48 HE=4	-2.551	18.05	.02	20.78	15.66	2.612	5.170
.3162	2571	8.412	-.461	NA=66 K =31 CA=3	-2.909	H2=51 HE=3	-2.547	18.17	.07	20.78	15.67	2.447	5.067
.3981	2648	8.427	-.332	NA=73 K =23 CA=3	-2.760	H2=55 HE=3	-2.545	18.28	.17	20.78	15.67	2.292	4.950
.5012	2700	8.436	-.208	NA=76 K =20 CA=3	-2.647	H2=58 HE=37	-2.544	18.38	.34	20.79	15.67	2.142	4.810
.6310	2757	8.444	-.107	NA=76 K =20 CA=3	-2.547	H2=60 HE=37	-2.544	18.47	.77	20.79	15.67	2.017	4.618
.7943	2819	8.450	.006	NA=78 K =17 CA=4	-2.514	H2=62 HE=33	-2.542	18.56	2.40	20.79	15.67	1.877	4.297
1.000	2894	8.464	.130	NA=80 K =14 CA=5	-2.366	H2=65 H=31	-2.544	18.66	3.03	20.78	15.67	1.724	0.000
1.259	2974	8.472	.256	NA=83 CA=7 K =4	-2.213	H2=68 H=27	-2.548	18.77	3.68	20.78	15.67	1.565	4.250
1.585	3068	8.478	.394	NA=83 CA=9 K =7	-2.040	H2=70 H=27	-2.555	18.88	4.40	20.76	15.66	1.392	4.524
1.995	3167	8.484	.526	NA=79 CA=11 K =7	-1.882	H2=73 H=23	-2.562	18.99	5.11	20.75	15.65	1.223	4.672
2.512	3272	8.488	.649	NA=79 CA=11 K =4	-1.724	H2=75 H=23	-2.580	19.10	5.82	20.73	15.63	1.062	4.772
3.162	3395	8.494	.780	NA=74 H=18 CA=17	-1.557	H2=77 H=21	-2.593	19.22	6.61	20.73	15.62	.889	4.846
3.981	3523	8.496	.898	H=64 MG=20 SI=7	-1.403	H2=81 H=17	-2.608	19.33	7.37	20.71	15.60	.728	4.902
5.012	3645	8.499	.997	H=70 MG=22 CA=6	-1.274	H2=81 H=17	-2.621	19.43	8.16	20.70	15.57	.593	4.902
6.310	3767	8.502	1.084	MG=64 CA=23 SI=3	-1.157	H2=83 H=16	-2.635	19.52	8.71	20.68	15.55	.336	5.029
7.943	3919	8.505	1.181	CA=57 CA=26 SI=1	-1.027	H2=84 H=16	-2.654	19.63		20.66	15.55		
10.00	4100	8.507	1.287	NA=47 CA=26 MG=16	-.880	H=86 H2=13	-2.676	19.74	10.29	20.63	15.53	.191	5.063
12.59	4303	8.509	1.448	NA=37 MG=35 CA=23	-.730	H=88 H2=10	-2.730	19.85	11.13	20.59	15.50	.036	5.092
15.85	4513	8.511	1.702	NA=27 CA=18 MG=10	-.564	H=89 H2=10	-2.731	19.95	11.89	20.55	15.48	-.125	5.118
19.95	4762	8.513	1.898	MG=18 CA=12	-.369	H=90 H2=9	-2.765	20.05	12.65	20.49	15.46	-.308	5.140
25.12	5000	8.514	2.110	MG=44 H=18	-.157	H=92 H2=7	-2.802	20.14	13.30	20.43	15.42	-.503	5.158
31.62	5238	8.515	2.404	H=64 MG=20 SI=9	.073	H=93 H2=7	-2.838	20.22	13.84	20.36	15.37	-.712	5.172
39.81	5520	8.516	2.736	H=74 MG=22 SI=3	.379	H=94 H2=6	-2.881	20.28	14.33	20.27	15.28	-.987	5.182
50.12	5820	8.516	3.077	H=92 MG=5 SI=1	.716	H=94 H2=4	-2.927	20.36	14.75	20.17	15.13	-1.285	5.189
63.10	6137	8.516	3.410	H=96 MG=2 SI=1	1.051	H=95 H2=4	-2.974	20.36	15.11	20.05	14.92	-1.575	5.193
79.43	6472	8.516		H=96	1.367	H=96 H2=3	-3.019	20.39	15.44	19.91	14.66	-1.839	5.196

TABLE 6

2500 DEGREE LOG G=5 H2O+H2 PID=.01 METALS
EFFECTIVE TEMPERATURE = 2500 LOG(SURFACE GRAVITY) = 2500 OPACITY AT 10000 ANGSTROMS = 5.000 OPACITY AT 5.000 TEMP = 3 233

TAU	TEMP	LOG P GAS	LOG P FLE	PRIMARY SOURCES OF FREE ELECTRONS	LOG KAPPA	PRIMARY SOURCES OF 10000A OPACITY	LOG RHO	LOG H	LOG H+	LOG H2	LOG H2O	LOG ADIAB GRADIENT	LOG DEPTH
.0010	1317	6.168	-6.015	K =.94 NA= 3 RB= 2	-3.917	PD=84 SI=15 K = 2	-4.504	13.03	0.00	18.83	13.68	5.327	6.330
.0013	1332	6.224	-5.954	K =.94 NA= 2 RB= 2	-3.870	PD=85 SI=13 K = 2	-4.450	13.09	0.00	18.88	13.73	5.226	6.318
.0016	1359	6.280	-5.755	K =.94 NA= 4 RB= 2	-3.827	PD=87 SI=12 K = 1	-4.403	13.28	0.00	18.93	13.78	5.134	6.305
.0020	1375	6.334	-5.676	K =.93 NA= 4 RB= 2	-3.781	PD=88 SI=11 K = 1	-4.352	13.40	0.00	18.98	13.83	5.037	6.291
.0025	1370	6.391	-5.628	K =.94 NA= 4 RB= 2	-3.732	PD=89 SI= 9 K = 1	-4.295	13.37	0.00	19.04	13.89	4.930	6.277
.0032	1359	6.446	-5.668	K =.94 NA= 4 RB= 2	-3.680	PD=90 SI= 8 K = 1	-4.237	13.37	0.00	19.09	13.95	4.821	6.263
.0040	1359	6.500	-5.646	K =.94 NA= 4 RB= 2	-3.631	PD=91 SI= 7 K = 1	-4.183	13.39	0.00	19.15	14.00	4.718	6.249
.0050	1366	6.553	-5.575	K =.94 NA= 4 RB= 2	-3.583	PD=92 SI= 7 K = 1	-4.131	13.46	0.00	19.20	14.05	4.618	6.234
.0063	1390	6.607	-5.398	K =.93 NA= 5 RB= 2	-3.538	PD=93 SI= 6 K = 1	-4.085	13.63	0.00	19.25	14.10	4.526	6.218
.0079	1420	6.661	-5.193	K =.93 NA= 5 RB= 1	-3.497	PD=94 SI= 6 K = 1	-4.040	13.82	0.00	19.30	14.14	4.435	6.202
.0100	1451	6.715	-4.994	K =.92 NA= 6 RB= 1	-3.449	PD=94 SI= 5 K = 1	-3.996	14.02	0.00	19.33	14.19	4.345	6.184
.0126	1509	6.624	-4.604	K =.91 NA= 7 RB= 1	-3.404	PD=95 SI= 4 K = 1	-3.950	14.19	0.00	19.38	14.23	4.252	6.165
.0158	1542	6.878	-4.428	K =.89 NA= 9 RB= 1	-3.356	PD=95 SI= 4 K = 1	-3.902	14.33	0.00	19.43	14.28	4.156	6.145
.0200	1562	6.932	-4.154	K =.86 NA=10 RB= 2	-3.313	PD=96 SI= 3 K = 1	-3.860	14.55	0.00	19.47	14.32	4.068	6.124
.0251	1592	6.987	-3.844	K =.86 NA=13 RB= 2	-3.269	PD=96 SI= 3 K = 1	-3.819	14.60	0.00	19.51	14.36	3.981	6.100
.0316	1653	7.042	-3.551	K =.83 NA=15 RB= 1	-3.227	PD=96 SI= 3 K = 1	-3.781	15.06	0.00	19.55	14.40	3.898	6.075
.0398	1716	7.097	-3.304	K =.81 NA=16 RB= 1	-3.185	PD=96 SI= 2 K = 1	-3.741	15.36	0.00	19.59	14.44	3.814	6.046
.0501	1771	7.153	-3.035	K =.78 NA=21 RB= 1	-3.140	PD=96 SI= 2 K = 1	-3.700	15.59	0.00	19.63	14.48	3.725	6.014
.0631	1836	7.208	-2.742	K =.73 NA=25 RB= 1	-3.096	PD=97 SI= 2 K = 1	-3.660	15.83	0.00	19.67	14.52	3.638	5.979
.0794	1913	7.633	...	K =.73 NA=25 RB= 1	-3.052	PD=97 SI= 2 K = 1	-3.623	16.10	0.00	19.71	14.56	3.553	5.938
.1000	2005	7.264	-2.426	K =.67 NA=31 RB= 1	-3.009	PD=96 SI= 2 K = 1	-3.587	16.39	.00	19.74	14.60	3.468	5.891
.1259	2093	7.319	-2.145	K =.61 NA=38 RB= 1	-2.961	PD=96 SI= 2 K = 1	-3.551	16.65	.00	19.78	14.64	3.377	5.836
.1585	2157	7.374	-1.948	K =.57 NA=43 CA= 1	-2.910	PD=97 SI= 2 H-= 1	-3.509	16.83	.00	19.82	14.68	3.279	5.772
.1995	2215	7.409	-1.783	K =.50 NA=49 CA= 1	-2.589	PD=97 SI= 2 H-= 1	-3.486	16.98	.00	19.84	14.71	2.931	5.726
.2512	2260	7.441	-1.611	K =.44 CA=11 K = 1	-2.563	H-=46 PD=45 SI= 2	-3.466	17.31	.00	19.86	14.73	2.879	5.676
.3162	2352	7.476	-1.428	NA=.84 CA= 1 K = 1	-2.532	H-=65 PD=34 H2= 1	-3.444	17.31	.00	19.88	14.75	2.849	5.617
.3981	2481	7.513	-1.294	NA=.82 CA=15 K = 3	-2.495	H-=25 PD=25 H2= 9	-3.421	17.47	.00	19.91	14.79	2.670	5.537
.5012	2501	7.552	-1.072	NA=.78 CA=20 AL= 1	-2.452	H-=77 PD=18 H2= 2	-3.394	17.72	.36	19.93	14.81	2.582	5.412
.6310	2575	7.592	-.905	NA=.71 CA=24 AL= 5	-2.401	H-=81 PD=14 H2= 3	-3.364	17.76	.36	19.96	14.84	2.486	5.244
.7943	2649	7.633	-.745	NA=.63 CA=24 K =15	-2.345	H-=85 PD=10 H2= 4	-3.337	17.90	.91	19.99	14.87	2.486	4.948
1.000	2731	7.674	-.576	NA=.55 MG=26 CA= 4	-2.277	PD=73 H-=27 SI= 2	-3.310	18.05	1.66	20.02	14.90	2.378	0.000
1.259	2818	7.713	-.410	NA=.94 NA=25 CA= 7	-2.199	PD=65 H-=35 H2= 6	-3.283	18.19	2.43	20.04	14.93	2.259	4.947
1.585	2915	7.750	-.240	MG=.85 K = 2 CA=12	-2.106	PD=56 H-=35 H2= 7	-3.299	18.33	3.25	20.07	14.95	2.126	5.243
1.995	3027	7.785	-.072	NA=.84 CA=17 CA=11	-1.998	H-=46 PD=45 H2= 8	-3.250	18.48	4.14	20.07	14.96	1.982	5.411
2.512	3151	7.815	.094	CA=.82 CA=11 K = 4	-1.868	H-=55 PD=34 H2= 9	-3.239	18.63	5.06	20.08	14.97	1.815	5.608
3.162	3293	7.841	.252	CA=.78 CA=15 CA= 2	-1.724	H-=65 PD=25 H2= 9	-3.234	18.79	6.04	20.08	14.97	1.637	5.670
3.981	3440	7.864	.390	NA=.71 CA=20 AL= 1	-1.543	H-=77 PD=18 H2= 4	-3.234	18.4	6.99	20.07	14.97	1.466	5.722
5.012	3576	7.884	.502	NA=.63 CA=24 KG=15	-1.461	H-=81 PD=14 H2= 3	-3.235	19.06	7.80	20.07	14.96	1.318	5.766
6.310	3717	7.902	.606	NA=.55 CA=26 KG= 8	-1.343	H-=85 PD=10 H2= 2	-3.239	19.17	8.58	20.06	14.95	1.178	5.804
7.943	3887	7.918	.721	NA=.46 CA=26 KG=15	-1.207	H-=85 H2= 7 PD= 7	-3.249	19.30	9.44	20.04	14.93	1.024	5.837
10.00	4082	7.932	.856	NA=.35 KG=26 CA= 6	-1.048	H-=8R H2= 6 PD= 4	-3.267	19.42	10.33	20.01	14.90	.854	5.864
12.59	4292	7.944	1.011	MG=.39 NA=25 CA=16	-.869	H-=91 H2= 5 PD= 1	-3.290	19.53	11.16	19.97	14.86	.673	5.887
15.85	4511	7.954	1.177	MG=.48 NA=17 CA=11	-.679	H-=93 H2= 5 PD= 1	-3.319	19.63	11.92	19.93	14.81	.490	5.906
19.95	4754	7.961	1.367	MG=.62 K =13 NA=10	-.469	H-=95 H2= 4 PD= 1	-3.356	19.72	12.66	19.83	14.74	.297	5.920
25.12	4967	7.967	1.574	VG=.60 K =32 SI= 6	-.250	H-=96 H2= 4 K = 1	-3.393	19.79	13.24	19.75	14.66	.108	5.932
31.62	5230	7.971	1.830	MG=.57 MG=25 SI= 6	.011	H-=97 H2= 3 HE= 0	-3.433	19.89	13.64	19.64	14.53	-.128	5.940
39.81	5505	7.973	2.151	NA=.95 VG=12 SI= 4	.327	H-=97 H2= 2 HE= 1	-3.476	20.00	14.36	19.37	14.27	-.666	5.945
50.12	5197	7.975	2.488	MG=.63 SI= 6 K = 6	.644	H-=98 H2= 1 HE= 1	-3.523	19.93	14.87	19.20	13.75	-.899	5.949
63.10	6157	7.977	2.905	VG=.5 SI= 3 SI= 1	.946	H-=98 H2= 2 HE= 1	-3.565	19.93	14.87	19.20	13.75	-.899	5.949
79.43	6424	7.977	3.136	H =.97 VG= 1 SI= 1	1.221	H-=98 H2= 1 HE= 1	-3.603	19.93	15.18	19.03	13.40	-1.101	5.951

TABLE 7

1500 DEGREE LOG G=5 WITH H2O

EFFECTIVE TEMPERATURE = 1500   LOG(SURFACE GRAVITY) = 5.000   OPACITY AT 10000 ANGSTROMS   TEMP= 3 092

TAU	TEMP	LOG P GAS	LOG P ELE	PRIMARY SOURCES OF FREE ELECTRONS	LOG KAPPA	PRIMARY SOURCES OF 10000A OPACITY	LOG RHO	LOG H	LOG H+	LOG .H2	LOG H2O	LOG ADIAB GRADIENT	LOG DEPTH
.0010	900	6.659	-8.893	K =95 RB= 5 NA= 0	-4.586	S2=67 W =33 H2= 0	-3.838	9.29	0.00	19.48	16.33	5.353	6.207
.0013	918	6.742	-8.725	K =95 RB= 5 NA= 0	-4.552	S2=62 W =38 H2= 0	-3.764	9.57	0.00	19.56	16.49	5.245	6.190
.0016	921	6.824	-8.609	K =95 RB= 4 NA= 0	-4.546	S2=61 W =39 H2= 0	-3.683	9.65	0.00	19.64	16.57	5.208	6.172
.0020	923	6.909	-8.474	K =95 RB= 4 NA= 0	-4.542	S2=60 W =40 H2= 0	-3.601	9.72	0.00	19.81	16.66	5.070	6.154
.0025	925	6.995	-8.339	K =95 RB= 4 NA= 0	-4.537	S2=60 W =40 H2= 0	-3.514	9.80	0.00	19.81	16.75	4.979	6.134
.0032	926	7.084	-8.321	K =95 RB= 4 NA= 0	-4.515	S2=59 W =41 H2= 0	-3.426	9.88	0.00	19.90	16.83	4.889	6.112
.0040	936	7.152	-8.152	K =95 RB= 4 NA= 0	-4.513	S2=57 W =43 H2= 0	-3.328	10.02	0.00	19.98	16.91	4.785	6.089
.0050	948	7.249	-7.719	K =96 RB= 4 NA= 0	-4.458	S2=53 W =47 H2= 0	-3.286	10.22	0.00	20.06	16.99	4.677	6.066
.0063	953	7.322	—	K =96 RB= 4 NA= 0	-4.458	S2=53 W =47 H2= 0	-3.185	10.45	0.00	20.14	16.99	4.565	6.041
.0079	961	7.422	-7.471	K =96 RB= 4 NA= 0	-4.413	S2=45 W =55 H2= 0	-3.113	10.69	0.00	20.21	17.06	4.452	6.016
.0100	995	7.498	-7.261	K =96 RB= 4 NA= 0	-4.378	S2=59 W =41 H2= 0	-3.043	10.89	0.00	20.28	17.13	4.346	5.989
.0126	1011	7.573	-7.045	K =96 RB= 4 NA= 0	-4.339	S2=62 W =38 H2= 0	-2.975	11.11	0.00	20.35	17.20	4.239	5.962
.0158	1027	7.646	-6.835	K =96 RB= 4 NA= 0	-4.300	S2=65 W =35 H2= 0	-2.909	11.31	0.00	20.41	17.26	4.134	5.932
.0200	1045	7.716	-6.597	K =96 RB= 4 NA= 1	-4.253	S2=69 W =31 H2= 0	-2.847	11.54	0.00	20.48	17.33	4.024	5.902
.0251	1068	7.783	-6.332	K =96 RB= 3 NA= 1	-4.197	S2=73 W =27 H2= 0	-2.789	11.80	0.00	20.53	17.38	3.909	5.870
.0316	1093	7.846	-6.096	K =96 RB= 3 NA= 1	-4.137	S2=76 W =24 H2= 0	-2.735	12.07	0.00	20.59	17.44	3.794	5.836
.0398	1110	7.908	-5.865	K =96 RB= 3 NA= 1	-4.096	S2=78 W =22 H2= 0	-2.681	12.25	0.00	20.64	17.49	3.698	5.800
.0501	1127	7.969	-5.612	K =96 RB= 3 NA= 1	-4.005	S2=80 W =20 H2= 0	-2.626	12.44	0.00	20.70	17.55	3.602	5.760
.0631	1149	8.029	-5.450	K =96 RB= 3 NA= 1	—	S2=82 W =18 H2= 0	-2.574	12.66	0.00	20.75	17.60	3.499	5.716
.0794	1193	8.086	-5.047	K =96 RB= 3 NA= 2	-3.908	S2=86 W =14 H2= 0	-2.534	13.04	0.00	20.79	17.64	3.359	5.669
.1000	1255	8.134	-4.540	K =95 RB= 3 NA= 2	-3.782	S2=89 W =11 H2= 0	-2.508	13.53	0.00	20.81	17.67	3.204	5.623
.1259	1315	8.176	-4.095	K =94 RB= 4 NA= 2	-3.672	S2=92 W = 8 H2= 0	-2.486	13.96	0.00	20.84	17.69	3.071	5.576
.1585	1380	8.214	-3.658	K =93 NA= 4 RB= 3	-3.567	S2=93 W = 5 H2= 2	-2.469	14.37	0.00	20.85	17.70	2.946	5.526
.1995	1451	8.248	-3.219	K =92 NA= 6 RB= 2	-3.463	S2=95 W = 5 H2= 0	-2.457	14.78	0.00	20.87	17.72	2.827	5.473
.2512	1493	8.261	-2.972	K =91 NA= 7 RB= 2	-3.361	S2=95 W = 4 H2= 0	-2.437	15.01	0.00	20.89	17.74	2.749	5.412
.3162	1531	8.315	-2.758	K =90 NA= 8 RB= 2	-3.222	S2=96 W = 3 H2= 1	-2.415	15.22	0.00	20.91	17.76	2.677	5.335
.3981	1659	8.346	-2.139	K =88 NA=10 RB= 2	-3.151	S2=95 W = 4 H2= 1	-2.404	15.79	0.00	20.92	17.76	2.534	5.241
.5012	1731	8.378	-1.818	K =86 NA=13 RB= 1	-3.089	S2=95 W = 2 H2= 2	-2.404	16.08	0.00	20.92	17.77	2.450	5.127
.6310	1748	8.408	-1.541	K =83 NA=16 RB= 1	-3.023	S2=93 H2= 4 W = 2	-2.390	16.33	0.00	20.93	17.78	2.371	4.961
.7943	1869	8.438	-1.266	K =77 NA=21 RB= 1	—	S2=90 H2= 6 W = 2	-2.377	16.58	0.00	20.95	17.80	2.289	4.668
1.000	1952	8.468	-.967	K =73 NA=25 RB= 1	-2.944	S2=85 H2=10 W = 2	-2.365	16.85	.00	20.96	17.81	2.193	0.000
1.259	2052	8.497	-.613	K =69 NA=30 RB= 1	-2.838	S2=76 H2=24 HE= 2	-2.357	17.12	.00	20.95	17.82	2.010	4.665
1.585	2152	8.523	-.366	K =64 NA=35 RB= 1	—	W =64 H2=31 HE= 3	-2.353	17.40	.00	20.97	17.82	—	4.754
1.995	2251	8.555	-.115	K =59 NA=40 RB= 1	-2.500	H2=38 W =29 HE= 2	-2.351	17.63	.00	20.98	17.82	1.767	5.108
2.512	2347	8.580	—	NA=46 K =53 CA= 1	-2.271	H2=40 W =29 HE= 2	-2.342	—	.00	20.98	17.83	1.635	5.122
3.162	2398	8.595	.305	NA=49 K =49 CA= 1	-2.104	H2=41 W =29 HE= 2	-2.342	17.98	.00	20.98	17.83	—	5.290
3.981	2476	8.606	.493	NA=52 K =45 CA= 2	-1.903	H2=40 W =18 HE= 1	-2.347	18.10	.04	20.97	17.83	1.306	5.348
5.012	2574	8.615	.499	NA=52 K =41 CA= 3	-1.693	H2=48 W =18 HE= 1	-2.347	18.27	.04	20.97	17.83	1.104	5.391
6.310	2675	8.621	.898	NA=45 K =37 CA=15	-1.478	H2=... W =14 HE= 1	-2.355	18.43	.30	20.93	17.82	.895	5.422
7.943	2780	—	1.092	NA=59 K =37 CA= 3	—	H2=54 W =14 HE= 1	-2.365	18.59	1.03	20.96	17.81	.681	5.445
10.00	2867	8.626	1.276	NA=63 CA=33 K =...	-1.267	H2=59 H2=34 W =...	-2.376	18.73	1.91	20.94	17.80	.471	5.462
12.59	2994	8.630	1.447	NA=66 CA=29 K =...	-1.067	H2=63 H2=29 HE= 3	-2.388	18.87	2.75	20.93	17.79	.270	5.476
15.85	3113	8.633	1.624	NA=69 K =25 CA= 6	-.855	H2=67 H2=24 HE= 2	-2.401	19.01	3.63	20.92	17.78	.058	5.486
19.95	—	—	1.817	NA=72 K =20 CA= 8	-.622	H2=71 H2=20 HE= 2	-2.411	19.16	4.58	20.90	17.77	-.178	5.495
25.12	3443	8.635	2.049	NA=74 K =15 CA= 8	-.334	H2=75 H2=20 HE= 2	-2.448	19.33	5.75	20.87	17.74	-.465	5.500
31.62	3597	8.637	2.223	NA=75 K =12 CA=11	-.043	H2=78 H2=18 HE= 2	-2.468	19.46	6.60	20.84	17.72	-.682	5.505
39.81	3725	8.638	2.356	NA=76 CA=13 K = 8	.212	H2=80 H2=18 HE= 1	-2.483	19.56	7.26	20.83	17.70	-.849	5.508
50.12	3867	8.639	2.493	NA=75 CA=13 K = 8	.387	H2=82 H2=17 HE= 1	-2.503	19.66	7.94	20.80	17.68	-1.021	5.511
63.10	4028	8.640	2.636	NA=74 CA=15 K = 6	.563	H2=83 H2=15 HE= 1	-2.524	19.77	8.65	20.78	17.65	-1.197	5.514
79.43	4208	8.641	2.779	NA=71 CA=18 K = 5	—	H2=85 H2=14 HE= 1	-2.548	19.87	9.39	20.75	17.62	-1.373	5.517

TABLE 8

1500 DEGREE LOG G=5 WITH H2O AND H2 PID
EFFECTIVE TEMPERATURE = 1500   LOG(SURFACE GRAVITY) = 1500   OPACITY AT 10000 ANGSTROMS = 5.000   TEMP = 3 251

TAU	TEMP	LOG P GAS	LOG P ELE	PRIMARY SOURCES OF FREE ELECTRONS	LOG KAPPA	PRIMARY SOURCES OF 10000A OPACITY	LOG RHO	LOG H	LOG H+	LOG H2	LOG H2O	LOG ADIAB GRADIENT	LOG DEPTH
.0010	825	6.098	-10.325	K=94 RB=6 NA=1	-3.842	PD=86 S2=12 H=2	-4.362	7.46	0.00	18.96	15.81	5.134	6.165
.0013	828	6.153	-10.248	K=94 RB=6 NA=1	-3.734	PD=87 S2=11 K=2	-4.308	7.96	0.00	19.01	15.86	5.033	6.154
.0016	832	6.206	-10.154	K=94 RB=6 NA=0	-3.747	PD=88 S2=10 K=2	-4.255	8.05	0.00	19.07	15.92	4.933	6.143
.0020	836	6.263	-10.061	K=94 RB=5 NA=0	-3.747	PD=90 S2=9 K=1	-4.202	8.14	0.00	19.12	15.97	4.833	6.131
.0025	838	6.317	-10.002	K=95 RB=5 NA=0	-3.651	PD=91 S2=8 K=1	-4.149	8.20	0.00	19.17	16.02	4.731	6.119
.0032	842	6.371	-9.889	K=95 RB=5 LA=0	-3.602	PD=92 S2=7 K=1	-4.046	8.24	0.00	19.23	16.08	4.628	6.107
.0040	855	6.425	-9.684	K=95 RB=5 NA=0	-3.554	PD=92 S2=6 K=1	-3.997	8.32	0.00	19.28	16.13	4.528	6.095
.0050	865	6.471	-9.471	K=95 RB=5 NA=0	-3.506	PD=93 S2=5 K=1	-3.948	8.55	0.00	19.33	16.18	4.435	6.082
.0063	855	6.532	-9.219	K=95 RB=5 NA=1	-3.461	PD=93 S2=5 K=1	-3.902	8.72	0.00	19.38	16.22	4.335	6.068
.0079	860	6.586	-9.219	K=96 RB=5 NA=1	-3.415	PD=94 S2=5 K=1	-3.902	8.97	0.00	19.42	16.27	4.247	6.054
.0100	890	6.639	-9.047	K=95 RB=5 NA=0	-3.368	PD=94 S2=4 K=2	-3.853	9.14	0.00	19.47	16.32	4.150	6.039
.0126	910	6.693	-8.739	K=95 RB=5 NA=0	-3.323	PD=94 S2=4 K=2	-3.809	9.44	0.00	19.51	16.36	4.061	6.023
.0158	930	6.747	-8.442	K=96 RB=5 NA=0	-3.277	PD=94 S2=3 K=3	-3.765	9.73	0.00	19.56	16.41	3.970	6.007
.0200	955	6.801	-8.093	K=96 RB=4 NA=1	-3.232	PD=94 S2=3 K=3	-3.723	10.08	0.00	19.60	16.45	3.882	5.989
.0251	985	6.855	-7.701	K=96 RB=4 NA=1	-3.186	PD=93 S2=3 K=3	-3.682	10.46	0.00	19.64	16.49	3.796	5.970
.0316	1015	6.909	-7.330	K=96 RR=4 NA=1	-3.142	PD=93 S2=4 K=2	-3.641	10.82	0.00	19.68	16.53	3.708	5.949
.0398	1050	6.963	-6.925	K=96 RB=3 NA=1	-3.097	PD=93 S2=5 K=2	-3.602	11.21	0.00	19.72	16.57	3.622	5.926
.0501	1080	7.017	-6.594	K=96 RB=3 NA=1	-3.050	PD=92 S2=6 K=2	-3.560	11.53	0.00	19.76	16.61	3.533	5.902
.0631	1125	7.071	-6.139	K=96 RB=3 NA=1	-3.005	PD=90 S2=7 K=3	-3.523	11.97	0.00	19.80	16.65	3.449	5.875
.0794	1175	7.125	-5.675	K=96 RB=3 RH=1	-2.958	PD=90 S2=9 K=2	-3.488	12.42	0.00	19.83	16.69	3.365	5.845
.1000	1220	7.179	-5.283	K=95 RB=3 NA=1	-2.911	K=10 S2=1 H2=1	-3.450	12.90	.00	19.87	16.72	3.278	5.811
.1259	1275	7.233	-4.805	K=94 NA=21 RB=1	-2.864	K=14 S2=1 H2=1	-3.416	13.22	.00	19.91	16.76	3.194	5.730
.1585	1330	7.287	-4.655	K=74 NA=24 RB=1	-2.816	K=15 S2=1 H2=1	-3.380	13.61	.00	19.94	16.79	3.108	5.680
.1995	1385	7.340	-4.061	K=71 NA=27 RB=1	-2.769	K=16 S2=1 H2=1	-3.344	13.96	.00	19.98	16.87	3.022	5.621
.2512	1450	7.394	-3.799	K=68 NA=30 RB=1	-2.723	K=17 H2=17 PD=10	-3.278	14.30	.00	20.01	16.90	2.936	5.551
.3162	1500	7.447	-3.353	K=61 NA=37 RB=1	-2.677	H2=15 PD=5	-3.236	14.64	.00	20.04	16.94	2.855	5.450
.3981	1560	7.500	-3.021	K=61 NA=40 PR=1	-2.581	H2=13 HE=1	-3.198	14.94	.00	20.08	16.98	2.679	5.344
.5012	1620	7.554	-2.709	K=55 NA=43 RB=1	-2.535	H2=13 HE=1	-3.161	15.23	.00	20.12	17.02	2.593	5.176
.6310	1680	7.607	-2.416	K=55 NA=43 RB=1	-2.488	H2=11 HE=1	-3.161	15.50	.00	20.16	17.05	2.523	4.883
.7943	1740	7.661	-2.140	K=52 NA=46 CA=1	-2.440	H2=10 HL=1	-3.123	15.76	.00	20.20	17.05	2.506	
1.000	1805	7.714	-1.860	NA=50 K=47 CA=2	-2.444	H2=11 H2=14	-3.085	16.01	.00	20.24	17.09	2.420	0.000
1.259	1865	7.768	-1.614	NA=60 K=36 CA=2	-2.398	H=25 H2=17 PD=5	-3.046	16.23	.00	20.28	17.13	2.333	4.899
1.585	1925	7.822	-1.379	NA=60 K=29 CA=4	-2.352	H=59 H2=28 PD=7	-3.006	16.44	.00	20.32	17.17	2.244	5.207
1.995	1995	7.875	-1.175	NA=71 K=22 CA=4	-2.305	H=67 H2=17 PD=10	-2.968	16.67	.00	20.36	17.21	2.154	5.391
2.512	2055	7.929	-.914	NA=77 K=12 CA=5	-2.255	H=75 H2=17 PD=5	-2.927	16.86	.00	20.40	17.25	2.061	5.522
3.162	2120	7.982	-.698	NA=77 K=12 CA=6	-2.202	H=80 H2=15 PD=3	-2.888	17.05	.00	20.43	17.29	1.966	5.625
3.981	2180	8.030	-.455	NA=77 CA=12 K=8	-1.911	H=80 H2=15 HE=1	-2.856	17.26	.00	20.47	17.32	1.638	5.704
5.012	2260	8.060	-.285	NA=40 PR=1	-1.877	H=81 H2=12 HE=1	-2.837	17.41	.01	20.48	17.34	1.582	5.750
6.310	2325	8.094	-.103	NA=14 K=3	-1.835	H=86 H2=10 HE=1	-2.816	17.56	.09	20.51	17.36	1.515	5.796
7.943	2407	8.129	.101	NA=16 K=1	-1.774	H=89 H2=10 HL=1	-2.795	17.74	.81	20.53	17.38	1.433	5.840
10.00	2500	8.164	.321	NA=50 K=47 CA=2	-1.703	H=64 H2=25 H2=11	-2.776	17.92	.01	20.55	17.40	1.330	5.882
12.59	2610	8.197	.561	NA=60 K=36 CA=2	-1.611	H=25 H2=14 H2=10	-2.762	18.12	.09	20.56	17.41	1.196	5.919
15.85	2752	8.225	.839	NA=60 K=29 CA=4	-1.417	PD=25 H2=28	-2.761	18.39	.81	20.56	17.42	1.001	5.949
19.95	2922	8.246	1.134	NA=71 K=22 CA=4	-1.171	H=67 PD=10	-2.739	18.29	2.19	20.55	17.42	.735	5.971
25.12	3105	8.260	1.417	NA=71 K=12 CA=5	-.882	H=75 H2=17	-2.790	18.81	3.59	20.49	17.38	.431	5.986
31.62	3300	8.268	1.677	NA=77 CA=12 K=6	-.582	H=80 H2=15 PD=3	-2.824	19.08	5.92	20.46	17.36	.119	5.990
39.81	3485	8.273	1.901	NA=77 CA=12 K=8	-.314	H=80 H2=15 HE=1	-2.821	19.18	7.01	20.46	17.30	-.119	6.002
50.12	3672	8.277	2.103	NA=77 CA=12 K=9	-.067	H=81 H2=12 HE=1	-2.872	19.36	8.02	20.43	17.30	-.472	6.007
63.10	3820	8.280	2.301	NA=75 CA=14 K=8	.177	H=86 H2=10 HE=1	-2.872	19.48	8.02	20.43	17.30	-.672	6.010
79.43	4092	8.281	2.476	NA=71 CA=18 K=6	.364	H=89 H2=10 HL=1	-2.901	19.61	8.95	20.39	17.25	-.892	6.013

Model atmospheres of M dwarf stars*

TAKASHI TSUJI†

Mount Wilson and Palomar Observatories
Carnegie Institution of Washington and California Institute of Technology

Abstract

In order to construct model atmospheres for M dwarf stars, molecular opacities in cool stellar atmospheres are discussed in detail. In this study, the molecular line absorption is characterized by two parameters, the mean absorption coefficient and the mean line separation. The mean absorption coefficients of the vibration–rotation bands as well as of electronic bands are evaluated in the approximation that they just overlap. A simple formula is derived to calculate the mean line separation as a function of wavelength for electronic bands of diatomic molecules. For water vapor bands, the mean line separation as well as the mean absorption coefficient is obtained directly from laboratory data. As sources of opacity, electronic bands of TiO, MgH, CaH and SiH, vibration–rotation bands of CO and H_2O, and pure rotation bands of H_2O are taken into account. The absorption coefficient of the collision-induced vibration–rotation band of H_2 is also evaluated, and is included in the opacity code. Based on the approximation which corresponds to the Elsasser band model, the harmonic mean absorption coefficient is evaluated and the Rosseland mean opacity at relatively high pressures is discussed. Atomic lines are omitted.

A pure radiative equilibrium model with $T_e = 3000°K$, $\log g = 4.8$, and solar composition is calculated. The flux and the mean intensity are evaluated for a statistical line opacity which is characterized by the mean absorption coefficient and the mean line separation. The model is characterized by a very low surface temperature and rapid temperature rise inward. In addition, because of the strong absorption by H_2O and CO bands in the infrared, the shortward wavelength region is more transparent than the infrared spectral region. These two facts necessarily give considerable flux excess in the region shortward of 1μ. Thus, even if molecular opacities in the blue and red spectral regions are taken into account, the flux calculated from our model is too blue,

*This research was sponsored by the U.S. Air Force under Grant AFOSR 68-1401, monitored by the Air Force Office of Scientific Research of the Office of Aerospace Research.
†Present address: Department of Astronomy, University of Tokyo.

as compared with black body radiation at the temperature corresponding to the effective temperature. If this fact is taken into account in the
analysis of Johnson's infrared photometric data, the effective temperature scale for *M*-type main-sequence stars seems to be more than
200°K lower than the empirical temperature scale currently in use. Thus,
the effective temperature of M4V star can be below 3000°K rather than
about 3200°K. The effect of blanketing by molecular lines on some of
the infrared colors is discussed.

I. INTRODUCTION

THE MODEL atmospheres of cool dwarf stars have recently been studied by
several authors. The radiative equilibrium models with only continuous
opacity sources show a surface temperature near to that of the grey model
and, sometimes a temperature inversion (Linsky, 1966; Gingerich, Latham,
Linsky, and Kumar, 1967). However, when the infrared molecular opacities
by H_2O and CO are taken into account, the model shows very low surface
temperature (Linsky, 1966; Linsky and Gingerich, 1967; Tsuji, 1967). The
convective atmosphere has been studied by Kandel (1967).

With these models, however, it is still impossible to carry out quantitative
analysis of observational material on cool dwarf stars. Some interesting
problems which require more realistic atmospheric models are, for example:
the determination of the temperature scale for late-type main-sequence stars
from the analysis of the infrared multi-color photometric data by Johnson
(1965a); the correction for the effect of differential line-blanketing by molecular bands on the colors and magnitudes among the stars of different
populations; the analysis of modern photoelectric scanning photometry,
and also the determination of chemical composition as well as atmospheric
parameters through the analysis of high-dispersion spectra.

For these purposes, it seems to be important first to have more knowledge
of the opacity sources in the atmosphere of cool stars. Obviously, TiO, MgH,
and other molecular bands which are not considered in the model atmospheres
mentioned above might be important in addition to H_2O and CO, especially
for the analysis of the emergent flux. Accordingly, we will extend our study of
molecular opacity to the electronic bands of several diatomic molecules.
Also, we will try to take into account the effect of the fine structure of molecular bands rather than to assume that the rotational fine structure is completely smeared out as has been done before (Tsuji, 1966a, 1967). The effect
of the convection may be important sometimes even for the analysis of the
emergent flux and it would be desirable to examine the effects of molecular
opacity and convection simultaneously. However, this analysis is deferred
to a future study.

II. MOLECULAR OPACITIES

(a) *General principles*

The total number of molecular lines in the spectra of cool stars is so numerous that it seems to be reasonable to adopt some statistical approximation. The most simple case is to assume that the rotational fine structure is completely smeared out, and to use an approximation such as the 'just-overlapping line' approximation (Tsuji, 1966a). However, this is a serious oversimplification especially at low temperatures where the line density decreases because of the weakening of the high excitation lines. For more detailed analysis of the emergent flux and the spectra of cool stars, it is desirable to overcome this restriction even by a rough approximation.

In simple radiative transfer problems, such as in the laboratory, the emissivity or absorption of a molecular gas is evaluated easily by simple formulae based on some band models, such as the Elsasser model, statistical model, or quasi-random model. However, radiative transfer in stellar atmosphere is much more complicated and the flux as well as the mean intensity is influenced by the structure of all the layers. Accordingly, it seems difficult to find simple formulae for these quantities by the use of band models. What we can do is to characterize the absorption coefficient so that the line characteristics of molecular line absorption can be taken into account. For this purpose, we must introduce a parameter which is a measure of the smearing effect. This can be the ratio of line width to line separation. However, to be applied for any physical condition, it is convenient to have the mean line separation as a function of wavelength. The evaluation of the mean line separation is sometimes difficult, because of the overlapping of many bands, or the rotational fine structure may become too complicated in the polyatomic molecules. On the other hand, another parameter, the mean absorption coefficient, is easily evaluated on the basis of the just-overlapping line approximation. This is a sufficiently good approximation even for a diatomic molecule such as CO (Kunde, 1967).

In what follows, we will try to obtain two parameters, namely, the mean absorption coefficient and the mean line separation within spectral interval in which these two parameters do not change appreciably. Then, based on these two parameters, a simple statistical method to take into account the effect of rotational fine structure of the molecular bands on radiative transfer is discussed in the next section.

(b) *Electronic bands of diatomic molecules*

The formula for a mean absorption coefficient based on the just-overlapping line approximation for electronic bands of a diatomic molecule has been given by Patch, Shackleford and Penner (1962) and improved by Golden (1967). The formula is

$$k_\omega{}^{v\,v''} = \frac{\pi e^2}{mc^2}\frac{\omega_{v'v''}}{\omega_{00}}\,f_e q_{v'v''}\frac{1}{Q_{v''}(T)}\frac{hc}{kT}\frac{B_{v'}}{|B_{v'}-B_{v''}|}\,F_{e''}$$

$$\times \exp\left[-\frac{hc}{kT}\left\{\omega_{v''}+\frac{B_{v''}}{B_{v'}-B_{v''}}(\omega-\Omega_{v'v''})\right\}\right]\left[1-\exp\left(-\frac{hc\omega}{kT}\right)\right], \quad (1)$$

with

$$F_e'' = \frac{g_e'' \, Q_v'' \, Q_R}{Q}\exp\left(-T_e''/kT\right), \tag{2}$$

$$\omega'' = \omega_e'(v''+\tfrac{1}{2}) - \omega_e'' x_e''(v''+\tfrac{1}{2})^2, \tag{3}$$

$$\Omega_{v'v''} = \omega_{v'v''} - \tfrac{1}{4}\frac{(B_{v'}+B_{v''})^2}{B_{v'}-B_{v''}}, \tag{4}$$

where $\omega_{v'v''}$ is the wavenumber of the band origin, f_e is the electronic oscillator strength, $q_{v'v''}$ is the Franck–Condon factor, and other notations have their usual meanings.

A simple formula for the line separation for electronic bands of diatomic molecules is obtained in the following way: the wavenumbers of lines of the P- and R-branches are given by (Herzberg, 1950)

$$\omega = \omega_{v'v''} + (B_{v'}+B_{v''})\,m + (B_{v'}-B_{v''})\,m^2, \tag{5}$$

where $m = J + 1$ for the R-branch while $m = -J$ for the P-branch. The line separation at wave number ω in a given band is

$$d = \frac{d\omega}{dm} = (B_{v'}+B_{v''}) + 2\,(B_{v'}-B_{v''})\,m. \tag{6}$$

We solve equation (5) for m and introduce it into equation (6). Then, the line separation for the P- and R-branches together is

$$d_{P+R} = \sqrt{[(B_{v'}-B_{v''})(\omega-\Omega_{v'v''})]}, \tag{7}$$

where $\Omega_{v'v''}$ is the wavenumber of the band head, given by Eq. (4).

By a similar consideration, the line separation for the Q-branch is

$$dQ = 2\sqrt{[(B_{v'}-B_{v''})(\omega-\omega_{v'v''})]}, \tag{8}$$

in which we have $\omega_{v'v''}$ in place of $\Omega_{v'v''}$ in Eq. (7). If P-, Q-, and R-branches overlap, the mean line separation is determined from $d^{-1} = d_{P+R}^{-1} + dQ^{-1}$. When several bands overlap, similar consideration should be applied with the proper selection criterion.

The above formulae are now applied to the electronic bands of TiO, MgH, CaH, and SiH. Unfortunately, intensity data as well as spectroscopic constants of these molecules are still far from being complete, and many spectroscopic constants, as well as Franck–Condon factors, had to be estimated. The most critical factor in Eq. (1), however, is the electronic oscillator strength f_e. The sources for f_e in the present study are the following:

MgH: Schadee (1964) determined the oscillator strength of the (0,0) band to be 0.008 from analysis of the solar spectrum. Main, Carlson, and DuPuis (1967) obtained an upper limiting value of 0.002 from a rocket engine experiment. The reason for the difference is not explained, and we used the average value of 0.005.

TiO: No laboratory measurement of f_e-value is known yet for any bands of TiO. Makita (1967) analyzed the α-system of TiO in the spectrum of a sunspot and estimated the $f_e = 0.01$. While the model of a sunspot is not as firmly established as that of the photosphere, the empirical excitation temperature for the TiO bands determined by Makita agrees well with the temperature of the layer of the line formation for the model adopted.

Weltner (1967) kindly measured the intensity peak of several bands of the α-, γ-, and γ'-systems of TiO from his spectra of matrices. From his measurement, the logarithms of the peak intensities are plotted against $q_{v'v''}\omega_{v'v''}$ for each system. Then the relative gradient near zero intensity gives the relative f_e-values, which turned out to be 1:3:4 among the α-, γ-, and γ'-systems. In general, the f_e-value in a matrix phase is different from that in the gas phase. However, it is expected that the relative f_e-values among different systems of the same molecule are less affected by the matrices than the f_e-values themselves.

The relative f_e-values between the α- and β-systems is estimated from the integrated intensity on the high-dispersion spectrograms of the M giant star, μ Gem. The ratio of f_e-values between the α- and β-systems seems to be 1:0.2. The f_e-values for the δ- and φ-systems are assumed to be the same as for the β-system.

SiH: Schadee (1964) estimated the value of $f_e = 0.002$ for the $A^2\Delta - X^2\Pi$ system from an analysis of the (0,0) band in the solar spectrum.

CaH: Makita (1967) is working on the sunspot spectrum in the region of the B-band. However, the final result is not yet available. We have used the values $f_e = 0.01$ for the A-band and $f_e = 0.005$ for the B-band.

The intensity data used are summarized in Table I. Then, the mean absorption coefficient and the mean line separation have been calculated for hundreds of wavelength meshes with 10–100 Å interval in the range of $\theta = 5040/T = 0.6$ to 6.0. An example is shown in Figures 1(a) and 1(b) for TiO. In Figure 1(b) the total half-width of the lines for $v = 2$ km/sec is also shown. It is to be noted that the mean line separation is smaller than the line width in some wavelength regions.

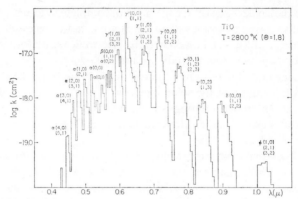

FIGURE 1(a)   Logarithms of the absorption cross section of TiO electronic bands for $\theta = 5040/T = 1.8$.

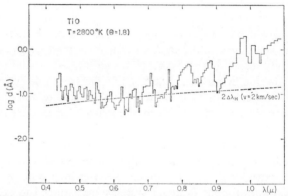

FIGURE 1(b)   Logarithms of the mean line separation of TiO electronic bands for $\theta = 5040/T = 1.8$. The dashed line indicates the line half width for turbulent velocity of 2 km/sec.

(c) *Vibration–rotation bands and pure rotation bands*

Water vapor and carbon monoxide are the most important sources of opacity in the infrared. For H_2O, Auman (1966) has calculated the harmonic mean by taking 2.3 million lines. However, as the line profile is assumed to be a pure Doppler shape, it seems difficult to apply his results to dwarf stars, in which pressure broadening is more important. The author has used the just-overlapping line approximation, which assumes that the rotational fine structure is completely smeared out (Tsuji, 1966a). Linsky and Gingerich (1967) have also treated the water vapor absorption as a pseudo-continuum. Recent laboratory measurements by Ludwig (1966) have shown that the complete smearing-out is truly the case at relatively high temperaures, say above 3000°K. However, at lower temperatures, complete smearing-out of the rotational fine structure is not observed at laboratory conditions (high pressure).

TABLE I Intensity data for electronic bands of diatomic molecules.

Molecule	Transition	ω_{00} (cm^{-1})	f_e	Source of f_e	Source of $q_{v'v''}$
MgH	$A^2\Pi - X^2\Sigma^+$	19,224	0.005	(1) (2)	(3)
TiO	$C^3\Delta - X^3\Delta$ (α)	19,433	0.01	(4)	(5)
	$c^1\Phi - a^1\Delta$ (β)	17,891	0.002	(6)	(7)
	$A^3\Phi - X^3\Delta$ (γ)	14,166	0.03	(8)	(5) (9)
	$?^3\Pi - X^3\Delta$ (γ')	16,248	0.04	(8)	(10)
	$b^1\Pi - c^1\Phi$ (δ)	11,322	(0.002)	(6)	(7)
	$b^1\Pi - d^1\Sigma$ (φ)	9106	(0.002)	(6)	(11)
SiH	$A^2\Delta - X^2\Sigma$	24,393	0.002	(1)	(11)
CaH	$A^2\Pi - X^2\Sigma$ (A)	14,420	(0.005)	(12)	(3)
	$B^2\Sigma - X^2\Sigma$ (B)	15,766	(0.01)	(12)	(13)

(1) Schadee (1964)
(2) Main, Carlson and DuPuis (1967)
(3) Ortenburg (1960)
(4) Makita (1967)
(5) Ortenberg and Glasko (1963)
(6) Est. (see text)
(7) $q_{\Delta v=0} = 1.0$
(8) Weltner (1967)
(9) Phillips (1967)
(10) Assumed to be the same as that of γ-system
(11) Estimated by Nicholls (1965) method
(12) Est.
(13) Pathak and Singh (1966)

Fortunately, so far as the atmospheres of *M* dwarf stars are concerned, the laboratory data can be used directly, because the line broadening mechanism is mainly collision broadening in both cases. Ludwig (1966) has measured the absorption coefficient and fine structure parameter, which is the ratio of line half width to mean line separation, for most of the water vapor bands. From these data we can obtain the mean absorption coefficient and the mean line separation.

For the H_2O 0.94 μ band, for which empirical opacity data is not available, the absorption coefficient is evaluated on the basis of the just-overlapping line approximation using intensity data of Burch and Gryvnak (1966). The mean line separation has been assumed to be the same as that of the corresponding part of the 1.4 μ band.

Q

For the far infrared region beyond 10 μ, the value discussed in the previous study (Tsuji, 1966a) has been used. The line separation has been assumed to be the same as the value at about 8 μ.

For CO bands, the mean absorption coefficient evaluated on the basis of the just-overlapping line approximation has sufficient accuracy for our purpose. The mean line separation is calculated from the formula given by Ludwig (1966).

(d) *Collision-induced vibration–rotation band of* H_2

Usually, a homonuclear molecule such as H_2 has no permanent dipole moment and hence shows no vibration–rotation bands or pure rotation bands. However, at high density, collision-induced transitions are observed in the laboratory (Welsh, Crawford, and Locke 1949) and the possible importance of the H_2 collision-induced band around 2.4 μ has been suggested before (Tsuji, 1966b; Gingerich, Latham, Linsky, and Kumar 1967).

The absorption coefficient of the collision-induced transition has been measured under laboratory conditions (Hare and Welsh 1958). However, no experiment has been done at a high temperature corresponding to the physical condition in stellar atmospheres. Accordingly, we must know the dependence of the intensity in temperature and pressure. The integrated absorption coefficient of the collision-induced transition is given by

$$\int k_\omega \, d\omega = \frac{8\pi^3\omega}{3hc} N_{12} \left| \int \psi_1\psi_2\mu\psi_1^*\psi_2^* d\tau \right|^2, \tag{9}$$

where ψ_1 and ψ_2 are the vibrational wave functions of the two colliding molecules, μ is the induced dipole moment, and N_{12} is the number of pairs of colliding molecules per cm^3. N_{12} is given by

$$N_{12} = N_1 N_2 \int \exp\left[-V_{12}(r)/kT\right] 4\pi r^2 \, dr, \tag{10}$$

where N_1 and N_2 are the number densities of the two colliding molecules, and $V_{12}(r)$ is the potential energy of interaction. Now we define

$$\psi(T) = \frac{\int \exp\left[-V_{12}(r)/kT\right] 4\pi r^2 \, dr}{\int \exp\left[-V_{12}(r)/kT_0\right] 4\pi r^2 \, dr}. \tag{11}$$

where T_0 is the room temperature for which the experimental absorption coefficient is known. The integrations in Eq. (11) have been carried out on the basis of the theory developed by Kranendonk (1958). For $T_0 = 298°K$, we have $\psi(T = 1008°K) = 1.85$, $\psi(T = 2520°K) = 3.14$, and $\psi(T = 5040°K) = 4.72$.

Now, we can evaluate the integrated absorption coefficient or, in other words, the effective oscillator strength for any temperature. For the fundamental band,

$$f_{\text{eff}} = 2.8 \times 10^{-14} \frac{P}{T} \psi(T). \tag{12}$$

Then, we apply the same theory of the just-overlapping line approximation as for the usual dipole transition.

The individual lines are highly broadened because of the short life of the intermolecular interaction inducing the dipole moment. The life time of the interaction is inversely proportional to the thermal velocity of the molecules and hence is proportional to the square root of temperature. Experiment (Crawford, Welsh, MacDonald, and Locke 1950) shows that the line half-width is given by

$$\Delta \omega_H = 320 \sqrt{(T/273)}, \tag{13}$$

and this is appreciably larger than the line separation. Accordingly, the rotational fine structure is completely smeared-out and the collision-induced transition works as a true continuum. For this reason, the collision-induced transition of H_2 molecule is an important source of opacity at high pressure.

III. OPACITY OF COOL STELLAR ATMOSPHERES AT RELATIVELY HIGH PRESSURE

(a) *Monochromatic absorption coefficients*

On the basis of the material presented in the preceding section, the absorption coefficients per gram of stellar material for the physical conditions representative of the cool dwarf stars have been calculated. Dissociation and ionization equilibria have been solved for the mixture of 16 elements and molecular concentrations for H_2, H_2O, CO, TiO, MgH, SiH and CaH are determined. The sources of continuous absorption are negative hydrogen ion (bf + ff), negative molecular hydrogen ion (ff), and hydrogen atom (bf + ff), together with Thomson scattering and Rayleigh scattering by H_2 and H. The sources of the cross section for these processes are the same as in the previous study (Tsuji, 1966a) except for H_2^-(ff) for which the cross section is assumed to be the same as that of H^-(ff), by considering the result of Dalgarno and Lane (1966).

An example of the result is shown for the case of $\Theta = 5040/T = 2.0$, log $P_g = 6.0$, and solar chemical composition (Goldberg, Müller and Aller 1960). The absorption features between 0.35 and 1.1 μ are shown in Figure 2(a) and extended to 12 μ in Figure 2(b). Inspection of these figures shows that the infrared opacity is still larger than the opacity in the short wavelength region even when we take into account the effect of electronic bands of several diatomic molecules. This fact has an important effect upon the atmospheric structure and emergent flux. It is interesting to note that the absorption coefficient of the collision-induced H_2 band around 2.4 μ is comparable with that of H_2^-, and much larger than the scattering coefficient of H_2 molecules, though it is considerably smaller than the molecular absorption by H_2O and CO. In the ultraviolet and blue parts of the spectrum, our survey of opacity sources is far from being complete. The source of the so-called Lindblad depression (Lindblad 1935), which is quite conspicuous in the spectra of cool dwarf stars, is not yet identified. We have not considered the atomic line opacity. Also, the photoionization continua of atoms, molecules, and negative ions are not taken into account.

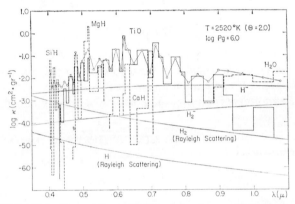

FIGURE 2(a)   Logarithms of mass absorption coefficients for the individual molecules and atoms versus wavelength for $\theta = 5040/T = 2.0$ and $\log P_g = 6.0$. The chemical composition is the solar mixture of Goldberg, Müller and Aller (1960), and $\log P_e = -0.75$, $\log P_H = 4.45$, $\log P_{H_2} = 5.99$, $\log P_{CO} = 3.02$, $\log P_{H_2O} = 2.88$, $\log P_{MgH} = -0.10$, $\log P_{TiO} = -1.18$, $\log P_{CaH} = -1.60$, and $\log P_{SiH} = -1.65$. The open circles connected by line are the harmonic mean of all the opacity sources for the small wavelength mesh, the center of which is the open circle.

(b) *Rosseland mean opacity*

An approximate calculation of the Rosseland mean opacity is carried out on the basis of the opacity data such as shown in Figures 2(a) and 2(b), together with the information on mean line separation for molecular bands. For each wavelength mesh, we evaluate the integrated mean absorption coefficient, $\bar{k}_{mol}$, and averaged mean line separation, d, if more than one molecular absorption overlap at a given wavelength mesh. On the other hand, for the

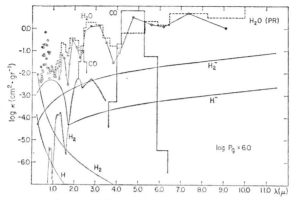

FIGURE 2(b) See legend to Figure 2(a). This figure shows the infrared opacity sources, for $T = 2520°$ K ($\Theta = 2.0$). The dotted line labelled H₂ is the collision-induced vibration–rotation band of H₂.

given physical conditions, including the turbulent velocity, we can evaluate the shape of spectral line if we assume that it is represented by a Voigt profile. Then, we assumed that lines of the same intensity are distributed uniformly within each wavelength mesh. This corresponds to the so-called Elsasser band model, although we have assumed a Voigt profile and we will not follow the mathematical formulation of the original Elsasser theory (Elsasser 1938).

Based on this model, the harmonic mean absorption coefficient at each wavelength mesh is evaluated in the following way: For the given physical condition, the damping line width is given by

$$\gamma = 0.05 \sqrt{\frac{273}{T}} \frac{P_g}{1.01325 \times 10^6}, \tag{14}$$

where we have assumed that the line half-width at normal condition ($T = 273°$K and $P = 1$ atm) is 0.05 cm^{-1} for any colliding particle and for any transition. This is a typical value for non-resonant molecular collisions. On the other hand, the Doppler half-width is given by

$$\Delta\omega_D = \frac{v}{c} \omega, \tag{15}$$

where v is the turbulent velocity. Then the Voigt profile is characterized by the parameter a, which is

$$a = \frac{\gamma}{\Delta\omega_D}. \tag{16}$$

Now, we evaluate the absorption coefficients at n points within $d/2$ by

$$k_{\text{mol}}^i = n\bar{k}_{\text{mol}}\frac{H(a, u_i)}{\sum\limits_{i=1}^{n} H(a, u_i)} \qquad i = 1, \ldots, n, \qquad (17)$$

(see Figure 3) where

$$u_i = \frac{d}{2\Delta\omega_D}\frac{i - 1}{n - 1} \qquad i = 1, \ldots, n, \qquad (18)$$

and

$$H(a, u) = \frac{a}{\pi}\int_{-\infty}^{\infty}\frac{e^{-y^2}}{a^2 + (u - y)^2}\, dy \qquad (19)$$

is the Voigt profile and is evaluated by the method described by Armstrong (1967). Since we have assumed a uniform distribution of lines, the harmonic mean absorption coefficient at each wavelength mesh is given by

$$\frac{1}{\kappa_H} = \frac{1}{n}\sum_{i=1}^{n}\frac{1}{k_{\text{mol}}^i + k_{\text{cont}}}, \qquad (20)$$

where k_{cont} is the continuous absorption and scattering coefficient. The results are shown by open circles in Figures 2(a) and 2(b) ($n = 5$ throughout).

Based on this harmonic mean, the Rosseland mean opacity is easily calculated. The result for solar composition mixture and $\log P_g = 6.0$ is shown in Figure 4(a) and the effect of turbulent velocity is also investigated. The case of $v = \infty$ corresponds to the complete smearing out of the rotational fine structure. Since the rotational fine structure is really smeared out above $T = 3000°K$, as has been mentioned before, the Rosseland mean opacity is almost independent of the turbulent velocity. However, below this temperature, the effect of rotational fine structure is quite serious. In Figure 4(b), the Rosseland mean opacity is shown for $\log P_g = 5.0$, 6.0, and 7.0 for solar composition and $v = 2$ km/sec. As is expected, the Rosseland mean opacity is larger at higher pressure because of the increased pressure broadening. The Rosseland mean opacity at lower pressure cannot be evaluated by the present model, since the mean line separation of H_2O bands is based on the laboratory data obtained at a pressure of one atmosphere. At lower pressure, many lines which were within the damping wing of a strong line at high pressure, will have important effect on the Rosseland mean opacity. Accordingly, if the present model is used at low pressure, it will give an underestimate of the Rosseland mean opacity.

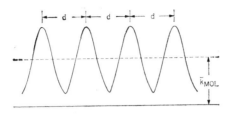

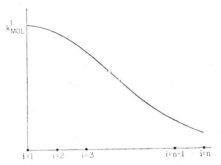

FIGURE 3 The molecular absorption for each wavelength mesh is characterized by the mean absorption coefficient $\bar{k}_{mol}$ and the mean line separation d. Then, the absorption coefficient for this wavelength mesh is approximated by the Elsasser band model, which consists of equidistant and equal intensity Voigt profiles, as is shown at the top, for $a = \gamma/\Delta\omega_D = 0.05$. Below, the probability distribution function corresponding to this type of band absorption is shown.

IV. MODEL ATMOSPHERE OF M DWARF STARS

(a) *Method for constructing the model atmosphere with molecular line-blanketing effect*

Model atmospheres of constant flux with line-blanketing effect included have been investigated recently by several authors. In A-type stars, strong hydrogen line profiles have been taken into account directly (Strom and Avrett 1964). Some strong metallic lines in the ultraviolet have also been taken into account directly in the construction of model atmosphere of B-type stars (Mihalas and Morton 1965). Strom and Kurucz (1966) have applied a statistical procedure to take into account more lines and the effect of blending of lines in F-type stars. For later type stars, accurate analysis of a line-blanketed model has not yet been carried out, even for the sun.

For M-type stars in which line absorptions are more complicated it is almost impossible to incorporate individual lines of atoms and molecules in the construction of model atmospheres. However, the very fact that the

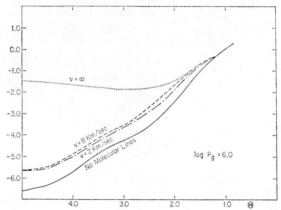

FIGURE 4(a)   The effect of molecular line absorptions on Rosseland mean opacity. Logarithms of the Rosseland mean absorption coefficient are plotted against $\theta = 5040/T$ for the case of no molecular line absorption (solid line), with lines and turbulent velocities of 2 km/sec (dashed-broken line) and 8 km/sec (dashed line), and also for the case of completely smeared out line absorption (dotted line).

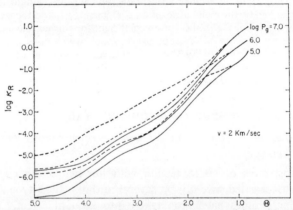

FIGURE 4(b)   Logarithms of Rosseland mean opacities with (dashed curve) and without (solid line) molecular lines are plotted against $\theta = 5040/T$ for $\log P_g = 5.0$, 6.0, and 7.0. The chemical composition is the solar mixture and the turbulent velocity is 2 km/sec.

number of lines is so great, and that the blending of lines is so serious, suggests the possibility of using some statistical method. In addition, the positions and intensities of molecular lines can be evaluated rather easily as compared with those of atomic lines and, moreover, molecular absorption will in some cases have the character of a pseudo-continuum, from the beginning. Thus, without carrying out any detailed counting or line statistics, we have already a kind of 'statistical' line opacity such as given by Eq. (17). This model is formally the Elsasser band model. However, this model is adopted only to facilitate the formulation, and Eq. (17) can also be understood as the probability distribution function of the statistical line opacity in each wavelength mesh. Now, by this method we can take into account statistically the effects of turbulent velocity, pressure broadening and change of line spacing throughout the atmosphere. It is to be emphasized, however, that this kind of approach is a reasonable one only when lines are actually at least partly overlapping, as in molecular bands.

Actual computation of a model atmosphere has been carried out on the assumption that the stellar atmosphere is in radiative equilibrium, in LTE, and in hydrostatic equilibrium. The computational procedure is the same as in previous work (Tsuji 1967), except that the source function, mean intensity, and flux have been evaluated for the absorption coefficient corresponding to several values of n in Eq. (17) for each wavelength mesh. This kind of analysis of the radiation field is essentially the same as that by Strom and Kurucz (1966). In general, line absorptions are smeared out into a pseudo-continuum in deeper layer because of high temperature and high pressure, and the effect of the fine structure of molecular bands is increasingly important towards the surface. The effect upon radiative transport is taken into account with reasonable accuracy if we take a sufficiently large number, n, in Eq. (17). The temperature correction procedure used is that developed by the author (Tsuji 1965): if the total flux $F(\tau_0)$ is found to be not the equilibrium value of $\sigma T_e^4/\pi$, the temperature gradient is corrected by

$$\left(\frac{dT}{d\tau_0}\right)_{\text{Rev}} = \frac{\sigma T_e^4}{\pi F(\tau_0)}\left(\frac{dT}{d\tau_0}\right).$$ (21)

The effect of convection is not considered at present, though it can easily be taken into account in Eq. (21). The surface temperature is determined so that the surface flux is the correct value of $\sigma T_e^4/\pi$. In the upper layer, the usual lambda iteration method is used, if necessary.

(b) *Pure radiative equilibrium model of* M *dwarf stars*

With the opacity sources discussed in Section II, the method outlined in the previous section has been applied to construct a model atmosphere of an M dwarf star characterized by the following parameters: $T_e = 3000°K$, $\log g = 4.8$, solar composition mixture, and turbulent velocity of 2 km/sec throughout the atmosphere. The number of optical depths considered is 75 between $\log \tau_0 = -6.0$ and 1.4, where τ_0 is the optical depth for the continuum at 0.8 μ. The number of wavelength meshes is 72 between 0.2 and

36.0 μ, and the flux and the mean intensity have been evaluated, based on the simple statistical model described by Eq. (17) with $n = 2$ or 3 at each wavelength mesh in which molecular line opacity is included.

The iteration has been started not from the grey model, but from the radiative equilibrium model which was constructed previously with the smeared-out opacity for H_2O and CO (Tsuji 1967). This model will no longer be in radiative equilibrium for our new opacity, since we have taken into account the effect of fine structure of molecular bands and also we have added several new sources of molecular line opacity. Also the absorption coefficient of $H_2^-(ff)$ has been increased by a factor of 3 to 5. Figure 5 shows the starting model and its flux error. To this model, we have applied 3 iterations, and the flux error for the resulting model is shown in Figure 5. Although the flux error of this model is still ± 2 per cent, the temperature structure does not show any significant change from the starting model. This indicates that the flux error of the kind shown in Figure 5 does not mean a serious error in the temperature structure. Accordingly, after 4 iterations, the model shown in Figure 5 is used in the analysis of radiation flux in the next section. A brief summary of the physical structure of this model is shown in Table II.

TABLE II   Model atmospheres.

$\log \tau_0$	$\log T$	$\log P_g$	$\log P_e$
−6.00	3.043	3.03	−8.33
−5.50	3.126	3.53	−6.23
−5.00	3.205	4.03	−4.52
−4.50	3.274	4.50	−3.19
−4.00	3.318	4.82	−2.42
−3.50	3.344	5.04	−1.97
−3.00	3.368	5.21	−1.58
−2.50	3.388	5.37	−1.25
−2.00	3.401	5.52	−0.91
−1.50	3.433	5.67	−0.58
−1.00	3.451	5.81	−0.22
−0.50	3.492	5.94	0.14
0.00	3.536	6.05	0.52
0.50	3.594	6.16	0.88
1.00	3.674	6.26	1.46
1.40	3.758	6.33	2.05

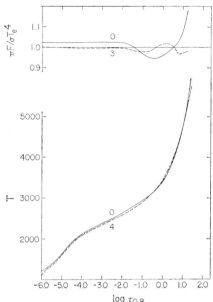

FIGURE 5 Temperature structures of the models plotted against log $\tau_{0\cdot8}$ where $\tau_{0\cdot8}$ is the optical depth of the continuum at 0.8 μ. The flux errors of the models are also shown. The number 0 means initial model, 3 and 4 the 3rd and 4th iterated model, respectively.

It is somewhat surprising to find that the temperature structures of the starting model and the new model show almost no change in spite of the considerable improvements in opacity. This fact, however, can be well understood in terms of the theory by Unno (1962) in the following way: Our model is characterized by high transparency on the short-wavelength side of the maximum of the Planck function. In such a case, the relative contribution of the violet compared to the red portion of the spectrum increases with depth. Since the violet side is more transparent, this model will show large excess flux. Then, to keep the total flux at the equilibrium value, it is necessary to reduce the total emissivity by lowering the temperature. On the other hand, heavy molecular absorption in the infrared increases the effective blanketing, and causes the temperature to rise in the deeper layer. This again strengthens the flux excess in the violet portion of the spectrum. The emergent flux indeed shows such a large flux excess in this spectrum region, as will be shown later. Thus, the simple theory of line blanketing based on a picket-fence model is not sufficient to account for the effect of line absorption in the case of *M*-type stars. Indeed, one of the important improvements in our new opacity over the previous one is the introduction of the effect of fine structure of the molecular bands which means the strengthening of the picket-fence characteristics of the absorption coefficient. However, this modification does not bring any serious effect upon the temperature structure of the model. Thus, the overall characteristics of the opacity and its interaction with the radiation field is more important in determining the physical structure of the model than the picket-fence feature of the individual line absorptions.

V. ANALYSIS OF THE OBSERVED STELLAR RADIATION

(a) *Comparison of the theoretical flux calculated from the model with the observed ones*

The emergent flux from the model discussed in the previous section is shown in Figure 6(a) for the spectral region between 0.3 and 1.1 μ, and in Figure 6(b) for the infrared region up to 12.0 μ. In each wavelength mesh, the flux from the transparent part and that from the opaque portion have been evaluated, based on the statistical line opacity discussed before, and then averaged. The characteristic features of the theoretical flux is its large flux excess on the short wavelength side of 1.2 μ, and the general flux deficiency in the infrared, as has been expected from the nature of opacity and the physical structure of our model. In what follows, we will examine the available observational data.

The result of scanning photometry of the M4V star, HD225213, by Willstrop (1965) is shown in Figure 6(a) with units $\log F_\lambda +$ const. The reduction to this scale has been carried out using three A0V stars, as comparison, together with the absolute scale for α Lyr given by Oke (1964, 1967). It is to be noted that some details of the molecular absorption bands are well reproduced in the theoretical flux. Quantitative discussions, however, are difficult on only the small observed portion of the spectrum. Moreover, the effect of atomic line opacity as well as some unknown source of opacity which is responsible to the so-called Lindblad depression is not considered in the present model. Accordingly, quantitative agreement is not to be expected, from the beginning, for the ultraviolet to visual spectral region (*U*-, *B*-, and *V*-regions).

In Figure 6(b), the results of wide band photometry by Johnson (1965a) are shown on an absolute scale for M4V and M6V stars. As representatives, Krüger 60A and 60B are used for types M4V and M6V, respectively. The reduction to absolute scale has been carried out with the sun as standard (Johnson 1964) together with the parallaxes (see Limber 1958) and radii given by Johnson (1965a) as a first approximation. It is to be noted that agreement between theoretical flux and observed ones is quite good for the M4V star as a whole, but not for the M6V star. This fact tentatively suggests

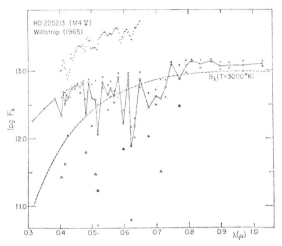

FIGURE 6(a) Logarithms of the emergent flux calculated from the model atmosphere of $T_e = 3000°$K, log $g = 4.8$ and solar composition is shown between 0.3 and 1 μ. The unit of F_λ is erg/cm²sec for $\Delta\lambda = 1$ cm. At each wavelength mesh, the flux is calculated at line center (triangle) and line wing (cross) and they are averaged to give the mean flux shown by the open circles. The black body radiation for $T = 3000°$K is shown by a dashed curve in the same units. The result of photoelectric scanning observation for the M4V star, HD 225213, by Willstrop (1965) is also shown, after reducing it to the same unit as the theoretical flux, from a constant factor which depends on the stellar radius and distance.

that the effective temperature of 3000°K corresponds to that of an M4V star rather than to M6V stars. This conclusion largely depends on the large flux excess below 1.2 μ, and especially near the *I* magnitude, in the theoretical flux. The physical reason for the large flux excess at the *I* magnitude is not only the general transparency below 1.2 μ, but also the fact that line absorptions are weakest around the *I* magnitude region, and also that the maximum of the Planck radiation for $T \simeq 3000°$K just comes around this region. For these reasons, the large flux excess near the *I* magnitude may remain unchanged even if we consider the effect of convective energy transport in the atmospheric layer. Now, if we assume that the effective temperature of an M4V star is about 3000°K, it shows large flux excess below 1.2 μ, especially at *I* magnitude in agreement with the theoretical flux. On the other hand, if we assume that the effective temperature of the M6V star is 3000°K, it is very difficult to understand its emergent flux on the basis of our model atmosphere. Probably convection cannot reduce the flux so much and we must introduce some unknown sources of opacity at *I* magnitude region

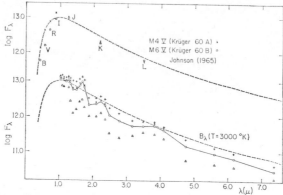

FIGURE 6(b)   Logarithms of the emergent flux in the infrared calculated from the pure radiative equilibrium model (see legend to Figure 6(a)). The result of multicolor photometry for M4V and M6V stars by Johnson (1965a) is shown together with the black body radiation curve for $T = 3000°K$. Johnson's result is reduced to absolute scale with the sun as the standard, together with the astrophysical data for Krüger 60 A (M4V) and 60 B (M6V).

The analysis just discussed, which is based on the absolute flux scale, is the most reasonable one in principle. On the other hand, it is sometimes dangerous if some parameters used are wrong. Accordingly, we discuss the same material on a relative scale. In Figure 7, the observed magnitude differences, with respect to the R magnitude, for M2V, M4V, and M6V stars obtained from Johnson (1965a) are compared with the theoretical magnitude calculated from the model with $T_e = 3000°K$. The computation of the magnitude has been carried out with the photometer response function given by Johnson (1965b), with the sun as the standard of calibration of color systems. For the sun, colors given by Johnson (1964) are used. In the analysis, the V magnitude is excluded for the reason noted previously. As far as R, I, J, K, and L magnitudes are concerned, the theoretical curve based on our model comes between the curves for M2V and M4V. This fact again suggests that the effective temperature of the M4V star is below $3000°K$. This is considerably lower than the effective temperature of $3200°K$ for the M4V star suggested by Limber (1958) and by Johnson (1965a).

(b) *Blanketing effect by molecular absorption on infrared two-color diagram, in terms of $R - I$ and $I - K$*

Infrared colors and magnitudes have been used to investigate the population effect among late-type dwarf stars by several authors (Kron, 1956; Eggen and Greenstein, 1965; Greenstein, 1965). These authors suggested that infrared

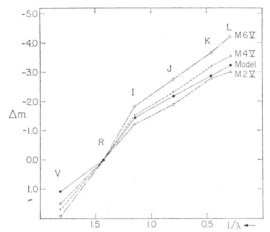

FIGURE 7 The observed spectral energy curves for M2V, M4V, and M6V stars by Johnson (1965a) are compared with the one predicted from model atmosphere of $T_e = 3000°K$. The ordinate scale is the magnitude difference with respect to the R magnitude, and the abscissa is $1/\lambda(\mu)$.

colors may be more useful, if the effect of differential line blanketing by TiO bands can properly be corrected for. Accordingly, we will discuss this problem briefly based on our model atmosphere. In Figure 8, $R - I$ colors are plotted against $I - K$ colors obtained from Johnson (1965a) for M and K dwarf stars. This two-color diagram is expected to be analogous to the $U - B$, $B - V$ color diagram in that $I - K$, like $B - V$, is a temperature indicator, while $R - I$, like $U - B$, can be a measure of metal abundance. This is because Greenstein (1965) found cool stars which show MgH bands but have no TiO bands, and he suggested that they may be highly metal deficient M dwarf stars. Such a star may show $R - I$ deficiency because the R magnitude is most strongly affected by TiO.

Inspection of Figure 8 at once shows that $I - K$ cannot be a good temperature index and, probably, that $I - L$ would be a better temperature index if the L magnitude is measured for many M dwarf stars. It is also noted that spectral sub-class is well correlated with $R - I$ color rather than with $I - K$ color. This is because the M sub-class is based on the intensity of TiO bands. Accordingly, if there is a considerable variation in metal abundance, the M sub-class cannot be a pure temperature sequence. The scattering of the points in Figure 8 is also very large and this may be another expression of the fact pointed out by Wilson (1962) that spectral sub-class is not a unique function of color in late-type dwarfs.

In this way, our two-color diagram may not be the best one to investigate population effect. However, we will tentatively try to interpret this diagram. In Figure 8, the colors corresponding to black-body radiation of $T = 3000°$K and $5000°$K are shown together with the theoretical colors for our model atmosphere of $T_e = 3000°$K. The theoretical color shows reasonable agreement with the $R - I$ color of M4V stars but $I - K$ appears too blue for M4V although it is displaced in the right direction from black-body colors. It is hoped that this disagreement between observed and theoretical $I - K$ color may be reduced by considering the effect of convection or by improved opacity data. In Figure 8, Barnard's star (M5V), which is believed to be metal deficient, occupies the lower left portion among the M5 stars. On the other hand our model with normal composition also appears on the left-hand side. Accordingly, it seems that the situation is more complicated than is expected from the simple blanketing theory. Our two-color diagram, however, seems to support our previous suggestion that the effective temperature of M4V stars is about 3000°K or lower. If the effective temperature of an

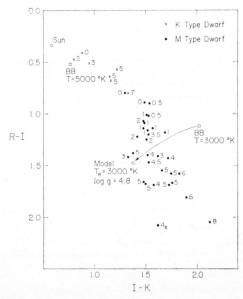

FIGURE 8 $R - I$ color is plotted against $I - K$ color for K and M dwarf stars based on the data given by Johnson (1965a). The crosses represent K dwarfs, while filled circles indicate the position of M dwarf stars. The number attached to each symbol is the spectral subclass of K and M types. The colors of black body radiation for $T = 3000°$K and $5000°$K are shown by squares. The colors calculated from the model atmosphere for $T_e = 3000°$K, log $g = 4.8$ and solar composition is shown by the open circle.

M6V star is about 3000°K, the blanketing line is almost straight downward from the black-body colors. It is quite difficult to understand this behavior, if we remember the physical properties of the atmosphere of cool dwarf stars discussed previously.

VI. DISCUSSION AND CONCLUSION

The infrared radiation flux of M-type main sequence stars can reasonably be understood based on the model atmosphere which properly incorporates the effect of molecular line absorptions. The essential features of the molcular line opacity can approximately be represented by the simplified statistical line opacity characterized by the mean line separation and the mean absorption coefficient. Since the radiation flux of cool stars is seriously distorted by the molecular absorption bands, it is most important to analyze the observed radiation on the basis of realistic model atmospheres. The direct comparison of the stellar radiation with the theoretical flux calculated from a line-blanketed model atmosphere is easier in M-type stars than in G- or K-type stars because of the predominance of molecular absorptions, for which a simple approximation is possible to evaluate the line opacity.

The essential character of the atmosphere of M-type stars is that it is more opaque in the infrared, because of the strong absorption by H_2O and CO. This is in marked contrast to that of G- and K-type stars in which the violet side of the spectrum is more or less opaque because of the predominance of H^-(bf) transition. The radiation flux from M-type stars necessarily shows considerable flux excess in the violet side of the spectrum even if molecular opacity, such as that due to TiO shortward of 1 μ, is taken into account. This fact has important bearing upon interpretation and analysis of the observed stellar radiations or colors. There is a possibility that the effective temperature scale of M dwarf stars is more than 200°K lower than estimated on an empirical basis without considering the model effect. The author's result in a previous study (Tsuji 1967) suggesting that the effective temperature of an M6V star may be 3000°K is also wrong. This is because the previous analysis was limited to the infrared flux beyond 1 μ, and was based on a too simplified, smeared out line opacity. On the other hand, all the known sources of opacity beyond the red spectral region are taken into account in the present study and it is expected that our model provides reasonable theoretical flux, except for U, B, and V regions. However, complete quantitative agreement at every color between theory and observation is not yet obtained and more examination of opacity sources is required.

One weak point in our model is that the effect of convection is not considered. However, so far as emergent flux is concerned, the effect of opacity sources may be more important than that of convection. This is the reason why we have examined the opacity problem first. Kandel (1967) also concluded that the effect of convection on the emergent flux is not important. It is necessary, however, to examine the effect of convective energy transport on emergent flux in the case where molcular opacity plays an important role.

Inspection of the theoretical flux, such as shown in Figures 6(a) and 6(b), also suggests that Johnson's infrared observations correspond to the peak radiations of the whole spectral energy curve except for M magnitude at which CO absorption is very strong. This fact, together with the general flux excess in the violet spectral region, suggests a possibility that the bolometric corrections by Johnson (1965a) are too large. If this is the case, then the luminosities are decreased from the proposed values, and the lowering of the temperature scale does not necessarily mean a serious change of radii of M dwarf stars.

Finally, it should be noted that our result on effective temperature is still a preliminary one and it is not a conclusion but rather a critical remark on the present situation. Before a final conclusion is reached, more careful examination of the opacity problem is necessary, and more detailed study of a sequence of models with convective energy transport should be carried out. Also the effect of chemical composition and turbulent velocity should be examined. These problems will be discussed in the near future.

The author would like to thank Dr. Jesse L. Greenstein for his invaluable advice and suggestions, and also for making available computing time at Caltech Computing Center. His thanks are also to Drs. D. E. Burch and C. B. Ludwig for sending valuable preprints on molecular intensity data, to Dr. R. P. Main for discussions on molecular intensity problems, to Dr. M. Makita for making available his result on the analysis of TiO spectrum of sun spot in advance of publication, to Dr. J. G. Phillips for sending unpublished Franck–Condon factors of TiO, and to Dr. W. Weltner, Jr. for measuring the relative intensities of TiO bands from his matrix spectra.

References

ARMSTRONG, B. H. 1967, *J.Q.S.R.T.*, **7**, 61.

AUMAN, J. Jr. 1967, *Ap. J. Suppl.*, **14**, 171.

BURCH, D. E., and GRYVNAK, D. A. 1966, *Aeronutronic Rep.*, *U*-3704, Aeronutronic Division of Philco Corporation.

CRAWFORD, M. F., WELSH, D. L., MACDONALD, J. C. F., and LOCKE, J. L. 1950, *Phys. Rev.*, **80**, 469.

DALGARNO, A., and LANE, N. F. 1966, *Ap. J.*, **145**, 623.

EGGEN, O. J., and GREENSTEIN, J. L. 1965, *Ap. J.*, **141**, 83.

ELSASSER, W. M. 1938, *Phys. Rev.*, **54**, 126.

GINGERICH, O., LATHAM, D., LINSKY, J., and KUMAR, S. S. 1967, *Proc. Trieste Colloq. on Late-type Stars* p. 291.
GOLDBERG, L., MÜLLER, E. A., and ALLER, L. H. 1960, *Ap. J. Suppl.*, **5**, 1.
GOLDEN, S. A. 1967, *J.Q.S.R.T.*, **7**, 225.
GREENSTEIN, J. L. 1965, *Galactic Structure* ed. by A. Blaauw and M. Schmidt, p. 361.
HARE, W. F. J. and WELSH, H. L. 1958, *Can. J. Phys.*, **36**, 88.
HERZBERG, G. 1950, *Spectra of Diatomic Molecules*, Princeton, N.J., Van Nostrand, p. 111.
JOHNSON, H. L. 1964, *Bull. Tonantzintla and Tacubaya Obs.*, **3**, 305.
JOHNSON, H. L. 1965a, *Ap. J.*, **141**, 170.
JOHNSON, H. L. 1965b, *Ap. J.*, **141**, 923.
KANDEL, R. 1967, *Ann. d'Astr.*, **30**, 439.
KRANENDONK, J. van. 1958, *Physica*, **24**, 347.
KRON, G. E. 1956, *Proceedings of the Third Berkeley Symposium*, **3**, p. 39.
KUNDE, V. G. 1967, *NASA Document X*-622-67-336, Goddard Space Flight Center, Greenbelt, Md.
LIMBER, D. N. 1968, *Ap. J.*, **127**, 363.
LINDBLAD, B. 1935, *Stockholm Obs. Ann.*, **12**, No. 2.
LINSKY, J. 1966, *A. J.*, **71**, 863.
LINSKY, J. and GINGERICH, O. 1967, *Smith. Ap. Obs. Special Rep.*, No. 240.
LUDWIG, C. B. 1966, *Technical Report GDC-DBE* 66-017, General Dynamics Convair.
MAIN, R. P., CARLSON, D. J., and DU PUIS, R. A. 1967, *J.Q.S.R.T.*, **7**, 805.
MAKITA, M. 1967, private communication.
MIHALAS, D. M., and MORTON, D. C. 1965, *Ap. J.*, **142**, 253.
NICHOLLS, R. W. 1965, *J.Q.S.R.T.*, **6**, 647.
OKE, J. B. 1964, *Ap. J.*, **140**, 689.
OKE, J. B. 1967, *I.A.U. Draft Report*, p. 637.
ORTENBERG, F. S. 1960, *Opt. Spectry.*, **9**, 82.
ORTENBERG, F. S., and GLASKO, V. B. 1963, *Soviet A.J.*, **6**, 714.
PATCH, R. W., SHACKLEFORD, W. L., and PENNER, S. S. 1962, *J.Q.S.R.T.*, **2**, 263.
PATHAK, A. N., and SINGH, P. D. 1966, *Proc. Phys. Soc.*, **87**, 1008.
PHILLIPS, J. G. 1967, private communication.
SCHADEE, A. 1964, *B.A.N.*, **17**, 311.
STROM, S. E. and AVRETT, E. H. 1964, *Ap. J.*, **140**, 1381.
STROM, S. E. and KURUCZ, R. L. 1966, *J.Q.S.R.T.*, **6**, 591.
TSUJI, T. 1965, *Publ. A.S. Japan*, **17**, 152.
TSUJI, T. 1966a, *Publ. A. S. Japan*, **18**, 127.
TSUJI, T. 1966b, *Proc. Japan Acad.*, **42**, 258.
TSUJI, T. 1967, *Proc. Trieste Colloq. on Late-type Stars*, p. 260.
UNNO, W. 1962, *Publ. A. S. Japan*, **14**, 153.
WELSH, H. L., CRAWFORD, M. F. and LOCKE, J. L. 1949, *Phys. Rev.*, **76**, 580.
WELTNER, Jr. W. 1967, private communicition.
WILLSTROP, R. V. 1965, *Mem. Roy. Astro. Soc.*, **69**, 83.
WILSON, O. C. 1962, *Ap. J.*, **136**, 793.

Discussion

McCARTHY   I would like to ask you if any of the observers have been using silicon hydride as a luminosity criterion.

GREENSTEIN   The discussion of SiH is outside the realm of this paper. The main problem with silicon hydride is its very open structure. It could be computed, I think, in considerable detail, but it is not only temperature dependent but also abundance dependent.

Model atmospheres of red dwarf stars

JASON R. AUMAN, Jr.

Princeton University Observatory, Princeton, N.J.

Abstract

Model atmospheres have been calculated for red dwarf stars with effective temperatures of 3000°K and 4000°K in addition to a model atmosphere corresponding to the Sun. In order to determine the effect of convection on the structure of the surface layers of the atmosphere and upon the emitted flux, the model atmospheres have been calculated both including and not including the flux carried by convection. It is found that although the convection does not affect the surface layers or the emitted flux of the solar model atmosphere, it does affect the surface layers and emitted flux of the cooler, red dwarf atmospheres.

There is also a discussion of the validity of some of the approximations made in calculating these atmospheres.

UNTIL VERY RECENTLY little progress has been made in calculating theoretical model atmospheres of stars cooler than the Sun. One of the big reasons for this lack of progress was the inability to calculate the opacity due to the many molecular lines which can be seen in the spectra of late-type stars. In addition, it is known that convection carries a significant amount of the flux in the atmospheres of late-type stars with the result that the temperature gradients in these atmospheres will be altered. Therefore, convection has to be included in calculating model atmospheres of late-type stars in contrast to the early-type stars where it can be neglected, at least as far as determining the emitted flux in the visual part of the spectrum. There is some hope that the lack of theoretical calculations concerning the atmospheres of late-type stars will be remedied over the next few years. With the advent of the new, large computers it is now feasible to determine the opacity due to the molecules by summing the contributions of the individual spectral lines such as has been done by Tsuji (1966), Auman (1967), and Kunde (1968). For some molecules, however, a considerable amount of laboratory work still needs to be done in order to determine the fundamental constants and transition probabilities needed to make these calculations.

Model atmospheres have been calculated for $T_e = 4000°K$, $\log g = 4.7$ and $T_e = 3000°K$, $\log g = 5.0$. In addition, a model atmosphere corresponding to the Sun was calculated with $T_e = 5800°K$ and $\log g = 4.44$. In order to determine the effects of convection, model atmospheres were also calculated for the above effective temperatures and gravities with the flux carried by convection neglected. The basic program for calculating the model atmospheres was obtained from D. Mihalas. The reader is referred to Mihalas (1967) for a description of this basic program and to Auman (1969) for a description of the changes that were made in the program to adapt it to the calculation of late-type model atmospheres. All of the atmospheres were calculated assuming solar abundances.

When the flux carried by convection was included, it was calculated using the mixing-length theory developed by Böhm-Vitense (1958). The mixing length was assumed to be equal to the pressure scale height in all of the model atmospheres.

The opacity due to H_2O becomes important in the atmospheres of red dwarfs for $T_e \leqslant 4000°K$. The H_2O opacity is in the form of a series of vibration-rotation bands and is made up of millions of spectral lines distributed over the spectrum for wavelengths greater than 8000 Å. The opacity due to these lines varies over a wide range with a characteristic scale of variation in frequency equal to or less than the widths of the spectral lines. In order to treat the detailed spectrum by integrating over the individual lines a total of over a million frequency points would be required. Obviously, some method has to be found to represent the spectrum without integrating over the individual lines. The method that has been used here was to use the harmonic mean of the sum of the continuous opacity plus the H_2O spectral lines over intervals of 100 cm^{-1} around each of the frequency points in the infrared where the H_2O opacity has to be included. The harmonic mean has been calculated using the tables given by Auman (1967). A total of 60 frequency points were used to calculate the model atmospheres so that the variation of the H_2O opacity over the bands could be well determined. Figure 1 gives the opacity as a function of the wavelength for several optical depths in the convective atmosphere with $T_e = 3000°K$. The H_2O bands can be seen very clearly.

Figure 2 gives the temperature as a function of the optical depth for the three effective temperatures and gravities, both with and without the flux carried by convection. Consider first the model corresponding to the atmosphere of the Sun. The atmosphere becomes unstable to convection at $\tau = 1.0$ where τ is the monochromatic optical depth at $\nu = 0.855\ \mu^{-1}$. The convection is carrying approximately one per cent of the flux at $\tau = 1.5$, a little more than one-fourth of the flux at $\tau = 3.0$, and half of the flux at $\tau = 5.0$. As can

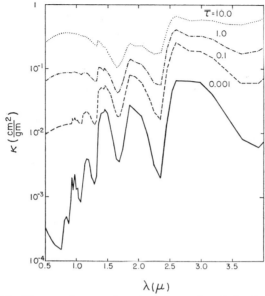

FIGURE 1 The opacity as a function of wavelength for four monochromatic optical depths at 1.17 μ in the convective model atmosphere wiht $T_e = 3000°$K.

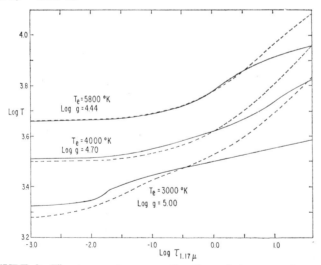

FIGURE 2 The temperature as a function of the monochromatic optical depth at 1.17 μ. The solid lines give the temperature in the atmospheres where the flux carried by convection has been included while the dashed lines give the temperatures in the atmospheres where the convective flux has been neglected.

be seen, the convection has very little effect on the temperatures in the surface layers with $\tau < 1.0$, at least within the approximations used to calculate these model atmospheres. This is not true, however, in the model atmospheres with lower effective temperatures. In the atmosphere with $T_e = 4000°K$, the atmosphere becomes unstable against convection at $\tau = 0.04$. The convection is carrying one per cent of total flux at $\tau = 0.1$, 10 per cent of total flux by $\tau = 0.4$, and over 25 per cent of the total flux by $\tau = 1.0$. For $\tau > 3.0$, the convection is carrying more than half of the total flux. Due to the fact that convection carries a significant amount of the flux for $\tau < 1.0$, the temperatures in the surface layers are affected by the convection. The temperature at the surface of the atmosphere including convection is about three per cent higher than the temperature in the atmosphere in which the flux carried by convection has been neglected. The effect of the convection is the greatest in the atmosphere with the lowest effective temperature. In this atmosphere the convection begins at $\tau = 0.015$. One per cent of the flux is carried by convection at $\tau = 0.025$. Approximately 10 per cent of the flux is carried by convection at $\tau = 0.1$, and 50 per cent of the flux is carried by convection at $\tau = 1.0$. As a result, the temperatures throughout the surface layers have been altered by quite a bit by the convection. The surface temperatures in the atmosphere with convection have been increased by approximately 10 per cent. In addition, the temperature gradients in the layers with $\tau > 0.02$ have been decreased due to the fact that less flux is being carried by radiation in these regions.

Figure 3 gives the emitted flux on a magnitude scale as a function of frequency for the six model atmospheres considered here. As would be expected due to the lack of change in the surface layers when convection is included the emitted flux in the solar model atmosphere changes very little when convection is included. The only effect which might be measurable is a change in the Balmer discontinuity by 0.045 magnitudes. The emitted flux is changed by a greater amount by convection at lower effective temperatures. In the atmosphere with $T_e = 4000°K$, the convection changes the $B - V$ colors by approximately 0.16. In addition, the convection decreases the flux in the infrared beyond $1.6\,\mu$ relative to the flux in the red and near infrared part of spectrum by approximately 0.1 magnitudes. The H_2O features in the infrared can be seen although they are not very strong. The emitted flux in the model atmosphere with $T_e = 3000°K$ is changed quite drastically when the flux carried by convection is included. The size of the H_2O features are considerably smaller in the convective model atmosphere due to the higher surface temperatures and the lower temperature gradients in this model. In addition, it should be noted that the H_2O opacity alters the emitted flux even in the regions between water vapor bands. In particular, the flux at 1.6 μ has been considerably reduced relative to the flux that would have been obtained if the water had not been there. As a result, care will have to be taken in using the infrared colors to determine the effective temperatures of late-type stars, particularly in the spectral region between 1 and 1.6 μ.

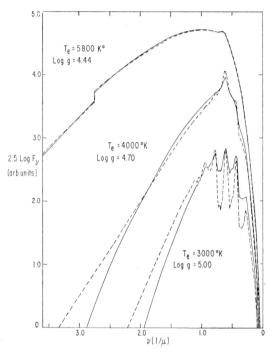

FIGURE 3 The emitted flux as a function of the frequency. The solid line gives the emitted flux by the atmospheres including convection while the dashed lines give the flux emitted by the model atmospheres neglecting convection.

Within the approximations used in calculating these atmospheres it is now possible to calculate a grid of red dwarf atmospheres. However, there must be considerable uncertainty concerning the extent to which these models may actually correspond to the actual atmospheres that exist in these stars. Before any theoretical model atmospheres can be used with confidence, several of the approximations that are at present being used may have to be improved.

The atmospheres have been assumed to be plane-parallel, homogeneous, and in hydrostatic equilibrium. These assumptions seem to hold quite well in the red dwarf atmospheres. Since it has to carry less flux, the convection is not as strong in the red dwarfs as it is in the Sun. The maximum calculated convective velocities are 0.6 and 0.15 km/sec for $T_e = 4000°K$ and $3000°K$ respectively compared to 1.9 km/sec in the Sun. As a result, the temperature fluctuations and the forces associated with the convection will be less than they are in the Sun. In addition, the depths of the red dwarf atmospheres compared to the radii of the stars are approximately one-half the corresponding quantity in the Sun.

The models have been calculated assuming that they are in local thermo-dynamic equilibrium. M.Ch. Pande (1968) has shown that the distribution of vibration-rotation states of the CO molecule are governed by collisional rather than radiative transitions. A similar analysis indicates that the vibration-rotation states of the H_2O molecule are also determined by collisional transitions. This means that the Boltzmann Equation can be used to calculate the distribution and that the assumption can be made that the H_2O spectral lines have been formed by absorption rather than by scattering. There is still the question of whether local thermodynamical equilibrium can be assumed when determining the abundance of the molecular and ionic constituents in the red dwarf atmospheres. Before this question can be answered, the appropriate radiative and collisional cross-sections will have to be determined as well as the ultra-violet radiation field in these stars.

One of the largest sources of uncertainty in these atmospheres is related to the proper method of treating convection. As has been stated above, this study has used the mixing-length formulation to calculate the convective flux. The uncertainties associated with this theory are quite large, mainly associated with how large the mixing length should be. The usual method for determining the mixing length is to assume that it is of the order of either the pressure scale height or the density scale height. The density scale height is always larger than the pressure scale height and can be considerably larger when a major constituent is being dissociated or ionized. While the mixing length may be of the order of one of these scale heights, the exact proportionality factor is still unknown. Since the flux carried by convection is proportional to the square of the mixing length, the convective flux cannot be considered to be well determined. In addition, the radiative transfer in these atmospheres are affected by the motions generated by convection since the opacity is due to a large part to spectral lines which can be broadened by the convective motions.

A great deal of work still needs to be done in determining the sources of opacity in the red dwarf stars. In the model atmospheres discussed here, the opacity due to H^-, He^-, H_2^-, H_2^+, the bound-free and free-free absorption due to H and the metals, Rayleigh scattering by H and H_2, and Thomson scattering by electrons was included in addition to the H_2O opacity. There are several other sources of opacity that have been neglected in this study. Goon and Auman (1969) have calculated the abundances of various molecular and atomic constituents above the photospheres in late-type stars using the dissociation constants given by Vardya (1966), and these abundances can be used to estimate what other molecules besides H_2O may contribute to the opacity in the red dwarf atmospheres. In the infrared the vibration-rotation bands of CO at 4.7μ and 2.35μ should be included. The opacity due to these bands have already been calculated by Tsuji (1966) and by Kunde (1968) so there should be no problem in including the opacity due to this molecule in the future. In addition, the vibration-rotation bands of SiO, HCl, and OH should be included.

There are several molecules which have electronic transitions which may be important sources of opacity. The bands due to TiO will be important, as would be expected due to the great strength of these bands in the spectra of late-type stars. In addition, there are several diatomic hydrids which will be important: OH, NH, AlH, and HS will be important in the ultraviolet while MgH will be important in the visual and CaH will be important in the red part of the spectrum. In the hotter red dwarfs, CN may become important. As has been suggested by Swings (1967), the molecule NH_2 may be an important source of opacity from the blue to the near-infrared part of the spectrum. In addition, the molecule HCO may be important in the same region. Because they are asymmetric molecules, their bands will have many more lines than is the case with diatomic molecules with the result that they can be very effective in blocking the radiation. In addition, the HCO molecule has continuous absorption in the visual to near-infrared due to photo-dissociation. Vardya (1967) has suggested that the photo-detachment of Cl^- may be an important source of opacity shortward of 3400 Å. The abundance of Cl^- in the atmosphere with $T_e = 3000°K$ is roughly equal to H^- so that it should be an important source of opacity in the ultraviolet unless its photo-detachment cross-section is much smaller than that of H^-. Another source of continuous opacity in the cooler dwarfs should be the photo-dissociation of NaCl which begins at 3700 Å.

Another problem which needs to be considered is the proper way to include the opacity due to the molecular spectral lines. The H_2O opacity was included in this study by taking the harmonic mean of the sum of the continuous opacities plus the opacity due to the H_2O spectral lines. Further analysis (Auman 1969) shows that the harmonic mean is probably not the best type of mean to use and that in fact no single mean will correctly treat the opacity due to the many spectral lines.

One method that could be used to treat the molecular opacities would be the one proposed by Strom and Kurucz (1966) and Mihalas (1967). In this method each frequency interval would be divided into a group of sub-intervals with different opacities in each sub-interval. The equations of radiative transfer could then be solved for each sub-interval separately, and, finally, the flux from each frequency interval could be determined by summing the flux from each sub-interval. This method should work reasonably well although it would take more computer time, as long as it is possible to assume that a frequency point at which the molecular opacity is small (or large) at one level in the atmosphere will also be small (or large) at all other levels in the atmosphere. This assumption may be reasonably good where there is only one molecule which is the dominant source of molecular opacity in a frequency interval. If there is more than one molecule important in a spectral region, particularly where they may be important at different levels in the atmosphere, the assumption that the relative opacities due to the molecular spectral lines within a frequency interval will remain the same throughout the atmosphere can no longer be made. In this case, it will be much more difficult to determine the proper solution to the equations of radiative transfer.

In order to determine the distribution of the opacities, the profile of the spectral lines must be known. In calculating the model atmospheres discussed here, the assumption was made that the spectral lines were broadened by micro-turbulent velocities of 2 km/sec in addition to the thermal motions of approximately 1 km/sec of the individual H_2O molecules. This is probably an overestimation of the turbulent velocities since the models predict maximum convective velocities of 0.15 km/sec in the atmosphere with $T_e = 3000°K$ and 0.6 km/sec in the atmosphere with $T_e = 4000°K$, although there is a possi-bility that there could exist waves in the atmosphere which could create higher velocities. Pressure broadening has not been included in determining the absorption due to the spectral lines. This is a particularly bad approxima-tion in the dwarf atmospheres. Since the Doppler broadening is proportional to the frequency while the pressure broadening is not, the pressure broadening will become more important compared to the Doppler broadening as the frequency is decreased. Therefore, there will always be some frequency below which pressure broadening will be important. Neglecting pressure broadening will cause the harmonic mean to be underestimated. Pressure broadening will probably be significant in the red dwarf atmospheres throughout the infrared part of the spectrum.

As can be seen from the discussion above, there are still quite a few prob-lems which need to be solved before realistic model atmospheres of red dwarfs can be calculated. Until these problems can be solved, care will have to be taken in applying these models to observations.

The author would like to express his great debt to Dr. D. Mihalas for many illuminating discussions. In addition, he would like to express his appreciation to Dr. M. Schwarzschild for his very helpful suggestions and encouragement.

This work was supported in part by Project Stratoscope of Princeton University, Princeton, New Jersey, sponsored by NSF, ONR, and NASA, and in part by the NASA Grant NsG-414. It made use of computer facilities supported in part by the NSF Grant NSF-GP579.

References

AUMAN, J. R. 1967, *Ap. J. Suppl.* **14**, 171.
AUMAN, J. R. 1969, submitted to *Ap. J.*
BÖHM-VITENSE, E. 1958, *Zs. f. Ap.*, **46**, 108.
GOON and AUMAN, 1969, in preparation.
KUNDE, V. 1969, *Ap. J.*, **153**, 435.
MIHALAS, D. 1967, in *Methods in Computational Physics*, **7** (New York: Academic Press).
PANDE, M. Ch. 1968, *Soviet Astronomy*, **11**, 592.
STROM, S. and KURUCZ, R. 1966, *J.Q.S.R.T.*, **6**, 591.
SWINGS, P. 1967, private communication.
TSUJI, T. 1966, *P.A.S.J.*, **18**, 127.
VARDYA, M. S. 1966, *M.N.*, **134**, 347.
VARDYA, M. S. 1967, *Memoirs Royal Astronomical Society*, **71**, 249.

Discussion

ALLER   I would like to make a comment about using these models for abundance work. The conventional procedure of using a curve of growth is no longer available when you go to stars of this sort. It is something like the far ultraviolet spectrum of the sun, if you are interested, for example, in determining the abundance of silver or cadmium or some other element which is represented only by weak lines, you have to take into account the actual mode of formation, and try and reproduce that region of the spectrum where the lines are formed. And I think we are going to have to face exactly the same problem here. We call this the method of spectrum synthesis and Ross at U.C.L,A. has used this with luck in work on lead, silver, gold and a few other elements. What you have to remember is that one needs observational material of very high quality. You need to know not only the exact shape of the line profiles, but you need to know also the intensity in the continuous spectrum in absolute units. It turns out, as Mr. Greenstein remarked earlier, that what you think is the continuum is only a fraction, may be 20 or 30 per cent in some instances of the actual height of the continuum, which would exist were the sources of absorption not present. In addition to that, you need very accurate atomic and molecular data so that you can perform the calculations for the spectrum synthesis in some realistic fashion. Finally, of course, we need the type of model atmospheres that Dr. Auman described.

KRAFT I would like to say that Roger Bell at Maryland has also done that type of thing. His interest has been in F and G supergiants, and perhaps he could be induced to do something on M-type stars. I have a question concerning color indices for cool stars. What color index would you use if you wanted to get a color temperature that would be close to the effective temperature?

AUMAN I think I would use R and I.

GINGERICH It is perfectly obvious that the bands are chosen in the photometric system to go between the water vapor bands, and no matter which lines we take we are always going to be hitting the high spots of the star. So we will always tend to overestimate the flux.

AUMAN I realize that we are not supposed to talk about giants here, but there exists a luminosity effect in the giants, namely, that the flux coming out at 1.6 microns increases. Whatever you do, do not use the 1.6 micron band for determining your temperature. You can use it to determine the luminosity, but do not use it for temperature. In general, in the infrared all of your colors are poorly determined as far as temperature is concerned because you are on the Rayleigh–Jeans tail of the black-body curve and so any changes in your opacity will mess up your colors. You do want to get to as short wavelengths as you can, except when you get to short wavelengths you have other bands coming in, which get stronger the further in the ultraviolet you go. So I would suspect that the R and I are probably the best you can do.

Atmospheric structure of K dwarfs

RICHARD A. BERG, JOHN L. HERSHEY
and SHIV S. KUMAR

Leander McCormick Observatory, University of Virginia

Abstract

A representative model atmosphere is presented for a *K* dwarf with an effective temperature of 4000° and a log surface gravity of 4.5. The model is non-gray and radiative flux constancy is enforced to within 1 per cent. Ten sources of opacity and scattering are included, but water vapor opacity is omitted.

I. INTRODUCTION

DURING THE PAST FEW YEARS there has been a growing interest in cool stars as is evidenced in the colloquium edited by M. Hack (1967). Kumar (1964) and Gingerich, Latham, Linsky, and Kumar (1967) began the work of computing accurate non-gray and flux-constant radiative models for late-type stars. During the past year a computer program for model atmospheres has been written and tested at the University of Virginia.

This paper presents a representative model of a late dwarf in radiative equilibrium. The radiative atmosphere computation is one aspect of a general investigation of radiative, convective and scaled solar atmospheres for dwarfs. Twelve strong-line profiles have been computed for each one of a grid of model atmospheres for comparison with observation. Recently several authors have published one or more profiles derived from high dispersion plates for each of 4 early *K* dwarfs and one *M*2 dwarf. We have acquired a few additional profiles at the Kitt Peak National Observatory. The differences between the various atmospheres and the resulting profiles are being discussed and compared with available observations (Hershey 1969).

The model atmosphere presented here has an effective temperature of 4000° and log $g = 4.5$. This corresponds closely to generally accepted values for a *K*5 dwarf (Allen 1963). The composition is the same as given by Goldberg, Müller, and Aller (1960), with the exception that the He/H ratio was

493

taken to be 0.1 by number instead of 0.2. A total of thirty elements are included. A list of these elements and abundances may be found in the paper of Gingerich *et al.* (1967).

The equation of state procedure includes five states of hydrogen, H, H^+, H^-, H_2^+ and H_2. Only first ionizations have been included for the 28 metals. No molecules other than H_2 have been included; in particular, water vapor has not been included in the model. The equilibrium functions and the method for solving the equation of state are taken from Mihalas (1967).

The method of numerical integration of the differential equation of hydrostatic equilibrium, $dP/d\tau_0 = g/\kappa_0$, is the same as used by Gingerich *et al.* and is described by Ralston and Wilf (1960). The reference wavelength adopted was 10000 Å which lies much nearer to the wavelength of maximum surface flux than do wavelengths in the visible region.

The initial pressure was found by a method similar to that described by Mihalas (1967). This method assumes an electron pressure by trial and error which corresponds to a point well above the first optical depth of the model atmosphere and then integrates down to the first point. This in effect starts the integration above the first point in the table and provides a self-consistent-starting value for the solution of the differential equation.

II. OPACITIES

The opacities given here are not listed in order of importance. The importance of individual opacities often varies strongly with wavelength and depth in an atmosphere. The behavior of the opacities at $\tau_0 = 1$ can be seen in Figure 1.

1. Hydrogen. The bound–free and free–free absorption are the well known expressions given by Menzel and Pekeris (1936). Hydrogen is an insignificant opacity source in cool stars.

2. Negative Hydrogen Ion. This ion is the dominant opacity source in *K* dwarfs. Polynomials given by Gingerich *et al.* (1964) for the bound-free and free–free components have been used.

3. Positive Molecular Hydrogen Ion. The expression used includes a polynomial fit to the tables given by Mihalas (1967).

4. Negative Molecular Hydrogen Ion. No bound state is known to exist. Free–free opacity due to H_2^- is important, but no accurate cross sections are available. A recent suggestion by Vardya (1966) has been followed, which is to set the H_2^- cross section per H_2 molecule equal to twice the cross section of H_{ff}^- per H atom.

5. Rayleigh Scattering from Hydrogen and H_2. Gingerich (1964) has given a simple expression for the hydrogen scattering cross section per neutral H

atom. Mihalas (1967) has given an expression for the scattering cross section of the H_2 molecule.

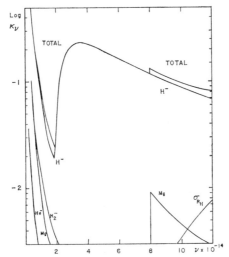

FIGURE 1 The monochromatic opacities at $\tau_{10,000} = 1$; at this depth $T = 3998°K$, $\log P = 5.525$ and $\log P_e = 0.396$.

6. Electron Scattering. This is completely negligible, being about 10^{-5} of total opacity in late-type dwarfs, but it was included as a matter of course.

7. Negative Helium Ion. The opacity computation of He^- interpolates in tables by McDowell *et al.* (1966). A computer library program was used to fit the table and produce the opacity as a function of θ and λ.

8. Magnesium and Silicon. Crude opacity expressions were formed following Unsöld (1955) and several absorption edges are included. The notable feature of the metal opacities is to block the small amount of black-body flux bluer than 2515 Å.

III. SOLUTION OF THE INTEGRAL EQUATION FOR THE SOURCE FUNCTION

Although scattering is not strong in any part of the model, the source function does differ from $B_\nu(T)$ in the visible region of the spectrum by 5 to 10 per cent at the surface. In the ultraviolet the differences are larger but very little flux is carried there. The integral equation for the source function was solved following the method given by Mihalas (1967). This is an iterative method which finds successively better values of $(J_\nu - B_\nu)$. The solution was checked by substituting S_ν into the integral equation independently. The error of the solution is well under 1 part in 1000.

R

IV. FLUX CONSTANCY

The Avrett–Krook procedure as modified to include scattering (Mihalas 1967), has been used to perform temperature corrections. In four iterations from an assumed $T(\tau)$ relation, flux errors of a few tenths of a per cent are usually achieved. The flux errors of this model are given in Table I. All of the remaining error could be removed with more iterations.

TABLE I The physical structure of the atmosphere.

$T_e = 4000°$				$\lambda_0 = 10,000$ Å				$\log g = 4.5$
Tau	Temp	Log p	Log pe	Log kap	Log rho	Depth	Conv.	Flux error
0.001	3241	3.923	−1.415	−2.000	−7.368	−2.18 @ +07		0.0%
0.001	3242	3.963	−1.383	−1.972	−7.327	−2.13 @ +07		0.0%
0.002	3243	4.005	−1.347	−1.943	−7.283	−2.07 @ +07		0.0%
0.002	3244	4.049	−1.310	−1.912	−7.236	−2.00 @ +07		0.0%
0.003	3246	4.096	−1.270	−1.879	−7.186	−1.94 @ +07		0.0%
0.003	3247	4.146	−1.228	−1.845	−7.134	−1.87 @ +07		0.0%
0.004	3249	4.197	−1.184	−1.810	−7.080	−1.80 @ +07		0.0%
0.005	3251	4.249	−1.138	−1.775	−7.025	−1.73 @ +07		0.0%
0.006	3254	4.303	−1.091	−1.739	−6.968	−1.65 @ +07		0.0%
0.008	3257	4.358	−1.042	−1.702	−6.909	−1.58 @ +07		0.0%
0.010	3260	4.414	−0.993	−1.665	−6.850	−1.50 @ +07		0.0%
0.013	3265	4.471	−0.941	−1.628	−6.789	−1.43 @ +07		0.0%
0.016	3270	4.528	−0.889	−1.590	−6.729	−1.35 @ +07		0.0%
0.020	3277	4.586	−0.836	−1.552	−6.667	−1.28 @ +07		0.0%
0.025	3284	4.645	−0.781	−1.513	−6.606	−1.20 @ +07		0.0%
0.032	3294	4.704	−0.725	−1.474	−6.544	−1.13 @ +07		0.0%
0.040	3305	4.763	−0.668	−1.435	−6.483	−1.05 @ +07		0.0%
0.050	3319	4.822	−0.610	−1.394	−6.423	−9.74 @ +06		0.0%
0.063	3335	4.881	−0.551	−1.353	−6.363	−8.98 @ +06		0.0%
0.079	3354	4.940	−0.490	−1.310	−6.305	−8.23 @ +06		0.0%
0.100	3377	4.999	−0.427	−1.265	−6.248	−7.48 @ +06		0.0%
0.126	3405	5.057	−0.363	−1.218	−6.194	−6.73 @ +06		0.0%
0.158	3438	5.114	−0.296	−1.170	−6.142	−5.98 @ +06		0.0%
0.200	3476	5.171	−0.227	−1.119	−6.092	−5.24 @ +06	C	0.0%
0.251	3522	5.226	−0.156	−1.065	−6.046	−4.49 @ +06	C	0.0%
0.316	3575	5.280	−0.080	−1.008	−6.003	−3.74 @ +06	C	0.0%
0.398	3638	5.333	−0.001	−0.948	−5.964	−3.00 @ +06	C	0.1%
0.501	3710	5.384	0.085	−0.884	−5.929	−2.25 @ +06	C	0.1%
0.631	3793	5.433	0.178	−0.815	−5.898	−1.50 @ +06	C	0.1%
0.794	3888	5.480	0.282	−0.741	−5.871	−7.45 @ +05	C	0.1%

TABLE I—(*cont.*)

$T_e = 4000°$				$\lambda_0 = 10,000$ Å			$\log g = 4.5$	
Tau	Temp	Log p	Log pe	Log kap	Log rho	Depth	Conv.	Flux error
1.000	3998	5.524	0.396	−0.661	−5.847	0.00 @ +00	C	0.1%
1.259	4122	5.566	0.523	−0.576	−5.826	7.37 @ +05	C	0.1%
1.585	4263	5.606	0.661	−0.487	−5.809	1.46 @ +06	C	0.1%
1.995	4423	5.643	0.807	−0.399	−5.794	2.18 @ +06	C	0.1%
2.512	4604	5.678	0.957	−0.315	−5.781	2.90 @ +06	C	0.1%
3.162	4807	5.712	1.105	−0.240	−5.769	3.63 @ +06	C	0.2%
3.981	5035	5.746	1.250	−0.174	−4.758	4.39 @ +06	C	0.2%
5.012	5289	5.780	1.395	−0.112	−5.747	5.20 @ +06	C	0.2%
6.310	5569	5.813	1.557	−0.036	−5.736	6.05 @ +06	C	0.3%
7.943	5877	5.845	1.761	0.081	−5.728	6.90 @ +06	C	0.3%
10.000	6212	5.872	2.017	0.251	−5.726	7.65 @ +06	C	0.3%
12.589	6576	5.893	2.310	0.457	−5.730	8.27 @ +06	C	0.2%
15.849	6973	5.908	2.617	0.680	−5.741	8.75 @ +06	C	0.4%
19.953	7406	5.919	2.929	0.911	−5.756	9.12 @ +06	C	−0.3%
25.119	7879	5.927	3.236	1.145	−5.776	9.40 @ +06	C	0.5%
31.623	8409	5.933	3.543	1.390	−5.799	9.62 @ +06	C	−0.1%
39.811	8980	5.937	3.836	1.643	−5.825	9.78 @ +06	C	−0.3%
50.119	9612	5.940	4.120	1.917	−5.855	9.90 @ +06	C	0.4%
63.096	10271	5.941	4.379	2.198	−5.888	9.98 @ +06	C	−0.8%
79.433	10962	5.943	4.615	2.485	−5.924	1.00 @ +07	C	0.3%

V. DISCUSSION

The essential parameters of the 4000° radiative model are given in Table I. The effective temperature is reached at almost exactly $\tau_0 = 1$. The gas pressure at $\tau_0 = 1$ is over twice its value in the sun at the same depth.

Using Schwarzschild's criterion, it is found that convection sets in at $\tau_0 = 0.2$ and thus the model cannot be very realistic for a K dwarf star. A flux constant radiative model is useful for comparison with convective and scaled solar models, for comparison with observation, and for testing the computer program.

Comparison of Figures 1 and 2 shows that the sharp H⁻ opacity minimum at 16,400 Å causes a sharp flux maximum which is accentuated by the peak of the black-body curve falling in that region. This was pointed out a few years ago by Kumar (1964) and Gingerich and Kumar (1964).

For the effective temperature of 4000° used here, the black-body peak on the frequency scale is much less sharp and is about 3000 Å further to the blue than the flux peak in Figure 2. The opacity minimum thus holds the flux peak at a constant wavelength over a range of effective temperatures. Observational tests of the reality of the theoretical H^- opacity minimum should be possible.

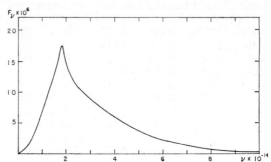

FIGURE 2   The monochromatic flux at the surface for $T_e = 4000°$ and $\log g = 4.5$.

We are grateful to K. Doremus, P. Knappenberger, J. Mast, J. Mischick, R. Myers, M. Hemenway and P. Shelus for their help in writing the model atmospheres program. A grant from the Computer Science Center at the University of Virginia is gratefully acknowledged.

References

ALLEN, C. W. 1963, *Astrophysical Quantities*, 2nd ed. (London: The Athlone Press).

ALLER, L. H. 1963, *The Atmospheres of the Sun and Stars* (New York: The Ronald Press).

ALLER, L. H. 1963, *ibid.*, 186.

GINGERICH, O. 1964, *Research in Space Science Report* no. 167, Smithsonian Astrophysical Observatory.

GINGERICH, O. J., and KUMAR, S. S. 1964, *A. J.*, **69**, 139.

GINGERICH, O., MIHALUS, D., MATSUSHIMA, S., STROM, S. 1965, *Ap. J.*, **141**, 316.

GINGERICH, O., LATHUM, D., LINSKY, J., and KUMAR, S. S. 1967, *Colloquium on Late-Type Stars*, ed. M. Hack, p. 291.

GOLDBERG, L., MÜLLER, E. A., and ALLER, L. H. 1960, *Ap. J. Suppl.*, **5**, 1.

HERSHEY, J. L. 1969, *Dissertation*, University of Virginia (unpublished).

KUMAR, S. S. 1964, *Mem. Soc. R. Sc. Liège*, **26**, 477.

McDOWELL, M., WILLIAMSON, J., and MYERSCOUGH, V. 1966, *Ap. J.* **144**, 827.

MENZEL, D. H., and PEKERIS, C. L. 1935, *M. N.*, **96**, 77.

MIHALAS, D. 1965, *Ap. J. Suppl.*, 92, **9**, 321.

MIHALAS, D. 1965, *Methods in Computational Physics*, **7**, (New York: Academic Press) p. 1.

RALSTON, A., and WILF, H. S. 1960, *Mathematical Methods for Digital Computers* (New York: John Wiley and Sons, Inc.) p. 99.

UNSÖLD, A. 1955, *Physick der Sternatmospharen*, 2nd ed. (Berlin: Springer Verlag), p. 173.

VARDYA, M. S. 1966, *M. N.*, **134**, 347.

Atmospheric structure of K dwarf stars

K. S. KRISHNA SWAMY*

Goddard Space Flight Center, Greenbelt, Maryland

Abstract

The structure of the atmospheric models constructed for $T_e = 4500°$K and $5000°$K and hydrogen-to-metal ratios of 1, 5, 10, 40, and 100 times the solar value has been discussed. The importance of the present investigation in relation to the study of metal deficient stars is pointed out.

I. INTRODUCTION

THE RADIATION emerging from a star carries considerable information about the physical condition in the star. One of the main aims in astrophysics is to deduce the physical structure and composition of the stars from the study of the observed stellar spectra. The general method employed is to compare the observed radiation with that of models of known characteristics. The degree of agreement between the observed stellar spectrum and the spectrum calculated from the model permits one to infer the physical structure and composition of stars.

A number of investigations have shown that there exist large variations in the metal abundances between stars of different population groups (Aller and Greenstein 1960, Baschek 1959, Pagel and Powell 1966). For example, subdwarfs have been found to be deficient in metals as compared to the sun by factors between 10 and 100. The variation in the abundances are not only useful in the interpretation of stellar evolution and the origin of the elements in stars, but they have an effect on the observable physical parameters such as temperature and luminosity obtained from spectroscopic and photo-electric observations.

*National Academy of Sciences–National Research Council, Postdoctoral Research Associate.

The present investigation deals with the study of the structure of the atmospheres of K dwarf stars. In particular, we would like to study the effect of reduced metal abundance on the structure of these stars. For this purpose we have constructed atmospheric models for $T_e = 4500K$ and $5000°K$ and hydrogen to metal ratios A of 1, 5, 10, 40, and 100 times the solar value. For surface gravity a value of 2×10^4 is used for all the models. We will show the importance of convection in the model atmosphere calculation and resonance damping in the calculation of hydrogen line profiles. We will first discuss briefly the calculation of model atmospheres and line profiles.

II. MODEL ATMOSPHERE CALCULATION

The equation of hydrostatic equilibrium is given by

$$\frac{dP}{d\tau} = \frac{g}{\bar{K}} \tag{1}$$

where τ is the optical depth, P the gas pressure, g the surface gravity and K the mean absorption coefficient per gram of the stellar material. We will now discuss briefly the various quantities needed to solve the Eq. (1).

(a) Composition: We take $X = 0.75$ and the models are calculated for $A = 1, 5, 10, 40$, and 100 times the solar value. The abundance of the individual elements are fixed using the solar abundances as standard (Goldberg, Müller and Aller 1960).

(b) Opacity: For the calculation of Rosseland mean opacity, we assume the continuous opacity to be due to absorption by negative hydrogen ions and scattering by hydrogen atoms and molecules (Krishna Swamy 1966).

(c) Equation of State: We consider hydrogen to be in the following Stages: H_2, H_2^+, H^-, H and H^+. Helium is considered only in the neutral state, since in the temperature range of interest the ionization is negligible. The metals are assumed to be only in the neutral and singly ionized states. We have included the detailed ionization equilibria of nine elements besides hydrogen. The calculation of the various partial pressures are described in Krishna Swamy (1966).

(d) T–τ: Relation: From an earlier analysis (Krishna Swamy 1966) it was found that the scaled solar model approach was quite adequate for solar-type stars. But for metal deficient stars, the scaled solar models are not very satisfactory. This is because of the fact that the role of convection is far more important in a metal deficient star than in a normal star of the same spectral type (Krishna Swamy 1967, 1968). Therefore, in the model atmosphere calculation it is necessary to take into account the effect of convection on the atmospheric structure. We divide the model atmosphere calculations into two parts; one radiative zone and the other superadiabatic convection zone. In the radiative zone, we use the following analytical temperature distribution (Krishna Swamy 1966, 1967).

$$T^4 = \tfrac{3}{4} T_e \left[\tau + 1.39 - 0.815 \, e^{-2.54 \, \tau} - 0.025 \, e^{-30\tau}\right] \qquad (2)$$

The high value of 1.39 in Eq. (2) arises mainly as a result of blanketing effect (Stewart 1964). The use of Eq. (2) for the temperature distribution in the radiative zone has the advantage of empirically allowing, to a first approximation, line blanketing effects.

In the superadiabatic zone, we do the calculations under the framework of mixing length theory (Böhm-Vitense 1958). The various thermodynamic quantities needed in the calculation of the superadiabatic convection zone are carried out in some detail (Henyey, Vardya and Bodenheimer 1965). In all the calculations we set the ratio of mixing length to pressure scale height to be unity.

III. LINE PROFILE CALCULATION

Here we would like to briefly mention a few essential points. (For details see Krishna Swamy 1966). We are interested in the wings of strong line profiles where the broadening plays an important role. Since the wings are formed deeper in the atmosphere than their centers, we assume that the line is formed by pure absorption. Radiation and collisional-damping have been included in the calculation of metallic lines. For the calculation of collisional-damping constant, collisions with neutral hydrogen atoms, neutral helium atoms, hydrogen molecules and electrons have been taken into account. The profile of Hα is calculated for Stark effect and for resonance broadening.

IV. DISCUSSION OF THE MODELS

(a) *Temperature distribution*

The temperature distributions of the model atmospheres for $T_e = 4500°$K and $5000°$K and for various hydrogen-to-metal ratios are shown in Figure 1. Figure 1 shows clearly the effect of metal deficiency on the temperature distribution of the model atmospheres. For the same A value the change introduced in the temperature distribution for $T_e = 4500°$K is larger than that for $T_e = 5000°$K. Also one can see clearly the branching of the curves at different optical depths, depending upon the A value. Therefore, it is clear that it is necessary to take into account the effect of convection in the model atmosphere calculations of metal deficient stars.

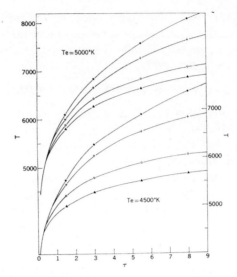

FIGURE 1 Temperature distributions for different hydrogen-to-metal ratios for $T_e = 5000°K$ and $T_e = 4500°K$, curves with dots, crosses, circles and triangles refer to $A = 1, 5, 40,$ and 100 times the solar value, respectively.

(b) *Pressure distribution*

The variation of gas pressure and electron pressure is plotted as a function of optical depth in Figure 2 for $T_e = 4500°K$ and for various A values. In Table I, we give ratios of the partial pressures of hydrogen molecule to hydrogen and hydrogen to total gas pressure for $T_e = 4500°K$ and for $A = 1,$ 40 and 100 times the solar value. It may be seen that although hydrogen is the dominant contributor to the total gas pressure, the contribution of hydrogen molecules is not negligible, particularly for metal deficient stars.

(c) *Logarithmic temperature gradients*

In Figures 3(b) and 4(b) we show the actual temperature gradient (∇), and the adiabatic temperature gradient (∇_{ad}) as a function of optical depth for $T_e = 4500°K$ and $5000°K$. Figures 3(a) and 4(a) show a plot of the fraction of hydrogen molecules with respect to the total hydrogen atoms as a function of optical depth. Near the surface the amount of molecular hydrogen can be quite appreciable and more so in metal deficient stars. The effect of the presence of hydrogen molecules in the top layers will be to lower the adiabatic temperature gradient through the dissociation of hydrogen molecules.

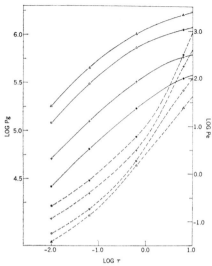

FIGURE 2 Gas and electron pressure distributions for different hydrogen-to-metal ratios for $T_e = 4500°$K. Solid curves refer to gas pressure and dashed curves to electron pressure. All the symbols have the same meaning as in Figure 1.

TABLE I Partial pressure ratios.

T_e	$\dfrac{A}{A_\odot}$	τ	$\dfrac{P(H_2)}{P(H)}$	$\dfrac{P(H)}{P(G)}$
4500°K	1	0.05	0.031	0.897
		0.1	0.022	0.906
		0.6	0.005	0.922
	40	0.05	0.129	0.812
		0.1	0.096	0.836
		0.6	0.025	0.898
	100	0.05	0.184	0.770
		0.1	0.137	0.804
		0.6	0.047	0.877

This is illustrated in Figures 3(b) and 4(b). This effect helps the convective instability to set in at shallow optical depths.

(d) *Absorption coefficients*

In Figure 5, we show the variation of absorption coefficient per gram of the stellar material of H^-, σ_H and σ_{H_2} as a function of λ for $T_e = 4500°$K. It

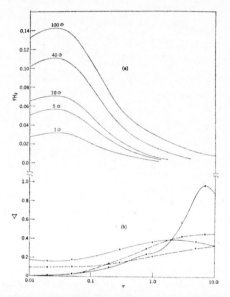

FIGURE 3   (a) Plot of the fraction of hydrogen molecules for different hydrogen-to-metal ratios for $T_e = 4500°K$. (b) Plot of ∇ (dots) and ∇_{ad} (crosses) as a function of optical depth. Solid line and dashed line refer to $A = 1$ and 100 times the solar value, respectively.

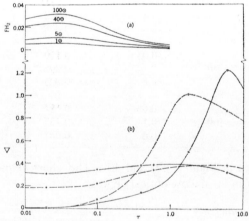

FIGURE 4   (a) and (b). Same as Figure 3(a) and (b) but for $T_e = 5000°K$.

TABLE II Damping constants of the lines λ3968 of CaII and λ5896 of Na I.

T_e	$\dfrac{A}{A_\circ}$	τ	λ3968			λ5986		
			Γ_H/Γ	Γ_{H2}/Γ	Γ_R/Γ	Γ_H/Γ	Γ_{H2}/Γ	Γ_R/Γ
4500°K	1	0.05	0.850	0.018	0.121	0.915	0.022	0.036
		0.6	0.932	0.003	0.054	0.956	0.004	0.016
	100	0.05	0.861	0.107	0.020	0.839	0.122	0.006
		0.6	0.956	0.030	0.009	0.932	0.035	0.003
5000°K	1	0.05	0.811	0.033	0.180	0.914	0.004	0.056
		0.6	0.921	0.001	0.070	0.953	0.001	0.021
	100	0.05	0.939	0.020	0.003	0.941	0.024	0.009
		0.6	0.966	0.004	0.020	0.972	0.003	0.006

TABLE III Residual intensity for Hα profile.

λ (A°)	$T_e = 4500°$K				$T_e = 5000°$K			
	$A = A_\circ$		$A = 100A_\circ$		$A = A_\circ$		$A = 100 A_\circ$	
	Stark effect	Resonance broadening	Stark effect	Resonance broadening	Stark effect	Resonance broadening	Stark effect	Resonance broadening
0.5	0.945	0.890	0.993	0.825	0.858	0.800	0.906	0.694
1.0	0.973	0.938	0.998	0.889	0.925	0.865	0.954	0.759
2.0	0.990	0.969	1.00	0.946	0.965	0.926	0.983	0.855
4.0	0.997	0.988	1.00	0.981	0.986	0.967	0.995	0.935
6.0	0.998	0.994	1.00	0.990	0.992	0.982	0.998	0.964

may be seen from Figure 5, that the scattering by hydrogen atoms contributes an appreciable amount to the total opacity for $\lambda \leqslant 0.3\,\mu$. The same result is valid even for the case $T_e = 5000°K$.

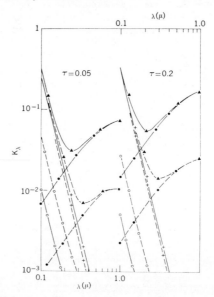

FIGURE 5 Plot of various absorption coefficients per gram of the stellar substance as a function of wavelength. Curves with dots, crosses, circles and triangles refer to contributions from H^-, σ_H, σ_{H_2} and total absorption coefficient respectively. Solid line and dashed line refer to $A = 1$ and 100 times the solar value, respectively.

(e) *Damping constants of metallic lines*

The various damping constants for the lines of λ 3968 of Ca II and λ 5896 of Na I are given in Table II. The damping due to collisions with hydrogen atoms is the main contributor to the broadening in these stars.

(f) *Resonance broadening of hydrogen profile*

We have calculated the H_α profile for two cases: (1) Stark effect only, (2) resonance broadening only. The resulting profiles are shown in Table III. One may see the importance of resonance broadening in the calculation of H_α profiles in the stars under discussion.

In conclusion, we would like to point out that the effect of convection on the stratification of the model atmospheres of metal deficient stars is quite important. Resonance broadening of H_α profile is also quite appreciable in metal deficient stars.

References

ALLER, L. H., and GREENSTEIN, J. L. 1960, *Ap. J. Suppl.*, **5**. 139.
BASCHEK, B. 1959, *Zs. f. Ap.*, **48**. 95.
BÖHM-VITENSE, E. 1958, *Zs. f. Ap.*, **46**, 108.
GOLDBERG, L., MÜLLER, E. A., and ALLER, L. H. 1960, *Ap. J. Suppl.*, **5**, 1.
HENYEY, L. G., VARDYA, M. S., and BODENHEIMER, P. 1965, *Ap. J.*, **142**, 841.
KRISHNA SWAMY, K. S. 1966, *Ap. J.*, **145**, 174.
KRISHNA SWAMY, K. S. 1967, *Ap. J.*, **150**, 1161.
KRISHNA SWAMY, K. S. 1968, *Ap. J.*, **152**, 477.
PAGEL, B. E. J., and POWELL, A. L. T. 1966, *Roy. Obs. Bull.* No. 124.
STEWART, J. C. 1964, *J. Quant. Spectrosc. Rad. Transfer*, **4**, 723.

The electron pressure in the atmospheres of late-type dwarfs

ROBERT S. KANDEL

Goddard Institute for Space Studies, NASA, New York, New York

Abstract

Observations of the Ca I resonance line at λ4227 Å, in late-type dwarfs, exhibit a constancy of the line profile going from spectral type K7 to M3, in disagreement with predictions from models. We discuss a non-LTE mechanism leading to excess electron pressure and H⁻ population, and conclude that while it probably cannot explain the observed discrepancy, it should be looked for in cooler stars.

SOME YEARS ago Vardya and Böhm (1965) examined the profile of the Ca I resonance line at λ4227 in the spectrum of the M dwarf HD 95735 and came to the conclusion that if the effective temperature of the star is lower than $3400°K$, unknown sources of opacity may be present in its atmosphere. This arises because the pressure-broadened profile predicted by the lower-temperature models is far broader than the profile actually observed. The introduction of an additional opacity changes the relation between gas pressure and optical depth, so that on the whole the line is formed in regions of lower pressure and therefore is not so broad.

In the course of observations which I made some years ago at the Observatoire de Haute-Provence, France, at the coudé spectrograph of the 193 cm telescope, I studied the profile of λ4227 line in several late-type dwarfs ranging in spectral type from K7 to M3. The stars observed were:

TABLE I

Star	Spectral type	Dispersion used (A/mm)
61 Cyg B	K7 V	10
HD 88230	dM0	10, 20
HD 95735	dM2	40
BD + 36°2393	dM2	40
HD 119850	dM3	40

511

S

The remarkable feature, as seen in Figure 1, is that the profile hardly varies among these stars; moreover the agreement with the published profile of Vardya and Böhm is quite good.

On the other hand, the profiles predicted by model atmospheres were far broader than those actually observed, so that the discrepancy suggested by Vardya and Böhm exists here too. The models used have been described by me elsewhere (Kandel 1967). Here I shall only remark that changing from a

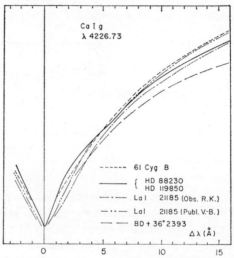

FIGURE 1   Smoothed observed profiles of the Ca I g line at λ4227 Å.

model in radiative equilibrium to one which goes over to an adiabatic at low optical depth has practically no effect on the predicted emergent profile. The Figures 2 and 3 illustrate the sense of the discrepancy, which, of course, becomes more serious as one goes to lower-effective-temperature models. A similar discrepancy exists for the region of Ca II *H* and *K* lines (Figure 4), but here the continuum level cannot be defined at all, so not much weight can be attached to this.

Now this discordance also manifests itself in the same sense for the Na *D* lines (Figure 5); therefore, if it is to be explained by an unknown absorber, it will mean that there is little wavelength dependence involved. In the atmosphere of a cool star, one way of obtaining such a nearly gray supplementary absorption would be to have an excess electron pressure and thus a higher concentration of H^-, the negative hydrogen ion.

It may seem that departures from local thermodynamic equilibrium (LTE) are unlikely in the dense atmospheres of late-type dwarfs. However, it

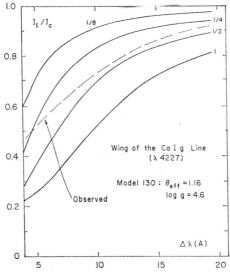

FIGURE 2 Comparison of observed and computed profiles. The curve numbered (1) is computed using the normal abundance: $A(Ca)/A(H) = 6.5 \times 10^{-6}$. For the other curves this abundance has been multiplied by the number shown.

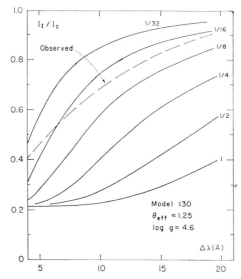

FIGURE 3 Same as Figure 2, but for a colder model: the disagreement becomes stronger.

should be noted that the *electron* densities are quite low ($N_e \sim 10^{12}$–10^{13}); thus, provided that collisions involving neutral particles, such as H atoms or H_2 molecules, are relatively ineffective in ionization or recombination, a strong radiation field could quite easily dominate the ionization equilibrium and lead to departures from the Saha equation.

The question then is: which metals are suitable candidates for over-ionization? Also, is the necessary radiation field available? In order for over-ionization to have an effect, the metal must not be already nearly totally

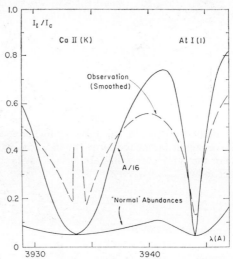

FIGURE 4   The blend of the Ca II K line and the Al I resonance line, between 3930 and 3945 Å, in the spectrum of 61 Cyg B. Again, observed and computed profiles.

ionized according to LTE. In the case of the stars observed, near spectral type M0, this excludes such metals as potassium, sodium and calcium, since an over-ionization will only affect the neutral population without changing the electron pressure. Examination of the LTE equation of state for these atmospheres (see e.g. Vardya 1966) shows that magnesium and silicon could furnish many electrons if there were departures from LTE. However, for these metals we cannot seriously expect such departures, since the bound-free absorption of the neutrals will practically cut off all the radiation in the relevant wavelengths (1400–1700 Å) (see e.g. Gingerich 1964). Thus, it seems that the suggested mechanism must be inoperative in these stars.

It may, however, come into play for cooler stars where according to the Saha equation, the metals Ca, K and Na are only partially ionized. These elements are not sufficiently abundant for their bound-free absorption to affect the radiation field. Furthermore, in the wavelength region for photo-ionization of these metals, especially K and Na, there may be significant chromospheric radiation available, including in particular the strong emission to be expected in the *H* and *K* lines of Mg II at λ2800 Å. Since there will be in any case very little ionization in the atmosphere, the blocking effect of ionized lines can be neglected.

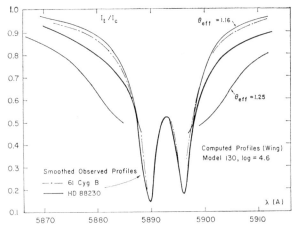

FIGURE 5   The Na *D* lines in the spectrum of 61 Cyg B and HD 88230, compared with computed profiles for θ_{eff} = 1.16 and θ_{eff} = 1.25. Note that H. L. Johnson (*Bol. Obs. Ton. y Tac.* **3**, 305, 1964) assigns θ_{eff} = 1.23 to 61 Cyg B, and HD 88230 is most probably cooler, so that a strong disagreement exists for Na *D*.

Thus, we suggest that while it still seems necessary to invoke unknown absorbants in the atmosphere of K6-M2 dwarfs, and while it is, of course, likely that such absorbants will continue to operate in cooler stars, one should take account of possible over-ionization of K and Na in the very coolest stars, leading to an excess electron pressure and higher H^- opacity.

This work was begun at the Observatoire de Paris-Meudon, France. It was completed at the Goddard Institute for Space Studies under the tenure of a NAS–NRC Resident Research Associateship, supported by the National Aeronautics and Space Administration. All of this support is gratefully acknowledged.

References

GINGERICH, O. 1964, *Proc. 1st Harvard-Smithsonian Conf. on Stellar Atmospheres,*
    S.A.O. Report No. **167**, 108.
KANDEL, R. 1967, *Ann. d'Ap.*, **30**, 439.
KANDEL, R. 1967, *ibid.*, **30**, 999.
VARDYA, M. S., and BÖHM, K. H. 1965, *M.N.R.A.S.*, **131**, 89.
VARDYA, M. S., 1966, *M.N.R.A.S.*, **134**, 347.

Discussion

GREENSTEIN   One of the problems of late-type stars is there is essentially no continuum, so that you should add to the computed continuum the effect of the solar lines increased by a large amount. Some years ago, Dr. Aller and his associates tried to measure the H and K ratio in an early K star and found that it fitted no possible prediction. From this they deduced that the continuum was 2 or 3 times as high as the apparent continuum. What you are in fact observing are just the inner cores of the extremely strong lines produced in a low temperature envelope. I do not think this is the whole effect, but it is a little dangerous to say that lines are independent of temperature ionization, the contrast may well be.

KANDEL H and K lines, of course, are too far in the ultraviolet to be used here. I just wanted to show what one could get there. But some sort of continuum may be available around 4227 Å, and the real test will be to look at the sodium D lines.

ALLER   In this context, I might remark that some of the work that Mr. Ross at UCLA is doing in synthesizing the solar spectrum in the region around 3000 Å to 3800 Å in looking for the profiles and the intensities of some of the lines of Si and other elements, shows that the spectrum is produced by overlapping effects of the crowded lines in that the continuous absorption has a very low influence. So, as Professor Greenstein remarked, you are looking at the very bottoms of all these profiles. I should imagine that the 3500 Å region of the sun might be comparable to what we see through most of the visible region in a K dwarf, as far as crowdedness is concerned.

Boundary conditions and late-type stars

M. S. VARDYA

Tata Institute of Fundamental Research, Bombay, India

P. GIANNONE, N. VIRGOPIA

Osservatorio Astronomico di Roma, Rome, Italy

Abstract

This paper consists of two parts. In the first part, the effect of pressure on the dissociation of molecular hydrogen, at low temperatures and high pressures, has been examined by exchange reaction approach as well as by radial distribution function approach. In the second part, we have tried to determine the molecular ratios which are sensitive to elemental abundance ratios of C/O, C/N, and C/H.

WE DISCUSS here not the boundary conditions *per se*, but rather two topics, somewhat independent, which may influence the boundary conditions for late-type stars. In Section I, we will take another look at the phenomenon of pressure dissociation, and in Section II, we will discuss the use of molecular ratios as indicators of elemental abundance ratios. The work on the two topics is still in progress and the results presented here are somewhat of a preliminary nature.

I. ANOTHER LOOK AT PRESSURE DISSOCIATION*

Traditionally, one considers in astrophysics, the phenomenon of pressure ionization, when dealing with the equation of state at high densities. However, when the temperature is not high enough, we encounter the analogous phenomenon of pressure dissociation. We considered this phenomenon and approximately incorporated the pressure dependence in the partition function, equilibrium constant and internal energy for molecular hydrogen

* M. S. Vardya.

(Vardya 1965; Paper I). This effect is bound to play an important role in the low luminosity and high density objects between red dwarfs and black dwarfs, as well as in the cool white dwarfs. It is worthwhile, therefore, to take another look at this phenomenon, which is bound to affect the boundary conditions.

As in Paper I, we will assume that the energy of any rotation-vibration level cannot exceed the effective dissociation energy, i.e. $\epsilon_v + \epsilon_J(v) \leqslant D_{\text{eff}}$, where ϵ_v and $\epsilon_J(v)$ are the energies of the vibrational level v and rotational level J, respectively, and D_{eff} is the effective dissociation energy.

In Paper I, the effective dissociation energy was determined by assuming that the atoms of the molecule and the perturbing hydrogen atom are collinear, and by applying the London equation. This essentially amounted to considering the exchange reaction:

$$\text{H—H} + \text{H} \longrightarrow \boxed{\text{H—H—H}} \longrightarrow \text{H} + \text{H—H}$$

This is really applicable if the gas consists mostly of hydrogen atoms with a small fraction as molecular hydrogen.

If the gas consists mostly of molecular hydrogen, then one should consider the following reaction:

$$
\begin{array}{c}
\text{H—H} \\
+ \\
\text{H—H}
\end{array}
\longrightarrow
\boxed{
\begin{array}{c}
\text{H—H} \\
|\quad| \\
\text{H—H}
\end{array}
}
\longrightarrow
\begin{array}{c}
\text{H} \\
| \\
\text{H}
\end{array}
+
\begin{array}{c}
\text{H} \\
| \\
\text{H}
\end{array}
$$

One can obtain in this situation also the effective dissociation energy by using the London equation for four nuclei (see, e.g., Glasstone, Laidler, and Eyring 1941) and by assuming that the four hydrogen atoms form a square. One can also assume that the four hydrogen atoms are collinear, but the square configuration is a more stable state than the collinear one. Table I gives

TABLE I Depression in the continuum.

r_0 (Å)	D_{3L} (ev)	D_{4L} (ev)	D_{4S} (ev)	N_H
2.0	0.794	1.314	1.472	2.984 (22)
2.5	0.263	0.436	0.505	1.527 (22)
3.0	0.082	0.136	0.161	8.841 (21)
3.5	0.025	0.041	0.049	5.568 (21)
4.0	0.007	0.012	0.014	3.730 (21)
4.5	0.002	0.003	0.004	2.619 (21)
5.0	0.001	0.001	0.001	1.090 (21)
	H—H—H	H—H—H—H	H—H | | H—H	

the depression of the continuum as a function of interparticle distance, r_0, for the cases when (i) three hydrogen atoms are collinear (D_{3L}), (ii) four hydrogen atoms are collinear (D_{4L}), and (iii) four hydrogen atoms form a square (D_{4S}). The number density of hydrogen *nuclei* is given by N_H. Table II compares the equilibrium constants for these three cases at three temperatures. As in Paper I, we have expressed the potential energy for the molecular hydrogen in the form of Rydberg potential.

TABLE II $\log_{10} K_p$ (dyn/cm²) [H₂ ⇌ 2 H]:
H—H—H
H—H—H—H
H—H
| |
H—H

r_0 (Å) \ $T \times 10^{-3}$°K	3	6	10
2.0	5.735	9.110	10.527
	6.608	9.552	10.814
	6.874	9.687	10.905
3.0	4.539	8.509	10.146
	4.630	8.554	10.174
	4.671	8.575	10.187
4.0	4.413	8.446	10.106
	4.421	8.450	10.108
	4.425	8.452	10.109

We can look at the problem of pressure dissociation from a somewhat different point of view also. The potential energy curve as a function of distance that one uses for a given molecule, is really valid in μ-space, i.e. in the molecule space. In the γ-space (the gas space), this potential energy function should be a good approximation if the density is low. However, it will undergo change by the presence of other particles, when the density is high. One can obtain the change in the potential energy by averaging $U(r)$ over the total volume, using a radial distribution function, g, such that Ng gives the probability per unit volume of finding a second particle at a distance r from a given particle. As N is the mean particle density, the average interparticle distance, r_0, can be written as:

$$(4\pi/3) r_0^3 N = 1.$$

Then

$$\bar{U} = (N/2) \int_{r^*}^{\infty} U(r) \cdot g \cdot 4\pi r^2 \, dr.$$

The distance r^* is the lower limit for which U goes to 0, and the factor 2 is to take into account the fact that the particles can interchange. We can express $\bar{U}$ as:

$$\bar{U} = (\tfrac{3}{2}r_0^3) \int_{r^*}^{\infty} U(r) \cdot g \cdot r^2 \, dr.$$

In general, g should depend on density and temperature. We will consider here four types of radial distribution functions.

Case (I): As the simplest case, one can take

$$g \equiv g_1 = 1$$

and then

$$\bar{U}_1 = (\tfrac{3}{2}r_0^3) \int_{r^*}^{\infty} U(r) \, r^2 \, dr.$$

This has recently been done by Penston (1967), while considering molecular hydrogen. This does not appear to be satisfactory as it does not take into account the local density variation.

Case (II): Let

$$g \equiv g_2 = \exp(-U/kT),$$

and the corresponding $\bar{U}_2$ can be written as before. This distribution function has the property that $g_2 \to 0$ as $r \to 0$, $g_2 \to 1$ as $r \to \infty$, and reaches its maximum value as $r \to r_e$, the mean equilibrium distance of the two atoms forming a molecule. Such a distribution function also implies that only pair interactions are involved. This distribution function takes into account the fact that the particles at a distance r_e have the greatest binding energy. Such a radial distribution function has been used in the literature (see, e.g. Hirschfelder, Curtiss, and Bird 1954). This distribution function, however, is independent of the number density.

Case (III): Let

$$g \equiv g_3 = (r/r_0)^n \cdot \exp\{-(r/r_0)^n + 1\},$$

n being an arbitrary positive number. Here r_0 is the mean interparticle distance. The behavior of g_3 is such that $g_3 \to 0$ with $r \to 0$ as well as with $r \to \infty$, and it reaches its maximum value with $r \to r_0$. This distribution function takes into account the fact that the maximum probability for the other particle is to be at a distance r_0. However, it does not take into account the effect of the interparticle interaction governed by the potential function U.

We have considered this distribution function with $n = 1$.

Case (IV): It will be best to combine g_2 and g_3 and obtain

$$g \equiv g_4 = \exp(-U/kT) \cdot (r/r_0) \cdot \exp\{-(r/r_0) + 1\}.$$

This distribution function may have peaks or maxima at $r = r_e$ as well as at $r = r_0$. The corresponding $\bar{U}_4$ is:

$$\bar{U}_4 = (\tfrac{3}{2}r_0^3) \int_{r^*}^{\infty} U \exp(-U/kT) \cdot (r/r_0) \exp\{-(r/r_0) + 1\} r^2 \, dr.$$

Table III gives the depression of the continuum for Cases (I) and (III), which are independent of temperature, and Table IV for the Cases (II) and (IV) for three temperatures. As expected, Case (I) gives the largest depression

TABLE III   Mean potential.

r_0 (Å)	$\bar{U}_1$ (ev)	$\bar{U}_3$ (ev)
2.0	1.307	1.177
3.0	0.387	0.309
4.0	0.163	0.113
6.0	0.048	0.026
8.0	0.020	0.009

TABLE IV   Mean potential.

$T \times 10^{-3}$°K	3		6		15	
r_0 (Å)	$\bar{U}_2$	$\bar{U}_4$	$\bar{U}_2$	$\bar{U}_4$	$\bar{U}_2$	$\bar{U}_4$
2.0	0.155	0.141	0.252	0.237	0.465	0.441
3.0	0.046	0.045	0.075	0.072	0.138	0.127
4.0	0.019	0.018	0.032	0.028	0.058	0.048
6.0	0.006	0.005	0.009	0.007	0.017	0.012
8.0	0.002	0.002	0.004	0.002	0.007	0.004

and Case (IV) the least. The differences between $\bar{U}_1$ and $\bar{U}_3$ as well as between $\bar{U}_2$ and $\bar{U}_4$ are small, but the differences between $\bar{U}_1$ or $\bar{U}_3$ to $\bar{U}_2$ or $\bar{U}_4$ are large. These differences decrease as r_0 increases. Further, $\bar{U}_2$ and $\bar{U}_4$ increase with increasing temperature.

We have so far not done it but it will be desirable to combine the exchange reaction approach and the radial distribution function approach.

II. MOLECULAR RATIOS AS INDICATORS OF ELEMENTAL ABUNDANCE RATIOS*

The determination of relative elemental abundances in cool stars is rather unsatisfactory. One generally uses solar photospheric abundances, in lieu of anything better known, for these stars. The relative abundance of elements in stars is generally determined using the equivalent width or profile of atomic lines. One can also use, in principle, the strength of molecular lines or bands, which abounds in late-type stars, to determine elemental abundance ratios. This in fact has been done by Spinrad and Vardya (1966) to estimate the relative abundances of H, C, N and O in M and S spectral type giant and supergiant stars. In practice, however, the use of molecular lines present certain difficulties. Some of these difficulties can be overcome with a modicum of accuracy, but the blending of lines and the fixing of the true continuum present serious problems.

Therefore, perhaps the best way to determine elemental abundances with certain precision at the moment is to determine not one but several molecular ratios, which are sensitive indicators of elemental abundance ratio of a given pair of elements. This is true more so, if the results are as

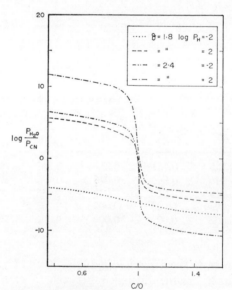

FIGURE 1 Variation of logarithmic ratio of partial pressure of H_2O to CN as a function of C/O ratio (H : N : O = 10^4 : 1 : 10.)

* P. Giannone, M. S. Vardya, and N. Virgopia.

startling as that obtained by Spinrad and Vardya, i.e. they obtained O/C = 1.05 for *M* giants and supergiants, when this ratio is about 1.6 for the Sun.

From our preliminary results, we have found that there are a very large number of molecular ratios, which can be used as sensitive indicators of O/C ratio. Note that the reason for obtaining such a high accuracy in pinpointing the O/C ratio, specially when O/C is close to unity, is due to CO having one of the highest binding energies for hetronuclear diatomic molecules of about 11 ev. Figure 1 demonstrates the variation of H_2O/CN ratio as a function of C/O ratio. Such a high accuracy should not be expected for other elemental abundance ratios. Table V gives the molecular ratios, which

TABLE V

Elemental abundance ratio	Sensitive molecular ratios	Remarks
C/O	H_2O/H_2 CO/CN H_2O/C_2 TiO/CN, H_2O/CN, CO/CH, C_2/CN, H_2O/HCN, $C_3/Si\ C_2$, SiC/SiS	C/O $\leqslant$ 1.1 C/O $\leqslant$ 1.2 Very good indicator
C/N	SiO/SiC_2 TiO/CN, NO/CO	C/O $\leqslant$ 1
C/H	H_2O/C_2, H_2O/CN TiO/CN, NO/CO	700 < H/C < 1200 and $\theta \geqslant$ 1.8

are sensitive to O/C, C/N and C/H ratios. We have also found that TiO/VO, ZrO/TiO, MgH/TiO and CN/CH are not good indicators of O/C ratio. These results have been obtained assuming thermodynamic equilibrium and the elements and molecules considered as in Vardya (1966). Note that some molecules used in forming the molecular ratios in Table V may not be visible in the spectrum of a certain star, because of its low abundance and/or the low value of the band strength. Incompatible values for a given elemental abundance ratio from several molecular ratios may perhaps indicate some kind of non-L.T.E. effect in the atmosphere of such stars. Once a satisfactory first estimate is made of the elemental abundances and one has a fairly good knowledge of the band strengths of the various features in the spectrum, one should try to compute a synthetic spectrum and check it with the observed spectrum.

Acknowledgements

One of us (M.S.V.) is thankful to Dr. P. K. Raju for several useful discussions on the problem of Pressure Dissociation and is grateful to Prof. Cimino for extending the hospitality during his stay at the Rome Astronomical Observatory.

References

GLASSTONE, S., LAIDLER, K. J., and EYRING, H. 1941, *The Theory of Rate Processes* (McGraw-Hill Book Co., Inc, New York), p. 67.
HIRSCHFELDER, J. O., CURTISS, C. F., and BIRD, R. B. 1954, *Molecular Theory of Gases and Liquids* (John Wiley and Sons, Inc., New York).
PENSTON, M. V. 1967, preprint.
SPINRAD, H., and VARDYA, M. S. 1966, *Ap. J.*, **146**, 399.
VARDYA, M. S. 1965, *M.N.*, **129**, 345.
VARDYA, M. S. 1966, *ibid.*, **134**, 347.

Discussion

GREENSTEIN   I, of course, am very interested in the C minus O (or C to O), and I am very concerned about the fact that this method always gives C minus O near zero, because of the predominance of the CO association. In fact, I have a strong skepticism as to whether there is something physically incorrect in such an approach, because I cannot believe that nature designed matter so that C and O should be nearly the same within one or two per cent. This has been an extreme puzzle. I do not know what is wrong, and I really think one must perhaps look through the whole star to see what the origin of this extreme sensitivity is. I am just worrying whether there is something peculiar in the molecular equilibria. Perhaps there exist regions where CO is not in fact the completely dominant molecule formed. Is there any region in a stellar atmosphere at these temperatures where CO does not consume all of the carbon?

VARDYA   If you go to very low temperatures, then instead of CO other molecules will start coming up. However, for stars with solar-type composition, CO will always be dominant.

AUMAN   One possible reason why the oxygen and carbon abundances seem to be so nearly equal in the M-stars is that people have been underestimating their effective temperatures. If I had to make a guess at the effective temperature of Betelgeuse, I would take something in the range 3500–3700° Kelvin, compared to 3000°K used by Vardya.

Summaries and discussion

Summary of theoretical papers

A. G. W. CAMERON

Belfer Graduate School of Science, Yeshiva University

IT IS difficult to know what the theme of this conference has been; nominally the subject is Low Luminosity, but that seems to include a great number of different types of objects. I think the most exciting discussion which we have had at the conference has been on the pulsars. These qualify as low luminosity objects perhaps only because the total energy emission in the observed wavelength bands is low. Whether or not they are intrinsically low luminosity objects remains to be seen. At any rate this has been an excellent excuse to bring Dr. Hewish over to this side of the Atlantic, and I congratulate the organizers of the Symposium for doing so.

Let me start with pulsars. In this connection I will report some discussion which took place yesterday afternoon (29 March) in a rump session held at NRAO. We have had one of those interesting situations that arise occasionally, whereby somebody comes to a conference and writes some numbers on the board, and then there is an immediate rush for the telephone. On the night of the day (28 March) that Hewish wrote down the coordinates of the other three pulsars, all three of them were observed at Arecibo. Typical pulse amplitudes at two frequencies for the three objects were written on the board yesterday afternoon. Although the pulsar with the quarter-second period appears to be closer to us than the others, its flux intensities are not as great as those from the one that we have been discussing for several weeks.

At the meeting yesterday afternoon Dr. Ostriker gave a very good review of some of the theoretical problems associated with the pulsars. I think one of the striking things that emerged from this is the indication that the pulsars are not pulsating. Of course they emit pulses of radiation, so the name is all right, but it may not have all of the implications that one might like to attribute to it. The reason for this is that we certainly cannot be observing the pulsations of a neutron star, and we cannot be observing the fundamental mode of the pulsations of a white dwarf. If we are observing some higher harmonic for the white dwarf then we should expect several harmonics to be excited, and it would be difficult to see why the pulsars would be such good time-keepers. Secondly, if we have a shock wave emerging in the atmosphere which is somehow pumping energy into modes which are radiating in the radio range, then it would be difficult to understand another feature of the radio pulses, which also was discussed yesterday afternoon. There seems to be frequency structure in some of the pulsars. The intensity is not a very smooth function of frequency, but has local minima and maxima, and that is a very difficult situation to understand. Perhaps we can still invoke nonradial pulsations of a white dwarf excited in some as yet unknown fashion. There was discussion of such nonradial vibrations by Miss Baglin earlier today, in connection with possible nova explosions, so I think that route is not yet completely explored. However, the radial pulsations look unpromising.

Ostriker also showed that if we want to find a good time-keeper, we can consider rotation. That can mean either orbital motions or rotations of a body. Orbital motions are difficult because of the period. There must be a very small separation between the things that are orbiting. Even if white dwarfs are orbiting in contact with each other then it is impossible for them to have a period as short as those which are observed. It may be possible to devise white dwarf models which can rotate fast enough, but it appears that these have to be Ostriker's highly flattened white dwarfs. Here there are problems of shear. Different latitudes rotate at different rates, and a good time-keeper must therefore be something which is emitting from a region on the surface not more than one centimeter wide in latitude in order to have the observed accuracy of time-keeping. That may be possible, but it is certainly difficult.

It appears to me that the easiest kind of rotating object which one might suggest for a time-keeper would be a very slowly rotating neutron star. This might solve time-keeping problems, but there is still the difficult problem of finding a suitable radiation mechanism. The very small time during which the pulse is emitted requires a radiating mechanism which is extremely directional. There are also energy problems. The radiation cannot be emitted from a region in which the plasma density is higher than 10^8 particles/cm³, or the plasma frequency would be higher than the frequencies of the observed radiation. If all of the allowed volume of the source is filled with plasma of the maximum allowed density, then the total amount of energy which has to be radiated in one of these pulses is very much larger than the total thermal energy of all of that plasma. That is a very severe problem for any mechanism in which the energy must pass into some mode from which it can be radiated with the necessary intensity.

In view of this I would say indeed that we are very, very far from having any understanding of the kind of object and the kind of radiation mechanism which can be involved in a pulsar. Presumably the telephonic communication between radio observers and optical observers will be continuing through the next few weeks, and we will have a lot more observations, but I suspect that they are going to leave us more puzzled than ever. Thus this is not a theoretical summary of the pulsar situation, but merely an indication of theoretical puzzlement.

The other objects discussed at the conference generally involved stars in the low mass range in various degrees of evolution. Let us start with the problem of stellar contraction toward the main sequence and the associated problem of star formation. Some very nice hydrodynamic calculations were presented by Bodenheimer and by Hayashi. I raised a question concerning these because of the neglect of angular momentum. That does not destroy the fact that these are very elegant examples of numerical hydrodynamic calculations.

One of the problems which Kraft mentioned, which perhaps clears up something in the realm of the relationship of observation to theory, is the large spread in the apparent positions of low mass stars in young clusters on the lower part of the main sequence. Kraft showed us that there is probably far too much dispersion in the points, and that the lower main sequence of a young cluster must be much narrower than it has usually been plotted. This is more satisfactory from the point of view of simultaneity of star formation. We are still left with some puzzlement that the contraction time for the indicated masses onto the lower main sequence seems to be longer than the nuclear burning ages given by the turnoff from the upper part of the main sequence. We may need large mass ejection during the contraction stage to allow low mass stars to arrive on the main sequence in a short enough time.

We heard also about the T Tauri stage in Kuhi's paper, where we saw that both dust and associated gas seem to vary from time to time in association with the T Tauri stars. I assume this represents gas and dust that has not yet been fully dissipated from the formation process, as well as gas ejected in the intense stellar winds associated with the T Tauri process.

Regarding low mass stars on or beyond the main sequence, we saw two very interesting features presented to us by Dr. Greenstein from the observational point of view, one of which is very nice from the theoretical point of view: the absence of large numbers of very red white dwarf stars. The other feature is a puzzle. The spread in luminosity at a given color on the lower main sequence appears to be a very real phenomenon. This point has also been discussed by Taylor, who raised the possibility of abundance variations to explain these luminosity differences. Presumably these would produce corresponding opacity variations, which might act in the right direction. I would caution him to be a little careful in proposing that most of the luminosity differences could arise from helium abundance variations, because other developments in astrophysics in the last year or two strongly suggest that helium may be a very constant component of the stars, both in our galaxy and in other galaxies, having been produced by cosmological nucleosynthesis. Therefore perhaps it is worth taking another look at the extent to which metal variations may be able to account for these luminosity differences.

We also had a discussion of the problems of constructing model atmospheres for these low mass stars. I was encouraged by the progress which is being made in a very difficult and messy subject. One has to put in the effects of molecules, particularly of water vapor at the lower temperatures, but also of many other molecules at the slightly higher temperatures. One must take into account the pressure broading of the bands and many other associated problems in order to produce a good model atmosphere. Even then there are serious remaining problems, in particular, the problem of convection.

On the subject of convection, I was intrigued with the very shallow convection zones which Böhm presented to us, although these referred to a very different kind of star. This discussion made me wonder if a further theoretical complication might eventually be worthy of consideration: are all the chemical reaction rates fast enough to maintain the equilibrium abundances assumed in these shallow layers?

The lower mass stars also produce flares. There was not much real discussion of flare stars, although several of the papers dealt with one or another aspect of them. In particular, the prospect of being able to discover unseen companions when they flare is a very exciting possibility. But it is an unforeseen way of tackling the problem and deserves to be followed up. The indication that some of the low mass stars are variable because of patches on the surface that may resemble giant sunspots seems to be a related phenomenon. Certainly the flares that occur in the flare stars are gigantic by solar standards. To have the patches or spots on the surface, which give rise to the flares, also gigantic by solar standards, would seem to be a nice correlation. Now all we need to do, as far as the theory is concerned, is to work out all of the problems of solar flares and to apply them to the flare stars, but that is still a substantial program of effort.

There was also some discussion of what we might call 'still-born' stars, and where they occur on the H–R diagram. These are stars unable to support their luminosity by nuclear burning. We wish to know the mass at which this failure occurs, below which stars will simply settle down as hydrogen white dwarfs. The mass is around one-tenth of a solar mass, probably a little bit less. But the detailed answer is going to depend on departures from the ordinary gas equation of state and on complicated surface opacities, and it will probably be some time before the critical mass is accurately located theoretically. However, this is a point of detail on which there is no immediate prospect of accurate observational confirmation.

Some of the stellar evolution calculations presented at this conference deserve comment. The recent failure of Davis to detect the high energy neutrinos from the sun raises unsettling questions about stellar evolution calculations carried beyond the initial main sequence. We must understand why we have not seen the high-energy neutrinos from the sun before we can carry the stellar evolution calculations into the advanced stages of evolution with any degree of confidence. As far as stellar evolution calculations dealing with the lower main sequence are concerned, there is probably no serious problem. But, the white dwarf stars represent the end point of stellar evolution for a wide mass range, and the detailed mechanisms by which the end point is approached must still remain speculative.

Much of the discussion at the conference dealt with the details of the equation of state for the white dwarfs. As the density of matter increases some fairly complicated Coulomb corrections are needed. Eventually much more difficult calculations are needed to determine how the interior of a white dwarf behaves as a crystalline solid. DeWitt remarked that in principle the Coulomb corrections might be determined to within an accuracy of about two per cent. I pinned him down at the cocktail party last night, and he was not willing to claim more than 4 or 5 per cent for the current state of the art. The accuracy with which one can go forward to the more complicated solid-state type of equation of state is very much less. Clearly this is a field in which a lot of research will be done for the next few years. Gradually astrophysics is attracting physicists of every breed and variety, and it is about time we got the solid state people involved too. I think that they will have quite a bit to contribute in this particular field.

How well do we know the mass of a white dwarf? The calculations of Matsushima indicated that a model atmosphere approach can, in the end, give better numbers than the simple *UBV* photometry approach. Unfortunately, there may not be too many systems in which this can be exploited to the limit.

Greenstein's observations of the paucity of white dwarfs in the red part of the H–R diagram seem to correlate very nicely with the concept of solid-state white dwarfs. Once these have cooled far enough to crystallize into the solid state, they cool very rapidly, so that their lifetime is then very short. However, the present theory predicts that only the more massive white dwarfs cool rapidly enough to fit Greenstein's observations. This may cast a little bit of doubt on the usual assumption of an average mass of $0.6 \, M_\odot$ that we use for white dwarfs. Perhaps the average mass should be closer to $0.9 \, M_\odot$.

Kippenhahn's calculations of the formation of white dwarf stars as members of binary pairs were very fascinating, subject to the reservation I made about calculating stellar evolution in advanced stages. I hope that studies involving many more different masses and mass ratios will be carried out. It would be particularly interesting to start with a very unequal mass ratio, for example a more massive star of two or three solar masses and a less massive one of only 0.2 or 0.3 of the solar mass. Then with mass transfer to the smaller companion, it does not move very far away. Consequently, after the companion evolves the mass will pass back to the original primary and the resulting models should be very interesting. They might have something to do with celestial X-ray sources.

Another interesting point discussed by Kippenhahn is the question of vibrational stability for white dwarfs with only one per cent of their mass of hydrogen sitting on the top, but which nevertheless occupies 50 per cent of their radius. The failure of Kippenhan to find vibrational instability for his models has a bearing on the discussion I gave regarding possible pulsar models. I think that the negative indication regarding instability in Kippenhahn's cases is also an argument against radial pulsations in some sort of white dwarf model that might be proposed for a pulsar. Hence Kippenhahn's remarks in that regard have been very timely.

In general, let me say that we have tried to absorb a tremendous amount of material at this conference. I have enjoyed it very much. I have tried to provide some connecting threads between the papers in this review. I will look forward eagerly to the next conference on this subject. Low mass stars are becoming much more interesting and a lot of research is going to be done on them in the future.

Summary of observational papers

O. J. EGGEN

Mt. Stromlo Observatory, Australia National University

I AM NOT certain as to what philosophy applies to a 'summary' of this kind. However, it is obvious that Dr. Cameron will need more time than I do. Why is there a dearth of observational results? I believe one factor is that the whole approach, from the observational side, is changing. It has been traditional that observational astronomers became masters of a technique—or two—and their success has been measured by their reading of the current theoretical structure and the application of their results to that structure. The newer approach is a more physical one in that the problem is conceived and the techniques required to solve it are mastered by an individual or a team. There is at least one small danger in this approach and that is we may lose the element of surprise—the serendipity that marks many of our advances—because I believe that in general we do not solve our problems; we discover them.

Three interfaces between theory and observation have been noticeable in this conference. In the discussions of the earliest stages of stellar evolution the theory is obviously chasing the observations. For various reasons, we do not have much observational data and as a result there are few boundary conditions to inhibit the theoretical output. On the other hand, the interplay between theory and observation in the field of the white dwarfs (or subluminous stars) is well balanced. Because this is Charlottesville, and we are near the Leander McCormick Observatory, a third interface looms large. Astrometrical knowledge is basic to nearly every paper presented here, but the astrometrists present here could easily feel that the conference very seldom degenerated into sense. Some years ago when the present upsurge of astronomical activity began, there was considerable worry that the glamour of astrophysics would drain the talent, techniques and telescopes from astrometry. One solution was to have a captive (and tamed) astrophysicist present at every astrometric conference. This grew, little by little, until today we have a nearly equal mixture. And at these mixed groups it has become traditional each time to once again plead for recognition and integration of the astrometrists. I think it is time to sound a warning that this integration can go too far. There are good reasons for not tying the astrometry at any one time too closely to the astrophysics of the time being. We would be in some difficulties if the astrophysicists of 1900 influenced too much the choice of double stars or proper motion stars then observed. Astrometrists must cast their nets widely. Also, in mixed groups such as this one they must chaff a little at the unfairness. It is *passé* and considered boring by astrophysicists to be subject to discussions of astrometric technique—while they inflict upon us long discussions of mathematical methods, which in the last analysis comes to the same thing.

General discussion

ALLER   Before we start the general discussion, I want to put in my nickel's worth on this abundance business, since I have been involved in it for many years. The point which I think could be made is that one has an opportunity to check out this apparent discordance between the solar carbon to oxygen ratio, which is of the order of 0.7 or something of this sort, depending on whose numbers you accept, and the value of around 0.95 which is deduced for the red M giant stars. Nature has provided us with an opportunity in the sun itself, where the spectrum of a good healthy sunspot imitates the spectrum of a K dwarf star, aside from the complexities introduced by a magnetic field, but with a modern computer you can dezeemanize the spectrum and examine what it would be like if these complexities did not exist. So with the use of a large solar telescope such as that which is available at Kitt Peak or at Mt. Wilson, one can get high dispersion spectra of the sun and high dispersion spectrum of a sunspot and compare the atomic lines in the sun with one another and the molecular lines in the spot with one another, and there you know that you have to come out with the same answer. And I think you could check out fairly quickly which of these ratios are the real ones. I now declare the meeting open for discussion.

SAVEDOFF   I do not recall hearing any discussion of the luminosity function here. But I was wondering if there was any knowledge of luminosity function down below say absolute magnitude 12 along the main sequence. I do not remember the precise number, but somewhere in the range 12 to 14 magnitude was the alleged peak in the number of stars and then from then on the numbers decrease.

EGGEN   I will say that some of the things that have been said at this conference should make it obvious that we have to look much closer at the kinds of stars involved. This is usually derived by just counting red stars. We now know red stars that are definitely below the main sequence, but they may or may not be degenerate, and they come in large numbers. How many of these are included when you count red stars? In other words, I think we are past the answer that you would get from a pure star-count, and the detailed investigations have yet to be made.

GLIESE   In the luminosity function derived by Luyten in 1967, the maximum is now at 15.7 photographic absolute magnitude. And this change from 14 photographic magnitude to 15.7 was caused by the inclusion of the 48 Schmidt plates from Mt. Palomar. The maximum corresponds to an absolute visual magnitude 14.

KRAFT   Where is the cut-off to the main-sequence?

KUMAR   The absolute visual magnitude at the end-point is approximately 17.

KRAFT   And, of course, below this point everything is degenerate. Isn't that so? The main sequence in the ordinary sense of the word does not exist.

SAVEDOFF   That is as far as we can push the mass distribution in the birth-rate function.

SALPETER   This is awfully important as far as the invisible mass is concerned, because if there was an error in any of these numbers, and those two things come together, then that means that there could be very, very many faint stars which we just have not seen. In other words the birth-rate function then could extend very greatly to very low masses and these stars would just not be visible because once there is no nuclear burning their life time is just so much less. So from the point of view of the invisible mass, all these quantitative details might be quite important.

TAYLOR   I would like to make two comments about Dr. Cameron's remarks on my paper. The first is that I was considering only effects on luminosity due to the effect of the abundance variation on the internal structure of the stars. This is the sort of thing that Bodenheimer, Faulkner, and Demarque have been considering. And the second thing is that I am not strongly committed to this. If we could prove that there was a given star with no helium, it would be a cosmological monstrosity. But, this is just thrown out as a possible explanation and I am amenable to any explanation anybody might offer, and I challenge anyone to come up with an explanation that does not burn his tongue while he is saying it.

KRAFT   Father McCarthy, how far down in the Pleiades do you continue to find flare stars or things that you are quite sure are members?

McCARTHY   The apparent magnitude of the faintest star which we have observed is about 15.4 visual.

KRAFT   That is only about 10 absolute. But, were there not more flare stars still farther. . . .

McCARTHY   There were more flare stars which were discovered by Haro and Chavira. And those go down to photographic magnitude of 20.9.

KRAFT   This means that they go down to absolute magnitude 11. But are not there now flare stars down to 15th absolute in the Hyades in Van Altena's discussion?

PESCH   Van Altena's results show that none of the flares from Hyades is a member.

VAN ALTENA   I would just like to add that the plates go down to a little above 20th magnitude photographic; it does appear that the main sequence tapers off before we get to the plate limit. So there is a possibility that the bottom of the main sequence is approximately reached.

McCARTHY   Professor Blaauw the other day asked about the availability of plates which have been taken in recent times since the Hertzsprung collection. I would like to mention that there exist two sets of plates. One set was taken with the Lick 20-inch astrograph and those are the ones which are discussed in the recent publication of Dr. Van Altena. The other is a set of plates including those taken at the time of the Palomar survey in 1945 with the 48 inch Schmidt and as compared with plates taken in 1964 and 1965 by Luyten. So both of these are available now and I understand from Dr. van Altena that he is setting up a measurement program of the Lick plates. This means that we would soon have more reliable astrometric data to separate the sheep from the goats or the birds from the bees as the case may be.

ALLER   Further discussion? If not, then the time has come to close this conference. Before doing that there are a couple of remarks that I would like to make. When Dr. Kumar asked me to attend this conference, I pointed out to him that I had not dabbled in this field for many years and his reply was "you can simply be here as a neutral observer." But I thought I would like to take advantage of the chairman's prerogative and mention a couple of things about the astrophysical end of the story as it was in the late 40's. At that time some of us tried to interpret the spectra of cool stars like K dwarfs by curve of growth methods, and quickly discovered that they would not work, and that the reason why they would not work was the extraordinary complexity of the spectrum. Some of us tried to dabble in the stellar structure; at that time the Cowling model was the gospel. Everyone believed in carbon cycle, the proton–proton reaction was not fully accepted. And it was shown that the carbon cycle would give you a nice Cowling model with a convective core and radiative envelope, but that there was no way of making a model with a proton–proton reaction that would fit the orthodox Cowling models. Therefore it was necessary to start out, as Strömgren pointed out, with a model which was convective in the exterior and perhaps radiative at the core. And listening to the discussions at this meeting, I have been very impressed by the enormous progress which has been made in this field. It might not be without some merit to take a minute to speculate on what to look forward to in the next 20 years. I should hope that by then we would have spectrograph- and telescopes operating above the earth's atmosphere so that the extremely important infrared portion of the spectrum could be studied in detail. The balloon observations have given us some idea of what can be accomplished, and it does not take much imagination to visualize what great contributions a good astrometric telescope, say on the moon, could make to the solution of many of these problems. But even before we can get above the Earth's atmosphere, it is worth noting that the advent of the modern computing machine, and of such highly sophistricated measuring devices, as is available, for example, at the University of California at Santa Cruz, make it possible to measure the astrometric plates about as rapidly as they can be taken. So that the great drudgery and backlog which has been the curse of so much of the astronomy can be dispensed with.

We have all been very fortunate in attending a conference which has been so well organized, both scientifically and otherwise, and I would like to call for a standing ovation to Professor Kumar, to Director Fredrick, and all the other members of the staff of the University of Virginia Astronomy Department, who have made our visit here so profitable.

Index

This index has been prepared from all the titles of papers in this volume, extensively cross-referenced. It is hoped that this will enable the reader to locate rapidly any papers in which he is interested.